BIOPOLYMERS

MOLECULAR BIOLOGY

An International Series of Monographs and Textbooks

Editors: BERNARD HORECKER, NATHAN O. KAPLAN, JULIUS MARMUR, AND
HAROLD A. SCHERAGA

A complete list of titles in this series appears at the end of this volume.

BIOPOLYMERS

Alan G. Walton and John Blackwell

Division of Macromolecular Science
Case Western Reserve University
Cleveland, Ohio

with a contribution by Stephen H. Carr

Department of Materials Science
The Technological Institute
Northwestern University
Evanston, Illinois

ACADEMIC PRESS New York and London 1973

A Subsidiary of Harcourt Brace Jovanovich, Publishers

ACADEMIC PRESS, INC.
111 Fifth Avenue, New York, New York 10003

United Kingdom Edition published by
ACADEMIC PRESS, INC. (LONDON) LTD.
24/28 Oval Road, London NW1

LIBRARY OF CONGRESS CATALOG CARD NUMBER: 79-182627

PRINTED IN THE UNITED STATES OF AMERICA

CONTENTS

8.   Electrical and Magnetic Field Effects

9.   Conformation of Polypeptides

10.   Fibrous Proteins and Biopolymer Models

PREFACE

The term biopolymer has been defined in a variety of ways by researchers in different disciplines. In strict analogy with the development of synthetic commercial polymers we might, for example, require that its use be limited to synthetic molecules fabricated from biological "monomer" units such as amino acids and sugars. However, if biological *in vivo* synthesis is encompassed by the definition, the full gambit of biological macromolecules—proteins nucleic acids, and polysaccharides—is included. We have chosen an intermediate ground in which emphasis is placed on the simpler synthetic (*in vitro*) biopolymers and native materials for which the molecular architecture is most fully understood. In this sense, the objectives of the biopolymer researcher would be, generally, to characterize the structure of biomolecules by using physical techniques and, by so doing, gain insight into physiological function.

The main theme of this book is concentrated on the methods of physical characterization and the principles which underly them. The contents stress the more quantitative aspects of sequence, conformation, and structure in both laboratory-synthesized and native biopolymers. Even in the area of characterization of biopolymers, the available techniques, the evaluation of underlying principles, and the experimental applications have mushroomed to such a degree in the past few decades that we have been forced to make some arbitrary limitations to the scope of the methods and applications discussed. To an extent, these limitations are imposed both by the nature of the audience to which this book is directed and the influence of our immediate academic environment. Thus, some of the newer methods are outlined, such as Raman spectroscopy, theoretical conformation analysis, and electron microscopy and morphology of laboratory-synthesized polymers, whereas others of recent vintage such as spin labeling, fluorescent spectroscopy, and electron spin resonance are excluded.

The book is directed to those who have relatively little background in the application of physical methods to the study of biological macromolecules— the undergraduate who wishes to familiarize himself (herself) with the area or the researcher who is faced with proceeding in a new direction. Thus, although much of the material is presented at an elementary level, an effort has been made to reference the more important aspects and to present a current account of the status of biopolymer research. In addition, the results and implications of physical characterization of native biopolymers are presented with emphasis on structure. Where possible, the principles of molecular conformational analysis are thus projected through model compounds to their native counterparts.

We are pleased to acknowledge the assistance of our colleagues in providing useful suggestions and material and the many investigators who provided us with photographs, original figures, and tables, several of which have not previously been published. We are particularly grateful for the contribution of Chapter 7 by Professor Stephen H. Carr of Northwestern University, for the diligent efforts of Drs. Elizabeth Simons and Barton Rippon, and for the unending patience, assistance, and cheerfulness brought to this project by Peggy Buccieri.

ALAN G. WALTON
JOHN BLACKWELL

STRUCTURAL UNITS OF BIOPOLYMERS

Introduction

A polymer is a large molecule comprised of many fundamental units joined together. If these units, called for present purposes the monomers, are identical, the result is a "homopolymer"; if they are of two or more kinds, the product is a hetero- or copolymer that is either random or sequential. In the latter case the monomers are present in the primary structure in a specific sequence. In the following text we shall be concerned with the primary structure (sequence and chemical structure), secondary structure (conformation or shape), and tertiary structure (arrangement and ultrastructure) of biopolymers, and it is well to realize from the onset that all are interrelated, although often in a manner which is not yet clear. This chapter deals with primary structure and specifically with the monomer units from which both synthetic and native biopolymers are generated. Three categories are dealt with—proteins and polypeptides, polysaccharides, and polynucleotides and nucleic acids.

Amino Acids

Proteins are broken down by hydrolysis into a mixture of about 20 different monomer units, known as amino acids. These are all α-amino acids,

i.e., the amino group is attached to the α carbon atom, and with two exceptions (proline and hydroxyproline) they are primary amino acids with the
general formula

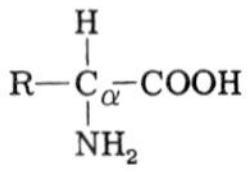

$$R-\underset{\underset{NH_2}{|}}{\overset{\overset{H}{|}}{C_\alpha}}-COOH$$

Amino acids fall into seven categories, depending mainly upon the nature
of the functional group R: aliphatic, hydroxylic, carboxylic, basic, aromatic,
sulfur-containing, and imino acids (proline and hydroxyproline). In the
imino acids the R group takes the form of a five membered ring containing
the α-carbon. A list of the naturally occurring amino and imino acids is
given in Table 1.1 (1). With such a large number of available building blocks,
nature is able to promote an extremely large variety of combinations, and can
thus control the subtle functions performed by proteins.

Apart from the simplest amino acid, glycine where R = H, all the amino
acids are optically active because of the asymmetry of the α-carbon atom.
We shall see later that this optical anisotropy is a useful basis for studying
the biopolymers that are comprised of these amino acids. With very few
exceptions the amino acids are found to be in the L configuration, where the
distribution of groups bonded to the α-carbon is as shown in Fig. 1.1. The

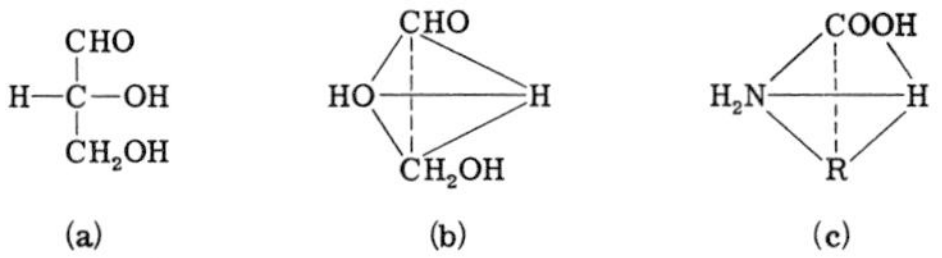

Fig. 1.1. Alternative representations of L-glyceraldehyde (a and b) and an L-amino
acid (c). Groups are at the corner of a tetrahedron.

exceptions are the occasional D forms which are found in certain bacterial
peptides. (The distinction between proteins, polypeptides, and peptides, is
mainly one of size and will be defined later.)

Polymerization of amino acids, either in the laboratory or in physiological
systems, involves the formation of peptide units,

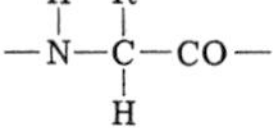

$$-\underset{\underset{H}{|}}{\overset{\overset{H}{|}}{N}}-\overset{\overset{R}{|}}{C}-CO-$$

by a condensation process, and the properties of the functional R groups
will, to a large extent, control the properties of the polymer. In biological
systems the polypeptide chain is always straight, i.e., there is no branching,
and is almost always made up of a variety of amino acid monomers. An
interesting exception to the latter generalization is poly-γ-D-glutamic acid, an

TABLE 1.1

Amino Acids That Commonly Occur in Proteins[a]

Name	Structure	pK_a (Side chain)
I. Aliphatic amino acids		
Glycine (Gly)	$H_2N-CH_2-CO_2H$	
Alanine (Ala)	CH_3 $\mid$ $H_2N-CH-CO_2H$	
Valine (Val)	$H_3C\diagdown\underset{C}{\overset{H}{}}\diagup CH_3$ $\mid$ $H_2N-CH-CO_2H$	
Leucine (Leu)	$H_3C\diagdown\underset{C}{\overset{H}{}}\diagup CH_3$ $\mid$ CH_2 $\mid$ $H_2N-CH-CO_2H$	
Isoleucine (Ile)	CH_3 $\mid$ CH_2 $\mid$ $CH-CH_3$ $\mid$ $H_2N-CH-CO_2H$	
II. Hydroxyamino acids		
Serine (Ser)	CH_2OH $\mid$ $H_2N-CH-CO_2H$	9.15
Threonine (Thr)	CH_3 $\mid$ $CH-OH$ $\mid$ $NH_2-CH-CO_2H$	10.43
III. Dicarboxylic amino acids and amides		
Aspartic acid (Asp)	CO_2H $\mid$ CH_2 $\mid$ $NH_2-CH-CO_2H$	3.86
Asparagine (AspNH$_2$ or Asn)	$CONH_2$ $\mid$ CH_2 $\mid$ $NH_2-CH-CO_2H$	

[a]Adapted in part from "Biological Chemistry," second edition; by Henry R. Mahler and Eugene H. Cordes. Harper and Row, New York, 1971.

TABLE 1.1 (*continued*)

Name	Structure	pK_a (Side chain)
Glutamic acid (Glu)	CO_2H — CH_2 — CH_2 — NH_2—CH—CO_2H	4.25
Glutamine (GluNH$_2$ or Gln)	$CONH_2$ — CH_2 — CH_2 — NH_2—CH—CO_2H	
IV. Amino acids having basic functions		
Lysine (Lys)	$(CH_2)_4$—NH_2 — NH_2—CH—CO_2H	10.53
Hydroxylysine (Hylys)	CH_2—NH_2 — CH—OH — $(CH_2)_2$ — NH_2—CH—CO_2H	9.67
Histidine (His)		6.0
Arginine (Arg)	$(CH_2)_3$—N—C—NH_2 — NH_2—CH—CO_2H	9.04
V. Aromatic amino acids		
Phenylalanine (Phe)		
Tyrosine (Tyr)		10.07

TABLE 1.1 (*continued*)

Name	Structure	pK_a (Side chain)
Tryptophan (Try)		
VI. Sulfur-containing amino acids		
Cysteine (CySH)	CH$_2$—SH NH$_2$—CH—CO$_2$H	8.33
Cystine (CyS—SCy)	$\left(\begin{array}{c}\text{CH}_2\text{—S}\text{—}\\ \text{NH}_2\text{—CH—CO}_2\text{H}\end{array}\right)_2$	
Methionine (Met)	(CH$_2$)$_2$—SCH$_3$ NH$_2$—CH—CO$_2$H	
VII. Imino acids		
Proline (Pro)		
Hydroxyproline (Hypro)		

insect wax that is not only a homopolymer, but consists of D residues linked through the γ- rather than the α-carboxyl group. However, no proteins are homopolymers and we may accept as a general rule that they are α-linked L-amino acids. In the laboratory the most straightforward synthetic process is the formation of a homopolymer, i.e., a poly-α-amino acid. Nevertheless, recent developments have made possible the preparation of mixed and sequential polypeptides that are, perhaps, realistic protein models.

As previously mentioned, the properties of the polypeptide are, to an extent, governed by the (R-)side group or groups, and some general rules may be pointed out here. Polymers of the aliphatic amino acids are "non-polar" because of the lack of ionizable side groups and are thus usually insoluble in aqueous solution. In proteins the apolar side groups interact unfavorably with the surrounding aqueous medium and tend to fold inside globular structures. The hydroxylic functional groups of serine and threonine are not ionized at neutral pH and, although inert, do tend to form hydrogen bonds when possible. Carboxylic and basic side-chain groups are ionized at neutral pH and interact strongly with aqueous media. The pH at which the side groups are 50% ionized is shown in Table 1.1 (pK_a). The values quoted are for the amino acids themselves, although little modification of this value is experienced in the polymer. Below the quoted pH value, the side chains are protonated ($-COOH$, $-NH_3^+$) and above, they are deprotonated ($-COO^-$ and $-NH_2$). The hydroxylic peptides deprotonate at high pH ($-OH \rightarrow O^-$). The acidic and basic peptides generally occupy surface positions in proteins.

In the sixth category, that of sulfur-containing peptides, cystine plays a very important part in protein structure by bridging (cross-linking) chains with the $-S-S-$ moiety, a feature that is found in almost all globular proteins. On the other hand, as we shall see later, nature often chooses to cross-link fibrous proteins by other means. Finally, the imino acids, which are secondary amino acids, possess no hydrogen atoms on the peptide nitrogen and have unusual structure-directing properties because of the steric restraints imposed by the pyrrolidine ring. Thus L-proline often occurs at "corners" in folded globular protein chains and imposes the specific structure of the fibrous protein collagen. L-Hydroxyproline is, in fact, unique to the collagen class of proteins. (Because of the all-pervading nature of the L residues in proteins, residues referred to in this book will be of the L-configuration unless otherwise noted.)

Naturally occurring amino acids that are generally found to be present only in conjunction with a specific process or polypeptide are listed in Table 1.2. Several of these acids have been polymerized with some interesting conformational results.

Although proteins, polypeptides, and peptides usually consist of covalently linked single chains, there are a number of small cyclic peptides occurring in nature. The physiological activity of some of these peptides has been determined, but the function of others, particularly those associated with cell membranes, is unknown. Whereas branched-chain polypeptides do not, as far as we know, exist in nature, it is possible to synthesize branched or multichain polyamino acids in the laboratory. This is possible because of the second carboxyl or amino group present in the R side chain of some of the amino acids. So far only lysine and glutamic acid seem to have been used for this purpose.

Nomenclature

The polymerization product of amino acids contains, as previously noted, peptide units $[-NH-CHR-CO-]_n$ joined together, n being the degree of polymerization (DP) or number of joined units. For homopolymers, when n is

TABLE 1.2

Important Nonprotein Amino Acids

Name	Structure	Occurrence and function
β-Alanine	$NH_2-CH_2CH_2-COOH$	Occurs in natural peptides carnosine and anserine
γ-Aminobutyric acid	$NH_2-(CH_2)_3-COOH$	A constituent of plant tissue and of brain tissue in mammals
α, γ-Diaminobutyric acid	$NH_2-(CH_2)_2-\underset{\underset{NH_2}{\mid}}{CH}-COOH$	Antibiotics
Sarcosine	N-Methylglycine	A constituent of the actinomycine
Betaine	N, N, N-Trimethylglycine	A constituent of plant and animal tissues; an intermediate in the metabolism of lipids
O-Diazoacetylserine (azaserine)	$\underset{NH_2-CH-COOH}{\overset{CH_2O-COCHN_2}{\mid}}$	Antimetabolite isolated from certain molds; inhibits the synthesis of nucleic acids
Homoserine	$\underset{NH_2-CH-COOH}{\overset{CH_2CH_2OH}{\mid}}$	An important intermediate in metabolism; occurs in plant and animal tissues
Ornithine	$\underset{NH_2-CH-COOH}{\overset{(CH_2)_3-NH_2}{\mid}}$	An important intermediate in the biosynthesis of urea
Citrulline	$\underset{NH_2-CH-COOH}{\overset{(CH_2)_3-NH-\overset{O}{\overset{\|}{C}}-NH_2}{\mid}}$	An important intermediate in urea biosynthesis; the immediate precursor of arginine
Thyroxine		In thyroid hormones
Pipecolic acid		In higher plants
α-Aminoadipic acid	$HOOC-(CH_2)_3-\underset{\underset{NH_2}{\mid}}{CH}-COOH$	In higher plants and antibiotics
Penicillamine	$\underset{NH_2-CH-COOH}{\overset{(CH_3)_2C-SH}{\mid}}$	A component of the antibiotic penicillin

Arg-Pro-Pro-Gly-Phe-Ser-Pro-Phe-Arg   Lys-Arg-Pro-Pro-Gly-Phe-Ser-Pro-Phe-Arg

(a)   (b)

Fig. 1.2. Amino acid sequence of (a) bradykinin and (b) kallidin.

small (2–25), the product is known as an oligomer; the same is true for very low molecular weight polymers of saccharides and nucleotides. Sequential or random polymers, either synthetic or native, with n in the same range, form the class of compounds known as peptides; for present purposes we may characterize the range of n for polypeptides $25 < n < 500$ and that for proteins as $n > 100$. These limitations are somewhat arbitrary since some proteins are known to have rather low molecular weights (e.g., certain enzymes). The average molecular weight of a peptide unit is close to 100 so that as a simple rule of thumb, $100 \times n$ is the molecular weight of the polymer. Synthetic polyamino acids, including the sequential and mixed varieties, almost invariably fall in the polypeptide range. For the more complicated materials a DP of <200 is usual.

Primary Structure

The primary structure, or sequence, of a peptide, polypeptide, or proteins dictates the secondary and tertiary structure and function of the entity in question. The experimental procedures involved in sequence determination represent a major area of research, but it is an area that is outside the scope of the present text. However, a few results will be quoted here so that some salient features may be noted. Peptides of two or more basic units are known and may be synthesized. With such small molecules, thorough conformational analysis may be accomplished by such methods as nuclear magnetic resonance (nmr) spectroscopy. However, the relationship between sequence, conformation, and function is not easily determined, and the linking of these three properties is certainly one of the major research areas in molecular biophysics. Among the simple straight-chain peptides that are physiologically active are bradykinin and kallidin (Fig. 1.2), which are nona-, and decapeptides, respectively. Both are muscle-tensing agents, the similarity in sequence being evident. 6-Glycylbradykinin, a synthetic analog of bradykinin, is also

Human   Ala-Glu-Lys-Lys-Asp-Glu-Gly-Pro-Tyr-Arg-Met-Glu-His-Phe-Arg-Try-Gly-Ser-Pro-Pro-Lys-Asp

Monkey   Asp-Glu-Gly-Pro-Tyr-Arg-Met-Glu-His-Phe-Arg-Try-Gly-Ser-Pro-Pro-Lys-Asp

Horse   Asp-Glu-Gly-Pro-Tyr-Lys-Met-Glu-His-Phe-Arg-Try-Gly-Ser-Pro-Arg-Lys-Asp

Beef   Asp-Ser-Gly-Pro-Tyr-Lys-Met-Glu-His-Phe-Arg-Try-Gly-Ser-Pro-Pro-Lys-Asp

Pig   Asp-Glu-Gly-Pro-Tyr-Lys-Met-Glu-His-Phe-Arg-Try-Gly-Ser-Pro-Pro-Lys-Asp

Fig. 1.3. β-Melanocyte-stimulating hormone.

strongly active, suggesting at an elementary level that a specific sequence is not entirely necessary for a specific function; rather, the overall molecular geometry would seem to be all-important.

The modification of an amino acid sequence during evolution can be traced for certain polypeptides and proteins isolated from different species (2). A comparison of sequence for the β-melanocyte-stimulating hormone (pigment-activating hormone) for five animal species is shown in Fig. 1.3 (3). It can be seen that the bulk of the peptide is very similar in each case, but an extra tetrapeptide is linked at the N-terminal end for the human hormone (peptide sequences are written with the NH_2-terminus at the left-hand end).

Many of the small peptides are cyclic compounds, joined through the cystine disulfide bridge. An example that has attracted unusual attention is the cyclic peptide gramicidin S (Fig. 1.4), for which the conformation has been studied on an entirely theoretical basis (4–6). (This approach is discussed in Chapter 2.) Even peptides of this size present awesome theoretical problems

L-Pro-L-Val-L-Orn-L-Leu-D-Phe
| |
D-Phe-L-Leu-L-Orn-L-Val-L-Pro

Fig. 1.4. Gramicidin S.

1 Val-Asp-Asp-Asp-Asp-Lys-Ile -Val-Gly-Gly-Tyr-Thr-Cys-Gly-Ala-

Asn-Thr-Val-Pro-Tyr-Gln-Val-Ser-Leu-Asn-Ser-Gly-Tyr-His-Phe-

31 Cys-Gly-Gly-Ser-Leu-Ile -Asn-Ser-Gln-Trp-Val-Val-Ser- Ala - Ala -

His-Cys-Tyr-Lys-Ser-Gly-Ile -Gln-Val-Arg-Leu-Gly-Gln-Asp-Asn -

61 Ile -Asn-Val-Val-Glu-Gly-Asn- Gln-Gln-Phe-Ile -Ser - Ala- Ser-Lys -

Ser-Ile -Val-His-Pro-Ser-Tyr-Asn-Ser-Asn-Thr-Leu-Asn-Asn-Asp-

91 Ile-Met-Leu-Ile - Lys-Leu-Lys-Ser-Ala-Ala -Ser-Leu-Asn-Ser-Arg -

Val-Ala-Ser-Ile -Ser-Leu-Pro-Thr-Ser-Cys- Ala-Ser -Ala- Gly-Thr -

121 Gln-Cys-Leu-Ile -Ser-Gly-Trp-Gly-Asn-Thr-Lys-Ser-Ser- Gly-Thr -

Ser-Tyr-Pro-Asp-Val-Leu-Lys-Cys-Leu-Lys-Ala-Pro-Ile-Leu-Ser-

151 Asn-Ser-Ser-Cys-Lys-Ser- Ala-Tyr-Pro-Gly- Gln -Ile-Thr-Ser-Asn-

Met-Phe-Cys-Ala-Gly-Tyr-Leu-Glu-Gly-Gly-Lys-Asn-Ser-Cys-Gln-

181 Gly-Asp-Ser-Gly-Gly- Pro-Val- Val-Cys-Ser-Gly-Lys-Leu- Gln-Gly-

Ile-Val-Ser-Trp-Gly- Ser-Gly-Cys -Ala-Gln-Lys-Asn-Lys-Pro -Gly-

211 Val-Tyr-Thr-Lys-Val-Cys-Asn-Tyr-Val-Ser-Trp- Ile-Lys-Gln-Thr-

Ile- Ala-Ser-Asn

Fig. 1.5. Primary sequence of bovine trypsinogen.

in conformational analysis, although continual progress is being made. An additional complication in this case is the presence of D residues.

A number of low molecular weight proteins (globular), as well as fragments of fibrous proteins, have been sequenced. The sequence of the enzyme precursor trypsinogen (bovine) (7) is shown in Fig. 1.5. It can be seen that there is apparently no regular repeating pattern of the fundamental peptides, a feature common to globular proteins. On the other hand, fibrous proteins do frequently appear to have specific sequences, as is demonstrated in Chapter 10.

Sugars

The constituent monomer units of polysaccharide chains are known as sugars or monosaccharides; there is considerable variety in the types of sugars that occur in nature, as well as the way in which they can be linked together to form polymers. One of the most common sugar monomers is α-D-glucose, which has a six-membered ring structure in the solid state, as shown in Fig. 1.6a. This ring is known as a pyranose ring in view of its similarity to tetrahydropyran. Similarly the five-membered ring of β-D-ribose (Fig. 1.6b) is analogous to tetrahydrofuran and is known as a furanose ring. The suffix "ose" is general for all sugars. In solution, ring structures of this type exist in equilibrium with aldehyde isomers. The aldose structures for D-glucose and D-ribose are shown in Figs. 1.6c and d.

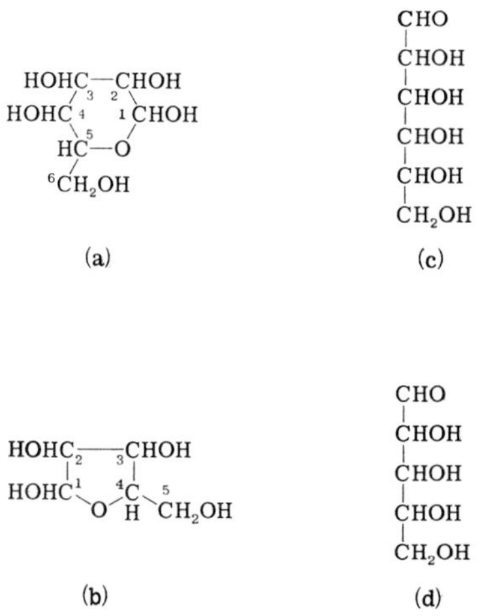

(a)

CHO
$CHOH$
$CHOH$
$CHOH$
$CHOH$
CH_2OH

(c)

(b)

CHO
$CHOH$
$CHOH$
$CHOH$
CH_2OH

(d)

Fig. 1.6. (a) α-D-Glucose in the pyranose form; (b) β-D-ribose in the furanose ring form; (c) Aldose structure of D-glucose; (d) Aldose structure of D-ribose.

The nomenclature for sugar monomers is based on the number of carbon atoms in the aldose chain. Glucose and ribose consist of chains of six and five carbons, respectively, and are thus a hexose and a pentose. The smallest molecule in the aldose series is glyceraldehyde (Fig. 1.7a), which is a triose.

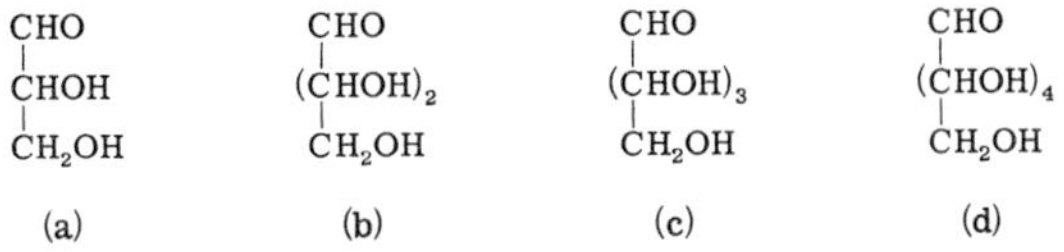

$$
\begin{array}{cccc}
\text{CHO} & \text{CHO} & \text{CHO} & \text{CHO} \\
| & | & | & | \\
\text{CHOH} & (\text{CHOH})_2 & (\text{CHOH})_3 & (\text{CHOH})_4 \\
| & | & | & | \\
\text{CH}_2\text{OH} & \text{CH}_2\text{OH} & \text{CH}_2\text{OH} & \text{CH}_2\text{OH} \\
\text{(a)} & \text{(b)} & \text{(c)} & \text{(d)}
\end{array}
$$

Fig. 1.7. Sugar monomers. (a) Triose: glyceraldehyde; (b) tetrose: threose, erythrose; (c) pentose: ribose, arabinose, xylose, lyxose; (d) hexose: allose, gulose, altose, idose, galactose, mannose, glucose, tallose.

This molecule has one asymmetric carbon atom and the two optical isomers are the D and L forms. By convention, all sugars with the same configuration of the carbon adjacent to the CH_2OH group, as in D- (or L-) glyceraldehyde, are labeled D (or L). Addition of further

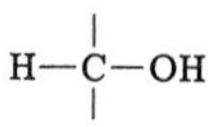

$$
\begin{array}{c}
| \\
\text{H}-\text{C}-\text{OH} \\
|
\end{array}
$$

asymmetric centers to the chain, doubles the number of possible structural isomers. Thus there are two D-tetroses, four D-pentoses, eight D-hexoses, etc. (see Figs. 1.7b–d). These structural isomers are denoted by trivial names. When the pentoses and hexoses are crystallized they take up furanose or pyranose ring structures and an additional asymmetric center is generated. The two isomers produced are designated α and β anomers, the former term being used for the structure with the highest positive optical rotation $[\alpha]_D$.

In polysaccharide structures the monomers are almost invariably hexoses and pentoses in the closed ring form. A large variety of sugar monomers is found in nature in polysaccharides. However, most of these occur only in small quantities, and the great majority of the monomers that make up the natural polysaccharides are D-glucose, D-galactose, D-mannose, D-xylose, L-arabinose, and D-ribose or derivatives of these, notably the acetamide, sulfate or uronic acid derivatives. Some of the most common sugar monomers in natural polysaccharides are shown in Fig. 1.8 (8).

These monomers are linked to form polysaccharide chains by condensation involving two of the hydroxyl groups. For D-glucose (Fig. 1.6a), for example, there are five hydroxyls, and thus there are several ways in which this linkage can be achieved. For reasons described in Chapter 11, the hydroxyl on C-1 is always involved in the formation of the so-called glycosidic

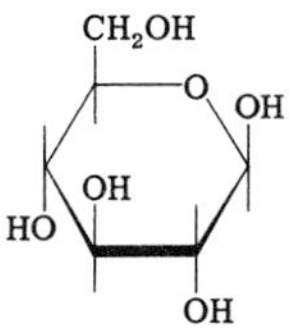

β-D-Glucopyranose
(glucose)

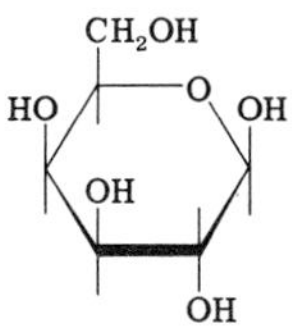

β-D-Galactopyranose
(galactose)

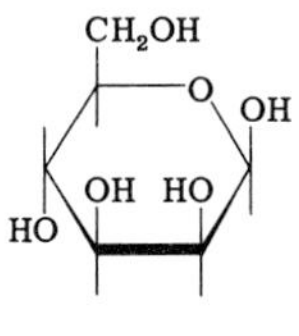

β-D-Mannopyranose
(mannose)

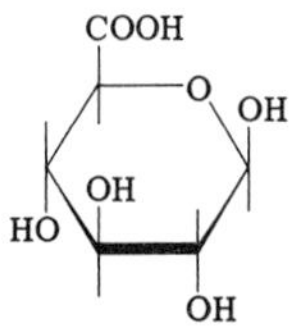

β-D-Glucopyranosyl-
uranic acid
(glucuronic acid)

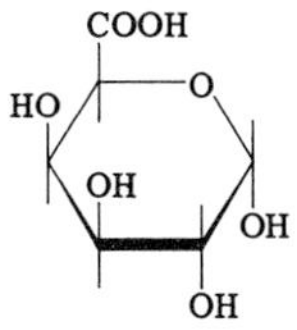

α-D-Galactopyrano-
syluronic acid
(galacturonic acid)

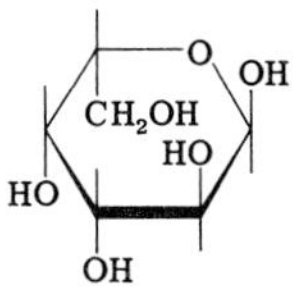

α-L-Galactopyranose
(galactose)

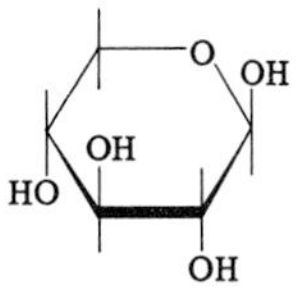

β-D-Xylopyranose
(xylose)

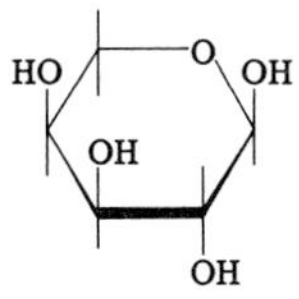

α-L-Arabinopyranose
(arabinose)

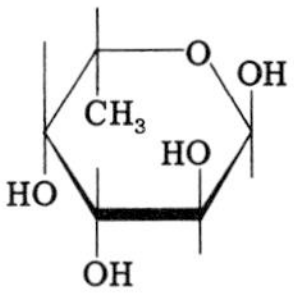

α-L-Fucopyranose
(fucose)

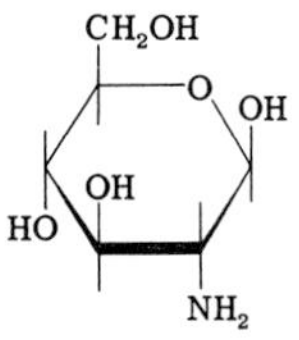

2-Amino-2-deoxy-
β-D-glucose
(glucosamine)

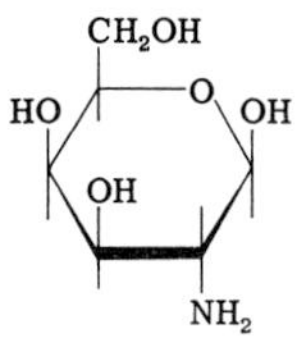

2-Amino-2-deoxy-
β-D-galactose
(galactosamine)

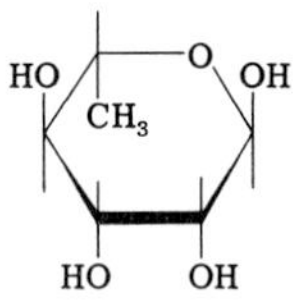

α-L-Rhamnopyranose
(rhamnose)

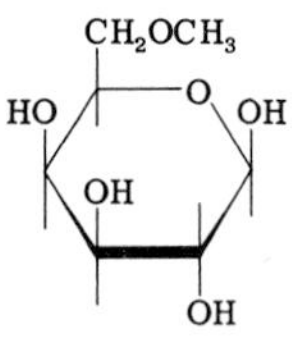

Methyl ether sub-
stituent (6-O-
methyl-D-galactose)

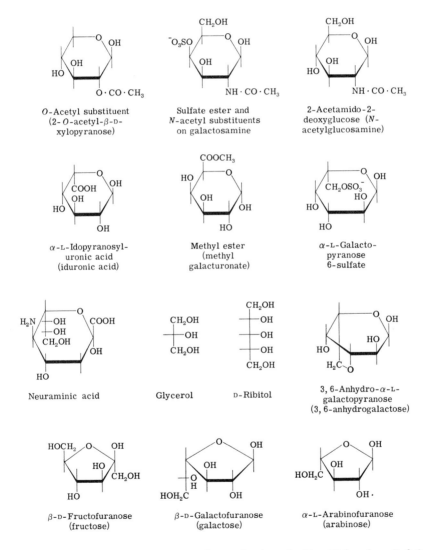

Fig. 1.8. Common sugar monomers of natural polysaccharides. [Taken from Ref. 8, with permission.]

bonds. Thus in the case of D-glucose there are four other O—H groups which can be involved in the condensation, at positions 2, 3, 4, and 6 on the ring. Most of the common sugar monomers are found in nature in a variety of of linkages. The type of linkage naturally affects the conformation adopted by the chain. Cellulose consists of long chains of β-D-glucose residues with 1,4-glycosidic linkages. Positions 1 and 4 are at opposite ends of the β-D-glucose residues and the respective C—O bonds are almost parallel. Consequently cellulose can exist in long extended chains. However, the xylan found in certain seaweeds has β-D-xylose residues in the pyranose ring form with 1,3 linkages. In this case the linkage does not favor straight chains, and a coiled helical conformation results.

In spite of the variety of monomers and linkages available, most of the common polysaccharides are homopolymers, e.g., cellulose, amylose, xylan, and mannan, or they consist of repeating sequences of two sugar monomers with the same repeating linkage such as mycodextran and the mucopolysaccharides in connective tissue. These polydisaccharides are often known as the A—B polysaccharides. Many heteropolysaccharides have been characterized chemically but probably do not in general have a regular conformation and will not be discussed in this book.

Nucleotides

The monomer units of nucleic acids are known as nucleotides. Nucleic acids and synthetic polynucleotides consist of a repeating chain of sugar residues linked by phosphate units, with organic bases as side chains on the sugars. The repeating sequence can be written

```
    base            base            base
     |               |               |
sugar–phosphate–sugar–phosphate–sugar–phosphate–
```

Sugars

The two types of natural nucleic acids are classified by the type of sugar they contain. Ribonucleic acid (RNA) contains exclusively β-D-ribose, while the sugar in DNA is β-2-deoxy-D-ribose. Polynucleotides have been synthesized as model compounds almost exclusively from ribose or deoxyribose, although a few other furanose ring sugars have been used, such as arabinose. The structures of β-D-ribose and β-2-deoxy-D-ribose are shown in Fig. 1.9.

Fig. 1.9. Structures of β-D-ribose and β-2-deoxy-D-ribose the sugar components of RNA and DNA respectively.

Bases

The bases in natural nucleic acids are derivatives of pyrimidines and purines. For RNA, the four most common bases are adenine, guanine, cytosine, and uracil, and the situation is very similar in DNA, except that uracil is replaced by thymine (5-methyluracil). These common bases (Fig. 1.10) account for approximately 95% of the base content in most nucleic

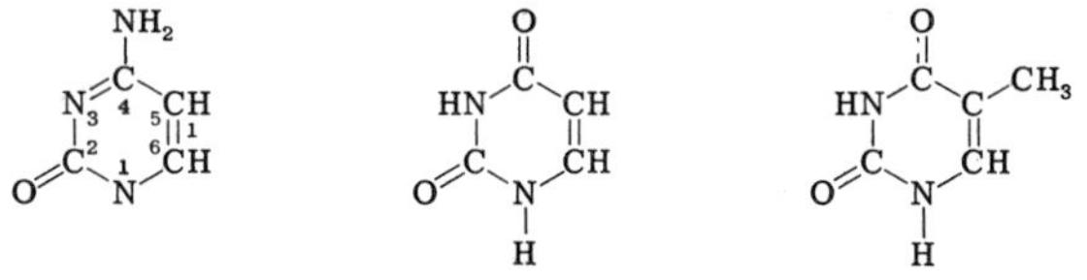

Cytosine Uracil Thymine

(a)

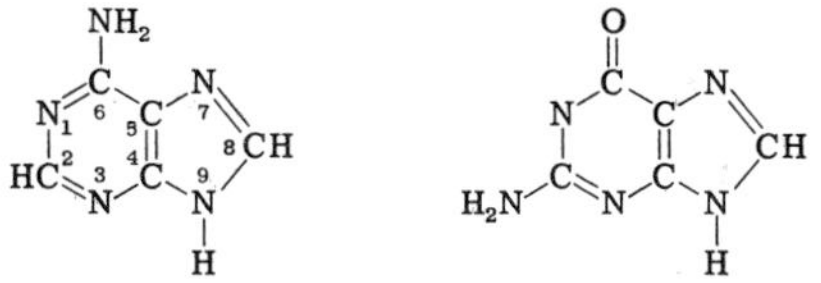

Adenine Guanine

(b)

Fig. 1.10. Common base of nucleic acids: (a) pyrimidines, (b) purines.

acids. The remaining 5% are the so-called minor bases, which are derivatives of the common bases. Table 1.3 lists some of the more common minor bases (9). Many of these are found almost exclusively in transfer RNA (see Chapter 12).

Condensation of a purine or pyrimidine base with ribose or deoxyribose produces a *nucleoside*. Linkage to the sugar occurs at position 1' on the sugar ring. Thus condensation of ribose with adenine, guanine, cytosine, and

TABLE 1.3

Some Minor Bases

Pyrimidines	Purines
5-Methylcytosine	Hypoxanthine
5-Hydroxymethylcytosine	Xanthine
	Uric acid
	2-Methyladenine
	6-Methylaminopurine
	6-Demethylaminopurine
	6-Hydroxy-2-methylaminopurine
	1-Methylguanine

uracil forms adenosine, guanosine, cytidine, and uridine (Fig. 1.11). For
the common bases, linkage takes place at N-3 on the pyrimidines and N-9
on the purines. The corresponding nucleosides involving deoxyribose are
known as deoxyadenosine, deoxyguanosine, etc.

For the unsaturated bases in nucleosides, various tautomeric forms are
theoretically possible due to resonance. Possible resonance structures for

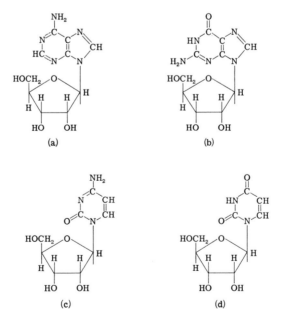

Fig. 1.11. Structure of (a) adenosine, (b) guanosine, (c) cytidine, and (d) uridine.

uridine are shown in Fig. 1.12. The infrared spectrum for uridine in D_2O solution shows two bands assigned to carbonyl stretching, and further evidence for the third form in Fig. 1.12 is that the spectra of substituted

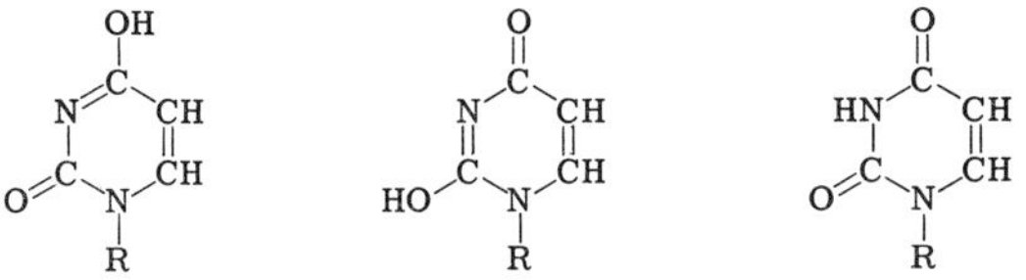

Fig. 1.12. Possible resonance structures for uridine.

uridines for which resonance is not possible are similar only to that of uridine itself if the derivatives are predominantly in the diketo form. In a similar way the predominant solution structures for the bases in the ribonucleosides adenosine, guanosine, cytidine, and uridine are shown in Fig. 1.11. The nonpredominant tautomers affect the hydrogen-bonding properties of the bases and may be involved in the formation of mutants.

Phosphate

Formation of a phosphate ester by addition of a phosphate unit to one of the OH groups on the sugar component of a nucleoside produces a *nucleotide*. For D-ribose there are three possible nucleotide monophosphates, e.g., for

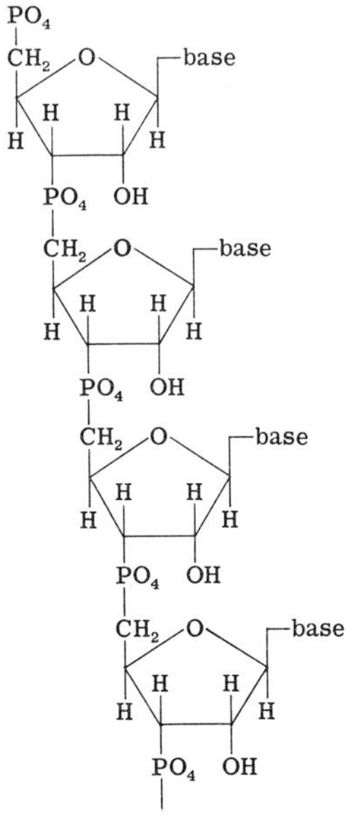

Fig. 1.13. Chain sequence typical of a deoxyribonucleotide, i.e., a DNA segment.

adenosine there are adenosine 5'-phosphate, 3'-phosphate, and 2'-phosphate, which are also known as adenylic acids. For deoxyribose only the 5'- and 3'-phosphates occur. For both DNA and RNA, the phosphates utilize positions 3' and 5' to form the polynucleotide chain. Figure 1.13 shows the structure of a section of the DNA chain.

REFERENCES

1. From H. R. Mahler and E. H. Cordes, "Basic Biological Chemistry." Harper, New York, 1966.
2. See, for example, E. Margoliach and W. M. Fitch, *Ann. N.Y. Acad. Sci.* **151**, 359 (1968).
3. From R. V. Eck and M. O. Dayhoff, "Atlas of Protein Sequence and Structure." National Biomedical Research Foundation, Silver Spring, Maryland, 1966.
4. A. Liquori, P. DeSantis, A. L. Kovacs, and L. Mazzarella, *Nature* **211**, 1039 (1966).
5. G. Vanderkooi, S. J. Leach, G. Némethy, and H. A. Scheraga, *Biochemistry* **5**, 2991 (1966).
6. D. Balasubramanian, *J. Amer. Chem. Soc.* **89**, 5445 (1967).
7. D. L. Kauffman, *J. Mol. Biol.* **12**, 929 (1965).
8. From D. A. Rees, "The Shapes of Molecules: Carbohydrate Polymers," Contemporary Science Paperbacks No. 14. Oliver & Boyd, Edinburgh, 1967.
9. J. N. Davidson, "The Biochemistry of the Nucleic Acids," 6th ed. Methuen, London, 1969.

2

CONFORMATION

Introduction

Although biological macromolecules may sometimes function in an entirely chemical sense, i.e., as a storehouse for material, in almost all cases the actual shape of the molecule is vitally important. Shape (or conformation) and function are closely related. In some cases, subtle conformational changes allow a chemical function to be performed, as with enzymatic degradation or the transport of oxygen and carbon dioxide in hemoglobin. In other cases, the conformation is important in a structural relationship, as with the fibrous proteins. It is, therefore, of some importance that we attempt to delineate the factors effecting the conformation of biological macromolecules. The most likely place to start is with synthetic homopolymers, particularly polypeptides, and this is indeed the area that has received most attention. The forces dictating the conformation of a polypeptide can emanate from essentially three sources—(a) from within the molecule itself, i.e., intramolecular forces, (b) from other similar molecules (intermolecular forces), or (c) from the solvent or surrounding dissimilar medium, which represents a special type of intermolecular interaction. Even if the detailed nature of the conformation-directing forces is not understood, some simple rules may be postulated. If a polymer or biopolymer consists of backbone atoms B and side groups S in the form shown in Fig. 2.1, then the ultimate shape will

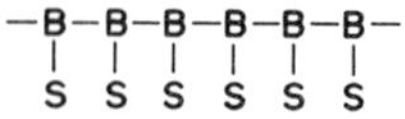

```
 -B-B-B-B-B-B-
  |  |  |  |  |  |
  S  S  S  S  S
```

Fig. 2.1

depend on allowable rotation about the B—B bond and must be such that the side groups do not overlap each other's space. These two criteria, however, still allow an infinite variety of shapes for such a chain. A postulate that simplifies the situation somewhat is that, in all likelihood, the shape of the molecule will be determined by the interaction between neighboring B—S pairs and, thus, the conformation of a B—S pair such as that shown in Fig. 2.2 is repeated down the chain. This postulate is known as the equivalence

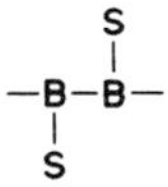

```
        S
        |
     -B-B-
     |
     S
```

Fig. 2.2. Schematic representation of a repeating unit of two backbone B and side-chain S groups.

condition and is believed to hold for most, if not all, homopolymers in the solid state. The result is a molecule extending down a linear path. Clearly, for polysaccharides, polynucleotides, and polypeptides which are not homo-polymers, the equivalence condition does not apply. Even here, however, if the biological molecule has a definable repeat sequence, such as ABABAB or ABCABC or even ABXABYABZ, etc., there is a strong probability that such a molecule will extend down a "linear" axis and will therefore be uniform, often helical in conformation and will typically be found in fibrous proteins. On the other hand, random sequences may, and generally do, lead to globular or random structures. The effort of molecular biologists has constantly been directed toward providing a unified understanding of the relationship between primary and secondary structure (conformation) and the function of the biological macromolecule.

The Peptide Bond

The peptide backbone consists of a repeated sequence of three atoms and their associated substituents. In principle, then, rotation may occur about any of the three bonds connecting them, as shown in Fig. 2.3. However,

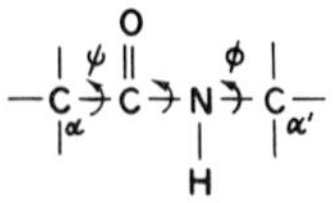

$$
\begin{array}{ccccc}
 & & \overset{\displaystyle O}{\underset{\displaystyle \quad}{|}} & & \\
 \overset{\displaystyle |}{\underset{\displaystyle |}{}} & \overset{\psi}{} & \overset{\displaystyle \|}{} & \overset{\phi}{} & \overset{\displaystyle |}{}
\end{array}
$$

$$
-\underset{|_\alpha}{C} \overset{\psi}{\to} \underset{}{C} \overset{}{\to} N \overset{\phi}{\to} \underset{|_{\alpha'}}{C} -
$$

with H below the N.

Fig. 2.3. Bonds in a peptide unit about which rotation might occur.

because of some specific properties of the peptide group (Fig. 2.4), the central C—N bond is not freely rotatable.

x-Ray crystallographic studies of simple peptides have shown that whereas the N—C_α bond length is 1.47 Å, as expected for a single bond, the C—N peptide bond is 1.32 Å. This short bond compares with the value of 1.25 Å for the average C=N bond length in model compounds and is taken as strong evidence for double-bond character in the peptide bond. (For comparison purposes the C—C bond length is ~1.53 Å, whereas C=C is 1.33 Å.) In addition, all of the x-ray work on the peptide group has shown that the six atoms C_αNHCOC are, or are very close to being, coplanar. Similar planarity is shown by other compounds possessing a double bond, such as ethylene. We should therefore modify our picture of the peptide group in Fig. 2.3 to that in Fig. 2.4, which shows the delocalization of an

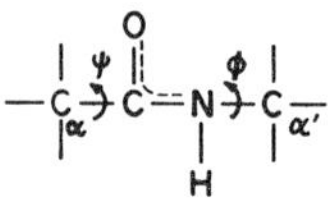

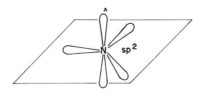

Fig. 2.4. Delocalization of electrons in the peptide group leads to coplanarity of O, C, N, H atoms and prevents rotation about the CO—NH bond.

electron from the carbonyl double bond. The double-bond character in the peptide bond has several very important implications and is worthy of further study. The first step is to examine the electron configuration of the three atoms directly involved in the bond, O, C, and N. All three atoms are involved in double bonding and, thus, the atomic orbitals of C and N have sp^2 hybridization. The nitrogen atom has an electron in each of the three sp^2 orbitals and a lone pair in the p_x orbital lying perpendicular to the plane of the sp^2, as shown in Fig. 2.5. In the double-bonded configuration, the carbon

Fig. 2.5. Spatial configuration of electron orbitals of nitrogen.

atom may also be regarded as having three electrons in hybridized sp^2 orbitals and a fourth bonding electron in the perpendicular p_x orbital. The oxygen atom, which normally contributes one electron to the carbonyl double bond, may be regarded as having one electron in the bonding sp_z orbital, one in the perpendicular p_x orbital, and two in the nonbonding p_y orbital, as shown in Fig. 2.6.

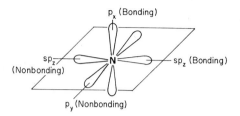

Fig. 2.6. Spatial arrangement of electron orbitals of oxygen.

When these three atoms are bonded together, one O(sp$_z$) and one C(sp^2) overlap to form a single σ bond, and as do C(sp^2) and N(sp^2). Since the N—H bond also involves a σ bond in the sp^2 plane of the nitrogen, all six atoms of the peptide group would be expected to be coplanar. The three p$_x$ orbitals are parallel to each other and form a conjugated π bonding system through which the four p$_x$ electrons [O(1), C(1), N(2)] traverse. The top and side view of the bonding scheme are shown in Figs. 2.7 and 2.8.

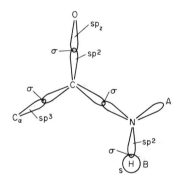

Fig. 2.7. Electron configuration of the atoms in the peptide group.

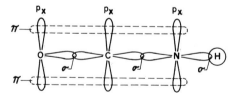

Fig. 2.8. Representation of the delocalized π electrons in the peptide group.

Although we shall examine the orbital situation more carefully when considering the spectroscopic properties of polypeptides, the geometrical features will be discussed here. Not only are the six peptide atoms coplanar, but because of the C=N double-bond character the O and H atoms can lie *cis* (position A in Fig. 2.7) or *trans* (position B) to this bond. Furthermore, the bonding angles may be specified more closely. The most precise angles can be arrived at by extrapolation from x-ray measurements of small peptides; as a result of a quantum mechanical analysis of the bonding scheme, the peptide bond is believed to contain approximately 40% double-bond character. On this basis the bond angles are established as shown in Fig. 2.9.

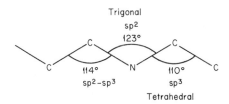

Fig. 2.9. Bond angles and electron hybridization of the polypeptide backbone.

The $NC_\alpha C$ angle is expected to be close to the tetrahedral angle of 110° because of the sp^3 hybridization of the C_α atom. This is, in fact, the observed value. The CNC_α angle is expected to be 120° based on the sp^2 hybridization of the nitrogen atom; the experimental value is 123°. Finally, the $C_\alpha CN$ angle should lie between sp^3 tetrahedral and sp^2 trigonal and is, in fact, found to be 114°.

A careful study of the configuration of the peptide group has shown that a nonplanarity (rotation about the C—N bond) of up to 6° or so is tolerable and that, in most cases, the O and H atoms should lie *trans* to the peptide bond. There are no known examples of the *cis* peptide bond occurring in polyamino acids, polypeptides, and proteins. We will see later, however, that for polyimino acids, the steric configuration is such that the energetics of the *cis* and *trans* forms are similar. Both have been shown to exist experimentally.

The steric restrictions on rotation about the peptide bond leave only two unique rotatable bonds per peptide in the backbone. These are the C_α—C and N—C_α bonds; rotations about these bonds are defined by the angles ψ, and ϕ, respectively. Since these rotation angles have become the basis for modern conformational analysis of polypeptides, we shall consider them in more detail shortly. First, however, it is appropriate to consider an alternative method of representing the peptide backbone, i.e., in terms of dihedral angles.

It was recognized some years ago that a convenient method of describing the polypeptide backbone was in terms of alternating planes drawn (a) through the plane of the peptide bond and (b) through the plane of the NC_α and $C_\alpha C$ bonds. The angles between these planes are called dihedral angles and are designated τ_{NC} and τ_{CC}. A schematic diagram of these alternating planes is

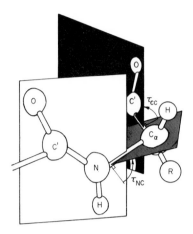

Fig. 2.10. Two planar peptide groups linked by an α carbon atom. The two dihedral angles τ_{NC} and τ_{CC} define the rotation about the single bonds $N-C_\alpha$ and $C_\alpha-C'$; $\tau = 0$ when the two planes defining the angle coincide; values of τ increase as the planes are rotated as indicated by the arrows. The figure shows a segment of a right-handed helix of L-amino acids. From J. A. Schellman and C. Schellman, *in* "The Proteins" (H. Neurath, ed.), Vol. 2, p. 10. Academic Press, New York, 1964, with permission.

shown in Fig. 2.10. Evidently, τ_{NC} and τ_{CC} represent rotations about the NC_α and CC_α bonds and are thus directly related to our previous backbone rotation angles ϕ and ψ. The specific relation* between the two is as follows:

$$\phi = -\tau_{NC} + 180° \tag{2.1}$$

$$\psi = -\tau_{CC} + 180° \tag{2.2}$$

It now remains for us to define either set of parameters so that backbone geometry can be analysed. Although several different standards were originally adopted, the definition now generally accepted is as follows: the backbone is "flat," with the $CONHC_\alpha$ atoms of successive residues coplanar, as shown in Fig. 2.11. In this conformation, both ϕ and ψ are defined as 0°.

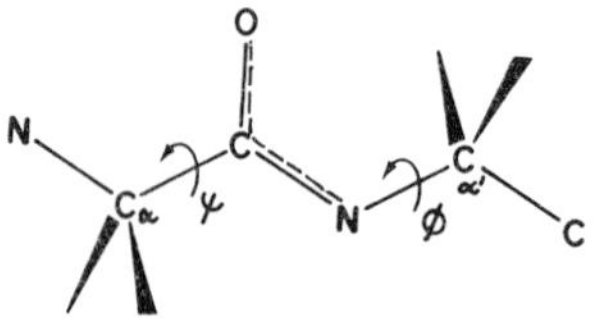

Fig. 2.11. Planar representation of the peptide backbone with ψ and $\phi = 0°$.

* Most of the literature relating to theoretical prediction of conformation has used this convention, although others have also been used.

Rotations of ϕ and ψ are then taken as positive in the clockwise direction looking in the $\overrightarrow{C_\alpha CN}$ direction. More specifically, looking along the $\overrightarrow{NC_\alpha}$ bond, as shown in Fig. 2.12, the configuration for $\phi = 0°$ can be seen. Note

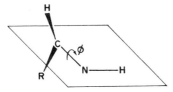

Fig. 2.12. Representation of increasing ϕ for L-residue; N—H bond rotating in the indicated direction.

that the hydrogen atom H and side chain R are above and below the plane of C_α—NH, respectively, thus defining the L configuration of the peptide. Similarly, looking along the C_α—C bond (Fig. 2.13), with the O, C, C_α, and

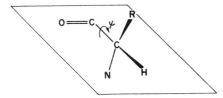

Fig. 2.13. The ψ rotation in positive direction (from zero); near group rotated.

N atoms in the same plane, $\psi = 0°$ and R and H lie above and below the plane, respectively, in the L configuration.

Factors Controlling Conformation of Polypeptides

We have already noted that, if the conformation of homobiopolymers is indicated by nearest-neighbor peptide interactions, the equivalence rule requires that successive monomer units be related by a symmetry axis, and, as a consequence, the polymer chain adopts an extended helical form.* If there is no identifiable peptide sequencing, then this postulate is not appropriate. The forces directing polypeptide conformation are essentially of three types: (a) electrostatic forces emanating from the dipolar electronic displace-

---

* In many biological texts, the helix is referred to as a folded form. The term folded is reserved here for a change in direction of the symmetry (fiber) axis.

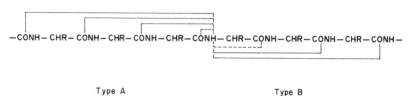

Fig. 2.14. Representation of the $N=H\cdots O=C$ hydrogen bond with the normally allowable atomic distances and angular distribution noted.

Type A   Type B

Fig. 2.15. Possible sites for hydrogen bonding in helical structures (not all are sterically feasible).

ment, (b) van der Waals forces, (c) hydrogen bonding. Of these forces, early workers supposed that the last two were dominant, mainly on the basis that the van der Waals interaction described the effective "volume" of the atoms involved and thus prevented atomic overlap, whereas hydrogen bonds were the driving force to conformational integrity. Although we now recognize several shortcomings of these ideas, nevertheless they did lead to very significant progress in conformational studies of polypeptides.

During the 1930's and 1940's, it became increasingly clear from x-ray studies of fibrous proteins that essentially two different conformations were present; these were termed α and β structures. A detailed historical account of the evolution of these concepts is outside the scope of this text, but the interested reader is referred to Ref. 1. Other structures have since come to light, but the α and β conformations are probably the most fundamental to polypeptide structure. It was recognized from these x-ray data that the β structure probably contained extended chains in a sheetlike arrangement, whereas the α form was probably helical. The nature of the helix, however, evaded detailed analysis for some years.

The driving force to formation of both these structures was believed to be hydrogen bonding. x-Ray analysis of the crystalline structure of many simple peptides, amides, and amino acids has shown (2) the presence of hydrogen bonds of the form $-NH\cdots O=C$, with an $H\cdots O$ distance of 2.7–3.1 Å and an $O\cdots HN$ angle (see Fig. 2.14) of 0°–30°.

The possible sites of hydrogen bonding in a single polypeptide chain (backbone) were suggested in 1950 by Bragg *et al.* (3) and are shown schematically in Fig. 2.15. Such intramolecular hydrogen bonds would cause the chain to twist into a helix. Note that two different hydrogen-bonded ring systems are apparently possible (Figs. 2.16 and 2.17), containing $3n' + 4$ (type A) and $3n' + 5$ (type B) atoms, respectively. Although we now know

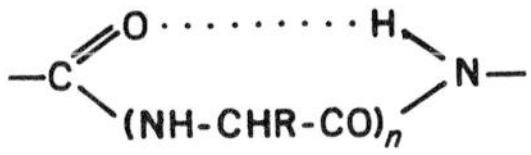

$$-C\overset{\displaystyle \nearrow O \cdots\cdots\cdots H}{\underset{\displaystyle (NH-CHR-CO)_n}{\diagdown}} \overset{\displaystyle }{\underset{\displaystyle }{N-}}$$

Fig. 2.16. Type A hydrogen-bonding scheme of Fig. 2.15. The number of atoms in a hydrogen-bonded ring is $3n' + 4$.

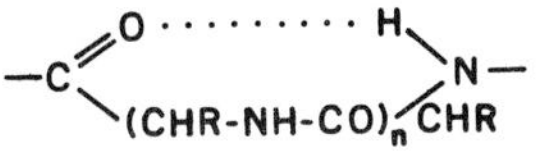

$$-C\overset{\displaystyle \nearrow O \cdots\cdots\cdots H}{\underset{\displaystyle (CHR-NH-CO)_n CHR}{\diagdown}} \overset{\displaystyle }{\underset{\displaystyle }{N-}}$$

Fig. 2.17. Type B hydrogen-bonding scheme of Fig. 2.15. The number of atoms in the hydrogen-bonded ring is $3n' + 5$. [Figs. 2.15–17 drawn according to the concepts of Bragg *et al.* (3).]

that many of these combinations are energetically unfavorable, others remain of considerable interest, including the $3n' + 4$ with $n' = 1$ and particularly $n' = 3$. The first of these structures with a $(3 \times 1 + 4 = 7)$ seven-membered hydrogen-bonded ring system, is a helix with a twofold screw axis; that is, successive peptide units are related by a translation along the axis, accompanied by a rotation through $180°$ about the chain axis (4). Bragg *et al.* (3) suggested that a helix should be defined by s_n, where s is the symmetry number and n the number of atoms in the hydrogen-bonded ring. On this basis the preceding structure would be a 2_7 helix. Two forms (left-handed and right-handed helices) of the 2_7 structure are shown in Fig. 2.18. Although the designation 2_7 is still sometimes used, this method of describing helices has fallen into disfavor, mainly because many helices, particularly in synthetic non-biopolymers, have no intramolecular hydrogen bonding. It is now customary

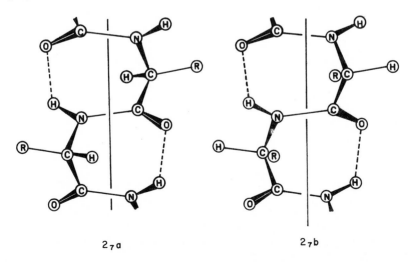

2_7a 2_7b

Fig. 2.18. Two forms of the 2_7 helix in a polypeptide of L-amino acid residues.

to define a helix by η_m, where η is the number of units that form a perfect repeat after m turns. Thus, the Bragg 2_7 helix has two peptide units in one complete turn of the helix and according to the current nomenclature would be a 2_1 helix.* It is not necessary for a helix to have an integral number of peptides per turn; for example, a structure with $3(3) + 4 = 13$ atoms in a hydrogen-bonded ring may have eighteen peptides in five complete turns and is then classified as an 18_5 helix.

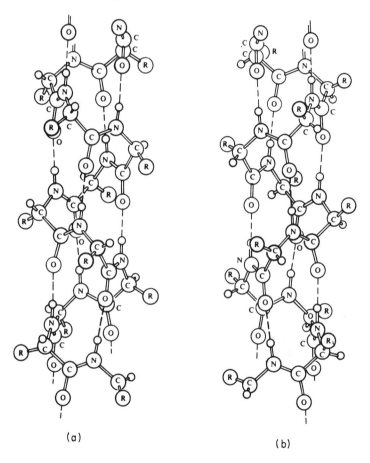

(a)   (b)

Fig. 2.19. Drawings of the left-handed (a) and right-handed (b) α-helical forms of a polypeptide chain containing L-amino acids. [From B. W. Low and J. T. Edsall, in "Currents in Biochemical Research" (D. E. Green, ed.), p. 398. Wiley (Interscience), New York, 1968, with permission.]

---

* It is also of interest to note that the β forms are degenerate, but different 2_1 helices, in the extended arrangement.

As we have seen, the number of possible intramolecularly hydrogen bonded helices is apparently very large. However, Pauling and Corey (5) defined three major criteria for the formation of a stable polypeptide chain conformation as (a) planar peptide groups, (b) nonoverlapping van der Waals radii, and (c) hydrogen bonds in which the OHN angle is acceptable (i.e., less than 30°). These criteria reduced considerably the number of possible structures to be examined. In addition, Pauling and Corey realized that chain molecules could adopt nonintegral helical conformations, and they thus provided a major breakthrough in the analysis of peptide structure. When this work was followed by detailed x-ray diffraction studies of the fibrous proteins (see Chapter 10), the α-helix and pleated-β-sheet structures were fully resolved and they now serve as a firm basis for the analysis of protein structure.

The α helix is the structure with eighteen peptides in five turns referred to previously. Each peptide N—H and C=O is involved in intramolecular hydrogen bonding, the hydrogen bonds lying essentially parallel with the fiber axis, as indicated in Fig. 2.19. Both left- and right-handed helices may be constructed from L residues or D residues (those composed of L residues are illustrated in the figure). Perhaps the easiest way of visualizing the left- and right-hand helices is to imagine that helix observation is made down the axis. If, on tracing the spiral, the tracing goes to the left, or counterclockwise, away from the observer, then the helix is left-handed; if the tracing goes to the right, away from the observer, the helix is right-handed. Diagrams of the first coil of the left- and right-handed helices are shown in Fig. 2.20. Of

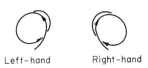

Left-hand   Right-hand

Fig. 2.20. Schematic representation of left- and right-hand helices; the former rotates to the left, away from the observer, and the latter to the right. Inversion of the helix does not affect the helix sense.

course, the helix has the same sense if viewed from either end. More details of the α structure will emerge when the forces directing chain conformation are examined in the following sections.

The other major structure presented by Pauling and Corey was that of the pleated-sheet or β conformation. If extended chains are placed side by side, they may be arranged such that efficient intermolecular hydrogen bonds are formed in the manner demonstrated in Figs. 2.21 and 2.22. In the first, the chains are antiparallel, i.e., the sequence $\overrightarrow{NC_\alpha CO}$ is of opposite direction

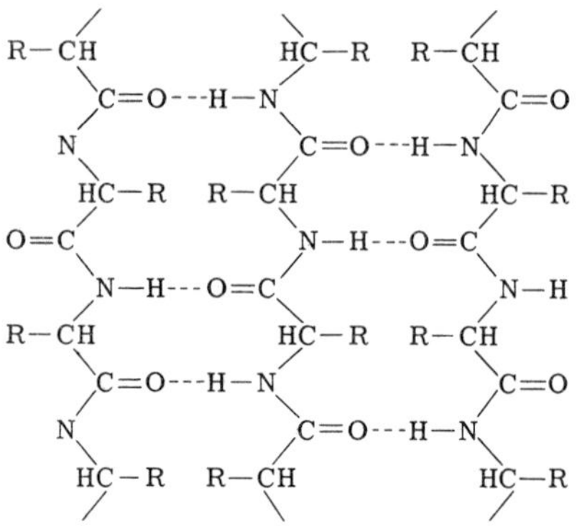

Fig. 2.21. Chain arrangement in the antiparallel β sheet.

in adjacent chains. In Fig. 2.22, the chains are arranged parallel. We shall see later that with modern characterization methods it is possible to distinguish between these possibilities with the result that the antiparallel structures are known to be the favored form. Pauling and Corey, however, pointed out that a pleated rather than sheet form was in accord with the x-ray data. The pleated form maintains the intermolecular hydrogen bonding, the pleat occurring along the molecular axis, as shown in Fig. 2.23. In addition to the commonly occurring α and β structures, model building showed several other structures to be possible from a purely stereochemical point of view. Pauling and Corey defined a so-called γ helix, while later workers have described the π and ω helices. The physical parameters of these structures are given in Table 2.1 along with those of other helices found for a number of synthetic polypeptides.

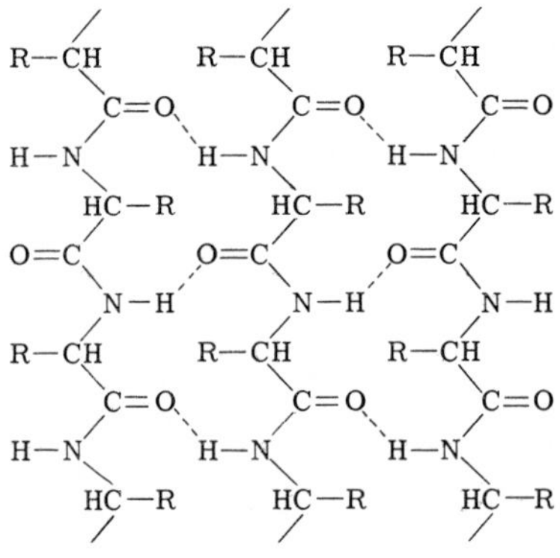

Fig. 2.22. Chain arrangement in the parallel β sheet.

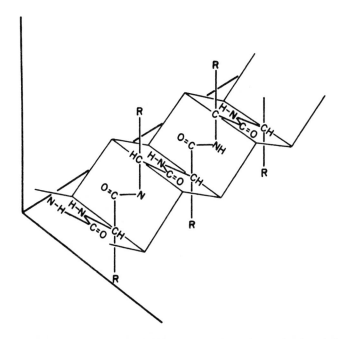

Fig. 2.23. Schematic three-dimensional representation of a molecular chain in the β conformation showing the "pleated sheet" effect. [From G. M. Barrow "Physical Chemistry" McGraw Hill, New York (1966) p. 813, with permission.]

TABLE 2.1

Properties of Some Standard Polypeptide Conformations

Structure	Symmetry[a]	Peptides/turn (n)[b]	Residue repeat (Å)
α Helix	18_5	3.60	1.50
β Parallel	2_1	2.0	3.25
Antiparallel	2_1	2.0	3.5
γ Helix		5.14	0.98
π Helix		4.40	1.15
ω Helix	4_1	-4.0	1.325
Polyglycine II	3_1	± 3.0	3.10
Poly-L-proline II	3_1	-3.0	3.12
2_7 Helix	2_1	2.0	2.80
Poly-L-proline I (*cis* peptide bond)	10_3	3.33	1.9

[a] Rational helices quoted.

[b] Positive values for right-hand, negative for left-hand helices.

Characteristics of Helices

Several important aspects of helical polypeptides have now emerged:
(a) the planarity of the peptide bond, (b) the importance of van der Waals
repulsive forces defining effective atomic volume, (c) the role of hydrogen
bonds, (d) the possible existence of various integral (e.g., 2_1) and nonintegral
(e.g., 18_5) helices. Evidently, some of the other factors that we have introduced,
i.e., backbone rotation angles and dihedral angles, must also bear some
definable relation to the characterization of helices. Furthermore, only intra
molecular hydrogen bonding has been considered as a major driving force to
helix formation. The number of peptides per helix turn, and the hydrogen-
bonding scheme, do not fully characterize the polypeptide helix since it can
be a narrow helix with gentle twist or a fat helix with strong twist.

A schematic polypeptide helix is shown in Fig. 2.24 with three major

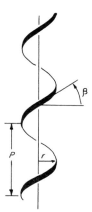

Fig. 2.24. Schematic diagram of a continuous helix showing parameters $P =$ pitch,
$r =$ radius, and $\beta =$ pitch angle which define the helix. [From R. E. Dickerson, *in* " The
Proteins" (H. Neurath, ed.), Vol. 2, p. 682. Academic Press, New York, 1964.]

characteristic parameters, the pitch of the helix P, radius r, and pitch angle β.
It is convenient to further extend the definition to take into account the peptide repeat d. The peptide repeat is the length (in angstroms) of the peptide
unit in the helix projected onto the helix axis. If all pairs of backbone rotation
angles ϕ and ψ are equal (i.e., the equivalence condition is obeyed), then the
peptide repeat d is identical for all peptides in the chain. This (equivalence)
criterion is valid for all homopolypeptides examined so far, although for
polydipeptides, polytripeptides, etc., the peptide repeat takes on the con-
notation of an average value for the individual peptide or an exact value for
the repeating group.

If a helix has three identical peptides per turn, i.e., 3_1 as in Fig. 2.25, then $d = P/3$. For the 18_5 α^* helix there are $n = 3.6$ peptides per turn and $d = P/3.6$. It is to be noted that the helix pitch P and the helix *repeat* are identical for integral helices, e.g., 3, but are different for nonintegral helices. The helix repeat (that is the translation parallel to the fiber axis necessary to bring the operator to an equivalent position) is, for an 18_5 helix, equal to $5P$.

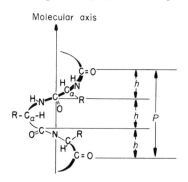

Fig. 2.25. Schematic diagram showing the relation between the peptide repeat h and helix repeat P for a threefold helix.

The two peptide parameters that define the helix are then the number of peptides per turn n and the peptide repeat d (in angstroms). In some cases the former is reduced to a rotation per peptide in the helix, $\theta = 360°/n$,; for the 3_1 helix, $\theta = 120°$. The helix could, of course, be equally well defined by ψ and ϕ or by τ_{NC} and τ_{CC}. It would be particularly useful if relationships between n, d, and the backbone rotation or dihedral angles were available. Miyazawa has presented such a relationship (6), which is as follows†:

$$\cos \tfrac{1}{2}\theta = 0.817 \sin \tfrac{1}{2}(\tau_{NC} + \tau_{CC}) + 0.045 \sin \tfrac{1}{2}(\tau_{NC} - \tau_{CC}) \qquad (2.3)$$

and

$$d \sin \tfrac{1}{2}\theta = 2.967 \cos \tfrac{1}{2}(\tau_{CC} + \tau_{NC}) - 0.664 \cos \tfrac{1}{2}(\tau_{CC} - \tau_{NC}) \qquad (2.4)$$

Therefore, with the aid of Eqs. (2.1)–(2.4), we are readily able to transpose between $n(\theta)$, d; τ_{NC}, τ_{CC}; and ϕ, ψ. The first two parameters may be extracted directly from x-ray fiber diagrams, as will be shown later. On the other hand, theoretical approaches to polypeptide conformation evolve around the calculation of allowable and most probable backbone rotation angles.

---

* Slight distortions are readily achieved which change the symmetry of the 18_5 helix. However, it is convenient to retain the designation 18_5 even for this distorted helix. In this context the helix is often referred to as an 18_5 rational helix.

† When using these equations it is necessary to note that $\theta \leq 180°$ and a quadrant correction must be applied to the trigonometric functions of the right-hand side.

Of course, not all polyamino acids are helical. However, planar extended structures may also be thought of as degenerate helices for some purposes. As an example, the extended zigzag form with $\phi = 0°$ and $\psi = 0°$ may be regarded as a degenerate 2_1 helix for which we can specify both d and $n (= 2)$.

Steric Maps of Dipeptides

If the van der Waals forces alone are considered, then the force field around the atoms might be regarded as centrosymmetric, and an effective radius is defined. The simplest approaches treat the effective volume of atoms as exclusion volumes in which two atoms cannot exist simultaneously. Ramachandran (7) has surveyed the available crystallographic data for model compounds and has tabulated the minimum approach distances for various atoms and small groups ($-CH_3$, CH_2, etc.). Any interatomic distance greater than this minimum value is known as "fully allowed." However, since in exceptional circumstances interatomic distances less than than the "fully allowed" value do occur, a "partially allowed" region is defined where the distance between the atoms can be less (usually by 0.3–0.5 Å) than the "fully allowed" separation. These data, which are summarized in Table 2.2 (7),

TABLE 2.2

Normally Allowed and Outer-Limit
Contact Distances (in angstroms)

Contact	Normally allowed	Outer limit
C···C	3.20	3.00
C′···C′	2.95	2.90
C···N	2.90	2.80
C···O	2.80	2.70
C···P	3.40	—
N···N	2.70	2.60
N···O	2.70	2.60
N···P	3.20	—
O···O	2.70	2.60
O···P	3.20	—
P···P	3.50	—
C···H	2.40	2.20
N···H	2.40	2.20
O···H	2.40	2.20
P···H	2.65	—
H···H	2.00	1.90

have been used to define regions on a plot of ϕ against ψ that correspond to sterically feasible structures. A model of two contiguous monomer residues is set up using a computer (in all but the earliest work), and all possible interatomic distances are calculated while ϕ and ψ are varied, usually in increments of 5°, 10°, or 20°. By comparing the results with the figures in Table 2.2 one can designate the various ϕ, ψ positions as fully allowed, partially allowed, or disallowed. The allowed regions can then be shown on a plot of ϕ against ψ. Figure 2.26 (8) shows such a map for L-alanylalanine. The fully

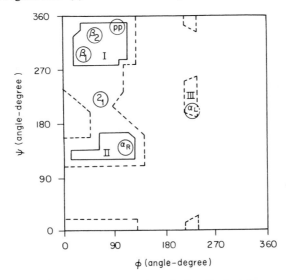

Fig. 2.26. Hard-sphere steric map of poly-L-alanine. Dashed lines represent allowed regions for outer contact distances, while the solid lines represent regions for normal contact distances.

allowed regions lie within the solid lines, while the dashed lines indicate the extent of the partially allowed regions. The ϕ, ψ angles corresponding to well-known polypeptide structures are also shown.

    Examination of Fig. 2.26 shows that there are two fully allowed regions. The larger, in the upper left portion of the figure, contains the antiparallel β (β_1), parallel β (β_2) and poly-L-proline II (pP) conformations; the smaller (center left) contains the right-handed α helix (α_R). The equivalent left-handed α helix (α_L) and 2_1 helix (2_7 in old nomenclature) occur in the partially allowed regions.

    Somewhat more informative is the map shown in Fig. 2.27 for various dipeptide pairs. It can be seen that L-glycylglycine has a wide range of conformational possibilities. As the peptide side chain increases in size the allowed regions are considerably decreased. All the combinations presented

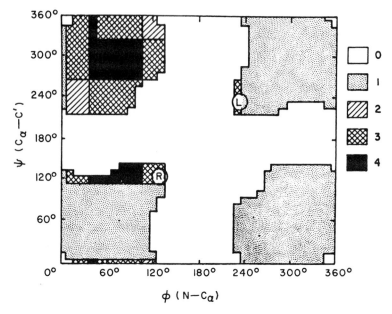

Fig. 2.27. Allowed areas of the steric map for various dipeptides. In area 0 no conformations are allowed. Conformations in areas 1–4 are allowed for glycylglycine, in areas 2–4 for glycyl-L-alanine, in areas 3–4 for higher straight-chain homologs, while only area 4 is allowed for glycyl-L-valine and glycyl-L-isoleucine. The circles marked R and L indicate the location of the standard right- and left-handed α helix. [From G. Némethy, S. J. Leach, and H. A. Sheraga, *J. Phys. Chem.* **70**, 998 (1966), with permission.]

which do not have branched C_β side chains may occupy the α-helix conformation, but glycyl-L-valine with the branched C_β on valine is sterically prohibited from forming α helices, the β-sheet conformation (top left) being a feasible alternative. Such steric maps are clearly limited in scope since they do not allow specific predictions to be made, and all allowed conformations appear equally probable.

The contact criteria listed in Table 2.2 have the drawback that they incorrectly assume angular independence for the van der Waals separation. In addition, the results for the partially allowed region can be misleading in that it is quite possible for a conformation with many partially allowed contacts to be less favorable than another with simply one contact just slightly less than the extreme limit. For these reasons, the simple steric maps have, to a large extent, been supplanted by steric *energy* maps in which the detailed nature of interatomic forces are taken into account. However, for more complex biopolymers, such as polynucleotides, steric maps still afford a particularly useful method of presenting conformational information. In a sense, steric maps, which are usually derived from direct computer analysis and printout, are directly analogous to model building in which the steric limitations are imposed by the physical volume of the atoms. The latter and older method does, however, allow contacts between atoms in nonneighboring peptide groups to be examined fairly readily. Often the steric maps compiled from computer analysis are presented for peptide pairs, although there is no practical reason for eliminating nonnearest neighbors from consideration except for ease of calculation. The inclusion of extra residues has the effect of decreasing the allowable rotation angles only slightly. One of the most interesting questions in biological chemistry is phrased, therefore, in terms of the extent to which interactions between peptide pairs, and hence their sequence, control the conformation and function of a polypeptide. To answer this question, at least in part, it is necessary for us to consider in more detail the nature of the interaction between atoms in the peptide chain.

Interatomic Interaction Potentials

The study of interatomic forces is, of course, one of the fundamental aspects of modern chemical physics, and many workers have presented theoretical expressions that are useful in calculating the forces stabilizing biopolymers. Since most of the approaches are based on quantum mechanics and are quite complex mathematically, only the salient aspects are mentioned here.

Nonbonded Potential Functions

Expressions for the potential energy due to steric nonbonded interactions between atoms take account of the electron–electron repulsion and electron–nucleus attraction forces. The two most commonly used potential functions are the Buckingham potential,

$$V = -A/r_{ij}^6 + B\exp(-\mu r_{ij}) \tag{2.5}$$

and the Lennard-Jones potential,

$$V = -A/r_{ij}^6 + B/r_{ij}^{12} \tag{2.6}$$

where V is the potential energy due to nonbonded interaction of two atoms (i and j) separated by a distance r_{ij}, and A, B, and μ are constants. These two potential functions have been used in most conformational analyses so far performed, although the various workers have differed in their choice of values for the constants A, B, and μ.

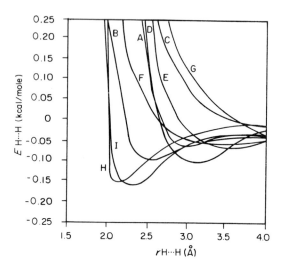

Fig. 2.28. Theoretical curves for hydrogen bond functions (from Ref. 8). A, M. L. Huggins, *in* "Structural Chemistry and Molecular Biology" (A. Rich and N. Davidson, eds.), p. 761. Freeman, San Francisco, California, 1967. B, A. R. Abe, R. L. Jernigan, and P. J. Flory, *J. Amer. Chem. Soc.* **88**, 631 (1966). C, R. Pauncz and D. Ginsburg, *Tetrahedron* **9**, 40 (1960). D, R. L. McCullough and P. E. McMahon, *J. Phys. Chem.* **69**, 1747 (1965). E, N. P. Borisova and M. V. Volkenstein, *Zh. Strukt. Khim.* **2**, 346 (1961). F, D. A. Dows, *J. Chem. Phys.* **35**, 282 (1961). G, J. O. Hirschfelder and J. W. Linnett, *J. Chem. Phys.* **18**, 130 (1950); P. DeSantis, E. Giglio, A. M. Liquori, and A. Ripaimonti, *J. Polym. Sci.* **A1**, 1383 (1963). H, A. J. Hopfinger and A. G. Walton, *J. Macromol. Sci. Phys.* **B3**, 171 (1969). I, R. A. Scott and H. A. Scheraga, *J. Chem. Phys.* **42**, 2209 (1965).

Figure 2.28 shows the results of nine different proposed functions for the H $\cdots$ H atom interaction. The derivation of potential energy curves relating the interaction energy of two atoms to the distance between their centers involves the assumption that the energy at a given distance is essentially independent of orientation of the atoms and the nature and strength of any bonds between the atoms. In the calculation of polypeptide conformation which began about 1963, four or five types of potential functions have been used, and they have been reviewed recently by Ramachandran and Sasise-kharan (9), Scheraga (10), and Hopfinger (8). As these authors have pointed out, the prediction of peptide geometry based on these functions is remarkably invariant, despite considerable differences in the form and magnitude of the potential functions. However, the predicted absolute energy of a conformation is not likely, at present, to be particularly accurate.

Equations of the form (2.5) and (2.6) relate to the hard-sphere model for the repulsive term becoming infinite at the van der Waals radius (r_0). Thus the first major step is achieved by softening the square well potential form for hard cores and adding in inverse sixth-power attractive potential.

Electrostatic Contributions

From the electronegativity concept it is evident that different atoms in a peptide share differently in the electron distribution. The interaction between these partially charged atoms is characterized by an electrostatic potential. The net result of the charge separation can be characterized either by a set of dipoles associated with the polar amide groups of the peptide backbone (and appropriate side groups) or by a set of residual charges associated with the constituent atoms. Descriptions of the dipoles associated with the peptide backbone are considered in Chapters 5 and 6; backbone dipoles form the basis of the Flory (11) method for assessing electrostatic contributions. Ooi and co-workers (12) have also calculated the partial charges of amide and side-chain groups in such a way that they reproduced the dipole moments (so-called monopole approximation). Hopfinger (8) has used the Del Re (13) method of calculating charge distribution on the assumption that σ bonding is the significant mode of bonding in polypeptides. The Del Re method may be considered as the quantum mechanical analog (for localized σ bonds) of the Hückel method for π (delocalized) electrons. The method has been applied successfully by Del Re to the problem of predicting first-order chemical shifts for proton magnetic resonance spectra (see Chapter 8). The results for a glycyl and a prolyl residue are presented in Table 2.3. Poland and Scheraga

TABLE 2.3

Partial Charges on the Atoms of Glycine and Proline Residues as Calculated by the Del Re Method[a]

Glycine		Proline (also Hypro)	
Atom	Charge (esu)	Atom	Charge (esu)
H	+0.201	N	−0.204
N	−0.353	C′	+0.197
C′	+0.207	O	−0.136
O	−0.135	H_α	+0.043
H_α	+0.046		

[a] For more details of the calculation see Chapter 8.

(13) have modified the Del Re technique by assuming that there is an additional contribution from the π bond. Their calculated values are shown in Table 2.4. Actually the sophistication of accurate charge calculation is at this time somewhat secondary to the requirement for a good criterion of dielectric

TABLE 2.4

Partial Charges[a] on Glycine and Proline Residues According to Poland and Scheraga

| Atom | Proline (Hypro) | | | Glycine | | |
	σ	π	Total	σ	π	Total
H				+0.204		+0.204
N	−0.188	+0.140	−0.048	−0.342	+0.140	−0.202
C′	+0.103	+0.208	+0.311	+0.110	+0.208	+0.318
O	−0.077	−0.348	−0.425	−0.074	−0.348	−0.422

[a] In fractions of unit electron charge.

constant in the region of the atoms. An examination of the electrostatic potential equation

$$V_{es} = -\frac{e_i e_j}{\varepsilon r_{ij}} \tag{2.7}$$

(e's are the residual charges separated by distance r, ε is the dielectric constant) readily shows the major error in magnitude of V_{es} arising from possible errors of severalfold in ε. Although no really good method of estimating an appropriate molecular dielectric constant exists, a value of 3.5 is commonly used, based on high-frequency dielectric constant measurements of solid amides and polyamides (14).

Hydrogen Bond Energy

Probably no other interaction in the field of molecular biology has received as much attention as the hydrogen bond. We have seen how the development of the concept led to the postulate of the α helix and other structures. Nevertheless, the nature of the bond is at best poorly understood. As with the other interactions, there are essentially two models, the classical Pauling (15) electrostatic model and the quantum mechanical model (16). For the purposes of conformational analysis it is not necessary to delve into the theoretical implications of the hydrogen bond, but rather to choose an appropriate model. Figure 2.29 (8) shows a plot of bond length versus bond angle for the experimentally measured N—H $\cdots$ O=C bond; the variability of the bond is obvious. In principle, the steric nonbonded plus the electrostatic point charge energies between the amide hydrogen and carbonyl oxygen should be the hydrogen bond energy. This approach applied to the peptide hydrogen bond gives a reasonable bond energy of approximately 3–7 kcal/mole

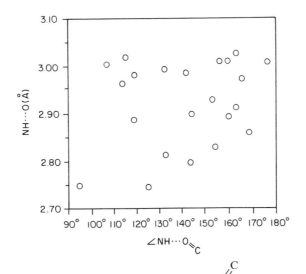

Fig. 2.29. A plot of NH⋯O distance versus NH⋯O angle in various crystals. (From Ref. 8.)

but gives a short optimum bond (2.2 Å) and is angular independent. Various angular-dependent equations have been used (8–10). The earliest (and most straightforward) function employed by DeSantis (17) used the Stockmayer relation for the interaction between polar molecules, i.e.,

$$V_H = -2V_m \frac{R^6}{r^6}\left(1 - \frac{R^6}{r^6}\right) - \frac{\mu_1 \mu_2}{r^3} g(\theta_1, \theta_2, \phi_1 - \phi_2) \qquad (2.8)$$

where the first term on the right corresponds to the minimum-energy Lennard-Jones equilibrium position $V_s = V_m$ and $r = R_1$ and represents the nonbonded interaction between the two dipolar groups. The second term is the electrostatic interaction between the dipoles μ_1 and μ_2 attached to the groups X—H and Y—Z. The directional nature of the dipole interaction is represented by the angular dependence of g_1 given by

$$g(\theta_1, \theta_2, \phi_1 - \phi_2) = 2 \cos \theta_1 \cos \theta_2 - \sin \theta_1 \sin \theta_2 \cos(\phi_1 - \phi_2) \quad (2.9)$$

where θ_1 and θ_2 are the angles that μ_1 and μ_2 make with the line joining them, and ϕ_1 and ϕ_2 are the azimuthal orientations of the two dipoles (see Fig. 2.30). Further, the assumption was made that the two dipoles N—H and O=C are centered on the H and O atoms, respectively, such that the first term on the right in Eq. (2.8) represents the nonbonded interaction between the H atoms and the O atom.

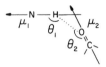

Fig. 2.30. Dipolar interaction parameters in hydrogen bond formation.

Torsional Potential Functions

A barrier to internal rotation of a single bond arises from orbital inter-actions of the substituents and of the bonding electrons. Brant and Flory (11) and Ooi *et al.* (12) have used torsional potential functions of the type

$$V_T = \frac{V_0}{2} [1 \pm \cos \alpha\chi] \tag{2.10}$$

where χ is the bond rotation angle. The choice of the constants V_0 and α has differed between the two schools. However, if the barrier is 0–1 kcal little difference is said to result in the peptide steric energy map. On the other hand, De Santis (17) has ignored the barrier altogether without apparently negating the validity of conformational predictions. Hopfinger (8) using a somewhat different approach, has pointed out that the bond energy changes as a func-tion of electronic distribution, which in turn results from environmental changes, irrespective of whether the bond is rotated. His torsional potential is calculated from the bond order, which is sensitive to rotation. In this latter approach the influence of hydrogen atoms is explicitly considered, whereas in most previous calculations the hydrogen atoms are implicitly included with the carbon atom to which they are attached.

Other Factors

Although the peptide bond is almost always treated, in theory, as planar, torsion about the peptide bond, and variability of bond lengths as a function of conformation, should be taken into account in refining the calculations. It is known that in simple peptides slight distortions from planarity occur (18, 19). Figure 2.31 shows the torsional energy about the peptide bond as a function of rotation. The calculations are those of Hopfinger (8) and are based on an extended Hückel quantum mechanical calculation. With this type of potential function the energy barrier to rotation is approximately 24 kcal/mole with the *trans* form being approximately 4 kcal/mole lower in energy than the *cis* form. Examination of this curve shows that a perturbation of 1 kcal/mole can accommodate a 7° rotation of the peptide bond.

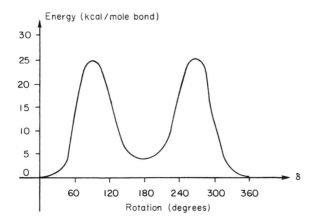

Fig. 2.31. The torsional potential about the backbone amide $C' \cdots N$ bond. (From Ref. 8.)

The extended Hückel method used for the Hopfinger calculations is thought to be the most accurate method for studying peptide conformation. Hoffmann (20) and Hopfinger have both used this method to examine the conformation of diglycine and dialanine, and quite recently the French school has extended this approach (21). However, despite the increase in sophistication of potential functions, the method is limited by the excessive computational time required for even the simplest units. Consequently, we shall concentrate on the results of the earlier work, which, in any case, are in fairly good qualitative agreement in the case of diglycine and dialanine.

All of the above-mentioned calculations relate to vacuum conformation, and clearly much is to be gained by consideration of the possible effects of solvents. Some very preliminary approaches to this problem have been attempted; notably, calculations of the free energy of solvation for various atoms in the peptide have been presented by Gibson and Scheraga (22). Somewhat more qualitative, but more useful from a practical viewpoint, are the calculations for helix/coil transition temperatures for collagenlike polypeptides presented by Brown *et al.* (23).

Conformational analysis based on energy criteria requires calculation of the interatomic distances, bond orientation angles, etc., for the different ϕ, ψ positions using the same approach as for the older hard-sphere method. These data are substituted into the various expressions for the potential energy, and the energy is summed for each position. Contours of equal energy are then plotted on the ϕ, ψ diagram, and the various energy minima are considered to indicate the most favorable structures.

We are now in a position to examine the most favorable conformation of some dipeptide pairs and polypeptides as deduced from potential energy maps.

Potential Energy Maps

Diglycine

The potential energy map for the diglycine pair is shown in Fig. 2.32. It was calculated using the Hoffman extended Hückel approach. The map is symmetrical due to the fact that two equivalent radicals (hydrogen atoms) are located on the C_α atom. This symmetry disappears for optically active amino acids with nonequivalent radicals. The potential energy contours, in kilocalories per mole above the deepest energy minimum, reveal regions of most favorable interaction. The steric energy map (Fig. 2.32) should be compared with the steric (exclusion) map (Fig. 2.27).

Several regions of particular stability are to be noted. In terms of absolute ranking of energy, the lowest-energy conformation of the dipeptide is given by $\phi = 120 \pm 5°$, $\psi = 120 \pm 5°$ and that of the corresponding symmetrical pair by $\phi = 240 \pm 5°$, and $\psi = 240 \pm 5°$. These values are very close to the values for the right- and left-hand α helix proposed by Pauling and coworkers. The right-hand α helix, for example, should have $\phi = 126°$ and $\psi = 130°$. It is to be noted that hydrogen bonds stabilizing the α-helix are formed between

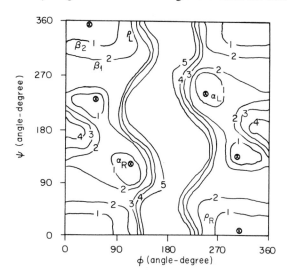

Fig. 2.32. Glycine conformational map using the extended Hückel technique (contours in kilocalories per residue per mole above deepest minimum). Encircled X's denote relative minima. Conformations denoted are α_L, left-hand α helix; α_R, right-hand α helix; P_L, left-hand polyglycine II; P_R, right-hand polyglycine II; β_1, parallel β sheet; β_2, antiparallel β sheet. (From Ref. 8.)

one peptide unit and its next-but-one-nearest neighbor and thus these bonds are not considered in the dipeptide calculation. The most favorable conformation for the dipeptide pair is located in a fairly sharp energy well and distortion of ϕ and ψ to the exact α conformation apparently destabilizes the conformation by approximately 0.5 kcal. However, it must be remembered that the precise planarity of the peptide bond has been assumed and may account for this slight discrepancy.

The second, and essentially equally favorable, conformation of the diglycine pair occurs broadly in the region of Fig. 2.32 labeled 2. The minimum in this region appears at $\phi = 50°$, $\psi = 360°$ with $E = 0.1$ kcal above the absolute minimum. This figure is certainly within the bounds of the accuracy of calculation. Very little energy is expended in changes from $\phi = 0°$, $\psi = 360°$ to $\phi = 110°$, $\psi = 340°$. In this region occurs the combination $\phi = 103°$, $\psi = 343°$, which is a left-hand, threefold (3_1) helix with $d = 3.1$ Å. Similarly the corresponding optical isomer with $\phi = 257°$, $\psi = 17°$ is the right-hand 3_1 helix. This conformation has been observed experimentally for polyglycine and is known as the polyglycine II (pGII) structure.

Also included in this energy region is the $\phi = 0°$, $\psi = 0°$ conformation corresponding to the extended chain. Other familiar structures such as the 2_1 helix (2_7, old nomenclature) and β structures are somewhat less stable on an intramolecular basis. A comparison of the energetics of the diglycine pair and known polymer conformation is made in Table 2.5.

Thus we see that by calculation based upon the vacuum conformation of a diglycine pair, the favored structures correlate well with information on polypeptides in the solid state, thus indicating that a major portion of conformation-directing forces does indeed emanate from nearest-neighbor interactions. If, however, this were the whole story, polyglycine would most likely

TABLE 2.5

Ranked Energy Minima for a Diglycyl Pair Using Extended Hückel Method

Rank	ϕ^a	ψ^a	Energy (kcal above min.)	Conformation
1	120	120		α_R
	240	240		α_L
2	245	235	~0.3	ω
3	100	330	~0.8	$pGII_L$
	260	30	~0.8	$pGII_R$
4	322	35	~1.0	*anti-parallel β*
5	299	67	~2.0	β

a Values given in degrees.

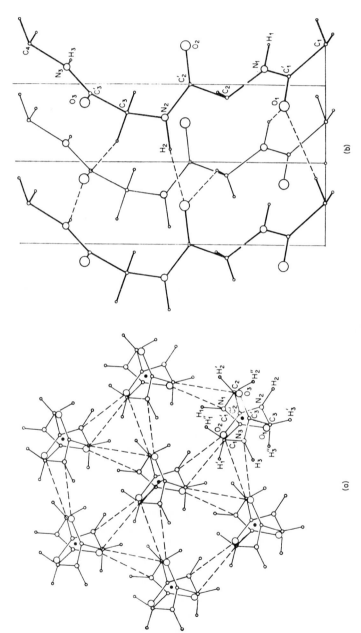

Fig. 2.33. Structures of polyglycine II. (a) Viewed along the axis, which is taken to be a left-handed threefold screw axis; (b) side view showing NH···O (and CH···O) hydrogen bonds. (From G. N. Ramachandran, ed., "Collagen," Academic Press, New York, 1967, with permission.)

have an α or pGII structure, with the other conformations being somewhat less likely. In fact the α, β, and pGII structures can all enter into efficient hydrogen-bonding schemes that have a modifying effect on the ranking of the energy minima. The α helix forms intra-molecular hydrogen bonds, and we must therefore suppose that an isolated strand of polyglycine in vacuum would most likely take on such a conformation.

In the solid state, α helices do not hydrogen bond with their neighbors and the intermolecular bonding is probably fairly weak. In contrast, the pGII and β structures are heavily hydrogen bonded in the solid state and therefore gain energetically on an intramolecular basis. Figure 2.33 shows the inter-molecular hydrogen bonding achieved in the polyglycine II conformation. Only the β sheet and pGII structures are thoroughly characterized experimentally for polyglycine, suggesting that the hydrogen-bonding schemes are at least 0.5 kcal/mole or so more efficient for these structures than for the α form. It should be noted, however, that biopolymer conformation in the solid state is also to some extent dependent upon the mode and conditions of crystallization. Since polyglycine is insoluble in most solvents and, as are nearly all biopolymers, is thermally unstable (i.e., does not melt), the methods of crystallization are strictly limited. In any event, glycine residues located in a peptide sequence can conveniently adapt to most peptide conformations. Since, on an absolute basis, the conformation of a biopolymer is determined by free-energy considerations alone, it is possible that the large number of plausible conformations for a glycine residue weighs heavily on entropy considerations.

Poly-L-Alanine

We have discussed the hard-sphere steric contour map for dialanine and it is instructive to examine a similar map that includes potential functions (see Fig. 2.34). The similarity between the two maps is obvious, although larger allowed regions are apparent in the potential energy map. An interesting extension of this calculation includes non-nearest-neighbor peptide inter-actions, thus incorporating the α-helix hydrogen-bonding scheme. Figure 2.35 shows the result of such an extension. The energy minima in the region of left- and right-hand α helices have deepened as expected, but the allowed areas for other conformations are much more limited. Detailed calculations in the region of the minima ($\phi = 240°$, $\psi = 240°$ and $\phi = 120°$, $\psi = 120°$) show the right-hand α helix to be very slightly (0.2 kcal/peptide) more stable* than

---

* The calculated energy difference between left- and right-handed α helices reflects the underlying differences in potential functions used. Thus Flory and Liquori with their "hard" potential functions obtained 2.0–2.5 kcal/peptide, whereas the "soft" functions of Scheraga and Kitaigorodsky give values of 0.2–0.5 kcal/peptide.

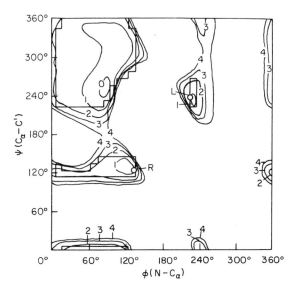

Fig. 2.34. Energy contours for an L-alanyl residue. The units of energy are kilocalories per mole. The circles marked R and L indicate the locations of the standard right- and left-handed α helix. Superimposed is the steric map. [Taken from Scheraga (10), with permission.]

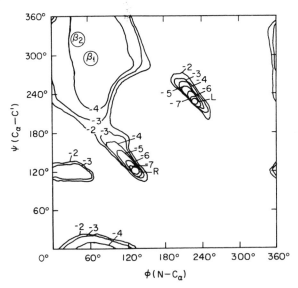

Fig. 2.35. Energy contours for poly-L-alanine helices in kilocalories per mole. The symbols R and L indicate the positions of the right- and left-handed α helices; β_1 and β_2 designate the positions of the parallel and antiparallel pleated-sheet structures. [From Ooi *et al.* (12), with permission.]

the left-hand. The symmetry of the diglycine steric energy map is now missing due to the asymmetric substitution on the C_α atom. On an intramolecular basis alone we would expect the poly-L-alanine helix to be a right-hand helix. High molecular weight poly-L-alanine is usually α helical, both in the solid state and in non-helix-breaking solvents, but the gain in energy achieved by the β structures through intermolecular hydrogen bonding renders these structures of comparable stabilization energy. Mechanical deformation is often sufficient to convert the α form to the β form experimentally, and end effects in low molecular weight poly-L-alanine often render the β form stable.

The left-hand polyglycine II (pGII$_L$) structure is approximately as stable as the β structures on an intramolecular basis, but the right-hand (pGII$_R$) helix is completely unstable. Stabilization of pGII$_L$ by intermolecular hydrogen bonding cannot occur in the solid state because the alanine methyl group causes helix separation that is too large to accommodate the interchain CO—HN hydrogen bonds. All other structures are considerably less favorable theoretically and have not been observed experimentally.

The potential energy diagram for poly-L-alanine is believed to be somewhat similar to those for other polyamino acids that are not substituted on the C_β atom, e.g., poly-L-leucine and the esters of poyl-L-glutamic acid. The complications of calculating the potential energy diagrams for the amino acids with longer side chains include the increased degrees of freedom produced in rotatable side-chain bonds and side-chain/side-chain interactions. The side-chain rotations are denoted by χ (see Fig. 2.36). In the Hopfinger nomenclature $\chi = 0°$ when the plane through the C_β and C_γ atoms bisects the NC$_\alpha$C' angle. The Scheraga nomenclature differs by 60°. In both cases the rotations are counterclockwise as observed from the peptide. Calculations by

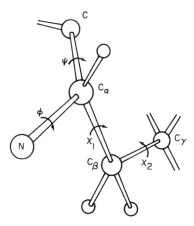

Fig. 2.36. Representation of side-chain rotation angles χ_1 and χ_2.

Hopfinger show that (excluding side-chain/side-chain interactions) poly-L-glutamic acid forms a stable α helix, and calculations of side-chain arrangement in the right-hand α helices of poly-L-tyrosine and poly-γ-methyl-L-glutamate have also been reported (24).

If the β carbon of a polyamino acid is substituted, the character of the potential energy diagram changes substantially.

Poly-L-Valine

The simplest β-substituted homopolypeptide is poly-L-valine. Scheraga and co-workers have presented a potential energy diagram for polyvaline, a modification of which is shown in Fig. 2.37. It can be seen that the left- and right-hand α helices are still stable conformations, but now the β conformations (top left of Fig. 2.37) are much more stable intramolecularly than are structures with non-β-substituted carbons. Thus with the additional stabilization of intermolecular hydrogen bonding in the solid state, the β structure is, as expected, the normal stable conformation. It is noteworthy that isolated molecules of poly-L-valine should be equally stable in α or β conformations. Thus it is possible that in very weakly interacting solvents polyvaline could

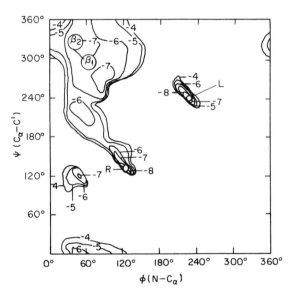

Fig. 2.37. Energy contours for poly-L-valine helices in kilocalories per mole. The energy at each value of ϕ and ψ represents the minimum value for side-chain–side-chain and side-chain–backbone interactions. The symbols R, L, β_1, and β_2 have the same meaning as in Fig. 2.35. [From Ooi *et al.* (12), with permission.]

be in α form. Most hydrogen-bonding solvents, however, are likely to promote a β-form. Although few calculations have been carried through for other β-substituted peptides, it seems likely that only minor modifications of the valine map will result.

Other Polyamino Acids

Another interesting example of a polyamino acid examined by theoretical conformational analysis is poly-L-serine. This biopolymer has an unsubstituted C_β atom and therefore might be expected to behave similarly to poly-L-alanine or poly-L-leucine. However, the side-chain hydroxyl opens up the possibility of specific side-chain interactions. Conformational calculations have shown (8, 25) that the most stable intramolecular chain structure is a left-hand α helix with the side-chain hydroxyl tending to form a hydrogen bond with the nearest backbone oxygen located in the N-terminal direction for $\chi_1 = 240°$. For $\chi = 150°$ the hydrogen bond is formed in the C-terminal direction, and for $\chi_1 = 0°$ the OH locates midway between the two nearest backbone oxygens.

The β structure (antiparallel) is only 0.8 kcal/mole per side-chain residue less stable than the left-hand α structure on an intramolecular basis and is clearly the most stable intermolecular conformation in accord with experiment. In this case $\chi_1 = 240°$. From the few cases examined for peptides with large side chains, the following seem likely:

1. The value of χ_1 is usually near 120° or 240°; $\chi_1 = 0°$ is a possibility in some cases.

2. The side-chain conformation energy is most strongly dependent upon χ_1.

3. For polar side chains the last rotation χ_h on a side chain usually assumes a value that allows maximum hydrogen bonding with a carbonyl in the backbone.

4. The left- and right-handed α helices and β-sheet structures normally produce the lowest backbone–side chain interaction energy.

Prolyl Oligomers

Proline, an imino acid, has conformational properties distinctly different from the amino acids. In globular proteins it is often found at chain kinks or "hairpin bends," whereas in fibrous proteins of the collagen type it has distinct helix-directing properties. Several groups have published work concerning proline–proline potential energy curves, but before considering them in more detail it is first necessary to recognize the new conformational aspects introduced by the secondary amine characteristics of proline. In Fig. 2.38

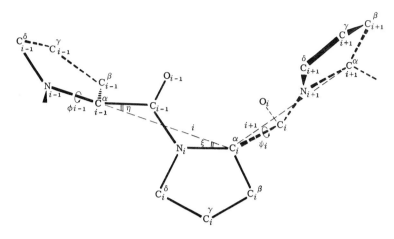

Fig. 2.38. Schematic representation of the poly-L-proline II chain with the imide groups in the planar *trans* conformation and the rotation angle ψ fixed ($\sim 103°$). Subscripts are serial indices of proline residues. [From Schimmel and Flory, *Proc. Nat. Acad. Sci. U.S.*, **58**, 52 (1967). With permission.]

(26) is shown a diprolyl radical indicating the two backbone rotation angles ϕ and ψ. It can immediately be seen that ϕ is fixed by the presence of the pyrrolidine ring, the usual value taken from x-ray data being 103°. Although this feature apparently simplifies the problem by restricting a geometric representation to one variable angle ψ, in fact, slight ring puckering and the possibility of *cis/trans* isomerism in the peptide bond add additional variables.

We have seen that although there is the possibility of *cis* and *trans* isomers (across the peptide bond) with the polyamino acids, the *trans* form is so much more stable that the *cis* form has never been identified. Calculations of the energetics of the peptide bond in *cis* and *trans* forms for diproline, however, indicate that both are approximately equally likely, and this feature proves to be unique to the polyimino acid class of polypeptides. An additional interesting feature of proline oligomers and poly-L-proline is that conformational stability must arise entirely from non-hydrogen-bonding forces since the peptide nitrogen is fully substituted and unprotonated.

Because ϕ is invariant, it is convenient to present the conformational energetics in terms of E versus ψ maps. Figure 2.39 shows such a diagram for the *trans* form (27). Included are the relative stabilization energetics (kilocalories per peptide) for di-, tri-, and tetraproline. It can be seen that a rather broad energy minimum occurs in the range $\psi = 280°–370°$, showing that the molecule is rather flexible. The curve for the polymer is shown in Fig. 2.40. The minimum-energy (most stable) form occurs at $\psi = 335°$ for the isolated (vacuum) conformation of the polymer, but an examination of the intermolecular interaction indicates a most stable conformation at $\psi = 343°$. This slight discrepancy arises because for $\psi = 343°$ the polyproline chain forms a

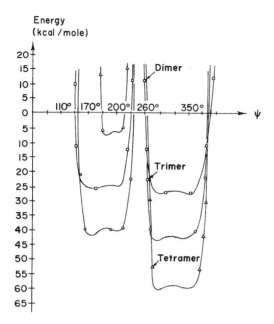

Fig. 2.39. Energy diagram ($\psi_2 = \psi_3 \cdots = \psi_m = \psi$) for *trans* form of di, tri- and tetra-L-proline. Indicated points are used to define individual curves and are not actual data points. The *total* conformational energy is plotted and has been redrawn from the data in Ref. 27 to conform with the nomenclature used throughout this book.

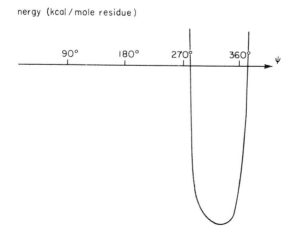

Fig. 2.40. *trans*-Poly-L-proline potential energy plot using the equivalence rule. (From Ref. 8.)

area. That which has been carried through has concentrated on the L-proline residue because of its known conformation-directing properties. In the first reported work (1968) on Gly-L-Pro and L-Ala-L-Pro, Schimmel and Flory (26) concluded that the right-hand α-helix conformation is disallowed for an amino acid containing a β carbon, which is followed by the prolyl residue. However, it has been pointed out (30) that the Lys(87)-Pro(88) combination in myoglobin is in an R_α (right-handed α-helix) conformation, and subsequently (31) several other proteins have been shown to have X-Pro pairs, where X is Leu, Ala, Val, Thr, Ser, or Ile, in the R_α form. A theoretical resolution of this problem (31), obtained by the quantum mechanical method, has demonstrated that R_α is allowed for X-Pro pairs if the proline ring is appropriately puckered. It would appear that the ring itself may have a different configuration in different biopolymers, and this additional subtlety must be taken into account in conformational analysis.

An extension of the theoretical method has also been attempted (32) for tripeptides of the form Gly-Pro-X and Gly-X-Pro because of their relevance to the collagen structure. In this case, because of the complexity of five degrees of rotational freedom in the peptide backbone, the assumption was made that the proline residue constricts the adjacent rotatable bonds. It was thus possible to calculate both the intramolecular and intermolecular energetics of various collagen models. It was clear, however, from these calculations that the relative stability of various conformers could not be derived entirely from a calculation of the interaction energy. A more recent calculation (23) of the *free* energy of the same (poly) tripeptides has shown that the relative stability is correctly predicted with the more refined approach. The success of this method augers well for the future of theoretical conformational analysis of biopolymers.

Maps for Globular Proteins

The ultimate objective of polypeptide conformation energetics is the *ab initio* theoretical prediction of protein structure. In practice we are still far from this objective since it is possible to obtain the complete conformation of only a simple decapeptide and since, if the peptide side chains are large, three or four peptides form the limit of direct calculation. Thus it is obviously impossible to calculate all possible conformations for, say, a globular protein. However, if the local conformation is dictated mainly by peptide pair interactions or if simplifying assumptions can be made in the calculation, possibly based on experience derived from complete x-ray determination of structure, then it becomes conceivable that the conformation of chain segments in a protein can be predicted.

If the peptide pair hypothesis is correct, the dipeptide pair angles must correspond with those predicted by theoretical calculation. Unfortunately, with 21 different amino acids there are $20^2 = 420$ combinations of dipeptides. As yet, very few dipeptide maps are available. We have seen that many of the amino acids seem to function similarly, e.g., as α-helix formers, and it is highly likely that the steric maps for many peptide pairs will be very similar. The ϕ, ψ angles for the various regular chain conformations have not been determined from atomic coordinates, but simply calculated, by means of the equations given in Chapter 3, from the observed n and h values obtained from the x-ray pattern. Direct determination of ϕ and ψ has so far been possible only for certain globular proteins whose structures have been determined by single-crystal methods (see Chapter 13). Even here the individual atoms are not properly resolved and the conformational angles can be measured only to within $\pm 20°$. Nevertheless, comparison of these angular combinations with a typical map for a peptide pair gives an indication of the extent to which the conformations in a globular protein are determined by immediate-neighbor interactions. In Figs. 2.43 and 2.44, data for rotation angles taken from

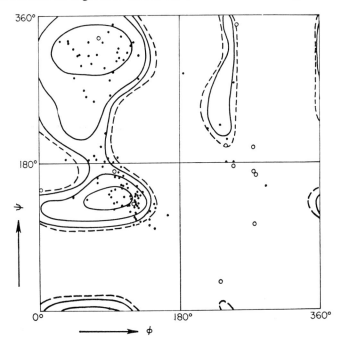

Fig. 2.43. The ϕ, ψ values of lysozyme along with (L-alanine) energy contours: ---, zero contour; —, contours at intervals of 1 kcal/mole going down to −3 kcal/mole. The points lying outside the highly favored regions are predominantly those involving glycyl residues. [From Ramachandran and Sasisekharan (9), with permission.]

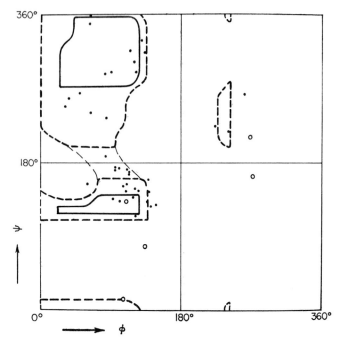

Fig. 2.44. A steric map (dialanine) showing areas of allowed conformation for minimum (dotted line) and maximum (full line) van der Waals radii. Superimposed are the conformations of the nonhelical peptide pairs for myoglobin. Open circles for glycyl residues. [From Ramachandran and Sasisekharan (9), with permission.]

complete x-ray structural analysis of lysozyme and myoglobin are plotted on the steric energy map for alanine. Liquori and co-workers (33) appear to have been the first to present such data.

As expected, no points lie in the forbidden steric region (5 kcal). But probably more significantly, with few exceptions the observed angles lie within 3 kcal of the minimum-energy conformation of diglycine pairs. There is a significant clustering around the right-hand α-helix minimum (74% of the myoglobin molecule consists of almost regular α-helical sections) and in the β-sheet region. It is notable, however, that the angles are widely distributed in the favorable energy regions, suggesting that the protein is conformed to satisfy the pair interaction criteria and long-range criteria (surface charge, hydrophobic interior) rather than any intermediate objectives of α-helix, β-sheet, etc., formation. The fact that several points lie near the forbidden (5 kcal) region shows that there is some distortion. The conformations adopted by globular proteins are such that the polar residues are found on the outside of the molecule, while the nonpolar side chains are more likely to be on the inside. The polar forces due to the solvent are thus thought to play a decisive role in determining the conformation. Such considerations also apply to the crystalline form since the crystals contain a high proportion (more than 40%) of water. The effect of hydrophobic forces on conformation has been discussed further by Gibson and Scheraga (22). Comparison of sequence and conformation has shown, as expected, that proline residues are generally found in the nonhelical sections of the chain. However, it is difficult at present to make further generalizations. Hemoglobins from different sources are found to adopt essentially the same conformation, but seemingly have such widely different sequences that a systematic relation to the conformation remains elusive.

Calculation of Tertiary Polypeptide Structure and Summary of Polypeptide Conformation

Several workers have published results of calculations concerning the secondary and tertiary structure of some small peptides and segments of proteins. Scheraga has reviewed the calculated conformations of the cyclic nonapeptides gramicidin S, oxytocin, and vasopressin (10). These calculations involve many shortcuts and approximations, and it will be most interesting to see how they compare with experimental conformations. Others have cal-

TABLE 2.6

Conformation Angles (in degrees) for Some Standard Polypeptide Structures[a][b]

Structure	C Bond angle	ϕ	ψ
α Helix	110	122.2	133.0
	109.5	133.0	122.8
β Sheet			
Parallel		61	293
Antiparallel		38	325
Polyglycine II			
Left-hand	110	100	330
Right-hand		260	30
Poly-L-proline II	110.0	102.8	325.9
ω Helix[b]	109.9	244.4	235.4
Poly-L-proline I[c]	114	97	338

[a] For *trans*, planar peptide units ($\omega = 0°$). See Appendix for complete tabulation of relation between ϕ, ψ, n, and h.
[b] Distorted peptide unit ($\omega = 5°$).
[c] The *cis* peptide unit ($\omega = 180°$).

culated the tertiary structure of α-helix fibers, crystal structure of some poly-imino acids and polytripeptides, as well as portions of the collagen structure. These results will be presented in the chapter on tertiary structure.

Although the application of theoretical methods to conformational analysis has been relatively successful, its role has so far been mainly confirmatory. This approach has many pitfalls, but it is a new method and probably its major triumphs lie ahead.

We have seen, then, that many of the structures arrived at by classical model building methods are substantiated by potential energy calculations, but relatively few are particularly stable. The most relevant of these structures and their corresponding bond rotation angles are listed in Table 2.6.

Conformational Analysis for Other Biopolymers

The work so far described has concentrated on the polypeptides, and most of the conformational analysis to date has been developed for the study of protein structure. This emphasis reflects the variety of possible chain con-formations for the proteins, the importance of these materials relative to other biopolymers, and the prospective use of conformational analysis in the pre-diction of the structure for globular proteins. In addition, the proteins are relatively simple in that the amide group is rigid and can be described by two parameters, ϕ and ψ. A similar rigidity is displayed by the polysaccharides; the monomer consists of a ring whose conformation is not expected to depend on the nature of the helix (for pyranose sugars at any rate), and thus the rela-tive orientation of the monomers is defined by the rotations about the linkage bonds. However, the backbone in other biopolymers such as the poly-nucleotides is not rigid and four parameters are required in addition to the two for the linkage in order to define the nucleic acid backbone conformation. Conformational analysis has complemented model building in the investiga-tion of polysaccharide and polynucleotide structures, and these two fields are reviewed below.

Polysaccharides

The details of polysaccharide structure, including the monomer con-figuration, nomenclature, and type of linkage, are described in Chapters 1 and 11. In view of the large number of monomers found in nature and the different types of linkage, the possibilities for variety in polysaccharide structure are enormous. Fortunately, however, a high degree of complexity is not found for most of the common examples. Most of the polysaccharides

serving skeletal, storage, and gel network functions are long-chain homo- ·
polymers or alternating copolymers of the type $[A-B]_n$. These are built up
mainly from a handful of monomer units such as D-glucose, D-mannose,
D-galactose, and D-xylose, and from some of their derivatives, among which
the $NHCOCH_3$-substituted and uronic acid compounds are very common.

In the structural work that has been reported so far, all the polysaccharides
have had monomer residues in the six-membered "pyranose" ring form
shown in Fig. 2.45a. Analogy is made with the cyclohexane ring in Fig. 2.45b.
If the latter were planar, the C—C—C bond angles would need to be 120°.
This strain is avoided by puckering the ring as shown in Figs. 2.45c and d to
give the two possible conformations known as the "chair" and "boat" forms.
Of these the "chair" is the more stable, where the six CH_2 groups are almost
perfectly staggered, while those on the "bottom" of the boat are in the eclipsed
position relative to each other. In addition, for each carbon atom in the chair
form, one C—H bond is almost perpendicular to the mean plane of the ring,
while the other is almost in this plane. These bond directions are known as
axial and *equatorial*, respectively, and this positioning minimizes the interaction between the hydrogens. In the boat conformation interaction also
occurs between hydrogens on the bow and stern. The situation is a little
more complex when the substituted pyranose rings are considered (34). The
eight possible pyranose conformations are shown in Fig. 2.45e. Some of these
are identical for simple tetrahydropyran but are not so for the substituted
compounds. For the sugars, the chair forms C1 and 1C are the most stable,
and Reeves (34) predicted which one would be favored for some 20 monomers
on the basis of the interactions of the side groups. For these sugar molecules,
each carbon in the ring is bonded to a hydrogen and a substituted group. In
principle, unfavorable interactions will occur when any of the substituted
groups are in the axial position. Any group that is axial in the 1C conformation will be equatorial in the C1 conformation, and the most favorable ring
is the one having the most hydrogens, and therefore the fewest substituted
groups, in the axial positions. Consider β-D-glucose, which is shown in Fig.
1.8. In the C1 conformation the hydrogens are all axial and the OH and
CH_2OH groups are equatorial. For the 1C ring, however, these bulky groups
are axial and unfavorable interactions are predicted. Thus it is no surprise
that the C1 conformation occurs in the solid state (35). β-D-glucose is an
extreme case and in most other cases, the stable ring is the one which has the
fewest large axial groups.

As with the work on polypeptides, conformational analysis of polysaccharides began with molecular models, and the development of computer
methods has made this model building more systematic and reliable. The
allowed angular ranges for the individual bond rotations in the cellulose chain
were first considered by Jones (36) using skeletal models. Calculation from the

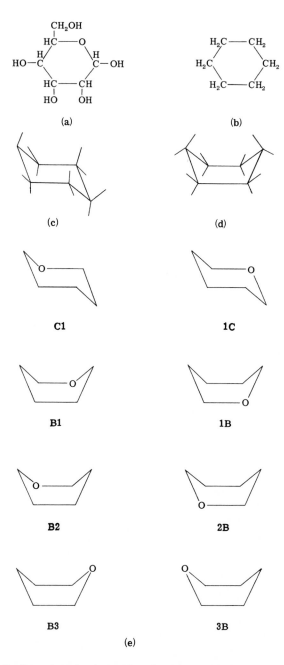

Fig. 2.45. Conformations for six-membered saccharide rings. (a) Pyranose, (b) cyclohexane, (c) chair form, (d) boat form, (e) eight possible conformations for pyranose.

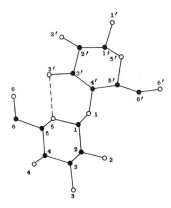

Fig. 2.46. Projection of the crystalline conformation of cellobiose. Note the formation of the O-5 $\cdots$ H—O-3′ intermolecular hydrogen bond.

atomic coordinates began with Ramachandran *et al.* (2), who considered the interactions between contiguous β-1,4-linked residues in cellulose and chitin. A model for the disaccharide unit is shown in Fig. 2.46. The zero positions for ϕ and ψ are taken when C-4, O-4, O-1, C-1′, and O-1′ are co-planar and the C-1—H-1 and C-4′—H-4′ bonds are directed to the same side of the molecule. Positive rotations are defined as follows. When looking along a particular bond, the rear groups rotate counterclockwise with respect to the front groups. The hard-sphere map obtained by Ramachandran *et al.* (2) for cellobiose is shown in Fig. 2.47a for the $(\phi, \psi)^*$ region for which they made calculations by hand. Contours of n and h are superimposed on those for the fully allowed and partially allowed regions. Calculation of n and h is more complicated in general for polysaccharides than for polypeptides, since the linkage bonds do not lie in the same plane. For the 1,4 linkages, the bonds can be made coplanar by a small adjustment of the coordinates, and this pro-cedure has been followed by some workers, without serious loss in accuracy.

---

* Unfortunately, different conventions have been used by the various research groups in order to define the zero positions for ϕ and ψ. Differences between the individual con-ventions and that used in the text are indicated in the legend to the figures.

    A common convention was adopted by many of the researchers in this field at a recent symposium on the conformation of polysaccharides, at the March 1971 meeting of the American Chemical Society in Los Angeles. The zero positions of ϕ and ψ for (1,4), linked residues are when C-1–O-1 is *cis* to C-4′–H-4′ ($\phi = 0°$), and O-1–C-4′ is *cis* to C-1–H-1 ($\psi = 0°$). A similar convention applies to (1,2), (1,3) and (1,6) linkages. In the case of (1,6) linkages, the additional rotation angle about the C-5′–C-6 bond is labelled ω, and the $\omega = 0°$ position is when O-1–C-6′ is *cis* to C-4′–C-5′. In linkages other than (1,6), the rotation angle for the C-6 side chain is labelled χ. Positive angles are defined when looking along a particular bond, as counterclockwise rotation of the front groups with respect to those at the back.

For other linkages, such as 1, 3, however, the full calculation has to be made. A degree of uncertainty remains concerning the C—O—C bond angle for the glycosidic linkage. Considerable variation is possible for this angle, although crystallographic work on the disaccharides indicates $117 \pm 2°$ as a reasonable value. Nevertheless, the four-degree range has a sizable effect on the limits for the allowed regions. Ramachandran *et al.* (2) chose 117° for the analysis shown in Fig. 2.47a.

The cellulose and analogous chitin chains are defined by the helical parameters $n = 2$ and $h = 5.15$ Å from x-ray fiber diagrams. This requires a buckling of the perfectly straight chain to avoid a bad contact between the hydrogens

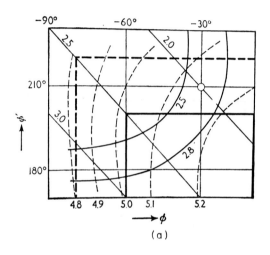

(a)

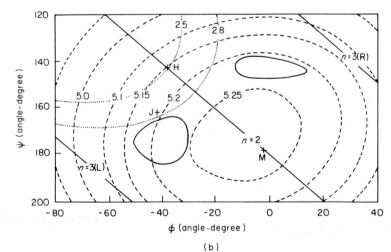

(b)

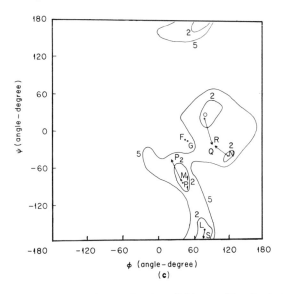

Fig. 2.47. (a) Hard-sphere steric map for the cellobiose residue due to Ramachandran *et al.* (2). [Thick broken line: outer limit; thick solid line: fully allowed region; straight thin solid lines: contours of equal n; curved thin dashed lines: contours of equal λ (Å); curved thin solid lines: contours for equal length O-3′—H···O-5 hydrogen bonds (in Å)]. (b) Potential energy map for the cellobiose residue, due to Rees and Skerrett (37). [Solid curved lines: zero energy contours (Kitagorodskii); solid straight lines: contours of equal n; broken lines: contours of equal λ (Å); dotted lines: contours for equal length O-3′—H··· O-5 hydrogen bonds (in Å).] N.B. Rees and Skerrett used a different convention for ϕ, ψ (ϕ, ψ) = (0°, 0°) when C-1, O-1, C-4′, H-1 and H-4′, are coplanar and when C-4′—H-4′, points in the same direction and C-1—H-1 in the opposite direction, as the apex of the glycosidic bond angle. Rotations are positive if the reducing residue (primed in Fig. 2.46) is held steady, and the nonreducing residue is rotated anticlockwise through ψ about C-4′—O-1 and clockwise through ϕ and O-1—C-1 when viewed from the nonreducing end. (c) Potential energy map for cellobiose from molecular orbital calculations due to Gracomini *et al.* (39). The contours are potential energy in kcals/residues. The calculations predict four minima labelled L, M, N, O. When the effect of hydrogen bonding is included, these minima move to P, R, and F. [Taken from the cited references, with permission.]

on C-1, and C-4′. There are two possible types of buckling, but one produces a bad contact between C-6 and O-2′ (Fig. 2.48a). The likely conformation, shown in Fig. 2.48b, has the added advantage of an intramolecular hydrogen bond. This conformation occurs at the intersection of the $n = 2$ and $h = 5.15$ Å contours in Fig. 2.47a and lies in the partially allowed region due to close contact between C-3′ and C-1(2). The map for cellulose has been recalculated by Rees and Skerrett (37) using potential functions, and the resultant energy contours are shown in Fig. 2.47b. Two minima are predicted, both of approximately the same energy, and one of these is close to the $n = 2$ axis. The observed conformation has an energy of +1 kcal/residue more than the minimum.

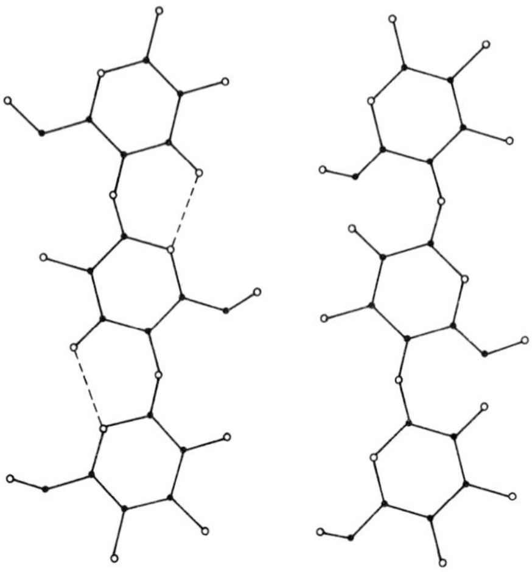

(a)   (b)

Fig. 2.48. Possible conformations of the cellulose chain when $n = 2$ and $\lambda = 5.15$ Å Note that an O-5$\cdots$H—O-3′ intramolecular hydrogen bond is formed in conformation (a) while in (b) there is a short contact between O-2 and C-6′.

Inclusion of terms for the intramolecular hydrogen bonding (38) moves this minimum onto the line for the twofold screw axis. In addition, two other subsidiary minima are observed. These are not possible conformations for a simple repeating polymer chain, but are possible for regions in which the chain might fold. Molecular orbital calculations by Gracomini *et al.* (39) for the cellobiose molecule predict four minima, which are shown as points L, M, N, and O on their conformational map (Fig. 2.47c). When the possibility of a hydrogen bond between O-3—H and O-5′ is allowed for, three minima are predicted, labelled P_2, R and F in Fig. 2.47c. Of these, conformation F is that observed for crystalline cellobiose and corresponds to the chain conformation of cellulose.

Adapting the above procedure to a consideration of β-1,4-xylan is extremely simple as the CH_2OH group need only be replaced by hydrogen. The net effect is that the allowed region is wider than that for cellulose. From a combination of x-ray work and hard sphere analysis, Settineri and Marchessault (40) proposed a left-hand 3_1 helical conformation, repeating in 14.85 Å. Rees and Skerrett (37) have computed the energy map for a β-1,4-xylobiose residue, and predict two energy minima of equal energy. The observed conformation has energy of 1.4 kcal/residue above the nearest minimum, but the effect of any intramolecular hydrogen bonding was not considered, and should account for the difference. However the energy map for β-1,4-xylobiose is very similar to that for cellobiose, and Rees and Skerrett could see no obvious reason, from intramolecular interactions alone, why the conformations of these two polysaccharides should be different. An energy map for β-1,4-xylobiose has also been obtained by Sundararajan and Rao (41). These workers used slightly different atomic coordinates and the observed conformation was found to have an energy of only 0.2 kcals/residue higher than the nearest minimum. The 0-3 $\cdots$ 0-5 distance in this conformation is 3.2 Å, which is too long for an intramolecular hydrogen bond, although a very weak bond could be formed by small variations in the coordinates. Sundararajan and Rao note that the crystalline structure is thought to contain one water molecule per residue which may stabilize the 3_1 conformation by hydrogen bonding.

Model building and computation is particularly useful when the n and h parameters cannot be defined fully by x-ray methods, and some of the helical structures for amylose are examples of this. The absence of meridional reflections for the V-amylose structure, due to low crystallinity, presented a difficulty in confirming the proposed helical conformation with parameters $n = 6$ and $h = \pm 1.33$ Å (the sense of the amylose helices is as yet undetermined). Rao et al (42) considered the nonbonded interaction for the maltose residue, the amylose dimer, that has the same skeletal formula as cellobiose (Fig. 2.46) except that the glycosidic linkage is α-(1,4) rather than β-(1,4). The potential energy map obtained with superposed n and h contours is shown in Fig. 2.49. It can be seen that the $n = 6$ contour lies within the allowed

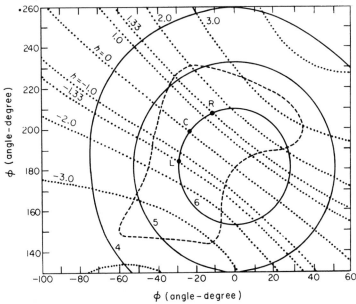

Fig. 2.49. Energy contour map for the conformation of the maltose residue, due to Rao *et al.* (41). [--- zero energy contour (Kitaigorodskii); $\cdots$ contour of λ in Å; — contours of n.] (From Ref. (41), with permission.)

region. Inclusion of hydrogen-bonding terms to allow for bonding between the O-2 and O-3′ hydroxyls produces the map shown in Fig. 2.50, where the energy minimum is moved onto the $n = 6$ contour (43). A similar allowed region has been obtained for maltose by Gracomini *et al.* in calculations based on quantum mechanics. Their results indicate a preference for O-2—H $\cdots$ O-3′—H intramolecular hydrogen bonding rather than the other possibility, H—O-2 $\cdots$ H—O-3′. The B-amylose helix has been shown to have six residues per turn with $h = \pm 1.73$ Å. The positions for the right- and left-hand V and B helices are marked in Fig. 2.50. The conformations of maltose and cyclo-hexa-amylose (the cyclic hexamer) also fall close to the $n = 6$ line; maltose adopts the conformation corresponding to a right-handed helix, which suggests a right-hand sense for amylose (44) (see Chapter 11). Sundararajan and Rao (45) have extended this type of analysis (nonbonded interactions only) to include interactions between successive turns of the helix. This produces a

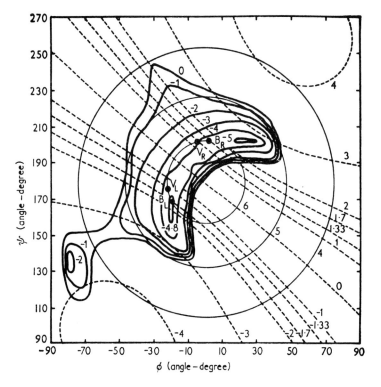

Fig. 2.50. Energy contour map for a maltose residue (42). [— potential energy in kcal/maltose residue; — contours of n; — contours of λ in Å; positive values of λ indicate right-hand helix; negative λ, a left-hand helix.]

region of high energy in the area containing conformations such as cyclohexa-amylose; the allowed region is thus separated into two areas for left- and right-hand helices respectively.

The salt complex KBr–amylose is believed to be a 4_1 helix with parameters $n = 4$, $h = 4.0(45)$. This conformation is not in the allowed region of Figs. 2.49 or 2.50, but this structure may be stabilized by the ionic forces of the complex (46). The n, h combinations of the 14_3 helix for amylose triacetate (47), $n = 4.67$ and $h = 3.77$ Å, and of the 7_1 helix proposed for an amylose hydrate (48), $n = 7$ and $h = 1.14$ Å, are not possible for the model analyzed. The maltose model had a bridge angle of $117°$, the most likely value from known structures of sugar dimers. Other values of the bridge angle are required for the structures in question, which could occur in the case of amylose triacetate in view of the presence of the bulky acetyl groups.

Little work has been done on polysaccharides other than the homopolymers. However, Rees (49) has considered the A—B copolymers that make up the carrageenans and connective tissue mucopolysaccharides. In a hard-sphere analysis he considered the A—B and B—A linkages in turn and the combined results show that less than 1 % of the total range of conformations is sterically possible. This 1 % still represents a large number of structures, but the double-helical conformation for the carrageenans (50) does lie in the allowed range.

Polynucleotides

Conformational analysis becomes more complex when the nucleic acids are considered. Their primary structure, which is described more fully in Chapter 1, is a repeating sequence,

$$\begin{bmatrix} \text{base} \\ | \\ -\text{sugar}-\text{phosphate}- \end{bmatrix}_n$$

and for present purposes it is appropriate to consider each component of the monomer unit in turn.

BASES

The common purine and pyrimidine bases that occur in the nucleic acids RNA and DNA are shown in Fig. 1.11. The bases are shown in their probable conjugated forms and would be expected to be planar by analogy with the planar amide group discussed in the Peptide Bond section. Detailed crystallographic work on the individual bases and nucleotides shows that in general the nucleotides are not perfectly planar, although the distortion is not large. It is not possible to generalize about these distortions, but several atoms can be displaced from the mean plane by up to 0.2 Å (51). Conformational analysis has generally been applied to models of specific nucleotides, taking the coordinates determined by x-ray crystallography and assuming the base conformations to be fixed. Complications are introduced by the changes in conjugation when, for instance, a hydrogen ion is picked up by the base.

SUGARS

The two sugar residues incorporated into native nucleic acids are β-D-ribose and β-2-deoxy-D-ribose,* which are analogous in conformation to cyclopentane. The skeletal formulas of these three molecules are shown in Fig. 2.51. A planar cyclopentane molecule would have C—C—C bond angles

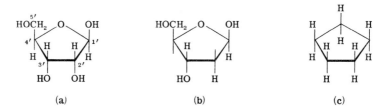

Fig. 2.51. Skeletal formulae of (a) β-D-ribose, (b) β-2-deoxy-D-ribose, (c) cyclopentane.

of 108°, and the CH_2 groups would be fully eclipsed. This strain is relieved when the ring adopts a puckered conformation. Energy calculations of Kilpatrick *et al.* (52) show that the puckered ring is 4 kcal/mole lower in energy than the planar structure. The planar conformation for ribose would be more unfavorable due to the OH and CH_2OH substituted groups. Crystallographic studies on ribose rings show that at least four conformations can exist (52). In these conformations, the ring atoms C-2' or C-3' are displaced either above or below the least-squares plane of the other four atoms. When the displacement is on the same side of the ring plane as C-5', the conformation is said to be *endo*; if it is on the side opposite C-5' it is termed *exo*. The four basic conformations for β-D-ribose, C-2' *endo*, C-2' *exo*, C-3' *endo*, and C-3' *exo*, are shown in Fig. 2.52. As an example, adenosine 5'-phosphate has the ribose ring in the C-3' *endo* conformation, and C-3' is displaced 0.66 Å from the least-squares planes of the other four ring atoms (53).

---

* The systematic name for this sugar is β-2-deoxy-D-erythrose, but the more common name has been retained here.

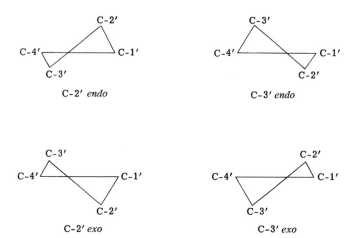

Fig. 2.52. Puckered conformations for β-D-ribose. The ring oxygen O-4, is collinear with C-4′ and C-1′ in projection.

PHOSPHATE

The phosphate unit has been shown to be approximately tetrahedral, although distortions of up to 10° are generally found in the bond angle (54).

Figure 2.53 (54) shows a skeletal diagram of the nucleotide monomer in which the angles that define the conformation are labeled The zero positions for the main chain dihedral angles are defined as *cis* arrangements of adjacent bonds in the backbone. For χ, which defines the orientation of the base relative to the furanose ring, the zero position is when C-1′—O-4′ is

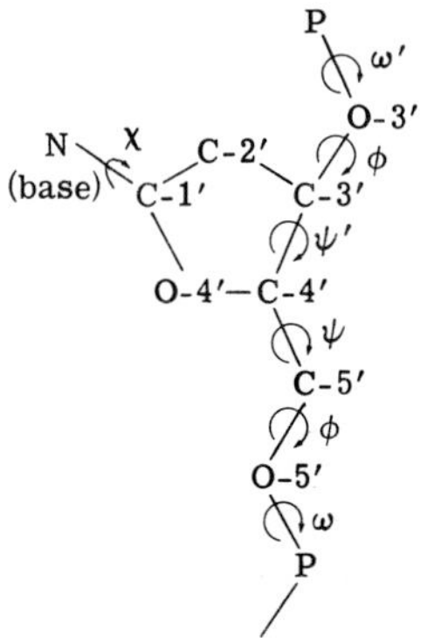

$$
\begin{array}{c}
\text{P} \\
\text{\textbackslash}\,\omega' \\
\text{O-3'} \\
\text{N}\quad\chi\quad\text{C-2'}\quad \phi \\
\text{(base)}\,\text{C-1'}\quad\text{C-3'} \\
\psi' \\
\text{O-4'}\!-\!\text{C-4'} \\
\psi \\
\text{C-5'} \\
\phi \\
\text{O-5'} \\
\omega \\
\text{P}
\end{array}
$$

Fig. 2.53. Skeletal diagram for the nucleotide monomer in which the rotation angles which define the conformation are labelled. [From Ref. (54), with permission.]

cis to N-6—C-1 for the pyrimidines and *cis* to N-9—C-6 for the purine bases. Looking along a particular bond, a positive rotation corresponds to a counter-clockwise rotation of the rear bond with respect to the front bond. Analysis of such a structure requires six parameters to describe the backbone conformation, plus the torsion angle χ for the base orientation. In addition four more torsion angles are required, along with ψ', to describe the conformation of the sugar ring, although this can be avoided by considering the four basic ring conformations in turn. It is therefore a complex problem to consider the possible conformations for a single nucleotide, let alone the complete chain.

Analysis was first directed at the arrangement of the bases. By means of molecular models, Donohue and Trueblood (55) described two ranges for the χ torsion angle, one centered at $\chi = 30°$, which was termed the *syn* conformation, and the other at $\chi = 210°$, which was labeled *anti*. In the original Watson and Crick structure the bases were in the *anti* conformation, as subsequently shown to be correct for DNA-A (56) and DNA-B (57). Further work showed that the allowed angular ranges for χ depended on the sugar ring conformation and the type of base. Haschemeyer and Rich (58) examined the rotational conformations for all permutations of the two types of base (purine and pyrimidine) with three of the four ring conformations (C-2' *exo* has not been observed in nucleotides, nor has C-3' *exo* been seen for pyrimidine com-

TABLE 2.7

Allowed Ranges of Torsion Angle (χ_{CN}) in Pyrimidine and Purine Nucleosides Based on Fully Allowed and Outer-Limit van der Waals Radii.[a]

Structure	Sugar conformation	Number of structures	Fully allowed range[b]	Outer-limit range[b]
Pyrimidine nucleosides	C-3' *endo*	4	+80 to +50	+105 to +25
	C-2' *endo*	5	None	+80 to +25
				+125 to −145
				−220 to −240
Purine nucleosides	C-2' *endo*	2	+85 to +160[c]	+260 to −60
			−110 to −155	
			+135 to +30[d]	+155 to −5
			−115 to −135	−105 to −150
	C-3' *endo*	2	+140 to +85	+150 to −20
			+60 to −10	−140 to −150
	C-3' *exo*	1	+85 to −25	+155 to −30

[a] From Ref. (57), with permission
[b] Values given in degrees.
[c] Based on deoxyguanosine in the 5-bromodeoxycytidine–deoxyguanosine complex.
[d] Based on adenosine in β-adenosine 2'β-uridine 5'-phosphate.

pounds, and these possibilities were not considered). Table 2.7 (Table 3 of Ref. (56)) gives the allowed ranges for χ based on hard-sphere criteria. For the purine bases and C-2' *endo* puckering, two ranges exist at $\chi = +60°$ to $+85°$ *anti* and $\chi = +205°$ to $+250°$ *syn*. These figures are for the fully allowed region; extension to the partially allowed contacts makes almost the whole range available. For the C-3' *endo* and C-3' *exo* rings the *syn* range is unlikely, although relatively large *anti* ranges are permitted. If a pyrimidine base is substituted, and if the sugar has C-3' *endo* puckering, then χ is restricted to an outer-limit *anti* range of 25°–105°. For C-2' *endo* rings, a similar *anti* range is allowed from 25° to 85°. Haschemeyer and Rich found good agreement between the calculated ranges for χ and the observed values for the 15 or so nucleotides, nucleosides, etc., whose structures had at that time been determined by single-crystal methods. This is borne out by the χ values for 20 β-pyrimidine and 18 β-purine nucleotides and nucleosides collected by Sundaralingam (54). It can be seen that there is excellent agreement between these and the theoretical ranges. A similar study has been made by Sasisekharan *et al.* (59) although only for the C-3' *endo* sugar ring structures. More recently, Lakshminarayanan and Sasisekharan (60) have determined the projected conformations for the sugar–base units in terms of potential energy due to nonbonded vibrational and electrostatic interactions. The ranges of χ for combinations of the four sugar conformations and the purine and pyrimidine bases are given in Table 2.8 (60). These results are very similar to those obtained from hard-sphere analyses and can be seen to agree with the observed values of χ from single-crystal analyses of nucleotides, nucleosides, etc.

In the work mentioned above, Sasisekharan *et al.* (59) also dealt with the arrangement of the main chain by considering the possible conformations for a monomer unit. In this case ω and ω' (Fig. 2.53) could be ignored, and ψ' was set constant by restricting the sugar to the C-3' *endo* conformation. Variation of the three remaining angles ϕ, ψ, and ϕ' indicated that these rotations were almost independent of each other. On the basis of hard-sphere criteria these workers showed that the allowed values for ϕ, ψ, and ϕ' had to be in the relatively small ranges given in Table 2.9. These ranges compared well with the few observed values for nucleotides then available. This study has been repeated using potential functions for nonbonded, torsional, and electrostatic energies (61). The ranges predicted for the three dihedral angles are given in Table 2.9. Sundaralingam (54) has also compared data for some 64 published structures of nucleotides, nucleosides, and related compounds with those predicted by hard-sphere methods, taking the torsion angles in successive pairs. The (ω, ϕ) map shows two allowed regions centered on $(285°, 170°)$ and $(65°, 170°)$. Most of the observed values are found to occur in the former region, which contains all the right-hand helices determined for the polynucleotides. The (ϕ, ψ) map contains five possible regions, although only two, at (170°, 60°) and (170°, 175°), are fully allowed. Again, most of the observed structures and helices occur in the first region. It is thought that the second region may contain conformations that occur in the non-base-paired loops of transfer RNA. Six allowed regions are predicted in the (ψ, ψ') map which are the possible combinations of the two values for ψ', which describe the sugar ring conformations, and the three staggered positions for ψ.

TABLE 2.8

Probable Ranges of χ (Energy Within 0.5 kcal/mole of the Minima), Allowed Ranges from Hard-sphere Model, and Observed Values in Crystal Structures[a]

	χ^b			
Sugar	Probable range (potential energy)	Allowed range (hard-sphere)	Observed value	Structure
		Purines		
3'-endo R[c]	−15 to +50	0–60	3.8	Adenylic acid
			12.4	Adenosine (in A + U complex)
			25.7	Adenosine 5'-phosphate
	200–220	200–220		
3'-endo D	−10 to +45	0–60		
	200–220	200–220		
2'-endo R	−5 to +75	0–50		
	215–245	210–230		
2'-endo D	−40 to +65	0–50		
	215–245	210–230	211.3	Deoxyguanosine (in G + C complex)
3'-exo R	−20 to +75	0–60		
	115–170			
	180–215	200–210		
	235–250			
3'-exo D	−35 to +60	0–60	5.0	Deoxyadenosine
	115–170			
	180–215	200–210		
	235–250			
2'-exo R	−15 to +20	0–30		
	120–170			
2'-exo D	−15 to +20	0–30		
	120–175			

[a] From Ref. (60), with permission.
[b] Values given in degrees.
[c] D and R denote β-2-deoxy-D-ribose and β-D-ribose respectively.

TABLE 2.8 (*continued*)

Sugar	χ^b		Observed value	Structure
	Probable range (potential energy	Allowed range (hard-sphere)		
		Pyrimidines		
3'-*endo* R	0–40	0–60	18.4	Cytidine
			20.0	5-Bromouridine (in A + U complex)
3'-*endo* D	0–40	0–60	43.5	Calcium thymidylate
2'-*endo* R	15–65	10–50	41.5	Cytidylic acid *b*
			43.3	Barium uridine 5'-phosphate
	220–240			
2'-*endo* D	−10 to +70	10–50	58.7	5-Bromodeoxycytidine (in G + C complex)
			59.1	5-Fluorodeoxyuridine
			63.2	5-Iododeoxyuridine
	220–240			
3'-*exo* R	10–70	10–60		
3'-*exo* D	−15 to +75	10–60		
2'-*exo* R	−15 to +25	10–40		
2'-*exo* D	−5 to +25	10–40		

TABLE 2.9

Ranges for the Nucleotide Backbone Conformation ϕ, ψ, and ϕ' Predicted by Sasisekharan et al. from a Hard-Sphere Analysis of a C-3' endo Nucleotide.[a]

$\phi = +150°$ to $+210°$
$\psi = +40°$ to $+70°$, $+170°$ to $+200°$, $+280°$ to $+320°$
$\phi' = +260°$ to $+270°$

Ranges for ϕ, ψ, and ϕ' Predicted by Lakshminarayanan and Sasisekharan (59) Based on the Potential Functions for Nonbonded, Torsional, and Electrostatic Energies (Cell Conformations for the Sugar Ring are Included).[b]

$\phi = +100°$ to $+260°$, with the region around $+180°$ being the most likely
$\psi =$ near $+60°$, $+180°$, and $+300°$
$\phi' = +180°$ to $+280°$

[a] From Ref. (58), with permission.
[b] From Ref. (59), with permission.

Almost all the observed structures occur in regions centered on (60°, 80°) (60°, 150°), and (175°, 150°). The (ψ', ϕ') map contains two allowed regions corresponding to the different sugar ring conformations, at (80°, 210°) and (150°, 210°). Both the B and C structures for DNA lie outside the fully allowed region, indicating probable strain on the sugar ring. Two regions are predicted on (ϕ', ω') and (ω', ω) maps in each case, corresponding to helices of opposite sense. The right-hand helices are in ranges centered on (210°, 280°) and (280°, 285°), respectively. A potential energy map (62) for the (ω, ω') angles in a dinucleotide, based on the known structure of dimethyl phosphate, shows that the most likely conformations are (60°, 60°) and (300°, 300°) with the latter corresponding to right-handed helices.

The above analysis reduces the conformational possibilities to two ranges for all the backbone angles except ψ, for which there are three possibilities. Further restriction to only right-handed helical structures leaves a total of six combinations, which are listed in Table 2.10 (ranges are represented by their approximate centers). Comparison with the proposed values for the known helical structures is effected in Table 2.11. Of the six theoretical structures, 1 and 2 were considered the most likely, and these differ only in that 1 has the C-3' *endo* conformation for the sugar, while 2 has a C-2' *endo* ring. There is close agreement between structure 1 and the proposed structures for DNA-A, RNA-11 and poly A. Undoubtedly these investigations will be advanced further, but the above agreement must be considered satisfactory at this stage.

TABLE 2.10

Preferred Conformations for the Sugar–Phosphate Chains of Right-Handed Double-Helical Nucleic Acids and Polynucleotides[a,b]

Conformation	ω	ϕ	ψ	ψ'	ϕ'	ω'
1	285	170	60	80	210	280
2	285	170	60	150	210	280
3	285	170	175	150	210	280
4	285	170	175	80	210	280
5	285	170	175	80	210	210
6	285	170	175	150	210	210
7[c]	60	200	200	150	200	285

[a] From Ref. (54), with permission.

[b] The torsional angles (in degrees) given here are within about ±20°.

[c] Conformation 7 has been included to indicate that double-stranded helices may be built with the above torsional angles for the sugar–phosphate backbone. However, it should be mentioned that the conformation of the phosphodiester in 7 is less preferred than any of the others (1–6). Hence, unless there is a compelling reason (some other factors stabilizing this conformation), conformation 7 will not be considered a possible in double-helical structures. It may, however, occur in the loops of tRNA's.

TABLE 2.11

Sugar-Phosphate Torsion Angles (in degrees) in the Proposed Models of Nucleic Acids and Polynucleotides[a]

Model	ω	ϕ	ψ	ψ'	ϕ'	ω'
DNA-W & C[b]	182	179	155	100	186	205
DNA-B	281	212	58	130	147	282
DNA-C	315	143	48	168	211	212
DNA-A	283	167	67	76	221	279
RNA-11[c]	282	165	74	95	216	273
Poly A	285	168	69	83	216	293
RNA-10[d]	257	188	88	80	203	285
A2′P5′U[e]	313	170	57(U)	84(U)	244	232
			45(A)	148(A)		

[a] From Ref. (54), with permission.
[b] Watson and Crick structure for DNA.
[c] The 11_1 helix for RNA.
[d] The 10_1 helix for RNA.
[e] Adenylyl 2′-phosphate 5′-uridine tetrahydrate (note 2′→5′ linkage).

REFERENCES

1. C. H. Bamford, A. Elliott, and W. E. Hanby, "Synthetic Polypeptides," p. 113. Academic Press, New York, 1956.
2. G. N. Ramachandran, C. Ramakrishnan, and V. Sasisekharan, *J. Mol. Biol.* **7**, 95 (1963).
3. W. L. Bragg, J. C. Kendrew, and M. F. Perutz, *Proc. Roy. Soc. London* **A203**, 321 (1950).
4. M. L. Huggins, *J. Org. Chem.* **1**, 407 (1937).
5. L. Pauling and R. B. Corey, *Proc. Nat. Acad. Sci. U.S.* **39**, 253 (1953).
6. T. Miyazawa, *J. Polym. Sci.* **55**, 215 (1961).
7. G. N. Ramachandran, *in* "Structural Chemistry and Molecular Biology" (A. Rich and N. Davidson, eds.), p. 77. Freeman, San Francisco, California, 1968.
8. A. J. Hopfinger, Ph.D. thesis. Case Western Reserve University, Cleveland, Ohio, 1970.
9. G. N. Ramachandran and V. Sasisekharan, *Advan. Protein Chem.* **23**, 238 (1968).
10. H. A. Scheraga, *Advan. Phys. Org. Chem.* **6**, 103 (1968).
11. D. A. Brant and P. J. Flory, *J. Amer. Chem. Soc.* **87**, 2788 (1965).
12. T. Ooi, R. A. Scott, G. Vanderkooi, and H. A. Scheraga, *J. Chem. Phys.* **46**, 4410 (1967).
13. G. Del Re, B. Pullman, and T. Yonezawa, *Biochim. Biophys. Acta* **75**, 153 (1963). D. Poland and H. A. Scheraga, *Biochem.* **6**, 3791 (1967).
14. W. O. Baker and W. A. Yager, *J. Amer. Chem. Soc.* **64**, 2171 (1942).
15. L. Pauling, *Proc. Nat. Acad. Sci. U.S.* **14**, 359 (1928).
16. R. Rein and F. E. Harris, *J. Chem. Phys.* **41**, 3393 (1964).
17. P. DeSantis, *Nature* **206**, 456 (1965).
18. A. V. Lakshminarayanan, *in* "Conformation of Biological Macromolecules" (G. N. Ramachandran, ed.), in press.
19. G. N. Ramachandran and V. Sasisekharan, *Advan. Protein Chem.* **23**, 296 (1968).
20. R. Hoffmann, *J. Chem. Phys.* **40**, 2474 (1964).

21. B. Pullman, B. Maijret, and D. Perakia, *Theor. Chim. Acta* **18**, 44 (1970).
22. K. D. Gibson and H. A. Scheraga, *Proc. Nat. Acad. Sci. U.S.* **58**, 420 (1967).
23. F. R. Brown, A. J. Hopfinger, and E. R. Blout, *J. Mol. Biol.* **63**, 101 (1972).
24. Y. Pao, R. Longworth, and R. L. Kornegay, *Biopolymers* **3**, 519 (1965).
25. K. P. Sarathy and G. N. Ramachandran, *Biopolymers* **6**, 461 (1968).
26. P. R. Schimmel and P. J. Flory, *J. Mol. Biol.* **34**, 105 (1968).
27. A. J. Hopfinger and A. G. Walton, *J. Macromol. Sci. Phys.* **83**, 171 (1969).
28. S. Krimm and J. E. Mark, *Proc. Nat. Acad. Sci. U.S.* **60**, 1122 (1968).
29. W. Hiltner, A. J. Hopfinger, and A. G. Walton, *J. Amer. Chem. Soc.* **94**, 4324 (1972).
30. A. Damiani, P. DeSantis, and A. Pizzi, *Nature* **226**, 542 (1970).
31. M. Maigret, B. Pullman, and J. Caillet, *Biochem. Biophys. Res. Commun.* **40**, 808 (1970).
32. A. J. Hopfinger and A. G. Walton, *Biopolymers* **9**, 29 (1970).
33. See A. M. Liquori, *Ciba Found. Symp. Principles Biomol. Organ.* 1965, p. 40 (1966).
34. R. E. Reeves, *J. Amer. Chem. Soc.* **72**, 1499 (1950).
35. S. S. C. Chu and G. A. Jeffrey, *Acta Crystallogr.* **B24**, 830 (1968).
36. D. W. Jones, *J. Polym. Sci.* **32**, 371 (1958); **42**, 173 (1960).
37. D. A. Rees and R. J. Skerrett, *Carbohyd. Res.* **7**, 334 (1968).
38. A. Sarko, in preparation.
39. M. Gracomini, B. Pullman, and B. Maigret, *Theor. Chim. Acta* **19**, 347 (1970).
40. W. Settineri and R. H. Marchessault, *J. Polym. Sci.* **C11**, 253 (1965).
41. P. R. Sundararajan and V. S. R. Rao, *Biopolymers*, **8**, 305 (1969).
42. V. S. R. Rao, P. R. Sundararajan, C. Ramakrishnan, and G. N. Ramachandran, *in* "Conformation of Biopolymers" (G. N. Ramachandran, ed.), Vol. 2, p. 271. Academic Press, New York, 1967.
43. J. Blackwell, A. Sarko and R. H. Marchessault, *J. Mol. Biol.* **42**, 379 (1969).
44. G. J. Quigley, A. Sarko, and R. H. Marchessault, *J. Amer. Chem. Soc.*, **92**, 5834 (1970).
45. P. R. Sundararajan and V. S. R. Rao, *Biopolymers* **8**, 313 (1969).
46. J. J. Jackobs, R. R. Bumb, and B. Zaslow, *Biopolymers* **6**, 1659 (1968).
47. A. Sarko and R. H. Marchessault, *J. Amer. Chem. Soc.* **89**, 6454 (1967).
48. B. Zaslow and R. L. Miller, *J. Amer. Chem. Soc.*, **83**, 4378 (1961).
49. D. A. Rees, *J. Chem. Soc. B*, p. 217 (1969).
50. N. S. Anderson, J. W. Campbell, M. M. Harding, D. A. Rees, and J. W. B. Samuel, *J. Mol. Biol.* **45**, 85 (1969).
51. M. Sundaralingam and L. H. Jensen, *J. Mol. Biol.* **13**, 930 (1965); see also references cited in this article.
52. J. E. Kilpatrick, K. S. Petzer, and R. Spitzer, *J. Amer. Chem. Soc.* **69**, 2483 (1947).
53. J. Kraut and L. H. Jensen, *Acta Crystallogr.* **16**, 79 (1963).
54. M. Sundaralingam, *Biopolymers* **7**, 821 (1969); see also references cited in this article.
55. J. Donohue and K. N. Trueblood, *J. Mol. Biol.* **2**, 363 (1960).
56. W. Fuller, M. H. F. Wilkins, H. R. Wilson, and L. D. Hamilton, *J. Mol. Biol.* **12**, 60 (1965).
57. R. Langridge, D. A. Marvin, W. E. Seeds, H. R. Wilson, C. W. Hooper, M. H. F. Wilkins, and L. D. Hamilton, *J. Mol. Biol.* **2**, 38 (1960).
58. A. E. V. Haschemeyer and A. Rich, *J. Mol. Biol.* **27**, 368 (1967).
59. V. Sasisekharan, A. V. Lakshminarayanan, and G. N. Ramachandran, *in* "Conformation of Biopolymers" (G. N. Ramachandran, ed.), p. 641. Academic Press, New York, 1967.
60. A. V. Lakshminarayanan and V. Sasisekharan, *Biopolymers* **8**, 475 (1969).
61. A. V. Lakshminarayanan and V. Sasisekharan, *Biopolymers* **8**, 489 (1969).
62. V. Sasisekharan and A. V. Lakshminarayanan, *Biopolymers* **8**, 505 (1969).

APPENDIX

Relation between ϕ, ψ, and Helical Parameters for Polypeptides[a]

$\phi\backslash\psi$	0	10	20	30	40	50	60	70	80	90	100	110
0	2.00	−2.09	−2.19	−2.29	−2.41	−2.54	−2.68	−2.83	−2.99	−3.17	−3.36	−3.56
	3.63	*3.63*	*3.61*	*3.58*	*3.54*	*3.48*	*3.41*	*3.32*	*3.20*	*3.07*	*2.90*	*2.69*
10	−2.10	−2.20	−2.31	−2.43	−2.56	−2.70	−2.85	−3.01	−3.19	−3.38	−3.58	−3.78
	3.63	*3.62*	*3.60*	*3.58*	*3.53*	*3.47*	*3.40*	*3.30*	*3.17*	*3.01*	*2.82*	*2.58*
20	−2.21	−2.32	−2.44	−2.57	−2.71	−2.87	−3.04	−3.22	−3.41	−3.61	−3.82	−4.03
	3.62	*3.61*	*3.59*	*3.56*	*3.52*	*3.45*	*3.37*	*3.26*	*3.11*	*2.94*	*2.72*	*2.45*
30	−2.33	−2.46	−2.59	−2.73	−2.89	−3.06	−3.25	−3.44	−3.65	−3.86	−4.07	−4.26
	3.60	*3.59*	*3.57*	*3.54*	*3.49*	*3.42*	*3.32*	*3.20*	*3.03*	*2.83*	*2.57*	*2.26*
40	−2.47	−2.60	−2.75	−2.91	−3.09	−3.28	−3.47	−3.69	−3.90	−4.11	−4.31	−4.49
	3.57	*3.56*	*3.54*	*3.51*	*3.45*	*3.37*	*3.26*	*3.11*	*2.92*	*2.68*	*2.39*	*2.04*
50	−2.62	−2.77	−2.93	−3.11	−3.31	−3.51	−3.73	−3.95	−4.17	−4.37	−4.55	−4.69
	3.53	*3.52*	*3.50*	*3.46*	*3.39*	*3.30*	*3.17*	*2.99*	*2.77*	*2.49*	*2.15*	*1.75*
60	−2.79	−2.96	−3.14	−3.34	−3.55	−3.77	−4.00	−4.23	−4.44	−4.63	−4.76	−4.83
	3.48	*3.47*	*3.44*	*3.39*	*3.31*	*3.19*	*3.04*	*2.84*	*2.57*	*2.24*	*1.86*	*1.42*
70	−2.98	−3.16	−3.37	−3.59	−3.82	−4.06	−4.29	−4.56	−4.71	−4.85	−4.92	−4.91
	3.42	*3.40*	*3.36*	*3.29*	*3.20*	*3.06*	*2.87*	*2.63*	*2.31*	*1.94*	*1.51*	*1.05*
80	−3.19	−3.40	−3.62	−3.86	−4.11	−4.36	−4.59	−4.80	−4.94	−5.07	−5.01	−4.92
	3.34	*3.31*	*3.25*	*3.17*	*3.05*	*2.87*	*2.64*	*2.35*	*1.99*	*1.57*	*1.11*	*0.64*
90	−3.43	−3.66	−3.91	−4.17	−4.43	−4.67	−4.89	−5.04	−5.12	−5.11	−5.01	−4.86
	3.23	*3.19*	*3.12*	*3.01*	*2.85*	*2.63*	*2.35*	*2.00*	*1.60*	*1.15*	*0.69*	*0.23*
100	−3.69	−3.94	−4.21	−4.49	−4.75	−4.98	−5.14	−5.23	−5.22	−5.12	−4.95	4.71
	3.10	*3.04*	*2.94*	*2.79*	*2.59*	*2.32*	*1.99*	*1.59*	*1.16*	*0.70*	*0.25*	*0.18*
110	−3.98	−4.26	−4.54	−4.82	−5.07	−5.25	−5.34	−5.34	−5.23	−5.03	4.78	4.52
	2.93	*2.84*	*2.70*	*2.51*	*2.26*	*1.94*	*1.55*	*1.13*	*0.68*	*0.24*	*0.17*	*0.55*
120	−4.30	−4.59	−4.88	−5.14	−5.35	−5.50	−5.50	−5.33	−5.14	4.89	4.59	4.29
	2.72	*2.59*	*2.40*	*2.16*	*1.85*	*1.47*	*1.07*	*0.62*	*0.19*	*0.21*	*0.57*	*0.89*
130	−4.64	−4.94	−5.22	−5.44	−5.57	−5.56	−5.46	−5.24	4.96	4.67	4.36	4.06
	2.45	*2.27*	*2.03*	*1.72*	*1.36*	*0.95*	*0.52*	*0.11*	*0.28*	*0.63*	*0.93*	*1.19*
140	−4.98	−5.26	−5.49	−5.63	−5.67	−5.56	5.34	5.06	4.75	4.42	4.11	3.82
	2.10	*1.87*	*1.56*	*1.21*	*0.81*	*0.39*	*0.01*	*0.38*	*0.71*	*1.00*	*1.24*	*1.44*
150	−5.30	−5.56	−5.70	−5.72	−5.63	5.42	5.14	4.82	4.49	4.16	3.86	3.59
	1.68	*1.39*	*1.03*	*0.64*	*0.23*	*0.16*	*0.52*	*0.83*	*1.10*	*1.33*	*1.51*	*1.67*
160	−5.61	−5.72	−5.77	−5.71	5.50	5.22	4.88	4.55	4.21	3.91	3.62	3.37
	1.18	*0.83*	*0.44*	*0.04*	*0.34*	*0.68*	*0.98*	*1.23*	*1.44*	*1.61*	*1.75*	*1.86*
170	−5.78	−5.84	5.80	5.55	5.25	4.93	4.60	4.26	3.95	3.66	3.40	3.17
	0.61	*0.23*	*0.16*	*0.54*	*0.86*	*1.15*	*1.38*	*1.57*	*1.72*	*1.84*	*1.94*	*2.02*
180	5.85	5.82	5.62	5.32	4.97	4.64	4.30	3.98	3.65	3.43	3.19	2.98
	0.00	*0.39*	*0.75*	*1.06*	*1.33*	*1.55*	*1.72*	*1.86*	*1.97*	*2.05*	*2.11*	*2.17*
190	5.78	5.60	5.33	5.01	4.66	4.32	4.01	3.71	3.45	3.21	3.00	2.81
	0.61	*0.97*	*1.28*	*1.53*	*1.73*	*1.89*	*2.01*	*2.10*	*2.17*	*2.22*	*2.26*	*2.29*

[a] Value of number of units per turn (n) and unit height (h in angstroms, shown in italics) of helical polypeptide chains corresponding to different values (in degrees) of (ϕ, ψ). (Taken from Ref. 8.)

APPENDIX (*continued*)

φ\ψ	0	10	20	30	40	50	60	70	80	90	100	110
200	5.61	5.35	5.02	4.67	4.34	4.02	3.73	3.46	3.22	3.01	2.82	2.65
	1.18	*1.49*	*1.73*	*1.92*	*2.06*	*2.17*	*2.25*	*2.31*	*2.34*	*2.37*	*2.39*	*2.39*
210	5.30	5.00	4.68	4.35	4.03	3.74	3.48	3.24	3.02	2.83	2.66	2.51
	1.68	*1.92*	*2.11*	*2.24*	*2.34*	*2.40*	*2.45*	*2.48*	*2.49*	*2.50*	*2.49*	*2.48*
220	4.98	4.66	4.34	4.03	3.74	3.48	3.25	3.03	2.84	2.67	2.52	2.38
	2.10	*2.28*	*2.41*	*2.50*	*2.56*	*2.60*	*2.62*	*2.62*	*2.62*	*2.61*	*2.59*	*2.56*
230	4.64	4.33	4.02	3.73	3.48	3.25	3.04	2.85	2.68	2.52	2.38	2.26
	2.45	*2.57*	*2.66*	*2.72*	*2.75*	*2.76*	*2.76*	*2.75*	*2.73*	*2.70*	*2.67*	*2.63*
240	4.30	4.00	3.73	3.48	3.24	3.04	2.85	2.68	2.53	2.39	2.26	2.15
	2.72	*2.80*	*2.86*	*2.89*	*2.90*	*2.91*	*2.88*	*2.85*	*2.82*	*2.78*	*2.74*	*2.69*
250	3.98	3.72	3.47	3.24	3.03	2.85	2.67	2.53	2.39	2.27	2.16	2.05
	2.93	*2.99*	*3.02*	*3.03*	*3.03*	*3.01*	*2.98*	*2.95*	*2.90*	*2.85*	*2.80*	*2.74*
260	3.69	3.45	3.22	3.02	2.84	2.68	2.53	2.39	2.27	2.16	2.06	−2.04
	3.10	*3.14*	*3.15*	*3.15*	*3.13*	*3.10*	*3.07*	*3.02*	*2.97*	*2.91*	*2.85*	*2.78*
270	3.43	3.21	3.01	2.83	2.67	2.52	2.39	2.27	2.16	2.06	−2.04	−2.14
	3.23	*3.25*	*3.25*	*3.24*	*3.21*	*3.18*	*3.14*	*3.09*	*3.03*	*2.97*	*2.90*	*2.82*
280	3.19	3.00	2.82	2.66	2.52	2.38	2.26	2.16	2.06	−2.04	−2.14	−2.24
	3.34	*3.34*	*3.34*	*3.32*	*3.28*	*3.24*	*3.20*	*3.14*	*3.08*	*3.01*	*2.93*	*2.85*
290	2.98	2.81	2.65	2.51	2.38	2.26	2.15	2.05	−2.04	−2.13	−2.24	−2.36
	3.42	*3.42*	*3.41*	*3.38*	*3.34*	*3.30*	*3.26*	*3.20*	*3.12*	*3.05*	*2.96*	*2.87*
300	2.79	2.64	2.50	2.37	2.25	2.14	2.05	−2.04	−2.14	−2.24	−2.36	−2.49
	3.48	*3.48*	*3.46*	*3.44*	*3.40*	*3.35*	*3.30*	*3.24*	*3.17*	*3.08*	*2.99*	*2.88*
310	2.62	2.48	2.36	2.24	2.14	2.04	−2.05	−2.14	−2.25	−2.36	−2.49	−2.63
	3.53	*3.52*	*3.51*	*3.48*	*3.43*	*3.39*	*3.34*	*3.27*	*3.19*	*3.10*	*3.01*	*2.89*
320	2.47	2.35	2.23	2.13	2.04	−2.05	−2.15	−2.26	−2.37	−2.50	−2.63	−2.78
	3.57	*3.56*	*3.55*	*3.52*	*3.48*	*3.43*	*3.37*	*3.30*	*3.22*	*3.12*	*3.01*	*2.88*
330	2.33	2.22	2.12	2.03	−2.06	−2.16	−2.26	−2.38	−2.50	−2.64	−2.79	−2.95
	3.60	*3.59*	*3.57*	*3.54*	*3.50*	*3.45*	*3.39*	*3.32*	*3.23*	*3.13*	*3.00*	*2.86*
340	2.21	2.11	2.02	−2.07	−2.17	−2.27	−2.39	−2.51	−2.65	−2.80	−2.96	−3.14
	3.62	*3.61*	*3.59*	*3.56*	*3.52*	*3.47*	*3.40*	*3.33*	*3.24*	*3.12*	*2.98*	*2.81*
350	2.10	2.02	−2.08	−2.18	−2.28	−2.40	−2.53	−2.66	−2.81	−2.98	−3.15	−3.34
	3.63	*3.62*	*3.60*	*3.57*	*3.53*	*3.48*	*3.41*	*3.33*	*3.23*	*3.10*	*2.95*	*2.77*

φ\ψ	120	130	140	150	160	170	180	190	200	210	220	230
0	−3.76	−3.96	−4.16	−4.33	−4.47	−4.55	4.59	4.55	4.47	4.33	4.16	3.96
	2.45	*2.15*	*1.81*	*1.42*	*0.97*	*0.50*	*0.00*	*0.50*	*0.97*	*1.42*	*1.81*	*2.15*
10	−3.99	−4.19	−4.39	−4.49	−4.55	−4.59	4.56	4.46	4.32	4.14	3.95	3.74
	2.31	*1.97*	*1.59*	*1.15*	*0.68*	*0.18*	*0.32*	*0.80*	*1.24*	*1.64*	*1.99*	*2.30*
20	−4.22	−4.38	−4.52	−4.59	−4.60	4.57	4.47	4.32	4.13	3.93	3.73	3.52
	2.13	*1.75*	*1.32*	*0.85*	*0.36*	*0.14*	*0.62*	*1.07*	*1.47*	*1.83*	*2.14*	*2.41*
30	−4.43	−4.56	−4.63	−4.65	−4.60	4.45	4.33	4.14	3.94	3.73	3.52	3.32
	1.90	*1.48*	*1.02*	*0.53*	*0.03*	*0.45*	*0.90*	*1.31*	*1.67*	*1.99*	*2.26*	*2.49*
40	−4.61	−4.68	−4.70	−4.63	4.52	4.35	4.16	3.95	3.73	3.52	3.32	3.12
	1.63	*1.17*	*0.69*	*0.20*	*0.29*	*0.74*	*1.15*	*1.51*	*1.83*	*2.11*	*2.35*	*2.55*

APPENDIX (*continued*)

φ\ψ	120	130	140	150	160	170	180	190	200	210	220	230
50	−4.75	−4.75	−4.69	4.56	4.39	4.19	3.97	3.74	3.52	3.32	3.12	2.94
	1.31	*0.83*	*0.34*	*0.14*	*0.59*	*1.00*	*1.37*	*1.69*	*1.96*	*2.21*	*2.41*	*2.59*
60	−4.83	−4.75	4.62	4.44	4.22	3.99	3.76	3.54	3.33	3.13	2.95	2.78
	0.95	*0.47*	*0.00*	*0.46*	*0.87*	*1.23*	*1.58*	*1.83*	*2.07*	*2.28*	*2.46*	*2.62*
70	−4.83	−4.69	4.48	4.26	4.02	3.79	3.56	3.34	3.14	2.94	2.78	2.63
	0.57	*0.10*	*0.35*	*0.75*	*1.11*	*1.43*	*1.71*	*1.95*	*2.16*	*2.34*	*2.50*	*2.64*
80	−4.77	4.55	4.31	4.07	3.82	3.58	3.36	3.15	2.96	2.79	2.63	2.49
	0.18	*0.26*	*0.66*	*1.02*	*1.33*	*1.60*	*1.84*	*2.04*	*2.22*	*2.38*	*2.52*	*2.64*
90	4.64	4.38	4.11	3.86	3.61	3.38	3.17	2.98	2.80	2.64	2.50	2.36
	0.21	*0.59*	*0.94*	*1.25*	*1.51*	*1.74*	*1.94*	*2.12*	*2.27*	*2.41*	*2.53*	*2.64*
100	4.45	4.17	3.90	3.65	3.41	3.19	2.99	2.81	2.65	2.50	2.37	2.25
	0.55	*0.89*	*1.19*	*1.45*	*1.67*	*1.86*	*2.03*	*2.18*	*2.31*	*2.43*	*2.53*	*2.62*
110	4.23	3.95	3.69	3.44	3.22	3.01	2.83	2.66	2.51	2.38	2.26	2.14
	0.88	*1.16*	*1.41*	*1.62*	*1.80*	*1.96*	*2.11*	*2.23*	*2.34*	*2.44*	*2.52*	*2.60*
120	4.00	3.73	3.48	3.24	3.04	2.85	2.68	2.53	2.39	2.26	2.15	2.05
	1.16	*1.39*	*1.59*	*1.77*	*1.92*	*2.05*	*2.16*	*2.26*	*2.35*	*2.44*	*2.51*	*2.57*
130	3.77	3.51	3.27	3.06	2.87	2.70	2.54	2.40	2.27	2.16	2.05	−2.04
	1.40	*1.59*	*1.75*	*1.89*	*2.01*	*2.11*	*2.21*	*2.29*	*2.36*	*2.43*	*2.49*	*2.54*
140	3.55	3.31	3.09	2.89	2.71	2.55	2.41	2.28	2.17	2.06	−2.04	−2.14
	1.62	*1.76*	*1.89*	*2.00*	*2.09*	*2.17*	*2.24*	*2.31*	*2.36*	*2.41*	*2.46*	*2.50*
150	3.34	3.11	2.91	2.73	2.57	2.43	2.30	2.18	2.07	−2.03	−2.13	−2.24
	1.80	*1.91*	*2.01*	*2.09*	*2.16*	*2.22*	*2.27*	*2.32*	*2.36*	*2.39*	*2.42*	*2.45*
160	3.14	2.93	2.75	2.59	2.44	2.31	2.19	2.08	−2.02	−2.10	−2.23	−2.36
	1.96	*2.04*	*2.10*	*2.16*	*2.21*	*2.25*	*2.29*	*2.32*	*2.34*	*2.36*	*2.37*	*2.39*
170	2.96	2.77	2.60	2.45	2.32	2.20	2.09	−2.01	−2.11	−2.22	−2.35	−2.48
	2.09	*2.15*	*2.19*	*2.23*	*2.26*	*2.28*	*2.30*	*2.31*	*2.32*	*2.32*	*2.32*	*2.31*
180	2.79	2.62	2.47	2.33	2.21	2.10	−2.00	−2.10	−2.21	−2.33	−2.47	−2.62
	2.21	*2.24*	*2.26*	*2.28*	*2.29*	*2.29*	*2.30*	*2.29*	*2.29*	*2.28*	*2.26*	*2.24*
190	2.64	2.48	2.35	2.22	2.11	2.01	−2.09	−2.20	−2.32	−2.45	−2.60	−2.77
	2.31	*2.31*	*2.32*	*2.32*	*2.32*	*2.31*	*2.30*	*2.28*	*2.26*	*2.23*	*2.19*	*2.15*
200	2.50	2.36	2.23	2.10	2.02	−2.08	−2.19	−2.31	−2.44	−2.59	−2.75	−2.93
	2.39	*2.39*	*2.37*	*2.36*	*2.34*	*2.32*	*2.29*	*2.25*	*2.21*	*2.16*	*2.10*	*2.04*
210	2.37	2.24	2.13	2.03	−2.07	−2.18	−2.30	−2.43	−2.57	−2.73	−2.91	−3.11
	2.47	*2.45*	*2.42*	*2.39*	*2.36*	*2.32*	*2.27*	*2.22*	*2.16*	*2.09*	*2.01*	*1.91*
220	2.25	2.14	2.04	−2.06	−2.17	−2.28	−2.41	−2.55	−2.71	−2.89	−3.09	−3.31
	2.53	*2.50*	*2.46*	*2.41*	*2.36*	*2.31*	*2.24*	*2.17*	*2.09*	*2.00*	*1.89*	*1.76*
230	2.15	2.04	−2.05	−2.16	−2.27	−2.40	−2.54	−2.70	−2.87	−3.06	−3.27	−3.51
	2.59	*2.54*	*2.49*	*2.43*	*2.36*	*2.29*	*2.21*	*2.11*	*2.01*	*1.89*	*1.75*	*1.59*
240	2.05	−2.05	−2.15	−2.26	−2.39	−2.53	−2.68	−2.85	−3.04	−3.24	−3.48	−3.73
	2.63	*2.57*	*2.51*	*2.44*	*2.35*	*2.26*	*2.16*	*2.05*	*1.92*	*1.77*	*1.59*	*1.39*
250	−2.04	−2.14	−2.26	−2.38	−2.51	−2.66	−2.83	−3.01	−3.22	−3.44	−3.69	−3.95
	2.68	*2.60*	*2.52*	*2.44*	*2.34*	*2.23*	*2.11*	*1.96*	*1.80*	*1.62*	*1.41*	*1.16*
260	−2.14	−2.25	−2.37	−2.50	−2.65	−2.81	−2.99	−3.19	−3.41	−3.65	−3.90	−4.17
	2.71	*2.62*	*2.53*	*2.43*	*2.31*	*2.18*	*2.03*	*1.86*	*1.67*	*1.45*	*1.19*	*0.89*
270	−2.25	−2.36	−2.50	−2.64	−2.80	−2.98	−3.17	−3.38	−3.61	−3.86	−4.11	−4.38
	2.73	*2.64*	*2.53*	*2.41*	*2.27*	*2.12*	*1.94*	*1.74*	*1.51*	*1.25*	*0.94*	*0.59*

APPENDIX (*continued*)

φ\ψ	120	130	140	150	160	170	180	190	200	210	220	230
280	−2.36	−2.49	−2.63	−2.79	−2.96	−3.15	−3.36	−3.58	−3.82	−4.07	−4.31	−4.55
	2.75	*2.64*	*2.52*	*2.38*	*2.22*	*2.04*	*1.84*	*1.60*	*1.33*	*1.02*	*0.66*	*0.26*
290	−2.49	−2.63	−2.78	−2.94	−3.14	−3.34	−3.56	−3.79	−4.02	−4.26	−4.48	4.69
	2.76	*2.64*	*2.50*	*2.34*	*2.16*	*1.95*	*1.71*	*1.43*	*1.11*	*0.75*	*0.35*	*0.10*
300	−2.63	−2.78	−2.95	−3.13	−3.33	−3.54	−3.76	−3.99	−4.22	−4.44	−4.62	4.75
	2.76	*2.62*	*2.46*	*2.28*	*2.07*	*1.83*	*1.58*	*1.23*	*0.87*	*0.46*	*0.00*	*0.47*
310	−2.78	−2.94	−3.12	−3.32	−3.52	−3.74	−3.97	−4.19	−4.39	−4.56	4.69	4.75
	2.75	*2.59*	*2.41*	*2.21*	*1.96*	*1.69*	*1.37*	*1.00*	*0.59*	*0.14*	*0.34*	*0.83*
320	−2.95	−3.12	−3.32	−3.52	−3.73	−3.95	−4.16	−4.35	−4.52	4.63	4.70	4.68
	2.72	*2.55*	*2.35*	*2.11*	*1.83*	*1.51*	*1.15*	*0.74*	*0.29*	*0.20*	*0.69*	*1.17*
330	−3.13	−3.32	−3.52	−3.73	−3.94	−4.14	−4.33	−4.45	4.60	4.65	4.63	4.56
	2.69	*2.49*	*2.26*	*1.99*	*1.67*	*1.31*	*0.90*	*0.45*	*0.03*	*0.53*	*1.02*	*1.48*
340	−3.32	−3.52	−3.73	−3.93	−4.13	−4.32	−4.47	−4.57	4.60	4.59	4.52	4.38
	2.63	*2.41*	*2.14*	*1.83*	*1.47*	*1.07*	*0.62*	*0.14*	*0.36*	*0.85*	*1.32*	*1.75*
350	−3.53	−3.74	−3.95	−4.14	−4.32	−4.46	−4.56	4.59	4.55	4.49	4.39	4.19
	2.56	*2.30*	*1.99*	*1.64*	*1.24*	*0.80*	*0.32*	*0.18*	*0.68*	*1.15*	*1.59*	*1.97*

φ\ψ	240	250	260	270	280	290	300	310	320	330	340	350
0	3.76	3.56	3.36	3.17	2.99	2.83	2.68	2.54	2.41	2.29	2.19	2.09
	2.45	*2.69*	*2.90*	*3.07*	*3.20*	*3.32*	*3.41*	*3.48*	*3.54*	*3.58*	*3.61*	*3.63*
10	3.53	3.34	3.15	2.98	2.81	2.66	2.53	2.40	2.28	2.18	2.08	−2.02
	2.56	*2.77*	*2.95*	*3.10*	*3.23*	*3.33*	*3.41*	*3.48*	*3.53*	*3.57*	*3.60*	*3.62*
20	3.32	3.14	2.96	2.80	2.65	2.51	2.39	2.27	2.17	2.07	−2.02	−2.11
	2.63	*2.81*	*2.98*	*3.12*	*3.24*	*3.33*	*3.40*	*3.47*	*3.52*	*3.56*	*3.59*	*3.61*
30	3.13	2.95	2.79	2.64	2.50	2.38	2.26	2.16	2.06	−2.03	−2.12	−2.22
	2.69	*2.86*	*3.00*	*3.13*	*3.23*	*3.32*	*3.39*	*3.45*	*3.50*	*3.54*	*3.57*	*3.59*
40	2.95	2.78	2.63	2.50	2.37	2.26	2.15	2.05	−2.04	−2.13	−2.23	−2.35
	2.72	*2.88*	*3.01*	*3.12*	*3.22*	*3.30*	*3.37*	*3.43*	*3.48*	*3.52*	*3.55*	*3.56*
50	2.78	2.63	2.49	2.36	2.25	2.14	2.05	−2.04	−2.14	−2.24	−2.36	−2.48
	2.75	*2.89*	*3.01*	*3.10*	*3.19*	*3.27*	*3.34*	*3.39*	*3.43*	*3.48*	*3.51*	*3.52*
60	2.63	2.49	2.36	2.24	2.14	2.04	−2.05	−2.14	−2.25	−2.37	−2.50	−2.64
	2.76	*2.88*	*2.99*	*3.08*	*3.17*	*3.24*	*3.30*	*3.35*	*3.40*	*3.44*	*3.46*	*3.48*
70	2.49	2.36	2.24	2.13	2.04	−2.05	−2.15	−2.26	−2.38	−2.51	−2.65	−2.81
	2.76	*2.87*	*2.96*	*3.05*	*3.12*	*3.20*	*3.26*	*3.30*	*3.34*	*3.38*	*3.41*	*3.42*
80	2.36	2.24	2.14	2.04	−2.06	−2.16	−2.26	−2.38	−2.52	−2.66	−2.82	−3.00
	2.75	*2.85*	*2.93*	*3.01*	*3.08*	*3.14*	*3.20*	*3.24*	*3.28*	*3.32*	*3.34*	*3.34*
90	2.25	2.14	2.04	−2.06	−2.16	−2.27	−2.39	−2.52	−2.67	−2.83	−3.01	−3.21
	2.73	*2.82*	*2.90*	*2.97*	*3.03*	*3.09*	*3.14*	*3.18*	*3.21*	*3.24*	*3.25*	*3.25*
100	2.14	2.04	−2.06	−2.16	−2.27	−2.39	−2.53	−2.68	−2.84	−3.02	−3.22	−3.45
	2.71	*2.78*	*2.85*	*2.91*	*2.97*	*3.02*	*3.07*	*3.10*	*3.13*	*3.15*	*3.15*	*3.14*
110	2.04	−2.05	−2.16	−2.27	−2.39	−2.53	−2.67	−2.85	−3.03	−3.24	−3.47	−3.72
	2.68	*2.74*	*2.80*	*2.85*	*2.90*	*2.95*	*2.98*	*3.01*	*3.03*	*3.03*	*3.02*	*3.00*
120	−2.05	−2.15	−2.26	−2.39	−2.53	−2.68	−2.85	−3.04	−3.24	−3.48	−3.73	−4.00
	2.63	*2.69*	*2.74*	*2.78*	*2.82*	*2.85*	*2.88*	*2.91*	*2.90*	*2.89*	*2.86*	*2.80*

APPENDIX (*continued*)

$\phi \backslash \psi$	240	250	260	270	280	290	300	310	320	330	340	350
130	−2.15	−2.26	−2.38	−2.52	−2.68	−2.85	−3.04	−3.25	−3.48	−3.73	−4.02	−4.33
	2.59	2.63	2.67	2.70	2.73	2.75	2.76	2.76	2.75	2.72	2.66	2.57
140	−2.25	−2.38	−2.52	−2.67	−2.84	−3.03	−3.25	−3.48	−3.74	−4.03	−4.34	−4.66
	2.53	2.56	2.59	2.61	2.62	2.62	2.62	2.60	2.56	2.50	2.41	2.38
150	−2.37	−2.51	−2.66	−2.83	−3.02	−3.24	−3.48	−3.74	−4.03	−4.35	−4.68	−5.00
	2.47	2.48	2.49	2.50	2.49	2.48	2.45	2.40	2.34	2.24	2.11	1.92
160	−2.50	−2.65	−2.82	−3.01	−3.22	−3.46	−3.73	−4.02	−4.34	−4.67	−5.02	−5.35
	2.39	2.39	2.39	2.37	2.34	2.31	2.25	2.17	2.06	1.92	1.78	1.49
170	−2.64	−2.81	−3.00	−3.21	−3.45	−3.71	−4.01	−4.32	−4.66	−5.01	−5.33	−5.60
	2.31	2.29	2.26	2.22	2.17	2.10	2.01	1.89	1.73	1.53	1.28	0.01
180	−2.79	−2.98	−3.19	−3.43	−3.65	−3.98	−4.30	−4.64	−4.97	−5.32	−5.62	−5.82
	2.21	2.17	2.11	2.05	1.97	1.86	1.72	1.55	1.33	1.06	0.75	0.00
190	−2.96	−3.17	−3.40	−3.66	−3.95	−4.26	−4.60	−4.93	−5.25	−5.55	−5.80	5.81
	2.09	2.02	1.94	1.84	1.72	1.57	1.38	1.15	0.86	0.54	0.16	0.00
200	−3.14	−3.37	−3.62	−3.91	−4.21	−4.55	−4.88	−5.22	−5.50	5.71	5.77	5.72
	1.96	1.86	1.75	1.61	1.44	1.23	0.98	0.68	0.34	0.04	0.44	0.00
210	−3.34	−3.59	−3.86	−4.16	−4.49	−4.82	−5.14	−5.42	5.63	5.72	5.70	5.50
	1.80	1.67	1.51	1.33	1.10	0.83	0.52	0.16	0.23	0.64	1.03	0.00
220	−3.55	−3.82	−4.11	−4.42	−4.75	−5.06	−5.34	5.56	5.67	5.63	5.49	5.26
	1.62	1.44	1.24	1.00	0.71	0.38	0.01	0.39	0.81	1.21	1.56	1.87
230	−3.77	−4.06	−4.36	−4.67	−4.96	5.24	5.46	5.56	5.57	5.44	5.22	4.94
	1.40	1.19	0.93	0.63	0.28	0.11	0.52	0.95	1.36	1.72	2.03	2.27
240	−4.00	−4.29	−4.59	−4.89	5.14	5.33	5.50	5.50	5.35	5.14	4.88	4.59
	1.16	0.89	0.57	0.21	0.19	0.62	1.07	1.47	1.85	2.16	2.40	2.59
250	−4.23	−4.52	−4.78	5.03	5.23	5.34	5.34	5.25	5.07	4.82	4.54	4.26
	0.88	0.55	0.17	0.24	0.68	1.13	1.55	1.94	2.26	2.51	2.70	2.84
260	−4.45	−4.71	4.95	5.12	5.22	5.23	5.14	4.98	4.75	4.49	4.21	3.94
	0.55	0.18	0.25	0.70	1.16	1.59	1.99	2.32	2.59	2.79	2.94	3.04
270	−4.64	4.86	5.01	5.11	5.12	5.04	4.89	4.67	4.43	4.17	3.91	3.66
	0.21	0.23	0.69	1.15	1.60	2.00	2.35	2.63	2.85	3.01	3.12	3.19
280	4.77	4.92	5.01	5.07	4.94	4.80	4.59	4.36	4.11	3.86	3.62	3.40
	0.18	0.64	1.11	1.57	1.99	2.35	2.64	2.87	3.05	3.17	3.25	3.31
290	4.83	4.91	4.92	4.85	4.71	4.56	4.29	4.06	3.82	3.59	3.37	3.16
	0.57	1.05	1.51	1.94	2.31	2.63	2.87	3.06	3.20	3.29	3.36	3.40
300	4.83	4.83	4.76	4.63	4.44	4.23	4.00	3.77	3.55	3.34	3.14	2.96
	0.95	1.42	1.86	2.24	2.57	2.84	3.04	3.19	3.31	3.39	3.44	3.47
310	4.75	4.69	4.55	4.37	4.17	3.95	3.73	3.51	3.31	3.11	2.93	2.77
	1.31	1.75	2.15	2.49	2.77	2.99	3.17	3.30	3.39	3.46	3.50	3.52
320	4.61	4.49	4.31	4.11	3.90	3.69	3.47	3.28	3.09	2.91	2.75	2.60
	1.63	2.04	2.39	2.68	2.92	3.11	3.26	3.37	3.45	3.51	3.54	3.56
330	4.43	4.26	4.07	3.86	3.65	3.44	3.25	3.06	2.89	2.73	2.59	2.46
	1.90	2.26	2.57	2.83	3.03	3.20	3.32	3.42	3.49	3.54	3.57	3.59
340	4.22	4.03	3.82	3.61	3.41	3.22	3.04	2.87	2.71	2.57	2.44	2.32
	2.13	2.45	2.72	2.94	3.11	3.26	3.37	3.45	3.52	3.56	3.59	3.61
350	3.99	3.78	3.58	3.38	3.19	3.01	2.85	2.70	2.56	2.43	2.31	2.20
	2.31	2.58	2.82	3.01	3.17	3.30	3.40	3.47	3.53	3.58	3.60	3.62

3

STRUCTURE DETERMINATION BY X-RAY DIFFRACTION

Introduction

Fundamental to all studies of the conformation of polymer chains is the determination of their structure in the solid state by the methods of x-ray crystallography. The wavelengths of x-rays are of the order of interatomic distances, and a regular array of atoms or molecules serves as a diffraction grating for this radiation. The x-ray diffraction pattern depends directly on the arrangement of the atoms within the crystal, and the refined techniques of x-ray crystallography are used to work back from this diffraction data to the crystal structure.

As with other techniques discussed in this book, it will not be possible to give more than brief coverage of the basic theory of x-ray diffraction, and the reader is referred to the excellent texts on this subject (1). This chapter will deal specifically with the theory and practice of determining the structure of polymers and macromolecules. It will be assumed that the reader is acquainted with the methods of application of Bragg's law. In addition, it is expected that he is accustomed to describing the three-dimensional lattice in terms of the basic repeating unit, the *unit cell*, and to defining equally spaced sets of planes in the lattice in terms of their *Miller indices, hkl,* where *h, k,*

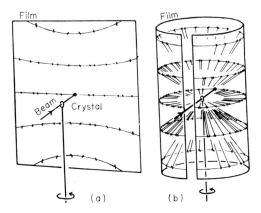

Fig. 3.1. Arrangements for recording single-crystal rotation photographs. (a) Flat-plate film, (b) cylindrical film. [From Bunn (1) with permission.]

and l define the number of intersections of the particular set of planes with the a, b, and c axes, respectively, of a single unit cell.

Figure 3.1a,b shows diagramatically the experimental setup to obtain the x-ray diffraction pattern of an oriented polymer. Monochromatic x radiation is collimated to give a fine beam of approximately parallel x rays, which is

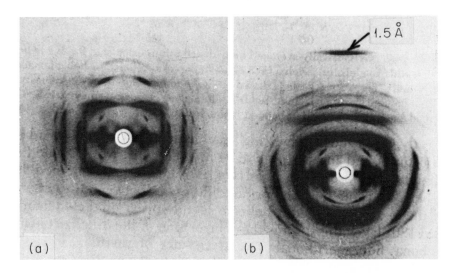

Fig. 3.2. The x-ray diffraction patterns of oriented crystalline fibers of α-poly-L-alanine. (a) Fiber axis perpendicular to the x-ray beam; (b) fiber axis tilted from the perpendicular position in (a) in order to record the meridional reflection at $d = 1.5$ Å. [From L. F. Brown and I. F. Trotter, *Trans. Faraday Soc.*, **52**, 537 (1956), with permission.]

scattered by the crystalline sample, and this diffraction pattern is recorded on a photographic film. A typical x-ray photograph is that of an oriented fiber of poly-α-L-alanine shown in Fig. 3.2. It consists of a number of spots, each of which is a result of reflection from different sets of *hkl* planes. The positions of these spots define the size and shape of the unit cell. The perpendicular distance, known as the *d*-spacing, between the planes that give rise to a particular reflection, is calculated from the Bragg equation (see below). The observed values for all reflections are compared with those calculated for postulated unit cells from which the cell dimensions are determined by trial and error. The relative intensities of the spots depend on the position of the atoms within the unit cell, and the main subject of this chapter will be the methods of working back from the intensity data to the detailed molecular structure.

Diffraction of x Rays by a Crystal. The Reciprocal Lattice

The traditional first approach to the diffraction of x rays by a crystal is to treat the spots on the x-ray photograph as reflections from sets of equally spaced *hkl* planes. Because of interference between the waves scattered by successive planes, reflection occurs only when a wave scattered by one plane of a set is an integral number (*n*) of wavelengths behind the wave scattered by the next plane of the same set. When the incident radiation is a parallel beam, this condition is fulfilled only when a set of planes is inclined at certain angles to that beam, defined by the Bragg equation,

$$n\lambda = 2d \sin \theta$$

where λ is the wavelength of the radiation, d is the perpendicular distance between the successive planes of the set, and θ is the angle between the planes and the beam.

When a single crystal is placed in the x-ray beam, in general only a very few sets of the planes that could reflect x rays are actually in the reflecting position. In order to obtain reflection from the other sets of planes, it is necessary to rotate the crystal. It is useful to mount the crystal in the beam so that it can be rotated about one of its unit cell axes, and the type of photograph that results is shown in Fig. 3.3 for magnesium silicate. Such a photograph is known as a *rotation photograph*, and all the reflections occur in lines known as *layer lines*. For rotation about the *c* axis, the spots on the same layer line are due to reflections from different sets of *hkl* planes that all have the same *l* index. It can be seen that certain sets of planes, such as those perpendicular to the *c* axis, are not brought into the reflecting position by this

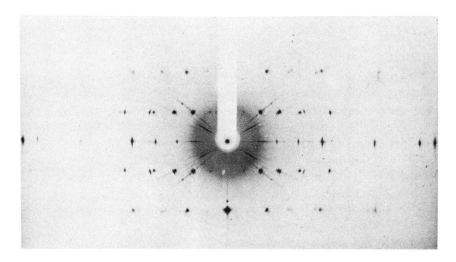

Fig. 3.3. The x-ray rotation diagram for single crystal of magnesium silicate, $MgSiO_3$. [From Alexander (1), with permission.] Note the horizontal layer lines consisting of *hkl* reflections with the same *l* index.

rotation and thus further rotation about another axis is necessary to record all possible reflections.

The way a reflection occurs can also be visualized in terms of the so-called reciprocal lattice. Every repeating lattice in real space can be described by a lattice in reciprocal space. A set of planes with spacing d in real space is equivalent to a point P in reciprocal space, a distance $1/d$ from the selected origin O, where OP is perpendicular to the set of planes. It is easily shown that the points for all *hkl* sets of planes form a three-dimensional lattice (2). The lengths of the axes of the lattice in reciprocal space, a^*, b^*, and c^*, and the angles between them are dependent on the axes and angles of the unit cell in real space. For a simple case of an orthorhombic unit cell, $a^* = 1/a$, $b^* = 1/b$, $c^* = 1/c$, and a^*, b^*, and c^* are mutually perpendicular.

A particular set of planes in real space reflects only when it is inclined at a certain angle θ to the beam. This set of planes becomes a point in reciprocal space, and it can be shown that reflection occurs only when this point lies on a sphere of radius $1/\lambda$, which has the direction of the incident beam as a diameter, and intersects the reciprocal lattice at the origin (3). This sphere is known as the sphere of reflection and is shown with a schematic reciprocal lattice in Fig. 3.4. When the crystal is rotated about the c axis, then a^* and b^* are perpendicular to the axis of rotation. The reciprocal lattice points then lie on planes perpendicular to the axis of rotation, and give rise to the reflections which lie on layer lines on the x-ray photograph. Full rotation of the crystal moves the reciprocal lattice points through the sphere of reflection as the various sets of planes in real space come into the reflecting position.

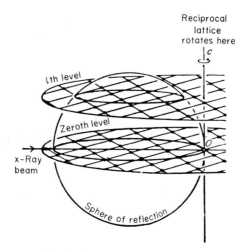

Fig. 3.4. Reciprocal lattice passing through the sphere of reflection as it rotates. [From Bunn (1), with permission.]

Crystalline Polymers and Macromolecules

In general, biological macromolecules do not form crystals large enough to be mounted separately in order to record the intensity data. Exceptions are the globular proteins which can be studied by single-crystal methods, and they are discussed later in this chapter. In addition, many polymers form microcrystals and their electron diffraction patterns often yield useful complementary information (see Chapter 4).

The most common polymer structure studied by x-ray methods is the partially crystalline fiber. This type of system consists of rod- or ribbonlike crystallites, which are arranged together with their long axes approximately parallel to the length of the fiber. The crystallites are generally surrounded by poorly crystalline regions of the same, or in some cases a different, material. Cross sections of idealized fibrous systems of this type are shown in Fig. 3.5 a–c. In the ideal fiber these crystallites are arranged parallel to the fiber axis, and this is the only orientation possessed by the structure. As shown in Fig. 3.5, the fibrils can take up any rotational orientation about the long axis and, in an ordinary fiber, crystalline regions are present in all such positions. Thus, the x-ray pattern of a fiber when the long axis is perpendicular to the beam is analogous to a single-crystal rotation diagram. Some of the crystallites are in the reflecting position for each set of planes and the fiber does not need to be rotated to record all possible reflections. From the point of view of the reciprocal lattice, the presence of this fiber disorientation of the crystallites converts the lattice points of Fig. 3.4 into circles in planes perpendicular to

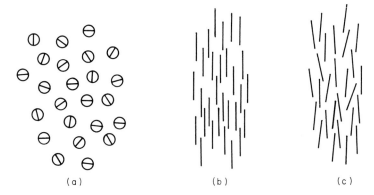

Fig. 3.5. Idealized arrangements of crystallites in fibrous specimens. (a) Cross section of the fiber showing crystallites with all rotational positions about the fiber axis; (b) parallel arrangement of crystallites along the fiber axis; (c) disorientation of crystallites about the fiber axis producing arcing of the reflections.

the rotation (fiber) axis. These circles intersect the sphere of reflection and thus all reflections are seen without moving the specimen.

Real specimens differ from the above idealized systems in that the orientation parallel to the fiber axis is not perfect, but the long axes are distributed as shown in Fig. 3.5c. The effect of this is to broaden the circles in reciprocal space (Fig. 3.6a), so that the reflections on the photograph are arcs instead of spots. The $00l$ points (Fig. 3.6b) are converted into "caps" by this disorientation, and if this effect is large enough, part of the cap intersects the sphere of reflection. Thus, disorientation about the fiber axis produces reflections from sets of planes that otherwise would not be in the reflecting position. Another difference between a fiber and a rotating single crystal is that the crystallites of the fiber are often very small, with the effect that the diffraction maxima are broadened, and the reflections have a measurable angular spread. In reciprocal space, the circle of Fig. 3.6a becomes a band with thickness varying over a range of $1/d$. The utilization of this effect to measure the apparent crystallite thickness is described below (Crystallite Size. Degree of Crystallinity). A further contribution to this spread of the reflections is the imperfection in the arrangement of the chains within the crystallites, i.e., the chains on the outside of the fibrils are probably not arranged as regularly as those near the center. Distortions and dislocations may also be present within the crystalline regions (4).

The reflections due to scattering by successive unit cells are termed the "wide-angle" pattern. Additional maxima are often observed at longer d-spacings known as the "small-angle" pattern and are related to the size and mode of packing of the crystallites themselves. Naturally, division of the pattern into small- and wide-angle regions is somewhat arbitrary.

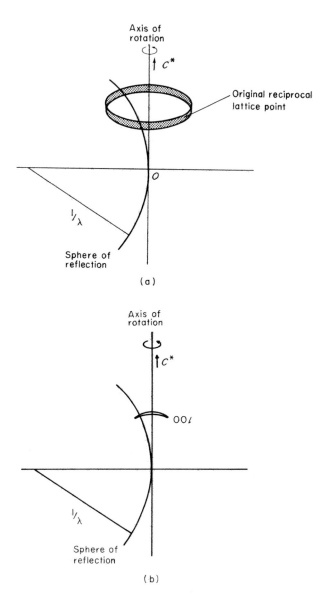

Fig. 3.6. (a) Single *hkl* point of the reciprocal lattice is converted into a circle by the complete rotational disorder about the fiber axis. This circle is broadened by the imperfect orientation parallel to the fiber axis and the small crystallite size. (b) Conversion of 00*l* point into "cap" as a result of fiber disorientation. Note that, in this case, the 00*l* cap intersects the sphere of reflection.

Meridional Reflections

As already described, reciprocal lattice points on or near the axis of rotation do not cut the sphere of reflection and so no reflection occurs. The axis of the unit cell parallel to the fiber axis is normally taken as c and thus the $00l$ reflections should not be observed. Disorientation with respect to the fiber axis enlarges the reciprocal lattice points and some of the lower $00l$ reflections are observed. It is particularly important to know whether a $00l$ reflection is absent because of insufficient disorientation, or whether the actual intensity is zero, since the latter property is related to the chain symmetry. Furthermore, on the higher layer lines, a general hkl point may be converted by the

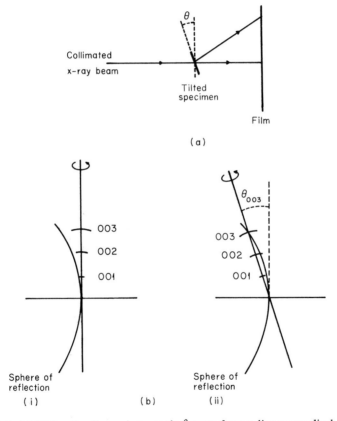

Fig. 3.7. (a) Tilting the fiber axis an angle θ away from a line perpendicular to the x-ray beam brings a meridional reflection with Bragg angle θ onto the x-ray photograph. (b) Effect of tilting the specimen in terms of the reciprocal lattice, showing how the $00l$ "caps" are moved so that they intersect the sphere of reflection. (i) Before tilting; (ii) after tilting.

disorientation into a broad circle that just touches the sphere of reflection, and may give rise to a reflection on the meridian, thus confusing the issue. These problems are resolved by tilting the fibre, which results in tilting the reciprocal lattice, as shown in Fig. 3.7a, so that a particular 00*l* lattice point touches the sphere of reflection. When the fiber axis is positioned so that, for example, the 004 point is touching the sphere of reflection, then any meridional reflection on the fourth layer line is indexed as 004. Any pseudo-meridional for the perpendicular fiber will now be split into two off-meridional reflections. An example of a diffraction pattern obtained with this technique is shown in Fig. 3.2b. for α-poly-L-alanine. When the fiber is tilted, a meridional reflection at $d = 1.5$ Å is observed which is not detected for the perpendicular fiber. Similarly, reflections on the lower layer lines that appear on the meridian for the perpendicular fiber are split into off-meridionals on tilting.

To determine which 00*l* reflections are absent, it is necessary to tilt the fiber for each layer line in turn. Figure 3.7b shows the lattice tilted for the 003 reflection. From the geometry it can be shown that the fiber has to be tilted at the Bragg angle θ to the perpendicular (i.e., with its axis making an angle of $90° - \theta$ to the x-ray beam) for the particular 00*l* reflection.

Crystallite Size. Degree of Crystallinity

For an infinite three-dimensional lattice, diffraction occurs at the Bragg angle θ and gives rise to a spot of infinitesimal thickness. For real crystals, however, this is definitely not the case. A densitometer trace of a line in a powder pattern, or a diffractometer scan of a reflection, shows a peak that is narrow or broad, depending on the specimen. The angular spread of a peak is defined by its half-width, i.e., the width at half the height, which is shown in Fig. 3.8. The crystallite width, B, perpendicular to the set of planes giving rise to the reflection is given by the Scherrer equation:

$$B = \frac{0.9\lambda}{L \cos \theta_0}$$

where L is the angular spread of 2θ for the reflection measured in radians, and θ_0 is the Bragg angle for the peak itself. The Scherrer equation is best applied for peak widths corresponding to 50–200 Å. With refined methods, crystallite widths can be measured to about 2000 Å (5), beyond which the lines are so sharp that differences cannot be detected.

For fibrous crystallites, the length is generally much greater than the width; consequently, reflections occurring on, or near, the meridian, which

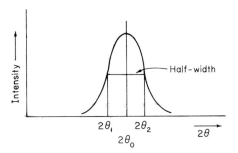

Fig. 3.8. Half-width of an x-ray reflection, i.e., width at half the height. $2\theta_0$ is twice the Bragg angle for the reflection. The angular broadening $L = 2\theta_2 - 2\theta_1$. The crystallite width B is given by the Scherrer equation (see text).

come from planes almost perpendicular to the fiber axis, are usually much sharper than those parallel to the axis, for which there are often only a few repeats.

This treatment assumes small but perfect crystallites packed together or in an amorphous matrix. This is a rather idealized picture since the degree of perfection in the center of the crystallite will be greater than that at the edges. At best the value of B determined from the Scherrer equation can be only an apparent crystallite size. Furthermore, it is clear that many crystals possess distortions and dislocation of different kinds which make more difficult the definition of the regular crystallites. Hosemann has considered the effects of these distortions in terms of his concept of the paracrystal (4). Nevertheless, values of B are useful for comparing specimens of similar materials.

In the same way, the concept of crystalline and amorphous regions is still used to give an idea of the degree of crystallinity. This requires separation of the diffraction pattern into the sharp reflections, due to the repeats of the crystalline regions, and the superimposed diffuse halos due to the amorphous material. It is important to have some measure of the degree of crystallinity, since other physical properties are so dependent on this factor.

The intensity of x rays scattered over all angles by a given assembly of atoms is independent of their state of order or disorder. The latter determines only where the scatter will be strong or weak. Thus, if we can separate the crystalline and amorphous sections of the pattern, then the ratio of proportions may be shown:

$$\frac{\text{crystalline}}{\text{amorphous}} = \frac{\text{total crystalline scattering intensity}}{\text{total amorphous scattering intensity}}$$

This type of approach has been used, for example, to determine a crystallinity index for comparison of cellulose specimens from different sources (6).

Symmetry of Chain Molecules

For any groups of identical atoms or molecules to form a regular crystal-line lattice, they must be arranged together in a symmetrical manner. This symmetry can be described in terms of a number of simple operations. The ways in which these basic symmetry operations can be combined form the subject of group theory. The possible symmetry in crystal lattices is described in terms of four operations: rotation, reflection, inversion, and translation. Combination of these elements leads to the 230 space groups that are the only ways in which identical units can be arranged to give a three-dimensional repeating lattice. The subject of space groups is covered in most general texts on x-ray crystallography (1). A description of each of the 230 space groups is given in "International Tables for X-ray Crystallography", Volume 1 (7).

In the case of biological polymers, the operations reflection and inversion are relatively rare. Proteins, nucleic acids, carbohydrates, etc., are all made up of optically active units, each with one specific configuration. Reflection or inversion of an L-amino acid produces the equivalent D-structure, but this does not occur in protein structures (except in very rare cases; see Chapter 1). Thus, for most biological work, we need consider only rotation and trans-lation as the symmetry operations. Combination of these two leads to the important symmetry element, the *screw axis*, used to describe a helical chain. It consists of a rotation about an axis through an angle $360°/n$, coupled with a simultaneous translation parallel to the axis of cm/n, where c is the repeat distance along the axis and m and n are integers such that $m \leqslant n - 1$. Such an axis is known as an n_m screw axis. Figure 3.9 shows examples of 2_1, 3_1, 3_2, 4_1, 4_2, and 4_3 screw axes. The coordinates for the positions generated by the operation of these axes on a general point x, y, z are referred to as *equivalent positions*. It can be seen that pairs such as 3_1 and 3_2, 4_1 and 4_3, and in

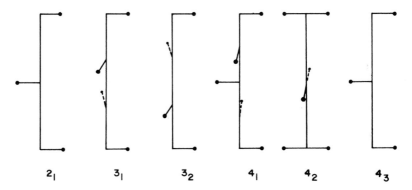

2_1 3_1 3_2 4_1 4_2 4_3

Fig. 3.9. Examples of screw axes.

general n_m and n_{n-m} ($m < n - m$), are very similar except that the first one in each case is a spiral with right-hand sense and the second has left-hand sense.

The symmetry of a chain molecule is precisely defined by the screw axis, although the description needs to be more specific than that given above. The 4_1 and 4_3 screw axes above refer simply to the arrangement of the asymmetric units and say nothing of the way these may be linked together. The symmetry of a helical molecule is normally defined by the number of monomer units M that repeat in N turns of the chain. For this N_M helix, it is necessary to define also the sense of the spiral, i.e., left- or right-hand. Thus, when we refer to a 4_1 left-hand helix or a 6_5 right-hand helix, we specify not only the symmetry of the arrangement of monomers in space, but also the way they are linked in a chain.

Structure Determination

Unit Cell Dimensions

When an x-ray pattern is obtained, the first requirement is to measure the d-spacings, which are obtained by calculating the θ angles for the reflections from the camera geometry and then applying Bragg's law. For a flat-plate fiber diagram, the four quadrants of the photograph are identical and the arcs are parts of circles (see Fig. 3.10). The diameter $2R$ of each circle is measured and used to calculate the Bragg angle θ in the relation

$$\tan 2\theta = R/D \qquad (3.1)$$

where D is the specimen-to-film distance. When D is not known accurately,

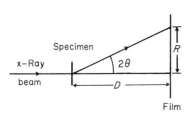
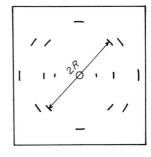

Fig. 3.10. Schematic of method of recording a fiber diagram on a flat-plate film and measurement of d-spacings, $\tan 2\theta = R/D$, $\lambda = 2d \sin \theta$. The specimen-to-film distance, D, is determined accurately by calibration with, e.g., calcite, for which the d-spacings are known; $\tan 2\theta$ is determined, and θ is substituted into the Bragg equation to give d.

it is calibrated by dusting the fiber with a powder such as calcite, which gives a strong diffraction ring at $d = 3.04$ Å. From this known spacing, D can be calculated accurately. Photographs for powder specimens, known as powder diagrams, can be obtained in the same way, but usually with a Debye–Scherrer camera which utilizes a cylindrical film. The radius of the camera is usually 57.3 mm, which conveniently allows 2θ to be measured from the film since the distance in millimeters from the center of the film to a particular reflection is numerically equal to 2θ in degrees. The measured d-spacings are then compared with those for the various sets of hkl planes of unit cells with different dimensions. Equations for orthorhombic and monoclinic unit cells are given below.

Orthorhombic

$$a \neq b \neq c \qquad \alpha = \beta = \gamma = 90°$$

$$d_{hkl} = \left(\frac{h^2}{a^2} + \frac{k^2}{b^2} + \frac{b^2}{c^2} \right)^{-1/2} \tag{3.2}$$

Monoclinic

$$a \neq b \neq c \qquad \alpha = \gamma = 90° \qquad \beta \neq 90°$$

$$d_{hkl} = \left(\frac{\dfrac{h^2}{a^2} + \dfrac{l^2}{c^2} - \dfrac{2hl \cos \beta}{ac}}{\sin^2 \beta} + \frac{k^2}{b^2} \right)^{-1/2} \tag{3.3}$$

The equivalent expression for the less symmetrical triclinic unit cell is more complicated than Eq. (3.3) since $\alpha \neq \beta \neq \gamma \neq 90°$, and it is more easily calculated in terms of the reciprocal lattice (8). Most indexing is usually done with the aid of the computer, which rapidly calculates the d-spacings for a number of trial unit cells. When the agreement is adequate, say within 1%, for the observed and calculated spacings, the hkl values are assigned to the observed data and the approximate unit cell parameters are refined using a least-squares procedure.

For many poorly crystalline polymer systems, often all that can be obtained is a powder pattern consisting of about 10 lines or less. In such a case, only a rough idea of the unit cell can be obtained. For polypeptide specimens it is usual to compare such data with spacings predicted for the standard protein structures made up of chains in the α, β, polyglycine II conformations, etc. The presence of certain d-spacings such as a 4.7 Å reflection leads one to suspect a β structure, while a 3.1 Å reflection might indicate a polyglycine II conformation, and the procedure is to check the other reflections against these possibilities. Helical chains often pack together in a hexagonal arrangement, and the unit cell dimensions are obtained from a special case of Eq. (3.3) where $a = c$ and $\beta = 120°$. Naturally, with such a small amount of data, one cannot be completely confident about the conclusions, and additional information is sought elsewhere. Ideally, if the polymer can be oriented, say by drawing a film or fiber, then the positions of the reflections in the fiber diagrams provide valuable information for indexing. Otherwise, other techniques such as electron diffraction and infrared spectroscopy are necessary to assist in definition of the conformation.

Intensities of Observed Reflections

Each *hkl* point in the reciprocal lattice has associated with it a term $\mathbf{F}(hkl)$, which is the value for the Fourier transform of the structure at that point in reciprocal space. The intensity of the observed reflection on the photograph is proportional to $g|F(hkl)|^2$ where $|F(hkl)|$ is the modulus of $\mathbf{F}(hkl)$ and g takes account of the Lorentz, polarization, and other correction factors (1). For a known or postulated structure, $\mathbf{F}(hkl)$ can be calculated from the following expression:

$$\mathbf{F}(hkl) = \sum_{j=1}^{n} f_j \exp[2\pi i(hx_j + ky_j + lz_j)] \tag{3.4}$$

where f_j is the atomic scattering factor, and x_j, y_j, z_j are the coordinates, of the jth atom, and the summation is over all atoms in the unit cell. This can be separated into real and imaginary parts,

$$\mathbf{F}(hkl) = A(hkl) + iB(hkl) \tag{3.5}$$

where

$$A(hkl) = \sum_{j=1}^{n} f_j \cos 2\pi(hx_j + ky_j + lz_j) \tag{3.6}$$

and

$$B(hkl) = \sum_{j=1}^{n} f_j \sin 2\pi(hx_j + ky_j + lz_j) \tag{3.7}$$

and $|F(hkl)|$ is given by

$$|F(hkl)| = [A(hkl)^2 + B(hkl)^2]^{1/2} \tag{3.8}$$

The term $\mathbf{F}(hkl)$ is a vector with magnitude $|F(hkl)|$ and phase angle α, which is given by $\tan \alpha = B(hkl)/A(hkl)$. This angle α defines the origin relative to the unit cell of the resultant wave produced by combination of all the waves scattered by the atoms within the unit cell.

From the properties of Fourier series, the expression equivalent to Eq. (3.4) describing the structure in terms of $\mathbf{F}(hkl)$ is

$$\rho(x, y, z) = \sum_{-\infty}^{\infty} \sum_{-\infty}^{\infty} \sum_{-\infty}^{\infty} |F(hkl)| \exp[-2\pi i(hx + ky + lz - \alpha)] \quad (3.9)$$

where $\rho(x, y, z)$ is the electron density at a point with coordinates x, y, z within the unit cell, and the summation is over all values of hkl; $\rho(x, y, z)$ varies continuously throughout the unit cell, but has maxima corresponding to the atomic positions. Thus, from knowledge of $|F(hkl)|$ and α we can determine the structure. Unfortunately, while we can obtain values of $|F(hkl)|$ from the intensities on the x-ray photograph, we have no way of measuring α. This is the essential problem of x-ray crystallography, known as the phase problem. The methods of solving this problem are the backbone of x-ray crystallography for single-crystal work, but they have had only limited application for most polymer work in view of the small amount of data generally available. An exception to this is the single-crystal work on the globular proteins; solution of the phase problem for these materials is discussed later, in the section on single crystals. Polymer structures are generally determined by the oldest method of crystallography: trial and error, in which a likely structure is postulated and the intensities are calculated using Eqs. (3.6), (3.7), and (3.8). These figures are compared with the observed data, and modifications of the proposed structure are made until the best agreement is obtained. (The criteria for satisfactory agreement between the observed and calculated data are discussed below.)

Systematic Absences

When a particular reflection is absent, i.e., has zero intensity, the atoms are arranged such that the scattering exactly "cancels out." If the unit cell contains a number of identical groups arranged in a symmetrical manner, then certain hkl reflections will be absent, for which there will be a systematic relationship in the h, k, or l indices, depending on the type of symmetry. Which particular set of hkl reflections will be absent can be predicted from the intensity equations [Eqs. (3.4)–(3.7)] and are tabulated for all the 230 space groups (7). For chain molecules the most important symmetry data to be determined are N and M for the N_M helix. The value of N can often be determined quite easily from the x-ray diffraction pattern. The coordinates of the equivalent positions along the helix axis will be

$$x_1, y_1, z_1; \ 3x_2, y_2, z_1 + c/N; \ 2x_3, y_3, z_1 + 2c/N; \ \text{etc.} \ \ldots.$$

For the $00l$ reflections, Eq. (3.4) will reduce to

$$F(00l) = \sum f_j \exp(2\pi i l z_j) \quad (3.10)$$

This equation is independent of x and y; thus $F(00l)$ has the same value as if all the equivalent positions had the same x and y coordinates. Thus, from the point of view of the $00l$ reflections, the c axis of the unit cell is reduced from c to c/N, and $00l$ reflections are observed only when $l = qN$, where q is an integer. Consequently, the absence of all reflections except 00ε, 002ε, 003ε, etc., suggests a helical structure with $N = \varepsilon$. The value of M can be determined from the spread of intensity between the layer lines as detailed below. A cautionary note here is that the presence of additional symmetry in the arrangement of the chains, such as the staggering of neighboring chains in special positions or coiling of the chains into multiple-strand helices, may modify this simple picture.

Diffraction by Poorly Crystalline Polymer Systems

The discussion so far has considered the diffraction from single crystals. The reciprocal lattice described above can be thought of as the product of the Fourier transform of a single unit cell and an interference function I, which takes account of the interference between the radiation scattered by all the cells in an infinite crystal. For such an infinite crystal, $I = 0$ at all hkl points in reciprocal space except when h, k, and l are all integers, in which case $I = 1$. However, for less perfect systems such as those found in polymers, the function I can be rather different. In order to consider the form I might take, it is necessary to examine the types of systems that are likely for polymer structures. Conformational analysis, described in Chapter 2, shows that for many polymer structures, the chain conformation is dictated mainly by intramolecular forces rather than by effects due to chain packing. Thus, it is common to find chains with an almost regular conformation but which are poorly packed together. This is particularly the case with many natural polymer systems where the chains may be grouped into thin rodlike fibrils. The diffraction patterns of such materials are best considered in terms of the scattering from a single polymer chain, which is subsequently modified by any regular packing that may be present. This approach was used by Cochran et al. (9), who derived the equations for the scattering by helical structures.

Scattering from an Isolated-Chain Molecule

The simplest model of an isolated chain is a line lattice where successive points are separated by a period c, as shown in Fig. 3.11. Diffraction by such a system would give a series of maxima at scattering angles (θ) predicted by

$$l\lambda = c \sin \theta \tag{3.11}$$

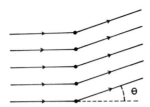

Fig. 3.11. Diffraction by a line lattice of period c. The diffraction maxima consist of straight lines of constant intensity, and their positions are predicted by $l\lambda = c \sin \theta$, where λ is the wavelength, θ is the scattering angle, and l is an integer.

where l is an integer, λ is the wavelength, and θ is simply the angle between the incident and diffracted beams. Equation (3.11) does not define the azimuthal angle ϕ which can take any value from 0 to 2π. Thus, the scattering consists of cones of semivertical angle $(90° - \theta)$, and the optical diffraction pattern is a series of lines. The spacing between the lines depends on c, while the thickness of the lines depends on the number of points forming the lattice. The intensity along the lines is constant.

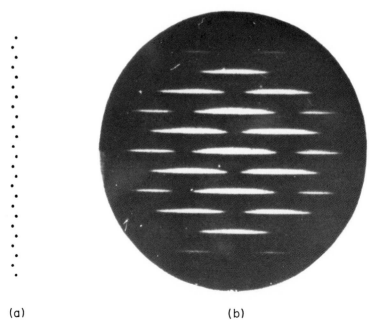

(a)   (b)

Fig. 3.12. (a) Optical diffraction mask representing a chain with a repeating unit of two points. (b) Optical diffraction pattern of the mask in (a). [From C. A. Taylor, *Pure Appl. Chem.* **18**, 533 (1969), by permission of the International Union of Pure and Applied Chemistry.]

For a polymer molecule, the scattering centers are atoms or asymmetric groups of atoms. For a chain of groups, each separated by c, the diffraction pattern is still a series of lines whose position is defined by Eq. (3.4). However, the intensity varies along the lines due to interference between the atoms. Figure 3.12a shows a simple optical diffraction mask for a chain of groups made up of two centers. The intensity variation along the layer lines of the diffraction pattern can be seen in Fig. 3.12b. This can be done for more complex repeat units, and greater fluctuations along the layer lines are obtained [for further details see references (10) and (11)]. The lines on these patterns arise from intersection of the Fourier transform in reciprocal space with the sphere of reflection. If the coordinates of a point in reciprocal space are X, Y, Z, we need to define the value of the transform at that point $\mathbf{F}(X, Y, Z)$. The transform is a series of planes separated by $1/c$; $\mathbf{F}(X, Y, Z)$ varies on these planes and is zero at all other points. Thus we need to define $\mathbf{F}(X, Y, l/c)$.

Diffraction by a Helical Chain

Almost all polymers with a regular sequence take up a helical-chain conformation in the solid state. The diffraction by a regular helix was first determined by Cochran *et al.* (9). This problem was solved by considering the diffraction by a continuous infinitesimally thin helical wire (Fig. 3.13a), which was then modified to a discontinuous helix, defined by the intersection of regularly spaced planes with the continuous helix (Fig. 3.13b). The most useful way of describing any helical structure is in terms of cylindrical polar coordinates, where the rectangular coordinates x, y, z, are written in terms of r, the distance from the helix axis, ψ, the angle of rotation about that axis, and z, which is defined as the distance along the helix axis, as shown in Fig. 3.13c. If R and Ψ are the equivalents in reciprocal space of r and ψ in real space, then both sets of coordinates are converted by the equations:

$$x = r \cos \psi \quad y = r \sin \psi \quad z = z \tag{3.12}$$

$$X = R \cos \Psi \quad Y = R \sin \Psi \quad Z = Z \tag{3.13}$$

For a continuous helical wire, z increases from 0 to p, the pitch of the helix, as ψ increases from 0 to 2π.

Thus

$$\psi = 2\pi z/p \tag{3.14}$$

Substituting Eq. (3.14) in Eq. (3.12), we have the polar coordinates of a continuous helix:

$$x = r \cos (2\pi z/p) \quad y = r \sin (2\pi z/p) \quad z = z \tag{3.15}$$

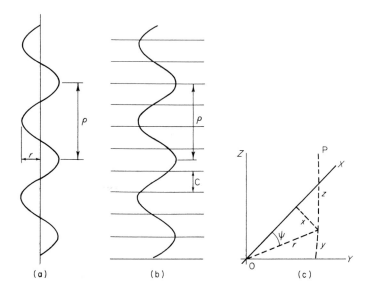

Fig. 3.13. (a) Projection of a helical wire of radius r and pitch p. (b) Discontinuous helix produced by intersection of a set of parallel planes of spacing C with the continuous helix in (a). (c) Conversion of rectangular coordinates to cylindrical polar coordinates for point p; $x = r \cos \psi$, $y = r \sin \psi$, $z = z$.

Equation (3.4), the expression for $\mathbf{F}(hkl)$ for a crystal made up of atoms, is a special case of the more general expression for the Fourier transform $\mathbf{F}(X, Y, Z)$ of a structure defined by an electron density function $\rho(x, y, z)$:

$$\mathbf{F}(X, Y, Z) = \int_{-\infty}^{\infty} \int_{-\infty}^{\infty} \int_{-\infty}^{\infty} \rho(x, y, z) \exp[2\pi i(xX + yY + zZ)] \, dx \, dy \, dz \qquad (3.16)$$

When the conditions for a continuous helix, Eq. (3.15), are substituted, then this equation rearranges to

$$\mathbf{F}\left(R, \Psi, \frac{n}{p}\right) = \int_0^p \exp\left\{2\pi i\left[Rr \cos\left(\frac{2\pi z}{p} - \Psi\right) + \frac{nz}{p}\right]\right\} dz \qquad (3.17)$$

The integral of the exponential of a cosine is solved by making use of the relationship

$$\int_0^{2\pi} \exp(iX \cos \psi)\exp(in\psi) \, d\psi = 2\pi i^n J_n(X)$$

where $J_n(X)$ is a Bessel function of order n and argument X. By analogy,

$$\mathbf{F}\left(R, \Psi, \frac{n}{p}\right) = 2\pi J_n(2\pi Rr)\exp\left[in\left(\Psi + \frac{\pi}{2}\right)\right] \qquad (3.18)$$

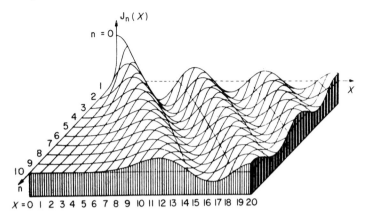

Fig. 3.14. Plot of Bessel functions with orders 0–10. (From E. Jahnke and F. Emde, "Tables and Functions with Formulae and Curves," Dover, New York, 1945, reprinted through permission of the publisher).

The intensity on the nth layer line of the diffraction pattern of a continuous helix is therefore

$$I = |2\pi J_n(2\pi Rr)|^2$$

Values of $J_n(X)$ can be computed and tables of values exist for this function. Figure 3.14 shows the curves for the first ten orders of Bessel functions. It can be seen that $J_0(X) = 1$ at the origin and fluctuates as X increases. All the rest are zero at the origin, and the first peak is progressively further from the origin as n increases.

A plot of the layer line intensities for a continuous helix is shown in Fig. 3.15; the intensity on the nth layer line depends on the square of $J_n(X)$.

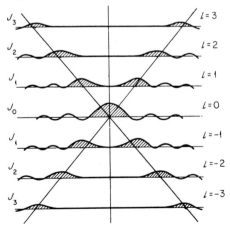

Fig. 3.15. Theoretical transform of a continuous helix. Taken from Vainshtein (1), with permission.

It can be seen that the first peak of the intensity moves further from the meridian on successive layer lines, and the result is an empty cross. The optical diffraction pattern for a mask of a projection of a continuous helix [Fig. 3.16 (12)] shows these predicted features. Figures 3.17 (13) shows the x-ray diffraction pattern of poorly crystalline DNA-B and, although this molecule is a discontinuous rather than a continuous helix, the same empty cross feature can be seen.

When this approach is extended to consider a discontinuous helix (Fig. 3.13b) where N points repeat in M turns of the continuous helix over a repeat distance of c, Eq. (3.18) becomes

$$\mathbf{F}\left(R, \Psi, \frac{l}{c}\right) = \sum_{n} J_{n}(2\pi Rr)\exp\left[in\left(\Psi + \frac{\pi}{2}\right)\right] \tag{3.19}$$

where the summation is over the values of n defined by the selection rule,

$$l = Nm + Mn \tag{3.20}$$

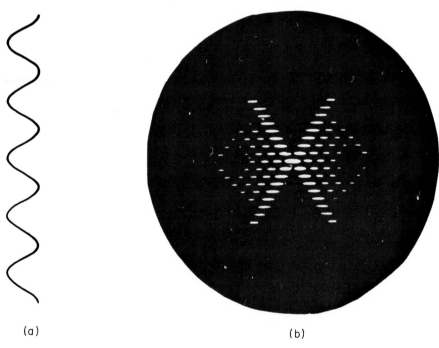

(a)   (b)

Fig. 3.16. (a) Mask of a projection of a continuous helix. (b) Optical diffraction pattern of the mask in (a). [Taken from Holmes and Blow (12), by permission of John Wiley and Sons, Inc.]

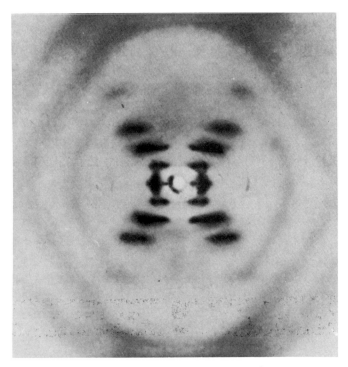

Fig. 3.17. The x-ray diffraction pattern of the sodium salt of DNA-B at 90% relative humidity. (Taken from Ref. 13, with permission.)

where l is the number of the layer line, and m is an integer. Equation (3.19) is for a repeat unit of one point only; for a repeat unit of atoms with coordinates (r_j, ψ_j, z_j) and scattering factor f_j

$$\mathbf{F}\left(R, \Psi, \frac{l}{c}\right) = \sum^n \sum^j f_j J_n(2\pi r_j R) \exp\left\{i\left[n\left(\Psi - \psi_j + \frac{\pi}{2}\right) + 2\pi \frac{lz_j}{c}\right]\right\} \quad (3.21)$$

where n is defined by Eq. (3.20) above.

Equations (3.19) and (3.21) indicate that a number of Bessel functions contribute to each layer line. A schematic representation of the regions of highest intensity on the diffraction pattern of a discontinuous helix is shown in Fig. 3.18. In this example, there are seven units every two turns. The "empty cross" for the continuous helix is replaced by "empty diamonds" for the discontinuous helix. The optical diffraction mask for a projection of a helical array of points and its resultant diffraction pattern are shown in Fig. 3.19 (12). In Fig. 3.17 the x-ray diffraction pattern for poorly crystalline DNA-B shows the predicted region of low intensity.

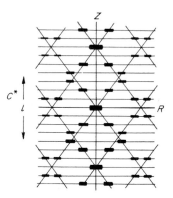

Fig. 3.18. Schematic showing the regions of highest intensity in the diffraction pattern of a discontinuous 7_2 helix. [Taken from Vainshtein (1) with permission.]

When the polymer chain is in a crystalline lattice, Eq. (3.21) can be used to calculate $F(R, \Psi, l/C)$ at the points on the layer line transform corresponding to the reciprocal lattice points. However, in general for polymer structures, the crystallinity is not high and the x-ray diffraction pattern is compared with the transform of the single chain. Even when the polymer is crystalline, this comparison is useful in the first stages of structure determination. The func-

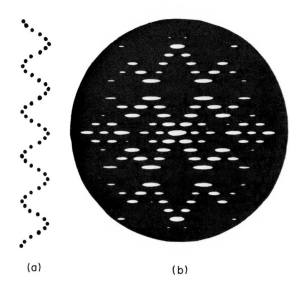

(a)   (b)

Fig. 3.19. (a) Mask of a projection of a discontinuous helix. (b) Optical diffraction pattern of the mask in (a). [Taken from Holmes and Blow (12), by permission of John Wiley and Sons, Inc.]

tion usually calculated is the cylindrically averaged transform, which is obtained by omitting Ψ from Eq. (3.21):

$$F\left(R, \frac{l}{c}\right) = \sum^{n}\sum^{j} f_j J_n(2\pi r_j R)\exp\left\{i\left[-n\left(\psi_j - \frac{\pi}{2}\right) + 2\pi\frac{lz_j}{c}\right]\right\} \quad (3.22)$$

The observed intensities are compared with curves for $|F(R, l/c)|^2$; after correction for the Lorentz and polarization effects, etc.

Identification of the Chain Conformation

The conformation of the polymer chain is the most important information that can be obtained from the x-ray photograph. Recognition of some of the simpler chain symmetries, i.e., integral screw axes, has been already discussed. This can now be extended to a consideration of the general chain with N units in M turns, the N_M helix.

The parameters N and M can often be determined from the intensity distribution on the x-ray photograph. The intense layer lines are predicted by Eq. (3.20) which determines which Bessel functions contribute to the intensity. As the order n increases, the position of the first peak moves outward, and the actual value of the Bessel function declines. For most polymer structures, only the lowest orders of Bessel functions need be considered and, generally, strong layer lines are predicted only when the contributing Bessel functions have orders 0, 1, 2, and perhaps 3, with the relative intensities decreasing as n increases. As an example let us consider the α helix that has 18 amino acids every 5 turns. Equation (3.20) becomes

$$5n + 18m = l$$

Solutions of this equation for successive layer line numbers l are given in Table 3.1, which also records the observed layer line intensities for α-poly-γ-methyl-L-glutamate based on a helix repeat of 27 Å. A meridional reflection can occur only when a contributing Bessel function is nonzero at $R = 0$. This is true only for $n = 0$ and so meridional reflections can occur when

$$l = 18m$$

Thus we are predicting meridional reflections of the 18th, 36th, 54th layer lines, etc., a conclusion which was arrived at in our discussion of the screw axes. The data for α-poly-L-alanine show that a meridional reflection occurs on the 18th layer at $d = 1.5$ Å. The next most intense layer lines should be those which have a J_1 contribution, predicted for the 5th and 13th layers. The 5th layer line is strong as predicted, although this is not true for the 13th

TABLE 3.1

Comparison of the Observed Layer Line Intensities (Iobs) and the Orders of Contributory Bessel Functions (n) Predicted for Poly-γ-Methyl-L-Glutamate, Assuming an 18_5 Helix with Repeat of 27 Å [a][b]

	Layer line (l)										
	0	*1*	*2*	*3*	*4*	*5*	*6*	*7*	*8*	*9*	*10*
Iobs	vvs	—	vw	vvw	—	m	—	—	vvw	—	w
n	0	7	4	3	8	1	6	5	2	9	2
	11	*12*	*13*	*14*	*15*	*16*	*17*	*18*	*19*	*20*	*21*
Iobs	—	—	vvw	—	—	—	—	vw	—	—	—
n	5	6	1	8	3	4	7	0	7	4	3
	22	*23*	*24*	*25*	*26*	*27*	*28*				
Iobs	—	vvw	—	—	vvw	—	vvw				
n	8	1	6	5	2	9	2				

[a] In this table and Table 1.3, the abbreviations are as follows: w, weak; m, moderate; s, strong; vw, very weak; vvw, extremely weak; vvs, extremely strong.
[b] From Bamford *et al.* (15).

layer. However, in the latter case, this is explained by interference between the waves from different atoms (9), a factor which is not considered further here. The other observed layer lines are the 8th and 10th, for which $n = 2$, and the 3rd, for which there is an $n = 3$ contribution. Thus N is found to be 18 from the meridional reflection on that layer and $M = 5$ is derived from the intensity distribution, since no other integer will give the same fit (except $M = 13$, of course, which is another way of describing the same structure, although not the path of the primary chain).

For the general case, the selection rule $l = Nm + Mn$ predicts:

1. A meridional reflection when $n = 0$, i.e., when $l = Nm$; thus, there will be meridionals on the Nth, $2N$th, $3N$th, layer lines, etc.
2. A strong layer line when $n = \pm1$, i.e., when $l = Nm \pm M$; thus, strong intensity should be present on the Mth, $N - M$th, $N + M$th, $2N - M$th layer lines, etc.

Therefore, if the x-ray photograph shows an 18th layer line meridional and a strong 5th layer line, then a good first guess is an 18_5 helix.

Once the layer lines have been indexed, then N and M can be determined relatively easily from the intensity distribution, provided that at least one meridional is observed. An example is the structure of amylose triacetate I, which has been shown (14) to form left-handed helices with 14 residues in 3 turns. The observed layer line intensities for this material are given in Table 3.2. Consideration of the unit cell dimensions and the requirements of stereochemistry rule out multiple-chain helices; N is thus limited to 14 by the meridional reflection on that layer line, and density measurements indicate that the asymmetric unit of the helix is a single sugar residue (anhydro-α-D-glucose 2,3,6-triacetate). It remains then to determine the likely value of M

TABLE 3.2

Order of Contributing Bessel Functions for Layer Lines 0–17 in the Diffraction Patterns of 14_1 14_3, and 14_5 Helices

Layer line l	Observed intensity[a] for amylose triacetate	Low orders of contributing Bessel functions		
		14_1	14_3	14_5
0	vs	0, ±14	0, ±14	0, ±14
1	—	1, −13	5, −9	3, −11
2	—	2, −12	−4, 10	6, −8
3	m	3, −11	1, −13	−5, 9
4	—	4, −10	6, −8	−2, 12
5	—	5, −9	−3, 11	1, −13
6	m	6, −8	2, 12	4, −10
7	—	±7	±7	±7
8	w	−6, 8	−2, 12	−4, 10
9	w	−5, 9	3, −11	−1, 13
10	—	−4, 10	−6, 8	2, −12
11	w	−3, 11	−1, 13	5, −19
12	—	−2, 12	4, −10	−6, 8
13	—	−1, 13	−5, 9	−3, 11
14	w(meridional)	0, ±14	0, ±14	0, ±14
15	—	1, −13	5, 9	3, −11
16	—	2, −12	−4, 10	6, −8
17	vw	3, −11	1, −13	−5, 9

[a] vs: very strong; m: medium; w: weak; vw: very weak.

from the distribution of intensity between the layer lines. This is done by substituting different values of M in Eq. (3.20), and thus predicting the orders of Bessel functions contributing to the individual layer lines for different 14_M helices. The value of M cannot contain a factor of 14 or the helix would reduce to a simpler structure. Thus the only values of M to be considered are $M = 1$, 3, 5, 9, 11, and 13. Solutions of Eq. (3.20) for the 14_1, 14_3, and 14_5, helices are given in Table 3.1 and apply equally to the 14_{13}, 14_{11}, and 14_9 helices, respectively. For a 14_1 helix, the layer lines should decrease in intensity up to $l = 7$ and then slowly increase in intensity up to $l = 14$. This does not occur, and the 14_1 (and 14_{13}) helix is therefore ruled out. The intensity match for the 14_5 (and 14_9) helix is also unacceptable. The $n = 1$ contributions are calculated for $l = 5$ and 9, predicting strong intensity on these layer lines. However, neither the 5th or 9th layer line is observed and this helix must be rejected. For the 14_3 (and 14_{11}) helix, the match of observed and predicted intensities can be seen to be quite good. A strong third layer line is observed which correlates with a J_1 Bessel function for this layer, and all the observed layer lines have contributions with orders $n = 0–3$. The match for the higher layer lines may not seem to be very good, but the geometrical factor (see Intensities of Observed Reflections section) has the effect of increasing the intensity of the lower layer lines relative to the higher ones. The 14_{11} helix is ruled out by stereochemical considerations; such a chain has almost one glucose residue per turn, which is easily shown to be impossible with space-filling models. The 14_3 helix is the only acceptable conformation, and for this only a left-handed helix is possible due to the intraresidue interactions (14). Also, this material is crystalline and the layer lines show reflections rather than the continuous transform. The unit cell is orthorhombic with dimensions $a = 10.87$ Å, $b = 18.83$ Å, and $c = 52.53$ Å, and it contains two antiparallel chains. The relative positions of these chains have an appreciable effect on the layer line intensities. Refinement of the structure was achieved by trial and error.

The polypeptide α helix proposed originally by Pauling and Corey has 18 amino acids repeating in 5 turns and is the approximate chain conformation found in the synthetic polypeptide α structure. The observed layer line intensities for poly-γ-methyl-L-glutamate (PMLG) are compared in Table 3.1 (15) with the orders of contributing Bessel functions predicted from the equation $l = 18m + 5n$. It can be seen that the observed layer lines correspond to the smaller-order Bessel function contributions. The weakness of the 13th layer line is explained by probable interference between waves scattered by atoms in the same asymmetric unit.

Closer study of the PMLG diffraction pattern shows that the layer lines are not exactly orders of 27 Å. Accurate measurement of the spacings indicates that 29_8 or 69_{19} helices may be better approximations. The rise per residue along the axes of these helices is still 1.495 Å, the d-spacing of the meridional reflection, and so the repeat distances must be 43.4 or 103.2 Å respectively. The values of l corresponding to the observed layer lines for these repeats and the predicted orders of contributing Bessel functions are given in Table 3.3. These are the only layer lines for which low orders are predicted, and the intensity distribution is essentially the same for all these helices. Similarly, the layer line spacings in the fiber diagram of α-poly-L-alanine are close to,

TABLE 3.3

Lowest Orders (n) of Contributing Bessel Functions for Observed Layer Lines (l) in X-Ray Diffraction Patterns of α-Poly-γ-Methyl-L-Glutamate; Comparison of 29_8 and 69_{19} Helices.[a]

l (29_8)	3	5	8	10	11	13	16
l (69_{19})	7	12	19	24	26	31	38
n	4	3	1	6	5	2	2
l (29_8)	18	19	21	24	26	29	
l (69_{19})	43	45	50	57	62	69	
n	5	6	1	3	4	0	

[a] From Bamford *et al.* (15).

but not quite exact orders of 27 Å, the repeat for the 18_5 helix (19). The best solution for the fiber repeat is $c = 70.3$ Å, which corresponds to a 47_{13} helix. The values of l that correspond to the observed layer lines are given in Table 3.4, and it can be seen that the solutions for n from the equation $l = 47m + 13n$ predict strong layer lines in approximately the same positions as for the 18_5 helix. [The pattern for α-poly-L-alanine contains a weak meridional on the 16th layer line for the 47_{13} repeat, which is not predicted for this symmetry; it is thought that this is due to distortions in the side-chain arrangement (15,16).] The conclusion to be drawn here is that helices with approximately the same N/M ratio, i.e., number of residues/turn, have very similar diffraction patterns, irrespective of the actual values of the integers N and M. These values can be determined only within the limits of accuracy of the indexing of the layer lines. The α-helix is defined as having approximately 3.6 residues per turn. The actual values of N and M are expected to depend on the intra-molecular forces, and slight differences for different polypeptides are quite reasonable.

TABLE 3.4

Observed Layer Line Intensities (I_{obs}) for α-Poly-L-Alanine and Orders of Contributory Bessel Functions (n) for a 47_{13} Helix

	Layer line (l)													
	0	*5*	*8*	*13*	*16*	*18*	*21*	*26*	*29*	*31*	*34*	*39*	*42*	*47*
I_{obs}	vvw	w	w	vw	vw	s	m	w	vw	ww	vw	vvw	vvw	m
n	0	4	3	1	6	5	2	2	5	6	1	3	4	0

TABLE 3.5

*Comparison of Observed (F_m) and Calculated (F_m') Structure
Amplitudes for the Refined α-Poly-L-Alanine Crystal Model[a,b]*

h	k	l	F_m	F_m'	h	k	l	F_m	F_m'
1	0	0	1002	1055	1	0	18	20[d]	12
1	1	0	459[c]	390	1	1	18	37[d]	46
2	0	0	566	498	2	0	18	37[d]	35
2	1	0	475[c]	464	2	1	18	139	85
3	0	0	372[c]	372	3	0	18	129	166
2	2	0	213	166	2	2	18	363	221
3	1	0	135	91	3	1	18	182	221
4	0	0	109	95	4	0	18	68[d]	189
3	2	0	259	193	3	2	18	124	138
4	1	0	189	205	1	0	21	301	329
5	0	0	134	151	1	1	21	138	156
3	3	0	199	106	2	0	21	43[d]	18
4	2	0	125	83	2	1	21	142	152
1	0	3	13[b]	1	3	0	21	225	133
1	1	3	23[b]	21	2	2	21	55[d]	82
2	0	3	27[b]	39	3	1	21	168	72
2	1	3	71[b]	77	1	0	26	256	204
3	0	3	82[b]	71	1	1	26	215	195
1	0	5	71	95	2	0	26	136	160
1	1	5	363	294	1	0	29	20[d]	9
2	0	5	394	306	1	1	29	28[d]	77
2	1	5	145[c]	163	2	0	29	31[d]	111
1	0	8	145[c]	202	2	1	29	205	126
1	1	8	200[c]	143	1	0	31	27[d]	1
2	0	8	104	23	1	1	31	65[d]	15
2	1	8	370	371	1	0	34	115	77
3	0	8	336	388	1	1	34	145	136
2	2	8	272	261	2	0	34	159	95
3	1	8	154[e]	210	2	1	34	136	66
1	0	13	274[e]	291	1	0	39	24[d]	15
1	1	13	119[e]	83	1	1	39	132	90
2	0	13	88[e]	63	2	0	39	152	124
2	1	13	168	131	2	1	39	135	137
3	0	13	156	152	1	0	42	23[d]	17
2	2	13	51[d]	66	1	1	42	108	71
3	1	13	38[d]	30	1	1	47	33[d]	7
4	0	13	152	50	2	0	47	37[d]	41
3	2	13	146	79	2	1	47	112	95
1	0	16	18[d]	4	3	0	47	149	110
1	1	16	85	91	1	1	52	23[d]	28
2	0	16	138	89	2	1	55	112	173
2	1	16	142	159	3	0	55	155	159
3	0	16	115	153					

A study of the detailed structure of α-poly-L-alanine was first made by Brown and Trotter (17). Both left- and right-handed Pauling and Corey α-helices were considered, but the agreement between observed and calculated intensities was poor in each case. Elliot and Malcolm (18) noted that the chain possesses a definite sense and can point either "up" or "down" the fiber axis. Much of the discrepancy between the observed and calculated intensities is removed when the model contains half "up" and half "down" chains with right-hand helical conformation arranged in a statistical manner. Arnott and Wonacott (16) have refined this structure, using a rigid-body, least-squares procedure. The bond lengths and angles of the amino acid residue were held constant at their probable values based on model compounds, while parameters defining the linkage (ϕ, ψ) and the position of the residue as a whole in the unit cell (rotation and translation) were allowed to vary until the best fit between the observed and calculated intensities was obtained. The comparison between the final calculated and the observed structure factors is given in Table 3.5. It must be remembered that the intensities are proportional to the square of $F(hkl)$. It can be seen that the match between observed and calculated intensities is very good by the standards of polymer crystallography.

This leads to the following question: what is the best agreement between the observed and calculated data? Small differences in postulated structures produce only slight differences in the calculated intensities. Refinement is often made in terms of the statistical "R value" given by

$$R = \frac{\sum \|F_o| - |F_c\|}{\sum |F_o|} \tag{3.23}$$

where $|F_o|$ and $|F_c|$ are the observed and calculated structure factors, and the summations are over all the hkl reflections predicted to occur in the range of the observed data. In single-crystal work, the best structure is usually the one for which R is lowest, and values for R between 0.05 and 0.10 are usually obtained for fully refined structures. In most polymer work, the errors in the intensity measurements are generally much higher than for single crystals, and there is very much less data. Once the approximate structure has been obtained, there is usually little further decrease in the value of R, and a good value is in the range 0.2–0.3. More attention is normally given to a

---

[a] The reflections were indexed in terms of a hexagonal single-chain unit cell with dimensions $a = b = 8.55$ Å, $c = 70.3$ Å, and $\gamma = 120°$. The structure consists of a statistical mixture of "up" and "down" chains, which are 47_{13} α-helices.

[b] Scale, $100 \times$ absolute.

[c] Corrected for overlap with β-phase reflection.

[d] Below threshold reflection, $F_m = \frac{2}{3} F_{threshold}$.

[e] Visually estimated because of presence of streaks.

sizable decline in R during the refinement than to its actual value at the end. In the work by Arnott and Wonacott (16), the value of R declined from 0.422 to 0.206 in the process of refinement of poly-L-alanine.

The value of N for a helix is determined from the meridional reflections. However, in some cases, no meridional reflections are observed since the diffraction pattern becomes diffuse before θ gets sufficiently large. Such a problem occurred with tobacco mosaic virus (TMV). Figure 3.20 shows the

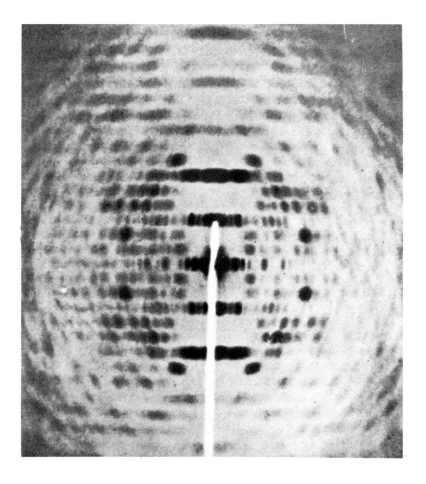

Fig. 3.20. The x-ray diffraction pattern of an oriented sol of tobacco mosaic virus (TMV). (From Ref. 19 by permission.)

x-ray diffraction pattern obtained from an oriented sol in a fine capillary, where all the rodlike particles are aligned approximately parallel to the tube, which in turn is perpendicular to the x-ray beam (19). Although the virus particles are packed approximately hexagonally, they have all possible orientations about their long axis (screw disorder), and the diffraction pattern shows the cylindrically averaged transform of a particle. The layer lines can be indexed as almost perfect orders of 69 Å.

First examination of the pattern showed that it was not possible to identify any meridional reflection, and thus N was not determined. However, the strongest layer lines are the third, sixth, and ninth, which suggests a non-integral helix repeating in three turns, i.e., N_3. The actual value of N, however, remained unknown for some years, until it was eventually determined from a study of mercury-substituted TMV (20). The separation between successive mercury atoms on the edge of the rodlike particle was determined from the equatorial scattering, and this distance was compatible only with a value of $N = 49$, which identified the particle as a 49_3 helix. Close examination of the layer line spacings allowed Franklin and Klug to show that 49 residues in 3 turns was in fact only an approximation, and a more accurate value was 49.02 ± 0.01 per 3 turns. This is an example of the accuracy with which certain parameters can be determined from x-ray diffraction.

The asymmetric units of the helix consist of separate globular protein molecules arranged together (presumably by hydrogen bonds) around the axis of the rod. This helical arrangement has been seen in the electron microscope in sections of the rod that are not too badly damaged by the high vacuum.

Multiple Helices

The chain conformations for many helical molecules, such as the protein α helix or V-amylose (21), are compact coils, and residues on successive turns are in contact. In some cases, as in the two examples given, the turns are linked through hydrogen bonds. For molecules that are not so compact, the pitch is greater, and a groove appears in the molecule. This groove has the same pitch as the chain helix, and if it is large enough, it is clear that another identical chain, or more than one, can be fitted into this groove. When the chains are simply intertwined helices, and both have the same axis, then such a system is known as a multiple-strand helix. In the case of some fibrous proteins, a number of helices are arranged so that their individual axes are coiled together around a common axis, and this arrangement is known as a coiled coil. The diffraction patterns of both systems will be discussed in turn.

Multiple-strand helices are found for a number of polynucleotides and polysaccharides. If we consider the possible arrangement of the chains, then identical chains are expected to pack together in a symmetrical manner so that they have identical intermolecular contacts. When two chains are coiled together this can be done symmetrically whether the two strands have the same or opposite sense. When the identical N_M helices are antiparallel, i.e., have opposite sense, they are related by a twofold rotation axis perpendicular to the common helix, and the symbol for this symmetry is $N_M/2$. On the other hand, if the two chains have the same sense, they can be arranged symmetrically with a two-fold rotation axis coincident with the helix axis, a situation which is denoted by $N_M 2$. Alternatively, the two parallel strands can be related by a screw axis. For three chains a symmetrical arrangement is possible only if they all have the same sense. The symmetry of multiple chain systems has been discussed by Klug *et al.* (22) and Vainshtein (1).

As an example of a parallel-chain system, three 5_1 helices could be arranged around a threefold axis, giving $5_1 3$ symmetry. The x-ray pattern would still show the same fiber repeat as a single strand structure with a meridional on the fifth layer line. The intensity distribution would be different from that for a single 5_1 chain, since the asymmetric unit now consists of an arrangement of three monomer units. A more obvious effect, however, occurs when there is an integral number of units per turn of a single chain and N is also a multiple of ρ, the number of strands forming the multiple-strand helix. Figure 3.21 shows a $6_1 2$ system in which two parallel 6_1 helices are coiled together around a twofold rotation axis. Units 1 and 7 on the first chain are identical with unit $4'$ on the second chain, just as 4 is identical with $1'$ and $7'$. Thus, the repeat for the two-chain complex is half that of the single chain. A single 6_1 helix with a repeat of c would show layer lines at spacings of c, $c/2$, $c/3$, $c/4$, $c/5$, $c/6$, etc., with the first meridional on the sixth layer line. However, with the repeat cut to $c/2$, the double helix would show layer lines at $c/2$, $c/4$, $c/6$, etc., with the meridional reflection on the third layer, although still at the same spacing. This is the same as saying that the odd layer lines for the single helix disappear, due to interferences from the second chain. The meridional on the third layer line implies a threefold screw axis and this is indeed how the arrangement over a repeat of $c/2$ is described. This illustrates the statement that the x-ray pattern shows how the asymmetric units are arranged irrespective of the covalent bonding for the polymer chain.

This type of $N_1 2$ multiple helix has been found for a number of poly-nucleotides and polysaccharides. In the fiber diagram of polyadenylic acid (poly A, see Chapter 12), the only observed meridional reflection is on the fourth layer line at $d = 3.8$ Å, suggesting a 4_1 helix. The unit cell could not be determined accurately due to the diffuseness of the pattern, but the strong equatorial at 17 Å suggested that the cell dimensions perpendicular to the

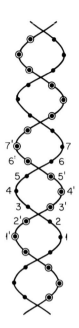

Fig. 3.21. Projection of a double helix formed by two 6_1 helices coiled together. Note that if the chains have the same sense and are exactly out of phase (as shown) then positions 1, 4', and 7 are identical, as are 1', 4, and 7', etc., and the overall repeat for the double helix is half that for the single chain.

fiber axis are probably ~ 17 Å. Such a cell is not compatible with a single-strand 4_1 conformation, and multistrand helices were therefore considered. The structure that seemed most probable on the basis of model building and intensity calculations was an $8_1 2$ structure, i.e., two parallel helices with eight nucleotides per turn coiled together around a twofold axis. Similarly, poly-ribocytidilic acid (poly C) is a double helix in which both parallel chains are 12_1 helices ($12_1 2$), and the first meridional reflection occurs on the sixth layer line (24). (The x-ray pattern of poly C is shown in Figure 12.9.)

One way of looking at the $6_1 2$ structure in Figure 3.21 is that a point on one chain is exactly halfway between identical points on the other chain, i.e. the chains are exactly (π radians) out of phase, and the overall repeat is cut to $c/2$. A similar effect can be obtained for two N_M helices, where N is not a factor of 2, provided these chains are also exactly out of phase. For example, if two 3_1 helices form such a double helix, then the odd layer lines predicted for a single strand are not observed, and the first meridional is on the sixth layer line with respect to the repeat for the single chain. The chains are no longer related by a two-fold rotation axis, but rather by a screw axis. Such a system has been found to occur in ι-carrageenan (see Chapter 11).

Similarly, if ρ chains with N_M symmetry form a multiple helix, where ρ is not a factor of N, they can be arranged symmetrically so that successive chains have a phase difference of $2\pi/\rho$. In this case only every ρth layer line predicted for the single chain repeat will be observed. Such a system is thought to occur in collagen, which is a three strand multiple of 10_3 helices.

Multiple-strand helices with $N_M/2$ symmetry are found for natural DNA, RNA, and base-paired complexes of synthetic polyribonucleotides, such as poly $(A + U)$ (see Chapter 12). If the two 6_1 chains coiled together in Fig. 3.21 are antiparallel, then unit 4' is no longer identical with units 1 and 7. Thus the fiber repeat remains at c and the same layer lines are seen for the two-chain complex, as would be predicted for the single 6_1 chain. For a chain such as a polynucleotide, however, an "up" chain is not very different in terms of electron density from a "down" chain, and thus while the odd layer lines do not disappear, they may be substantially weakened in intensity from those for the single chain.

The final symmetry arrangement for a combination of helical chains involves both a vertical rotation axis and a horizontal twofold axis for which the symbol is $N_M\rho/2$. Such a structure requires an even number of chains and the simplest example (other than the trivial $N_M/2$ case just considered) has four chains that are alternately parallel and antiparallel. This type of structure has not been observed so far, although coiled coils with this symmetry have been considered for α-keratin (see below). Symmetrical arrangements of the chains in multiple helices should be the most likely structures. Of course, it is possible that multiple helices could be formed with no symmetry between the chains, e.g., with parallel chains out of register or three chains with two up and one down. Such a two-up–one-down system has been proposed for the (Gly-Pro-Pro)$_n$ triple helix in single crystals (see Chapter 4), and Atkins has considered various models for a four-strand, coiled coil for honey bee silk (26), one of which has a three-up–one-down arrangement of the chains.

The Diffraction Pattern of a Coiled Coil

The theoretical α helix satisfactorily explains the diffraction patterns for many poly-α-amino acids. The x-ray patterns for most naturally occurring α proteins are similar to that for the 18_5 helix, but significant differences are observed. Examples of x-ray photographs for the kmef (keratin, myosin, epidermin, fibrinogen) group are shown in Fig. 3.22. For these structures, a meridional is observed at $d \simeq 1.5$ Å (not seen in Fig. 3.22), as is expected. However, a meridional is also observed at $d \simeq 5.1$ Å, unlike the theoretical α pattern, which has an off-meridional on the fifth layer line (for 18_5) at $d = 5.4$ Å. It was suggested by Crick (27), and by Pauling and Corey (28)

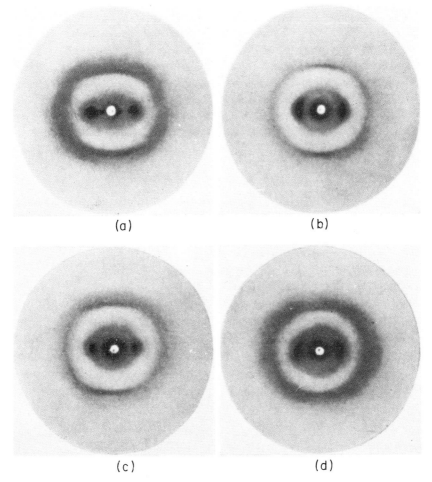

(a) (b)

(c) (d)

Fig. 3.22. The x-ray fiber diagrams of (a) keratin, (b) myosin, (c) epidermin, and (d) fibrinogen. [Taken from W. T. Astbury, *Proc. Roy. Soc., London* **B134**, 303 (1947), with permission.]

that this effect is due to supercoiling of the α helices into coiled coils, probably in order to facilitate better packing of the amino acid side chains.

Crick (27) considered the theoretical scattering from a coiled coil, which is defined by the following parameters. The coiled-coil repeat contains N units (amino acids) in a repeat distance c. Over this distance c, there are M turns of the major helix (supercoiling) and M_1 turns of the minor helix (in this case, the α helix); in this case, the major and minor helices have opposite sense. The Fourier transform at point $(R, \Psi, l/c)$ in reciprocal space for a set of points on a coiled coil is given by

$$F(R, \Psi, l/c) = \sum_p \sum_q \sum_s J_p(2\pi r_0 R) J_q(2\pi r_1 R) J_s\left(2\pi \frac{l}{c} r_1 \sin \alpha\right)$$

$$\times \exp\left\{i\left[p\left(\frac{\pi}{2} + \Psi\right) + q\left(\frac{\pi}{2} - \Psi\right) + s\pi\right]\right\} \qquad (3.24)$$

where r_0 and r_1 are the radii of the major and minor helices and α is the angle of slope of a tangent to the major helix. The selection rule for the orders of Bessel functions is

$$M + (M_1 - M)q + M_1 s = l + NM \qquad (3.25)$$

The supercoiling is proposed as a means of providing better packing for the side chains. The 18_5 helix has 7.2 residues every two turns. The projection in Fig. 3.23a shows that there is a lag of $20°$ between every seventh residue.

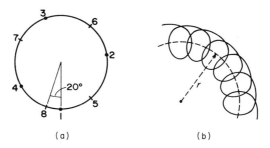

(a)   (b)

Fig. 3.23. (a) Projection of an 18_5 helix perpendicular to the chain axis: 3.6 residues per turn = $100°$ per residue; for seven residues, total rotation = $700°$, i.e., $20°$ short of two complete turns. (b) Supercoiling of the 18_5 helix to move every seventh residue into an equivalent position with respect to the axis of supercoiling.

There is no doubt that the packing would be better if these residues were in identical positions. Thus, if supercoiling is introduced, as in Fig. 3.23b, every seventh side chain is in an equivalent position on the inside of a coil. For this to occur, the stagger of $-20°$ has to be eliminated, which requires supercoiling of the opposite sense to the α helix. For a $20°$ rotation, there must be 18 such operations per full turn of the major helix ($18 \times 20° = 360°$) and thus, there are 36 turns of the minor (2 for every $20°$) for each single turn of the major. The repeat thus consists of $18 \times 7 = 126$ amino acid residues. From the stereochemistry, $r_0 = 5.2$ Å and $\alpha = 10°$ for the major helix. Thus, the repeat c is given by $c = 126 \times 1.5 \cos 10° = 186$ Å. The repeat observed for α-keratin from the small-angle region is 197 Å. The selection rule, Eq. (3.25), now becomes

$$p + 35q + 36s = l + 126m \qquad (3.26)$$

The expression for $F(R, \Psi, l/c)$ depends on three Bessel functions, J_s, J_q, and J_p. If we consider only the intensity on the meridian where $R = 0$ this depends on the product $J_p(0)J_q(0)J_s[2\pi(l/c)r_1 \sin \alpha]$. Now the first two terms are always zero unless p and q are zero. Thus for meridional reflections, the selection rule becomes

$$36s = l + 126m \qquad (3.27)$$

and the intensity depends on $J_s[2\pi(l/c)r \sin \alpha]$. The argument of J_s is not zero and thus the intensity is finite. However, the value of J_s is negligible except for small values of s. Thus, the relative intensities of the meridional reflections can be predicted from the smallest order of contributing Bessel function J_s. Solutions of Eq. (3.26) are given in Table 3.6, and the observed x-ray diffrac-

TABLE 3.6

Lowest Orders of J_s Bessel Functions Contributing to the Meridional Intensities for the Theoretical Coiled-Coil: Solutions of the Equations $36s = l + 126m$, where $l = 18u$, i.e. $2s = u + 7m$

u	0	1	2	3	4	5	6	7	8	9	10
s	0	-3	1	-2	2	-1	3	0	-3	1	-2
$d(\text{Å})$	∞	10.33	5.17	3.44	2.58	2.07	1.72	1.48	1.29	1.15	1.03

tion data is summarized in Table 3.7. It can be seen that the only values of l for which meridionals can occur are order of 18. This is not really surprising since the coiled coil can be described as having an 18_1 screw axis. Table 3.6 shows that the strongest meridional would be at $d = 1.48$ Å for a J_0 contribution. Thereafter, we would expect strongish meridionals on the layer lines at 5.17 Å and 2.07 Å. The former is observed, but the latter is always weak or absent. A similar effect occurs for the 13th layer line for the straight α helix (18_5) and is probably also due to interference between atoms in the same

TABLE 3.7

Solutions of the Equation $p + 35q + 36s = l + 126m$ and the Observed Intensities for the Coiled α Helix

Solution		Observed intensity
$p = q = s = m = 0$		Equator
p small	$q = s = m = 0$	Near equator
$s = 1$	$p = q = m = 0$	5.1 Å meridian
$m = 1$	$p = q = s = 0$	1.5 Å meridian
$q = 1$	$p = s = m = 0$	5 Å near meridian

residue. The strong off-meridionals near 5.0 Å and the near equatorials are explained by combinations of low values of p, q, and s in Eq. (3.26). It can be seen here that the intensity distribution is very similar to those for the various straight α-helical conformations (Tables 3.1 and 3.3) and illustrates that small conformational changes have little effect on the diffraction pattern.

These features have been predicted for a single-coiled coil. Combinations of two, three, and four chains have been considered in theoretical calculations (26,30) and optical diffraction analogs (29). In addition to the α proteins, the coiled-coil structure has been found for collagen. Here three chains in a 10_3 conformation are coiled together with each chain repeating after 10 turns and 30 residues (see Chapter 10).

It has been show by Parry (30) that the same intensity distribution as for the α coiled coil can be obtained from a straight α helix provided that it has a particular sequence of residues. Reasonable agreement was obtained for the polyheptapeptide sequence (lys-X-glu-X-asp-X-X)$_n$. The sequence of α-keratin is not known and there is no way of knowing if these sequences are reasonable, although they are compatible with amino acid analyses. Fibrous proteins are expected to possess some sort of regular amino acid sequence or, otherwise, they should be globular. The nature of the side chains is normally ignored in calculations and such an approximation may be less valid than has previously been assumed.

Single Crystals of Globular Proteins

Single crystals have been obtained for some globular proteins and the structures have been determined by more refined methods than trial and error. The proteins are usually crystallized from buffer solution and contain between 25 and 95% water; these crystals have dimensions of the order of 0.1 mm and are mounted in a capillary containing their mother liquor so that x-ray data can be recorded.

As indicated earlier, the main problem in x-ray crystallography is the lack of knowledge of the phase angle associated with the amplitude $|F(hkl)|$ of the scattered wave. The first method of overcoming this was to plot the *Patterson function*, $P(u, v, w)$, where

$$P(u, v, w) = \sum |F(hkl)|^2 \exp[-2i(hu + kv + lw)] \qquad (3.28)$$

and where u, v, w are coordinates of a point within the unit cell and the summation is over all values of hkl. This function is very similar to Eq. (3.9) except that $|F(hkl)|$ is replaced by its square and the phase angle α is no longer present. While Eq. (3.9) produces a series of peaks in the unit cell corresponding to atomic positions, the peaks of the Patterson function correspond to interatomic vectors, i.e., a peak at u, v, w indicates that two atoms are present at x_1, y_1, z_1 and x_2, y_2, z_2, such that $u = x_1 - x_2$, $v = y_1 - y_2$, and $w = z_1 - z_2$. For a structure containing n atoms, there will be n^2 interatomic vectors and this number of peaks in the Patterson synthesis. The Patterson function has been interpreted for many small molecules, but for a protein with $n \simeq 2000$, the peaks will overlap such that they cannot be resolved.

The intensity of the Patterson peak is dependent on the product of the atomic numbers of the two atoms separated by u, v, w. Thus, a peak for a vector between two carbon atoms will have intensity 36 relative to oxygen–oxygen and hydrogen–hydrogen vectors of 64 and 1, respectively. This is very useful when the structure contains a few heavy atoms such as mercury (atomic number 80). In this case the mercury–mercury peaks have intensity 6400 and stand out from the unresolved peaks for carbon–carbon, intensity 36, and other vectors between light atoms. From the interatomic vector components and knowledge of the unit cell symmetry, the positions of the heavy atoms can be determined.

The above method of finding the heavy atoms is used (in modified form) to determine the structures of globular proteins. This requires the crystallization of a number of heavy-ion–protein complexes that have the same unit cells and the protein in the same position, but that have the heavy ions, in some cases at least, in different positions. The reflections for these crystals occur in the same positions and the intensity differences are due to the different types of heavy ions and their positions in the unit cell. Consider a single *hkl* reflection observed for two different heavy-atom derivatives of this type for a globular protein. The structure factors for this *hkl* plane for the two derivatives $\mathbf{F_A}$ and $\mathbf{F_B}$ can be written

$$\mathbf{F_A} = \mathbf{F_L} + \mathbf{F_{HA}} \tag{3.29}$$

$$\mathbf{F_B} = \mathbf{F_L} + \mathbf{F_{HB}} \tag{3.30}$$

where $\mathbf{F_L}$ is the vector (amplitude and phase) due to the contribution of the light (protein) atoms, and $\mathbf{F_{HA}}$ and $\mathbf{F_{HB}}$ are the vector contributions for the heavy atoms of A and B, respectively. Subtraction of Eq. (3.30) from Eq. (3.29) leads to

$$\mathbf{F_A} - \mathbf{F_B} = \mathbf{F_{HA}} - \mathbf{F_{HB}} = \Delta \mathbf{F} \tag{3.31}$$

Now the positions of the heavy atoms for A and B can be determined from the Patterson function as described above. From these coordinates, $\mathbf{F_{HA}}$ and $\mathbf{F_{HB}}$ can be calculated, and thus the magnitude and phase of $\Delta\mathbf{F}$ are known. Since we know the magnitudes of $\mathbf{F_A}$ and $\mathbf{F_B}$, there are only two ways these can be combined in a vector diagram to give $\Delta\mathbf{F}$. These possibilities are shown in Fig. 3.24. The phase angle α can be measured from these diagrams, and there are two possible values of α for both $\mathbf{F_A}$ and $\mathbf{F_B}$, only one of which is

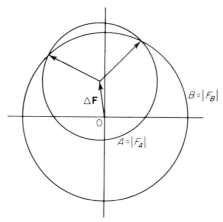

Fig. 3.24. Ambiguity in phase determination for a non-centro-symmetric structure. The magnitude and phase of ΔF is known; the ends of $\mathbf{F_A}$ and $\mathbf{F_B}$ must be at one of the points where the circles intersect. [From Stout and Jensen (1), with permission.]

correct for each. This process is repeated for structure A and a third heavy atom complex, C, for which the ions are in positions with respect to the protein that are different than they were in complex B; again two values of α emerge for $\mathbf{F_A}$. Only one of these will be the same for the two sets A, B and A, C and, thus, the phase of F_A is determined. This result will be accurate within the experimental error of the three structure factors. A more accurate value of α is determined if more than three ionic complexes are studied. For myoglobin, Kendrew *et. al.* (31) averaged α over the comparison of four heavy-atom derivatives. The solution of the structure then emerges from a Fourier synthesis of the electron density by substituting the values of F_A and α_A into Eq. (3.9). The x-ray data generally consist of about 2000–3000 reflections with d-spacings with $d > 2$ Å. In the Fourier synthesis of the structure, it is not possible to resolve individual atoms, although they can be positioned fairly accurately from a knowledge of the stereochemistry. It is usually possible to identify most of the R groups of the amino acids, although the sequence is generally known. For a description of the structures determined by this method for globular patterns see Chapter 13.

REFERENCES

1. General texts on x-ray crystallography:

   C. W. Bunn, "Chemical Crystallography," 2nd ed. The Clarendon Press, Oxford 1961.
   G. H. Stout and L. H. Jensen, "X-ray Structure Determination." Macmillan, New York, 1968.
   M. J. Buerger, "Crystal Structure Analysis." Wiley, New York, 1960.

H. Lipson and W. Cochran, "The Determination of Crystal Structures." Bell, London, 1959.

S. C. Nyburg, "X-ray Analysis of Organic Structures." Academic Press, New York, 1961.

L. E. Alexander, "X-ray Diffraction Methods in Polymer Science." Wiley, New York, 1969.

B. K. Vainstein, "Diffraction of X-rays by Chain Molecules." Amer. Elsevier, New York, 1962.

2. C. W. Bunn, "Chemical Crystallography," 2nd ed., p. 461. Oxford Univ. Press, London and New York, 1961.

3. G. H. Stout and L. H. Jensen, "X-ray Structure Determination," pp. 31–37. Macmillan, New York, 1968.

4. R. Hosemann and S. N. Bagchi, "Direct Analyses of Diffraction by Matter." North-Holland Publ., Amsterdam, 1962.

5. N. F. M. Henry, H. Lipson, and W. A. Wooster, "The Interpretation of X-ray Diffraction Photographs," p. 212. Macmillan, New York, 1960.

6. J. H. Wakelin, H. S. Virgin, and E. Crystal, *J. Appl. Phys.* **30**, 1654 (1959).

7. N. F. M. Henry and K. Lonsdale, eds., "International Tables for X-ray Crystallography," Vol. I. Kynoch Press, Birmingham, England, 1952.

8. G. H. Stout and L. H. Jensen, "X-ray Structure Determination," p. 31. Macmillan, New York, 1968.

9. W. Cochran, F. H. C. Crick, and V. Vand, *Acta Crystallogr.* **5**, 581 (1952).

10. C. A. Taylor and H. Lipson, "Optical Transforms." Cornell Univ. Press, Ithaca, New York, 1965.

11. C. A. Taylor, *Pure Appl. Chem.* **18**, 533 (1969).

12. K. C. Holmes and D. M. Blow, "The Use of X-ray Diffraction in the Study of Protein and Nucleic Acid Structure." Wiley (Interscience), New York, 1966. Reprinted from *Methods of Biochem. Anal.* **13**, 113 (1965).

13. M. F. H. Wilkins, *Cold Spring Harbor Symp. Quant. Biol.* **21**, 75 (1956).

14. A. Sarko and R. H. Marchessault, *J. Amer. Chem. Soc.* **89**, 6454 (1967).

15. C. H. Bamford, A. Elliot, and W. E. Hanby, "Synthetic Polypeptides." Academic Press, New York, 1956.

16. S. Arnott and A. J. Wonacott, *J. Mol. Biol.* **21**, 371 (1966).

17. L. Brown and I. F. Trotter, *Trans. Faraday Soc.* **52**, 537 (1956).

18. A. Elliott and B. R. Malcolm, *Proc. Roy. Soc. London,* **A249**, 30 (1959).

19. A. Klug and D. L. D. Caspar, *Advan. Virus. Res.* **7**, 225 (1960).

20. R. E. Franklin and A. Klug, *Acta Crystallogr.* **8**, 777 (1955).

21. J. Blackwell, A. Sarko, and R. H. Marchessault, *J. Mol. Biol.* **42**, 379 (1969).

22. A. Klug, F. H. C. Crick, and H. W. Wyckoff, *Acta Crystallogr.* **11**, 199 (1958).

23. A. Rich, D. R. Davies, F. H. C. Crick, and J. D. Watson, *J. Mol. Biol.* **3**, 71 (1961).

24. M. Spencer, W. Fuller, M. H. F. Wilkins, and G. L. Brown, *Nature* **194**, 1014 (1960).

25. R. Langridge, H. R. Wilson, C. W. Hooper, M. H. F. Wilkins, and L. D. Hamilton, *J. Mol. Biol.* **2**, 19 (1960).

26. E. D. T. Atkins, *J. Mol. Biol.* **24**, 139 (1967).

27. F. H. C. Crick, *Nature* **170**, 882 (1952); *Acta Crystallogr.* **6**, 689 (1953).

28. L. Pauling and R. B. Corey, *Nature* **171**, 59 (1953).

29. R. D. B. Fraser, T. P. MacRae, and G. E. Rogers, *Nature* **183**, 592 (1959); **193**, 1052 (1960).

30. D. A. D. Parry, *J. Theor. Biol.* **24**, 73 (1969); *J. Theo. Biol.* **26**, 429 (1970).

31. J. C. Kendrew, R. E. Dickerson, B. E. Strandberg, R. G. Hart, D. R. Davies, D. C. Phillips and V. C. Shore, *Nature,* **185**, 422 (1961).

4

TERTIARY STRUCTURE AND MORPHOLOGY
OF SYNTHETIC BIOPOLYMERS

Introduction

Whereas intramolecular forces play a large part in determining the shape or conformation of biopolymers, the intermolecular interactions and packing arrangement are also very important. Thus the most stable theoretical (vacuum) conformation of an isolated molecule is frequently not the form observed experimentally in the solid state or solution (poly-L-proline and α helices are major exceptions). The intermolecular forces, therefore, stabilize conformations which are often not the most stable on an intramolecular basis. Molecular arrangement can be determined by crystallographic methods as discussed in Chapter 3. The principles underlying the development of tertiary structure are important because of their relationship to the structural development in native tissue. Evidently, if we are trying to design a fibrous material, for example, it is important to resolve the features controlling fiber formation. In general, biopolymers that are capable of highly symmetrical arrangements and strong intermolecular interaction [β form (degenerate 2_1 helix) and polyglycine 3_1 helices] have their conformations determined, to a large extent, by intermolecular interactions. The poly-L-proline II helix also adjusts itself to the more symmetrical (3_1) arrangement in order to accommodate intermolecular interactions in the solid state. Less symmetrical forms, e.g., the 18_5 α helix, undergo less (nonbonding) interaction with neighboring molecules in the solid state. Very unsymmetrical molecules with nonrepeating primary structure often form spherical or amorphous tertiary structures.

Types of Molecular Arrangement

If we regard the homobiopolymer helix as a uniform cylinder or rod, then we might expect a number of these to pack in a hexagonal array in the solid state. In fact, most simple helical polypeptides do indeed pack hexagonally or pseudohexagonally when they crystallize. It is easily seen that models such as the 3_1 polyglycine II helix would be likely to pack in a "hexagonal" manner.

Fig. 4.1. Hexagonal packing of helices with a threefold screw axis.

Figure 4.1 shows an idealized schematic diagram in which the triangles represent the three peptides per turn and the hexagons represent the six planes of the peptide backbone. In fact, the dihedral angles are not identical ($\phi \neq \psi$) and some modification should be allowed for on this basis. However, the optimization of intermolecular interaction is clearly revealed. The α helix, on the other hand (Fig. 4.2), is obviously much less symmetrical and, by packing side by side, the molecules can achieve a maximum registry of one in six planes of the peptide backbone. We might suspect that other molecular arrangements are comparably favorable. In the preceding arrangements, the fiber axis and helix axis coincide, but this restriction is removed if groups of

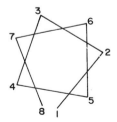

Fig. 4.2. Schematic diagram of the α helix; end view with 3.6 peptides/turn. The dihedral planes are imagined to be symmetrically related to the line representing the peptide.

molecules are twisted together in rope or cable form, otherwise known as a coiled coil. Figure 4.3 shows possible models for two-, three-, and four-strand coils. It seems at first glance that a simple homobiopolymer is unlikely to gain sufficient intermolecular energy by this maneuver to compensate for the loss of nearest neighbors from the hexagonal array. In practice the situation is complicated by the role of the solvent from which the biopolymer is precipitated, any one material being deposited in two or more different forms, as will be shown later in this chapter. However, a few simple polyamino acids (and esters) do precipitate with fibrous morphology which may be indicative of super-coiling, and, in these cases, the solvent effect seems all-important.

The original suggestion of the coiled coil arose from the observation that in fibrous α-proteins such as α-keratin, the 5.4 Å helix repeat decreased to ~ 5.1 Å. Crick (1) supposed that the pitch of the α helix was changed by supercoiling, the inclination angle of the chains being shown in Fig. 4.4. It is not usually possible to determine unambiguously the number of chains in the cable by x-ray methods.

Pauling and Corey (2) have proposed several coiling possibilities, including two-, three-, four-, and seven-strand versions. The most extensive theoretical conformational analysis of coiled coils has been performed by Parry and Suzuki (3). Taking a poly-L-alanine chain as a model, they have shown that the intramolecular energy of the α helix changes very slightly (an increase of less than 0.007-kcal/mole residue) on conversion to a coiled-coil structure of radius 5.5 Å and pitch 186 Å. The calculated intermolecular energies for two-, three-, and four-strand cables, consisting of both parallel and antiparallel

A

Fig. 4.3.

B

C

Fig. 4.3. Schematic diagrams of α-helix coiled coils with two, three, and four strands, respectively.

Fig. 4.4. Inclination angle θ of a supercoiled helix.

arrangements of chains, are shown in Table 4.1 along with the comparable intermolecular energies for the linear helices. Interestingly, the antiparallel arrangements are favored, as are the coiled coils. However, we note that the "hexagonal" crystal lattice should be the most stable (unless the coiled coil can derive additional stabilization energy in larger bundles).

In all probability, the coiled-coil arrangement of fibrous proteins arises not

TABLE 4.1

Minimum van der Waals Energy for the Rope Systems with the Corresponding Packing Parameters[a]

System	Radius r_0 (Å)	Rotation of reference residue β[b]	Total translation t (Å)	Minimum energy (kcal/mole residue)
Coiled coil				
Two-strand parallel	4.0	−1		−1.86
Two-strand antiparallel	3.8	13	−0.32	−2.75
Three-strand parallel	4.5	−2		−2.36
Four-strand parallel	5.7	−1		−1.92
Four-strand antiparallel	5.4	13	−0.40	−2.72
Straight α-helix analogs				
Two-strand parallel	4.8	8		−0.52
Two-strand antiparallel	4.0_5	−172	−0.83	−1.70
Three-strand parallel	5.1	16		−0.88
Four-strand parallel	6.6	1		−0.61
Four-strand antiparallel	5.7	−172	−0.83	−1.70
Two parallel strands related by crystallographic symmetry observed in α-poly-L-alanine	4.2_5	14		−1.03

[a] From Ref. (3).
[b] In degrees.

so much from energetics of the backbone conformation as from the favorable "knob-hole" packing for the side chains, which was the basis for the original proposal of the structure (1). If the near-cylindrical symmetry of a polyamino acid is destroyed by placing a large side group in a regular position down the chain, then a marked effect on the molecular arrangement results. For example, a 3_1 helix with a large residue in every third position may form an array such as those in Fig. 4.5a,b.

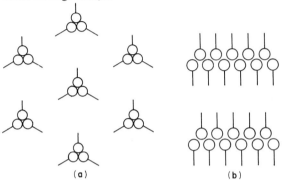

Fig. 4.5. Possible packing arrangements of threefold (screw) helices containing a bulky residue in every third position. (a) Clusters of triple helices, (b) sheet structure.

On the other hand, an α helix with a bulky group in any one regular position down the chain would probably not crystallize readily into a lattice with high symmetry. Thus, it is to be expected that the primary sequence plays a large part in fiber formation by rendering straight-chain single crystals unstable, but with an arrangement such that coiled coils are readily formed, Unfortunately, not enough is known of the primary sequence of amino acids in any known coiled coil to bear out the above hypothesis with the possible exception of collagen, which is discussed in a later chapter.

The Calculation of Crystal Structures

In principle, it should be possible to carry through the conformational calculations mentioned in Chapter 2 to include intermolecular interactions. In practice, it is much easier (but less exciting) to calculate the energies of various trial structures. Both approaches have been attempted only recently. The former presents very awesome mathematical problems, particularly if the peptide under consideration is anything other than a very simple sequential material. The first reported crystal structure of a biopolymer derived by *ab initio* conformational calculation appears to be poly-L-hydroxyproline (4).

In this case, the intramolecular conformation was calculated first, and then minor conformational changes about the most stable position were allowed as the other chains were placed around it, such that their separation and translation were controlled variables. The crystal structure calculated with parallel chains was in excellent agreement with the experimental data of Sasisekharan (5) (experimental unit cell: hexagonal, $a = 12.3$ Å, $c = 9.15$ Å; theoretical unit cell: hexagonal, $a = 12.5$ Å, $c = 9.1$ Å). It now seems likely that poly-L-hydroxyproline crystals have antiparallel chains (by analogy with other biopolymer single crystals, see Chapter 2). This change in packing arrangement does not, apparently, modify the crystal structure. The second and more complicated crystal structure presented by Hopfinger and Walton (6) was for the polytripeptide (Gly-Pro-Gly)$_n$. In this case, unlike poly-L-hydroxyproline, there were many favorable isolated molecular conformations, several of which provided a basis for possible intermolecular structures. Most of the conformations do not possess equivalent peptide rotation angles and are not, therefore, readily identifiable with known homobiopolymer structures. The conformation which is fourth in order of predicted (vacuum) stability was the first that could be packed into a compact three-dimensional structure. It is a left-handed helix with three residues per turn and a peptide repeat of 3.05 Å, which is reminiscent of the poly-L-proline II structure. However, its characteristics are quite different, as shown in Table 4.2. The next conformation that appears to pack reasonably ranks seventh in stability and is the familiar poly-L-proline II (PPII) conformation. It has been found that these two conformations can pack according to two different schemes each, as shown in Fig. 4.6 (6). Based on calculations of the total

TABLE 4.2

*Calculated Parameters for the Fourth Intramolecular
Energy Minimum (RM4) for (Gly-Pro-Gly)$_n$*

N^a	ϕ^b	ψ^b	Property^c	Value
1	340.0	110.0	t	3.05 ± 0.36 Å
2	102.5	340.0	r	1.22 ± 0.20 Å
3	105.0	330.0	ε	$-120.0° \pm 45.6°$
			E	-25.9 kcal/mole tripeptide

[a] Peptide residue number counting from the N-terminal end of the tripeptide.

[b] Bond rotation angles in degrees.

[c] The property t is the translation per residue, r is the radius of gyration using C_α, and ε is the rotation per peptide unit.

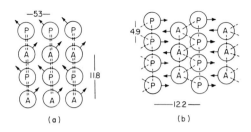

Fig. 4.6. A schematic representation of two possible crystal structures of (Gly-Pro-Gly)$_n$ of comparable energy based on theoretical calculation. (a) Unit cell of $a = 5.3$ Å, $b = 11.8$ Å, and $c = 9.3$ Å. (b) Unit cell of $a = 4.9$ Å, $b = 12.2$ Å, $c = 9.3$ Å. Dashed lines represent the hydrogen bonds and the arrows show the orientation of proline rings. Structure (a) uses RM4 chain conformation (see text) and (b) contains PPII chain conformation. [The RM4 conformation could be used with structure (b).] The letter P stands for parallel and A, antiparallel. Only structure (b) has been observed experimentally. [Reprinted from Ref. (6), p. 196 by courtesy of Marcel Dekker, Inc.]

TABLE 4.3

Comparison of Intramolecular and Intermolecular Interactions and Theoretical Unit Cells for (Gly-Pro-Gly)$_n$

Structure	Stabilization energy/tripeptide (kcal)			Unit cell
	Intramolecular	Intermolecular	Total	
RM 4 conformation	−4.72		} −8.0(9)	Orthogonal $a = 5.3$ Å, $b = 11.8$ Å
Type A packing		−3.37		$c = 9.3$ Å
RM 4 conformation	−4.72		} −7.9(1)	Orthogonal $a = 5.1$ Å, $b = 12.4$ Å
Type B packing		−3.19		$c = 9.3$ Å
Polyproline II conformation	−4.53		} −8.0(8)	Orthogonal $a = 4.9$ Å, $b = 12.2$ Å
Type B packing (Traub and Yonath structure)*		−3.55		$c = 9.3$ Å

* The Traub and Yonath structure involves an experimentally determined orthogonal unit cell of $a = 12.2$ Å, $b = 4.9$ Å and $c = 9.3$ Å. This is the same as the theoretical structure where the a and b axes have been reversed.

crystal energy (inter- and intramolecular), three possible crystal structures emerge, each with an orthorhombic unit cell. The energies, packing scheme, and cell parameters are given in Table 4.3. The data suggest that these forms and possibly others are likely to exist, but particularly significant is the fact that the polyproline II conformation, which is far from the most stable intramolecular form, yields a crystal of particularly stable nature. The reason is that a large intermolecular interaction results from the high symmetry of the PPII conformation. Under experimental conditions the solvent also plays an important role in crystal structure and morphology, as will be noted later. Traub and co-workers (7,8) have obtained one experimental form of poly (Gly-Pro-Gly) which apparently has the PPII conformation and characteristics listed in Table 4.3. It should be noted that the above-mentioned calculations of polypeptide crystal structure are based upon interaction energies controlling conformation and chain organization. In fact it is the free energy that should be considered in these calculations; such refinements are currently being introduced (9), although no crystal structures based on this modification have yet been published.

Observation of Morphology of Synthetic Biopolymers

Supercoiled* Aggregates

Because of the important role of morphology in the structural makeup of living tissue, it is of considerable value for us to understand the relation between the primary and secondary structure of biopolymers and their observable morphology. Since it is not possible to grow macroscopic crystals of most biopolymers, it is necessary to use electron microscopy to observe the morphology. The attendant advantage is that electron diffraction studies on the sample *in situ* provide important structural information which often allows an interpretation of morphology in terms of the underlying chain conformation and arrangement. For the biopolymers studied so far, the most important controlling factor is the solvent from which the material is precipitated. This feature points out the undoubtedly vital role played by peptide/solvent interactions in controlling the conformation and function of body proteins. It seems likely that the incorporation of solvent interaction factors into conformation calculations of the type presented in Chapter 2 will represent an area of significant advance in the next few years.

---

* The term supercoiled is used here to denote observed coiling in fibrous structures; such a structure may or may not reflect the Crick molecular supercoil described in the section on types of molecular arrangement.

The α helix

There has been much discussion over the years concerning the relation between the conformation of biopolymers in solution and in precipitates from the same solution. In general, the conformation as derived from optical rotatory dispersion (ORD) and x-ray techniques appears to indicate that generally the solution conformation is maintained in the solid state. However, more recently it has been realized that the generalization is not entirely valid. For example, the α helix, which is destroyed in strong hydrogen-bonding solvents, may be regenerated in the crystalline product of evaporation from such solvents. Alternatively, weak hydrogen-bonding solvents, which promote helix formation, do not promote crystalline precipitates, but rather favor fibrillar aggregates. The most thoroughly studied fibrillar materials are the esters of poly-L-glutamic acid. Ishikawa and Kurita (10) have reported that poly-γ-methyl-L-glutamate (PMLG) forms fibrillar structures from aged dimethylformamide (DMF) solutions. The fibrils contain α helices aligned parallel with the fibrillar axis, the fibrils being 45–67 Å in diameter.

The diameter of the PMLG helix is approximately 12 Å and, thus, a seven strand coiled coil would be expected to be about 36 Å in diameter. The appearance of these fibrils is highly suggestive of higher-order (more strands) supercoiling, even in the smallest observable units. Recent electron microscope work (11,12) has revealed more detail of the fibrous arrangement. Poly-γ-benzyl-L-glutamate, presumably a right-handed α helix in poor solvents (mesitylene, butoxy-, ethoxybenzene), precipitates as left-hand superhelical aggregates or fibrils, whereas poly-γ-benzyl-D-glutamate produces right-hand aggregates. Figures 4.7A and B show micrographs of the poly-L- and poly-D-esters. It can be seen that there is a wide variation in fiber thickness but not in the "handedness" of the supercoiling, which suggests that the molecular supercoiling is reflected in these micrographs. In fact, in the latest work (12), structures down to 30 Å or so have been observed which are probably two or three supercoiled molecules, although the supercoiling could not be discerned in such small structures. It seems, then, that molecular aggregation which occurs in poor solvents, often leads to superhelical aggregates in which the morphology probably reflects the underlying molecular geometry.

It is interesting to note that random copolymers of D- and L-glutamate esters form a distinctly different morphology where "block" fibers are produced, the blocks being approximately equal in length to the molecules (see Fig. 4.8).

A somewhat more general study of fibrillar polypeptide aggregates has been published by Blais and Geil (13). They observed fibrils of PMLG, poly-γ-benzyl-L-glutamate (PBLG), poly-β-benzyl-L-aspartate (PBLA), and

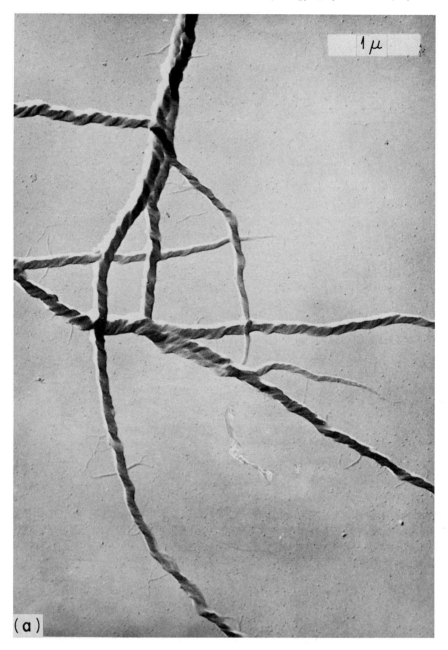

(a)

Fig. 4.7A. An electron micrograph of poly-γ-benzyl-ʟ-glutamate precipitated from tri-methylbenzene and butoxyethanol (1:2). The right-hand α helix becomes wound into ropes with a left-hand twist (1 μ insert).

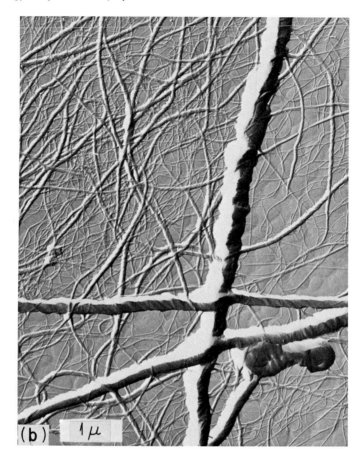

Fig. 4.7B. Poly-γ-benzyl-ᴅ-glutamate precipitated from trimethylbenzene and but-oxyethanol (1:1) showing right-hand twisted ropes based on a left-hand α-helix conformation. (Micrographs kindly supplied by F. Rybnikar and P. H. Geil.)

poly-ʟ-alanine (PLA) precipitated isothermally from various non-hydrogen-bonding solvents. The primary units were found to be fibrils approximately 50 Å in diameter. The fact that layer upon layer can form supercoils seems to be demonstrated when larger fibrils are examined, such as those for PBLG shown in Figs. 4.7A and B. The detailed nature of this "macroscopic" super-coiling is not fully understood; nevertheless, the helical pitch does not seem to change as a function of fiber thickness but only as a function of solvent. Data collected by Blais and Geil for PBLG are shown in Table 4.4. It is not known whether there are any concomitant changes in the x-ray diffraction pattern.

Fig. 4.8. An electron micrograph of copoly-γ-benzyl-DL-glutamate (50 : 50) precipitated from the same solvent mixture as in Fig. 4.7B showing blocks of structure approximately equal to the length of the average molecule, built into a fibrous structure. (Micrograph by F. Rybnikar and P. H. Geil.)

Although little is known about the basic structure of fibrous biopolymer precipitates, the evidence indicates that this type of morphology results from a situation in which the material is already aggregated in the poor solvent prior to precipitation. It is known that PBLG forms liquid crystalline arrays in solvents from which it may be precipitated. Luzzati and co-workers (14) have shown that several complex solution structures may be delineated by x-ray diffraction procedures. Common among solution aggregated forms are hexagonal or complex hexagonal forms. In the latter, it is suggested that three-strand coils make up the basic unit. Sedimentation and light-scattering studies of the parent acid (PLGA) have also indicated species which are thought to contain three or four molecules.

TABLE 4.4

Variation of Helical Pitch in PBLG[a]

Solvent[b]	Solvent ratio	Precipitation temperature (°C)	α^c
BE–Tetralin	2:1	70	(−)42–48
BEE–TMB	2:1	70	(−)40–45
Cyclohexanol–TMB	2:1	70	(−)20–25
BE–Xylene	2:1	70	(−)20–25
BE–Toluene	2:1	50	(−)15–20
BE–TMB	2:1	70	(−)15–20
MEE		70	(−) 5–10
MEE		30	0
TMB		30	(+)10–15
TMB–Xylene	2:1	70	(+)20–25
TMB–Xylene	2:1	50	(+)20–25
TMB–Xylene	2:1	30	(+)20–25
TMB–Xylene	1:1	30	(+)25–30
TMB–Xylene	3:1	70	(+)28–33
BE–TMB	1:1	70	(+), (−)30–40
TMB		70	(+)40–45

[a] From Ref. (13).
[b] Abbreviations: BE, butoxyethanol; TMB, trimethylbenzene; BEE, butoxyethoxyethanol; MEE, methoxyethoxyethanol.
[c] Helical pitch α in degrees.

Single Crystals

THE α HELIX

Whereas the predominant morphology of α-helical polypeptides precipitated from weak non-hydrogen-bonding solvents is either fibrillar or amorphous, a rather remarkable change is observed when α-helical polymers are slowly precipitated from certain strong hydrogen-bonding solvents. Optical rotatory dispersion measurements indicate that the solution conformation is random prior to precipitation, yet the product may be highly crystalline and gives rise to α-helical x-ray diffraction. Biopolymers crystallized under such conditions are remarkably similar to synthetic nonbiological polymers such as polyethylene, both in appearance and molecular orientation. In fact, most of the important observations of biopolymers have arisen from developments in the early 1960's with common commercial polymers. Although single crystals of biopolymers were observed in the 1950's, it was not until Padden and Keith's definitive report (15) in 1965 that many of the implications were recognized. Figures 4.9 and 4.10 (15) show single crystals of polyethylene and

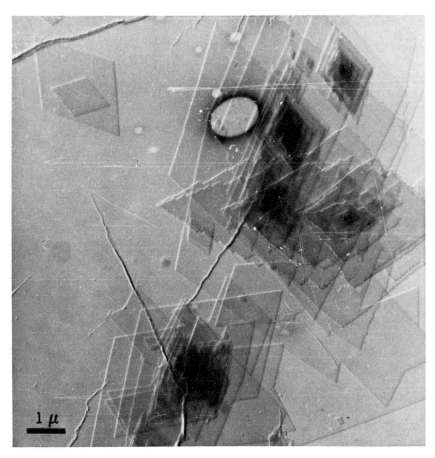

Fig. 4.9. Lathlike single crystals of polyethylene. The molecular axis lies perpendicular to the major face of the crystals, the molecules being folded many times to pack into the crystal. The height of the crystals, and thus the fold period of the molecules, is ∼100 Å. By Dr. P. H. Geil. "Polymer Single Crystals," (1963), John Wiley and Sons, Inc.

the α-helical poly-L-tyrosine, respectively. The former were grown by cooling a saturated xylene solution, the latter by thermal precipitation from DMF–*n*-heptanol solution. It can be seen that both are thin lathlike crystals of the order of 100 Å in thickness and 1 μ in width. These dimensions are typical of biopolymer single crystals and are ideal for single-crystal electron diffraction *in situ* with the reservation that it is often preferable to have two or more layers of crystals, and the crystals are quite unstable in the electron beam.*

---

* Actually, carbon-shadowed replicas are often used in electron microscopy. Platinum shadowing at a defined angle allows calculation of the crystal height. Electron diffraction can be achieved in areas where intact crystals remain adhering to the carbon film.

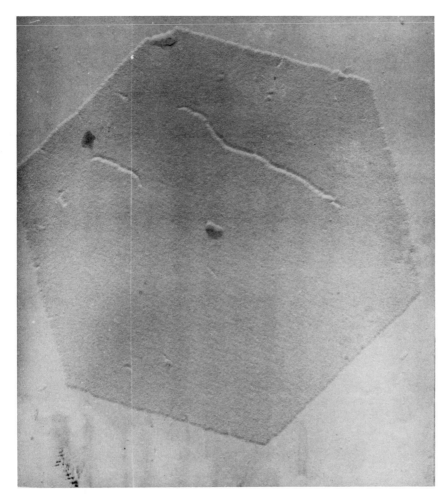

Fig. 4.10. Lathlike single crystal of poly-L-tyrosine grown from dimethylformamide and *n*-heptanol (1:2) at 45°C. Electron diffraction shows the α-helical molecules to be perpendicular to the major crystal face. Since the thickness of the crystal (~150 Å) is considerably smaller than the chain length of the poly-L-tyrosine (MW ~100,000) the molecules are believed to be chain folded (×22,000). [From Ref. (15), with permission.]

Some simple principles of electron diffraction are discussed in the Appendix.

Padden and Keith were able to deduce from their electron diffraction patterns that the α-helical molecular axis of poly-L-tyrosine is perpendicular to the lamellar crystal surface. (It should be noted that the electron diffraction from lamellar crystals lying flat yields reflections only from planes perpendicular to the chain axis. Thus, the measured *d*-spacings give information on the chain packing and separation, but not chain conformation. This two-dimensional representation must be supplemented either by tilting the crystal or by x-ray diffraction to obtain more definitive results.) Possibly most remarkable is the fact that since the material is of high molecular weight, and the molecular chain is several times longer than the crystal is thick, the chain must be folded. A schematic diagram of the folded lamellar crystal is given in Fig. 4.11. Although the precise conformation in the fold region is not

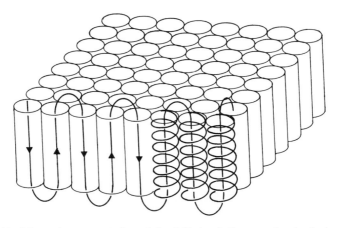

Fig. 4.11. Schematic representation of the folded α helix occurring in single crystals (such as that of poly-L-tyrosine in Fig. 4.10). Note that adjacent chains are antiparallel and that non-α-helical segments occur at the surfaces.

known, it is not difficult to arrive at various postulated structures. Since the conformation and environment must be different in these fold regions, the characteristic atomic vibrations will also be different and may lead to unique infrared and Raman absorption bands, similar to those reported for synthetic commercial polymers (16) and certain polysaccharides (17). No such effects have yet been observed. Apart from the fold structure, which may be quite relevant to fold structures in proteins, adjacent peptide chains must be antiparallel, i.e.,

$$\overrightarrow{-C_\alpha - CO - NH - C_\alpha -}$$
$$\overleftarrow{-C_\alpha - NH - CO - C_\alpha -}$$

A great deal of work on the nature of chain folding in commercial polymers has been published. Its relation to the mode of crystallization is pointed out in Ref. 18.

In many cases it is not easy to grow crystals by thermal precipitation since a careful and exhaustive examination of possible solvent systems is often required. More often, crystalline deposits may be obtained from film casting, i.e., by allowing a film of solution to dry on a suitable substrate, usually glass. The film is then stripped from the glass by flotation and mounted and treated (shadowed) in the usual manner. In this case a quite different morphology is produced, which is defined as bundlelike or "spherulitic." It is relatively easy to understand how this morphology arises. The initial molecules depositing from solution lie flat on the substrate and build up in lamellae which appear "edge on" when viewed from above. Further lamellae grow in close contact and, as they lengthen, there is a tendency to twist and circle. The details of polymer structure in spherulites are explained in Ref. 14. Figure 4.12 shows a schematic build-up of bundle and spherulite texture, and

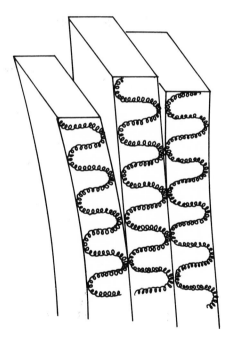

Fig. 4.12. Schematic diagram of helical polypeptide chains in a bundle or spherulitic structure (such as for poly-L-alanine in Fig. 4.13).

Fig. 4.13 shows a bundle of poly-L-alanine crystals, film cast from trifluoroacetic acid (TFA) plus trifluoroethanol (TFE) solutions. This time, electron diffraction does yield information concerning the chain conformation and orientation. As before, the lamellar thickness is of the order of 100 Å, which is much less than the molecular length, and the molecules are folded. The fold period (lamellar thickness) for synthetic nonbiopolymers

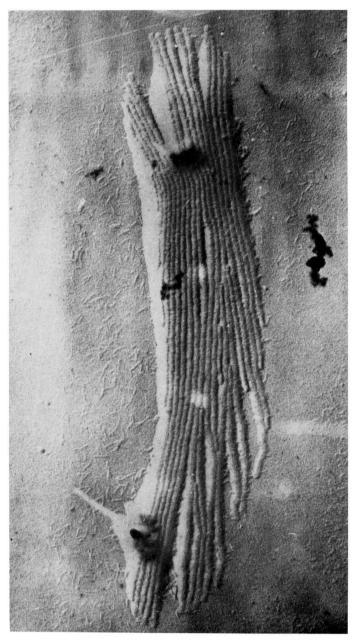

Fig. 4.13. Electron micrograph of a bundle of poly-L-alanine crystals film cast from TFA–TFE solution. Electron diffraction shows that the molecules lie perpendicular to the "fiber" axis and flat on the substrate. Molecular chain folding is believed to occur as shown schematically in Fig. 4.12. [From Padden and Keith (15), with permission.]

changes with crystallization temperature and solvent (16), but it does not change with molecular weight. These features have not, in general, been investigated for simple synthetic biopolymers. The substrate on which the biopolymer is deposited can alter the fold period and the morphology. These features are discussed later.

In addition to poly-L-tyrosine and poly-L-alanine, the acid salt of poly-L-lysine is known to show crystalline morphology with α-helix conformation (19). Poly-L-lysine HPO_4^{2-} and the magnesium salt of poly-L-glutamic acid may be deposited from aqueous solution at normal temperature and appropriate pH, the former by precipitation with $(NH_4)_2HPO_4$, the latter by addition of ethanol. Lamellar crystals are produced with thicknesses of 150Å and 100Å respectively.

THE β STRUCTURE

As pointed out in Chapter 2, the conformation of chains in the β-sheet arrangement may, in theory, be parallel or antiparallel. Although there was some indication of antiparallel β (cross-β) structures in some of the aged "fibers" observed by Ishikawa and Kurita, the discovery and definitive study of lamellar β crystals has not come until fairly recently. Keith *et al.* (20) have found that the calcium, strontium, and barium salts of poly-L-glutamic acid precipitated from alkaline solutions give very thin (25–60 Å) lamellar crystals. The chain orientation and crystallographic axes are shown in Fig. 4.14 (20) for CaPLG. The peptide repeat in the b direction (chain axis) is 3.4 Å, and thus between seven and seventeen peptide repeats occur in the crystal thickness depending on the salt and preparation method. Thus, at least in the observed cases, the peptide chains in the β conformation are folded into the antiparallel cross-β conformation. (This structure differs from the simple antiparallel structure only in that a single chain is folded such that the chain axis and fiber axis are perpendicular to each other in the cross-

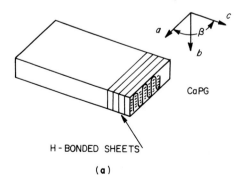

CaPG

H - BONDED SHEETS

(a)

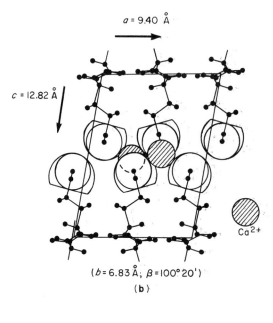

$a = 9.40 \text{ Å}$

$c = 12.82 \text{ Å}$

Ca^{2+}

$(b = 6.83 \text{ Å}; \ \beta = 100° 20')$

(b)

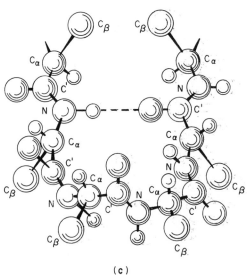

C_β C_β

C_α C_α

C N

N C'

C_α C_α

C' N

C_β C_α C_α C_β

N C N C'

C_β C_β

(c)

Fig. 4.14. Diagram showing orientation of molecules and unit cells in the calcium salt of poly-L-glutamic acid (β-CaPLG). (a) Chain-folded (cross-β) arrangement of chains in crystal of β-CaPLG and unit cell orientation. (b) The b-axis projection of the monoclinic unit cell showing interdigitation of side chains from two hydrogen-bonded pleated sheets. (c) Proposed molecular fold of β-CaPLG. [From Ref. (20) with permission.]

β conformation.) Astbury and co-workers (21,22) were the first to note and describe the cross-β conformation as a basis for structural units in certain proteins. A more recent consideration of peptide units in the β fold region (23) of *Chrysopa* silk has suggested that glycyl residues particularly facilitate a fold, and it is perhaps not surprising that most glycine-containing sequential peptides (except those containing proline) readily achieve a cross-β conformation. Other electron microscope studies of homopolypeptide crystals in the β conformation have been limited to the acid salts of poly-L-lysine, particularly the monohydrogen phosphate, produced by precipitation from aqueous solution at elevated temperatures. (The use of elevated temperatures is known to induce the $\alpha \rightarrow \beta$ transition for poly-L-lysine; see Chapter 8.)

POLY-L-PROLINE CONFORMATIONS

In both the α and β structures, hairpin folds produced at the crystal surface involve an irregularity in conformation that must arise from permissible rotation in the peptide backbone. The polyimino acids and their derivatives have only one degree of backbone rotational freedom, as noted in Chapter 2, and it therefore is of some interest to know whether these materials also form lamellar crystals. The first reported study (24) concerned poly-*O*-acetyl-L-hydroxyproline (POAHP). This biopolymer is said to have two solution conformations analogous to the poly-L-proline I (*cis*) and II (*trans*) forms. It was found that lamellar platelets could be produced from acetic anhydride by thermal precipitation [see Fig. 4.15 (24)]. Apparently, POAHP is in form I, as determined by optical rotation, in acetic anhydride at room temperature, but the crystalline deposit is in form II as determined by both x-ray and electron diffraction. This observation is unusual since, in general, there is believed to be a close relation between solution and solid-state conformations. Although the length of the POAHP molecules was, on the average 1.6 times the crystal thickness of 150 Å, this is rather short to establish definitively that the poly-L-proline II backbone conformation undergoes hairpin folds. However, shortly after the initial report, detailed studies of the morphology of poly-L-proline itself were presented (25). In this case, the morphology of poly-L-proline, precipitated by film casting from formic acid, was studied. If the "as polymerized" polyproline is dissolved in formic acid, its initial PPI conformation gradually changes over the period of a few hours to PPII. Film casting at various stages provides a morphological record of this conversion. Initially, very ill-defined spherulites are produced that give poor x-ray and electron diffraction, which correspond to form I. At a later stage, the lamellar spherulitic structure appears and may be correlated with form II. Micrographs of the two extreme stages are shown in Figs. 4.16A and B (25). The lamellar thickness in the micrograph of form II is 150 Å. Since the average

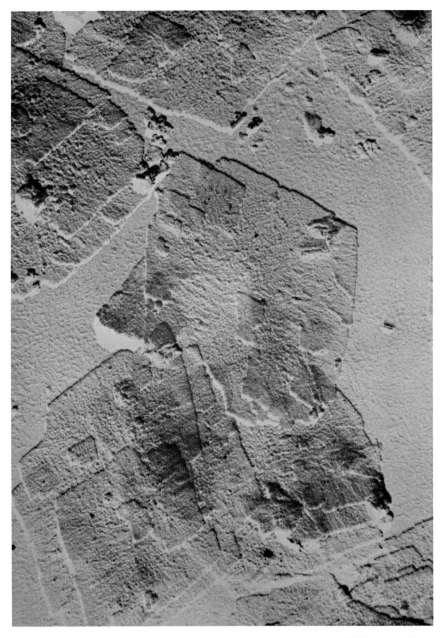

Fig. 4.15. Single crystals of poly-*O*-acetyl-L-hydroxyproline (POAHP) grown from acetic anhydride solutions. The molecular axis (in the PPII conformation) is perpendicular to the major face of the crystals, the heights of which are 150 Å ($\times 28,500$). [From Ref. (24), with permission.]

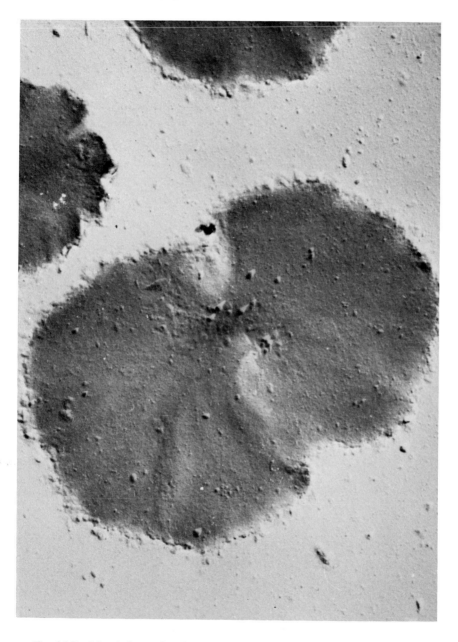

Fig. 4.16A. Morphology of poly-L-proline (MW 25,000) in the PPI conformation, film cast from a formic acid solution 5 min old. Only poorly formed spherulites are present with no evidence of chain-folded lamellar crystals (×31,700). [From Ref. (25), with permission.]

Fig. 4.16B. Morphology of poly-L-proline (MW 25,000) in the PPII conformation, film cast from a formic acid solution 5 hr old. Clearly defined lamellae are present which are ∼150 Å wide. Molecules are perpendicular to the long axis of the lamellae (i.e., perpendicular to the fiber axis (×35,000). [From Ref. (25), with permission.]

chain length is 815 Å, there can be little doubt that the PPII conformation undergoes hairpin folding. On the other hand, the *cis*-poly-L-proline I is an exceptionally rigid molecule with essentially no freedom of backbone rotation (the stable conformation lies in a deep, narrow energy trough) and is thus unable to fold to give well-formed spherulites; presumably, the chains are essentially extended. We can see, therefore, that the ability of the backbone chains to fold is an important crystallization requirement. The poly-L-proline II conformation lies in a broad, favorable energy trough (see Fig. 2.40), is a floppy molecule, and can fairly easily "bend" into a lamellar crystal structure.

MORPHOLOGY OF OTHER LINEAR POLYPEPTIDES

The morphology of other more complex synthetic polypeptides has recently been investigated. These polypeptides have been chosen because of their resemblance to specific protein structures, particularly silk fibroin and collagen. The more detailed aspects of chain conformation of biopolymer models for proteins are presented in Chapter 10, but, in brief, sequential polymers of the type $(Gly-X)_n$, $(Gly-Pro-Pro)_n$, $(Gly-Ala-Pro)_n$, and $(Gly-X-Y)_n$ have been studied. The first of the polytripeptides is said to be an excellent structural model for collagen (26,27), and its morphology is therefore of particular interest. It is possible to produce the usual spherulite morphology, with clearly defined lamellae when $(Gly-Pro-Pro)_n$ is film cast from dioxane/water mixtures (Fig. 4.17). Unfortunately, it is not possible to obtain electron diffraction from such samples, presumably because water molecules, necessary to preserve the crystal structure, are lost in the vacuum and electron bombardment of the microscope. However, there seems to be little doubt that the molecular chain axis lies perpendicular to the face of the lamellae and that once again hairpin chain folding is accomplished. From x-ray data, poly (Gly-Pro-Pro) is known to be triple helical, and the most rational explanation of the morphology requires the rodlike antiparallel triple helices to be stacked in register as shown in Fig. 4.18. In concurrence with this model, the lamellar thickness changes as indicated in Table 4.5.

It should perhaps be pointed out that, although poly (Gly-Pro-Pro) would appear to have antiparallel chains in the triple helices crystallized in the above manner, it does not follow that precipitation under different conditions will produce a similar result (in fact no lamellae could be crystallized from aqueous solutions alone). Data for another proline-containing polytripeptide $(Gly-Ala-Pro)_n$ has also been presented, although the molecular length was not great enough to cause chain folding (28).

The morphologies of two other polytripeptides have been investigated extensively. The first of these, poly [Gly-Ala-Glu(OEt)], may be precipitated from various solvents with the antiparallel β structure. In each case fibers are

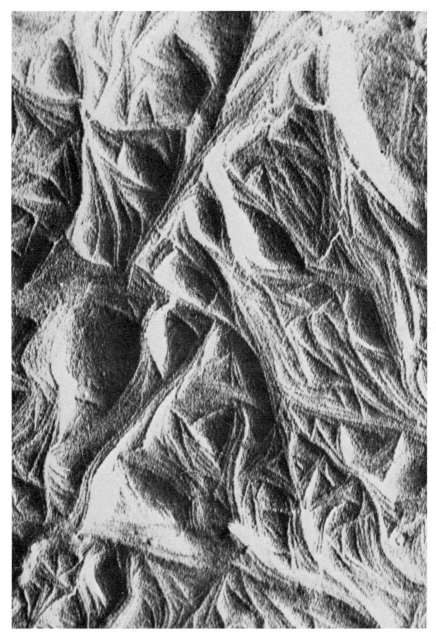

Fig. 4.17. Electron micrograph of poly (Gly-Pro-Pro) precipitated from dioxane–water mixtures. Clearly defined lamellae are present; the molecules probably lie perpendicular to the fiber axis (×75,000). (Micrograph by J. C. Andries.)

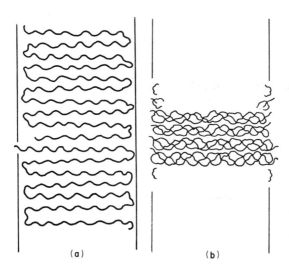

(a)   (b)

Fig. 4.18. (a) Schematic drawing of the folding of polyproline II helices. (b) Schematic drawing of the possible folding of antiparallel, triple-helical poly (Gly-Pro-Pro). [From J. C. Andries and A. G. Walton, *J. Mol. Biol.* **54**, 579 (1970).]

TABLE 4.5

Dependence of Lamellar Thickness on Molecular Weight of Polyglycylprolylproline

Mol. wt	Lamellar thickness (Å)
<6,000	No lamellae
7,000	70–90
8,800	80–100
12,000	110–140
Unfractionated, 6,000–14,500	70–150

composed of lamellae whose cross section is about 60 Å. Figures 4.19A and B (29) show the material precipitated from formic acid with water and ethanol, respectively. The molecules lie essentially flat on the substrate and are folded in the cross-β form; interestingly, however, an area in Fig. 4.19B seems to contain a twisted lamellar form which may be a coiled cross-β structure (29). Poly (Gly-Gly-Ala) may be crystallized in two different forms, both apparently fibrous, from different solvents. Acidic solvents produce a cross-β form similar to poly [Gly-Ala-Glu(OEt)], as shown in Fig. 4.20A (30); the second form, a folded polyglycine II conformation (26) is crystallized from an aged solution in water [Fig. 4.20B (30)]. In the latter case, oriented electron diffraction shows the chain orientation (30). Thus long fibers of crystalline cross-β and

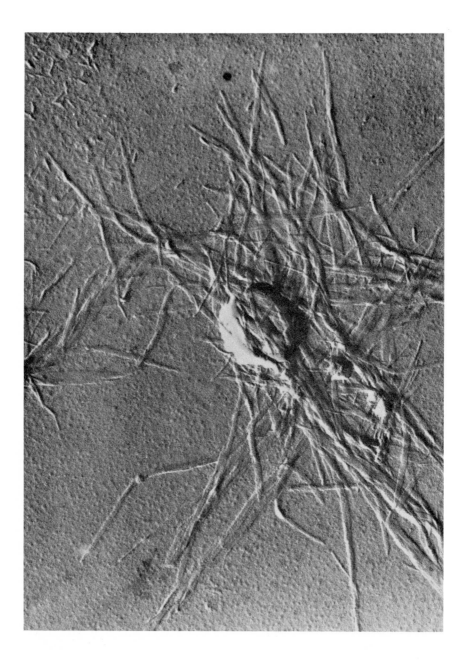

Fig. 4.19A. Thin lathlike crystals of poly [Gly-Ala-Glu(OEt)] precipitated from formic acid with water at room temperature (×25,000). [From J. C. Andries and A. G. Walton, Ref. (29).]

Fig. 4.19B. Morphology of poly [Gly-Ala-Glu(OEt)] precipitated from formic acid with ethanol at room temperature. In both Figs. 4.19A and B the polymer is in the β form. The differences in morphology are wrought by the solvent. Here, areas of twisted rods are apparent. Electron diffraction shows the molecules to be perpendicular to the fiber axis and in the cross-β conformation ($\times 100,000$). [From Ref. (29), with permission.]

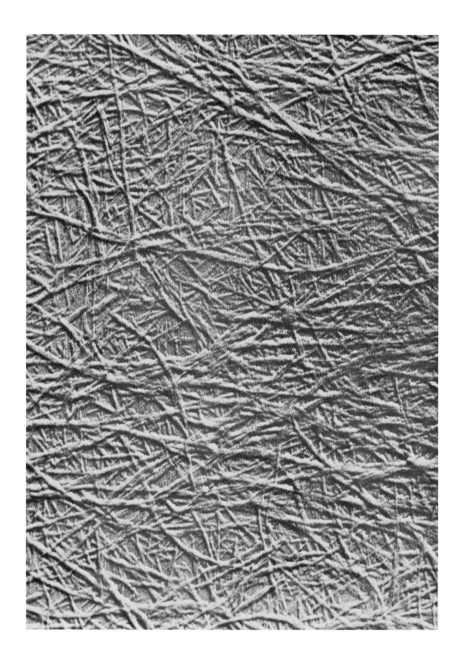

Fig. 4.20A. Fibers (lamellae?) of poly (Gly-Gly-Ala) I precipitated from an acetic acid solution by cooling slowly from 100° to 0°C (×53,800). [From Ref. (30), with permission.]

Fig. 4.20B. Fibers of poly (Gly-Gly-Ala) II crystallized from an aged solution in water. The molecular axis is perpendicular to the fiber axis (×25,800). [From Ref. (30), with permission.]

PGII forms are readily obtainable even from low molecular weight synthetic polypeptides (mol. wt., 10,000–20,000).

Polydipeptides $(Gly\text{-}Ala)_n$, $(Gly\text{-}Ser)_n$, and $[Gly\text{-}Glu(OEt)]_n$ have been examined (31–33). All published micrographs are of the cross-β form, examples of which are shown in Chapter 10. The lamellar thickness is in the range 50–80 Å. (The data for the morphology of the above-mentioned polypeptides are summarized in Table 4.6.)

TABLE 4.6

Morphology and Fold Period of Synthetic Polypeptide Crystals

Material M.W.	Conformation	Morphology	Crystal thickness	Method (Solvent)
1. Poly-L-Tyrosine 100,000	α	Platelet	180Å	20-50% D.M.F. & n- heptanol, slow cooling.
2. Poly-L-Alanine	α	Platelet		D.C.A.
		Spherulites	150Å	T.F.A.
		Sheaves		T.F.A.-T.F.E.
3. Polyglycine 10,000	pG II	Spherulites, Sheaves	60Å	T.F.A.-T.F.E.
4. Poly-L-lysine HPO_4 180,000	α	Platelets	150Å	Aq. solution $+Na_2HPO_4$
	β		n.a.	Same at 75°C.
5. Poly-L-glutamic acid 102,000				
Ca salt	β	Rect. lamellae	35-40Å	Heating Aq.
Sr salt	β	Poorly defined	25-30Å	solutions
Ba salt	β	Acicular	55Å	Excess in salt
Mg salt	α	Hexagonal plates	100Å	Addn. of ethanol
6. Poly-L-proline	pp I	Poorly formed spherulites	—	Formic Acid
Poly-L-proline	pp II	Spherulites	150Å	Formic acid
7. Poly-O-acetyl-L-hydroxyproline 11,800	pp II	Platelet	150Å	Acetic anhydride Thermal Pptn.
8. PolyGlyAla	β	Fibers	60Å	Formic acid
GlySer	β	Fibers	60Å	Formic Acid
Poly(gly-ala-pro)	pp II	Rods	150Å	Butanol Thermal Pptn.
9. Poly(gly-pro-pro)	Triple helix	Sheaves	80-120Å (M.W. dep.)	Dioxane-Water
10. Poly(gly-ala-glu(OEt)) 16,000	β	Spherulites	60Å	From formic acid
11. Poly(gly-gly-ala) 16,000	β	Fibers	60Å	Acetic acid
Poly(gly-gly-ala) 16,000	pG II	Fibers	80-120Å	Water

From the data currently available it appears that the lamellar thickness directly reflects the underlying molecular arrangement. The materials with strong intermolecular bonding and high lattice energy have the smallest fold period and vice versa. There is a theoretical approach to understanding such properties based on the nucleation or crystal-initiation step. Further comments on the theory are made in the Discussion of Morphology section.

AMORPHOUS POLYPEPTIDES

The previously discussed morphological observations are based upon polyamino acids or polypeptides that form more or less crystalline phases when deposited from appropriate solutions. In such cases, a well-defined biopolymer conformation is achieved which is representative of some segments of proteins. However, even fibrous proteins contain amorphous regions, and attempts have been made to produce biopolymer models for the amorphous or globular proteins. Although, in principle, random linear co-polymers of many peptide components would serve this purpose, and some progress has been made in the synthesis and study of these so-called "proteinoids," a simpler, but less realistic model is that of the multichain (branched) polypeptide. Several polypeptides of this type have been prepared at the Weizmann Institute, Rehovot, Israel, and some are available commercially. In general, amino acids are condensed in conjunction with poly-L-lysine or poly-L-ornithine such that the resulting polymer consists of oligopeptides condensed onto the free amino group of the polymer backbone.

It has proven exceptionally difficult to characterize the morphology of amorphous polymers and biopolymers. Some commercial polymers show rows of spherical structures, each sphere being approximately 80 Å (34). Replicas and stained examples of the multichain polypeptides also seem to show similar structure (32), although the possibility of artifacts arising in sample preparation cannot entirely be ruled out.

EPITAXY

Many instances of the interaction of inorganic solids with proteins and polysaccharides exist in nature. Calcified tissue, for example, consists of interspersed collagen and calcium phosphate. In the laboratory, it is not uncommon to cast films of biopolymers on quartz or salt plates for purposes of spectroscopic examination. Until recently, it was felt that such substrates were inert and would not modify the conformation or morphology of the deposited phase. However, it is now known that, under some circumstances, an intimate structural relation can exist between deposit and substrate. The orientational effect of the underlying lattice may be determined by morphological observation, the phenomenon of orientation overgrowth or

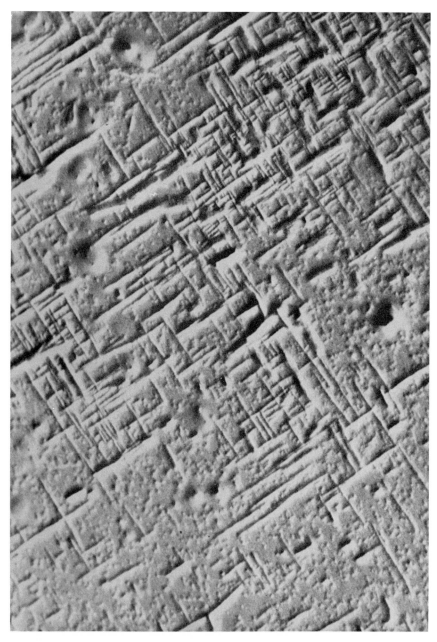

Fig. 4.21. Epitaxy of POAHP on sodium chloride, precipitated from acetic anhydride at 55°C with 30 min immersion. Needle direction is [100] and [010] ($\times$ 13,500). (Taken from Ref. 24., with permission.)

deposition being known as epitaxy. Although the effect is well characterized for commercial polymers (35), relatively few studies have been directed toward biopolymers. Optical microscopy by Seifert (36) has revealed that several homobiopolymers undergo epitaxy on quartz. Electron microscopy of poly-*O*-acetylhydroxy-L-proline (24) and poly-γ-benzyl-L-glutamate (37) on alkali halide crystals reveals the epitaxial effect (Figs. 4.21 and 4.22), and these morphologies should be compared with those produced in the absence of a substrate (Figs. 4.15 and 4.8). For POAHP, oriented lamellar crystallization is produced with the molecular axis lying flat on the substrate and perpendicular to the lamellae. The lamellar thickness appears to have increased considerably (~ 300 Å) compared with the free crystals. The morphology of

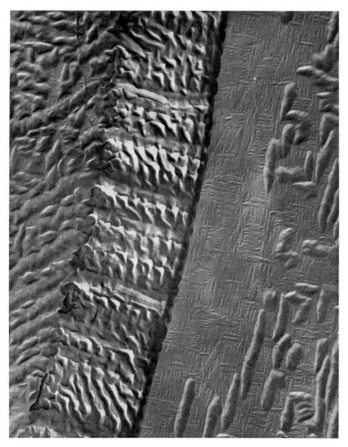

Fig. 4.22. Electron micrograph of poly-γ-benzyl-L-glutamate (PBLG) precipitated onto a potassium chloride surface. Orientation direction is ($\times 30,000$) (Micrograph by F. Rybnikar.)

PBLG is completely changed from the fibrillar aggregate to the lamellar crystal (using the same solvents). Here again the indications are that the lamellae are much thicker than those of the usual crystalline structures.

Discussion of Morphology

The development of chain folding in polymer and biopolymer crystals must result from a minimization of the free energy of the system. Two theoretical approaches based on nucleation and equilibrium thermodynamic arguments are outlined in Ref. 15. The nucleation approach predicts that the larger the driving energy to nucleation and crystal growth, the smaller will be the fold period. The equilibrium approach supposes that vibrational anisotropy leads to a minimum-free-energy conformation when the chain is of discrete length. In this case also, larger lamellar thickness is expected as the equilibrium temperature is raised. Since there are no data on the change of lamellar thickness with temperature or supersaturation for biopolymers, it can only be assumed that their behavior is similar to that observed for commercial polymers (which is in accord with the preceding concepts). However, the crystal thickness does appear to be correlated with the polypeptide backbone conformation and is, as far as we can tell, rather independent of the precise conditions of crystallization. It is notable that materials with the weakest "intermolecular" stabilization forces (α helix and PPII) have the largest fold period, whereas those with strong "intermolecular" forces (β sheet and PGII) have very small fold periods. It seems likely that the driving force to the formation of antiparallel folded β form originates from molecules that are already folded in solution, whereas the gain in lattice energy for α and PPII structures is minimal and may involve cooperative aggregation prior to crystallization. In any case it is evident that the process of chain folding is highly statistical and entropic in origin.

In most examples studied so far, the solid conformation seems to reflect that of the molecules in solution prior to precipitation. The esters of poly-L-glutamic acid, which are of complex agglomerated form in many weak solvents, form fibrils when deposited from those solvents. The $\alpha \rightarrow \beta$ conformation change of poly-L-lysine with temperature in aqueous solution is amply reflected by change in morphology of the solid precipitated from these solutions, as is the PPI $\rightarrow$ PPII conformation change in formic acid. On the other hand, poly-L-alanine precipitates from strong solvents [dichloroacetic acid (DCA) and TFA] in the α conformation, despite the fact that ORD evidence seems to indicate that the material is in the random conformation in solution. One clear anomaly is the POAHP, as previously noted. No explanation can be offered for this exception, nor are the actual structure and chain conformation in the crystal folds yet determined.

Perhaps the most important aspect of biopolymer crystallization is the choice of suitable solvent systems. Although biopolymers containing ester groups often precipitate from weakly polar organic solvents with fibrillar morphology, the choice of solvent for lamellar crystallization is usually problematical. It is commonly necessary to scan fifty or more solvents and then mixtures thereof. The morphology produced from most of these solutions will appear glassy or will have no definable structure. It may be that the molecule/solvent interactions which relate to the crystal/solution interfacial energy are not readily optimized to promote chain folding. The apparently amorphous material precipitated from such systems, nevertheless shows considerable long-range order by x-ray diffraction, although the extent of this order is not as great as in crystalline lamellar material. We may deduce, therefore, that the chain-folded lamellar structure is not necessarily the most common solid-state structure, which probably contains more extended and randomized chain orientations.

Perhaps the most startling aspect of the highly crystalline lamellar morphology of synthetic polypeptides discussed in this chapter is that the molecular axis lies perpendicular to the long axis in all cases. This effect causes the infrared dichroism to be the opposite of that normally expected for polypeptides, as explained in Chapter 5. Although there are numerous cases of "anomalies" in infrared and x-ray fiber diagrams resulting from the super-folded-chain conformation, the process of orientation in "wet" films, usually brings the molecular axis and fiber axis into alignment, thus breaking down the lamellar crystal structure if it existed. We must therefore differentiate between crystalline biopolymers in which both the chain conformation and tertiary (and quarternary) structure are well defined, and "ordered" structures in which the chain conformation is maintained but three-dimensional order is minimal.

Morphology of Polysaccharides and Synthetic Polynucleotides

The electron microscopy of polysaccharides appears to have been limited to the native species; these results are discussed in Chapter 11. The situation with polynucleotides is more complex. They usually crystallize with solvating (water) molecules and are generally not accessible to *in situ* electron diffraction in the vacuum of the electron microscope. There does, however, seem to be ample scope for the development of such research. It has, for example, been shown that sonicated fragments of DNA undergo lamellar crystallization with chain folding (38). The fold period was ∼ 100 Å and electron diffraction identified the chain direction. It would seem that the combination of morphology and electron diffraction (possibly using moisture-containing devices) can be of considerable potential value in studying synthetic polynucleotide structure.

Summary

In conclusion it may be stated that the vast majority of electron microscope studies of synthetic biopolymers have been concentrated on polypeptides. Materials that can be crystallized from strong solvents reveal morphology similar to that of commercial polymers but not to that of polypeptides or proteins in native tissue. On the other hand, it may be that if native materials, or fragments thereof, were treated in a similar manner, they also would show crystalline morphology. Such an assertion is supported by the observation of lamellar chain-folded DNA fragments. The crystalline polypeptides do have the peculiar characteristic of chain superfolds in which the backbone conformation is strongly perturbed. Studies of the superfold region may give insight into similar superfolds observed in globular proteins (by x-ray crystallography). A less common morphology is that of the supercoiled fibrillar aggregates. More detailed studies of such aggregates should reveal quantitative aspects of α-helix supercoiling.

Appendix: Electron Diffraction

Electron diffraction is highly suitable for investigation of the structure of thin films and polymer single crystals since electrons interact about 10^6 times more strongly with matter than do x-rays; thus the diffraction from very small specimens can be recorded at short exposure times. These are two appreciable advantages over the x-ray technique. The strong interaction of electrons with matter often produces problems in intensity measurement which can usually be neglected in x-ray work. However, the study of electron diffraction by polymers has been concerned mainly with the determination of the unit cell orientation for which only the d-spacings of the reflections are required. From this viewpoint, the theory for electron diffraction is the same as that detailed in Chapter 3 for x-ray crystallography. The position of the hkl reflections and their d-spacings are related by Bragg's law,

$$n\lambda = 2d \sin \theta$$

where θ is half the scattering angle, and λ is now the wavelength of the electrons (of the order of 0.05 Å).

The conditions for reflection in terms of the reciprocal lattice, (see Chapter 3, Diffraction of x-Rays by a Crystal, Reciprocal Lattice section) require the lattice point to be in contact with the Ewald sphere of reflection. The wavelength of the electrons is considerably less than the lattice repeat distances, with a result that the sphere of reflection, diameter $1/\lambda$ is approximately planar and almost coincides with a plane of reciprocal lattice points. As described in this chapter, the single crystals of polymers are frequently flat lamellae or hexagonal plates and lie flat on the surface of the electron microscope grid. The electron diffraction from such specimens represents a projection of a plane of the reciprocal lattice. Prior knowledge of the unit cell from x-ray measurements (or by analogy with similar materials) allows the reflections to be indexed; this identifies the reciprocal lattice plane giving rise to the reflections, usually the $hk0$, $h0l$, or $0kl$ reflections, for polymer structures. Hence, the orientation of the unit cell relative to the observed structures can be determined, and knowledge of the positions of the chains within the unit cell defines the direction of these chains within the crystal.

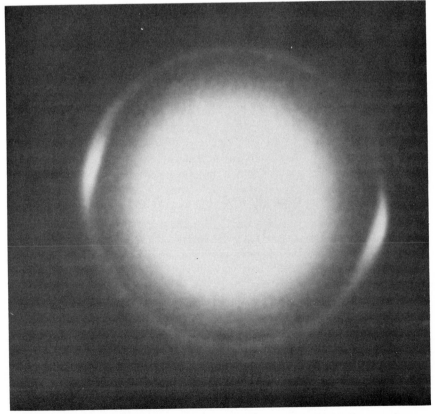

Fig. 4.23. Electron diffraction pattern of poly [Gly-Ala-Glu(OEt)] film cast from a formic acid–ethyl ether mixture at room temperature.

In the case of polymers that crystallize as fibrous lamellae, the electron diffraction pattern can be treated in much the same way as the x-ray pattern from a fibrous specimen. Regions can often be found where all the long axes of the fibrils are approximately parallel but where there is no apparent preference for them to lie down in a particular manner. Such is the case for the fibrous crystals of poly [Gly-Ala-Glu(OEt)], for which the electron diffraction pattern is shown in Fig. 4.23. The appearance of the 4.7 Å reflection corresponding to the hydrogen bond direction of the β-sheet structure on the meridian of the electron diffraction pattern identifies this structure as the cross-β conformation.

Electron diffraction is feasible for very small quantities (10^{-18} gm) but not for samples thicker than several hundred angstroms. With increasing specimen thickness, absorption of electrons becomes so high that the diffraction pattern cannot be observed. A further limitation is that a crystalline biopolymer specimen is usually destroyed by the beam in 3–4 sec. For a further discussion of electron diffraction see the book on this subject by Vainstein (39).

REFERENCES

1. F. H. C. Crick, *Acta Crystallogr.* **6**, 689 (1953).
2. L. Pauling and R. B. Corey, *Nature* **171**, 59 (1953).
3. D. A. D. Parry and E. Suzuki, *Biopolymers* **7**, 189, 199 (1969).
4. A. J. Hopfinger and A. G. Walton, *J. Macromol. Sci. Phys.* **B3**, 195 (1969).
5. V. Sasisekharan, *Acta Crystallogr.* **12**, 903 (1959).
6. A. J. Hopfinger and A. G. Walton, *J. Macromol. Sci. Phys.* **B4**, 185 (1970).
7. W. Traub and A. Yonath, *J. Mol. Biol.* **16**, 404 (1966).
8. W. Traub and A. Yonath, *in* "Conformation of Biopolymers" (G. N. Ramachandran, ed.), Vol. 2, p. 449. Academic Press, New York, 1967.
9. F. R. Brown, A. J. Hopfinger, and E. R. Blout, *J. Mol. Biol.*, **63**, 101 (1972).
10. S. Ishikawa and T. Kurita, *Biopolymers* **2**, 381 (1964).
11. T. Tachibana and H. Kambara, *Kolloid-Z.* **219**, 40 (1967).
12. F. Rybnikar and P. H. Geil, *Biopolymers*, **11**, 271 (1972).
13. J. J. B. P. Blais and P. H. Geil, *J. Ultrastruct. Res.* **22**, 303 (1968).
14. See, for example, P. Saludjian and V. Luzzati, *in* "Poly-α-Amino Acids" (G. D. Fasman, ed.), Chapter 4. Dekker, New York, 1967.
15. F. J. Padden and H. D. Keith, *J. Appl. Phys.* **36**, 2987 (1965).
16. J. L. Koenig and M. J. Hannon, *J. Macromol. Sci. Phys.* **B1**, 119 (1967).
17. J. L. Koenig and P. D. Vasko, *J. Macromol. Sci. Phys.* **B4**, 347, 369 (1970).
18. P. H. Geil, "Polymer Single Crystals." Wiley, New York, 1963.
19. F. J. Padden, H. D. Keith, and G. Giannoni, *Biopolymers* **7**, 793 (1969).
20. H. D. Keith, G. Giannoni, and F. J. Padden, *Biopolymers* **7**, 775 (1969).
21. W. T. Astbury, S. Dickinson, and K. Bailey, *Biochem J.* **29**, 235 (1935).
22. W. T. Astbury, E. Beighton, and K. D. Parker, *Biochim. Biophys. Acta* **35**, 17 (1969).
23. A. J. Geddes, K. D. Parker, E. D. T. Atkins, and E. Beighton, *J. Mol. Biol.* **32**, 343 (1968).
24. J. C. Andries and A. G. Walton, *Biopolymers* **8**, 465 (1969).
25. J. C. Andries and A. G. Walton, *Biopolymers* **8**, 523 (1969).

26. W. Traub and A. Yonath, *J. Mol. Biol.* **16**, 404 (1966).
27. A. Yonath and W. Traub, *J. Mol. Biol.* **43**, 461 (1969).
28. A. Schwartz, J. C. Andries, and A. G. Walton, *Nature* **226**, 161 (1970).
29. J. C. Andries and A. G. Walton, *J. Mol. Biol.* **56**, 515 (1971).
30. J. C. Andries, J. M. Anderson, and A. G. Walton, *Biopolymers* **10**, 1049 (1971).
31. B. Lotz and H. D. Keith, *Bull. Amer. Phys. Soc.* **15**, 305 (1970).
32. H.H. Chen, M.S. thesis, Case Western Reserve University, Cleveland, Ohio, 1971.
33. J. M. Anderson, H. S. Chen, and A. G. Walton, *Bull. Amer. Phys. Soc.* **16**, 320 (1971).
34. G. S. Y. Yeh and P. H. Geil, *J. Macromol. Sci. Phys.* **B1**, 235 (1967).
35. J. A. Koutsky, A. G. Walton, and E. Baer, *J. Polym. Sci. Part A-2* **4**, 611 (1966).
36. H. Seifert, *Kolloid-Z. Z. Polym.* **224**, 7 (1968).
37. S. H. Carr, A. G. Walton, and E. Baer, *Biopolymers* **6**, 469 (1968).
38. G. Giannoni, F. J. Padden, and H. D. Keith, *Proc. Nat. Acad. Sci. U.S.* **62**, 964 (1969).
39. B. K. Vainstein, "Structure Analysis by Electron Diffraction." Pergammon, New York, 1964.

5

INFRARED AND RAMAN SPECTROSCOPY

Introduction

Infrared (IR) spectroscopy has been recognized for many years as an important tool for probing molecular structure and conformation of biological macromolecules. With the recent development of the high-energy laser, Raman scattering has come to the fore as a useful complementary method. Both infrared and Raman spectra arise from absorption of energy, which produces an increase in vibrational or rotational energy of atoms or groups of atoms within the molecule. However, the two effects reflect different properties of the molecule. The infrared technique uses a variable wavelength source, and the molecules absorb the radiation at specific wavelengths provided that the vibrational mode produces a change in the permanent dipole moment ($\partial\mu/\partial q \neq 0$). Raman scattering utilizes incident light of a fixed frequency, which is in the visible range when the source is a laser. The molecule absorbs the incident radiation and then emits radiation characteristic of the vibrational modes for which there is a change in the polarizability of the molecule ($\partial x/\partial q \neq 0$). Two sets of "emission" lines are observable at frequencies higher and lower than the incident beam, and are known as the "Stokes" and the "anti-Stokes" lines, respectively (see Fig. 5.1).

The wealth of information that may be derived from infrared studies of polypeptides has been reviewed extensively (1–3). By comparison, there seems

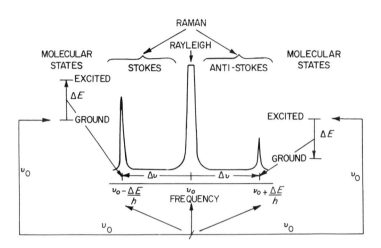

Fig. 5.1. Schematic diagram of the origin of the Raman effect (ν_0 is the frequency of the exciting source). Raman lines are usually measured in the Stokes region and thus correspond to vibrational excitation from the ground state. Although the Stokes lines lie in the visible region, the $\Delta E/h$ frequency *difference* corresponds to infrared and far-infrared frequencies. (Diagram by W. B. Rippon.)

to be no general review of Raman studies of biological macromolecules, although recent reviews concerning nonbiological polymers have been presented (4,5). Raman spectroscopy of polypeptides was, in fact, virtually impossible until the advent of the laser since the scattering intensity is only of the order of 10^{-6} of incident intensity. Furthermore, highly detailed spectra have been available only since 1969 with the use of more powerful lasers (e.g., Ar^+). The advantages of the laser are that it provides high-intensity monochromatic radiation which is also inherently polarized, with the result that orientational properties of the molecular vibrations may be readily characterized.

Because the atomic environment and directional motion are dependent upon the conformation and spatial array of the system we can expect that conformational information may be deduced from IR and Raman spectroscopy, the conformational effect making itself felt through the intensity and frequency of the various bands. An elementary treatment of the principles of infrared spectroscopy is given in Ref. 6, and the analysis of infrared and Raman spectra for organic molecules is discussed in Ref. 7.

Because of the different physical processes giving rise to the infrared and Raman spectra, the two effects are governed by different selection rules. These give rise to the mutual exclusion principle for molecules possessing a center of symmetry so that bands observed in the infrared are not seen in the Raman and vice versa. Centers of symmetry are not common in biopolymer structures, which are usually optically active. However, the nature of the two processes means that, in general, a highly asymmetrical vibrational mode will probably give rise to a strong infrared band, whereas a relatively symmetrical mode will give a prominent Raman line. Thus, polar groups with large dipole moments are strong absorbers of infrared radiation, whereas nonpolar bonds give rise to Raman scatter. Table 5.1 shows this effect qualitatively applied

TABLE 5.1

Representation of Intensity Relations
for Various Groups

$C{=}O$	$C{=}N$	$C{=}C$
Raman (polarizability) $\longrightarrow$		
$\longleftarrow$ Infrared (dipole moment)		
$O{-}H$	$N{-}H$	$C{-}H$

to various $C{-}X$ moieties. The vibrational bands characteristic of certain groups of atoms are often observed at a virtually constant wavelength, independent of the total composition of the molecules, and this fact, plus the relatively small changes that occur in biopolymer spectra due to conformational differences, provides us with a useful tool for probing molecular structure.

For a fundamental understanding of the intensity and position of the line or band in the Raman or infrared spectrum it is necessary to have a detailed picture of the vibrational modes, transition energies between excited states, and symmetry of the molecule in question. In practice, the experimental spectrum is usually observed and then efforts are made to relate models to the observed data. For a more extensive analysis it is necessary to introduce symmetry properties and the associated group theory. The symmetry elements and the resulting character tables have been used extensively to calculate the allowed bands for simple organic molecules with considerable success (8). The extension of these calculations in order to consider long-chain molecules has occurred more recently and is still largely incomplete. However, the only potentially active vibrations are those for which corresponding atoms in all unit cells move in phase. Thus, for example, the normal modes for a single polyethylene chain have been derived from consideration of a single unit cell (9). This approach is rigorous only for an infinite polymer chain in a crystalline lattice. Deviations from this ideal model, due for example to low crystallinity, are detected in the observed spectra as band broadening and as frequency shifts. Figure 5.2 (10) shows an infrared spectrum of nylon in which

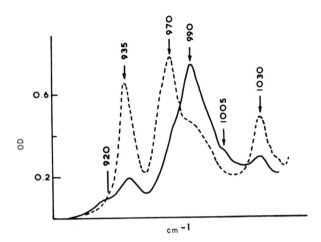

Fig. 5.2. An infrared spectrum (optical density versus reciprocal wavelength) for nylon 6. Solid line is for low-crystallinity material. The dashed line is for annealed and more highly crystalline material. [From Ref. (10), by permission of John Wiley and Sons, Inc.]

both poorly crystalline and annealed (more crystalline) samples have been examined. Nylon is a polymer with a peptidelike repeat group (Fig. 5.3) and is therefore of relevance to our consideration of biopolymers. Although there seem to be no reported studies of the crystallinity of biopolymers based on the width of the bands in the infrared or Raman spectra, it appears that such measurements should be at least feasible. It may be argued that Fig. 5.2 represents more a conformational change than that of a three-dimensional order. We shall see later that not only are infrared and Raman methods able to delineate conformational changes, but the nature of the chain packing in the crystal may also be deduced from these methods.

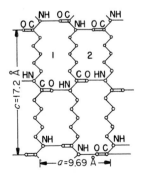

Fig. 5.3. The molecular arrangement of the α-form nylon 6 in the projection down the *b* axis.

Spectroscopy of Polypeptides

Most of the early infrared work on polypeptides was carried out in the wavelength range 4000–1400 cm^{-1}, and in this region there are several important bands. In more recent work this range has been extended down to approximately 200 cm^{-1}, where additional important bands are evident. Raman spectra, in particular, show interesting lines in this "low-frequency" region. In the high-frequency region, vibrations tend to be localized mainly to one, two, or perhaps three coupled bond vibrations, but at lower frequencies considerable delocalization along the peptide chain gives rise to cooperative modes.

For polypeptides and proteins we can divide the characteristic vibrations into those arising from the peptide backbone and those from the side chains. It is usual to define up to nine characteristic bands occurring in the infrared spectra of polypeptides, the so-called amide A and B and amide I–VII vibrational modes. Often only three or so of the amide I–VII bands are clearly observed.

Amide A and B Bands

Two bands are observed in the infrared spectra of amides, polypeptides, etc., in the 3300–3000 cm^{-1} region due to vibrations involving the amide group. For a long time this was a puzzling observation, since early workers expected only a single N—H stretching mode in this region. An apparent solution to this problem came (11) when it was realized that the first overtone of one of the lower-frequency amide bands, the so-called amide II, should also occur in this region. Ordinarily such an overtone would be very weak in intensity. However, this frequency falls close to the N—H stretching frequency, and resonance occurs between the two levels. This phenomenon, which is known as Fermi resonance, shifts the original frequencies so that we observe a new band at a frequency higher than the theoretical unperturbed N—H stretch and another one at a frequency lower than the first overtone of amide II. This effect is shown schematically in Fig. 5.4, where the frequencies are those for *N*-methylacetamide. The new bands are known as amide A and amide B. The perturbation also tends to equalize the intensities, so that while amide A is generally stronger than amide B, the latter band is more intense than the theoretical unperturbed overtone. At this point it should be noted that the frequency of amide B is frequently more than twice that for amide II, rather than less than twice this frequency as predicted. It has been argued that in fact the resonance occurs between the N—H stretch, and an IR-inactive component of complex amide II motions at a higher frequency than the IR-active components.

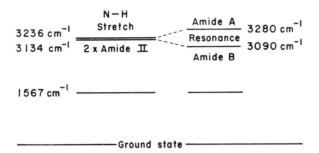

Fig. 5.4. Diagrammatic representation of Fermi resonance for liquid *N*-methyl-acetamide. [Reprinted from Miyazawa (2), p. 69, courtesy of Marcel Dekker, Inc.]

Amide I, II, III, and IV Bands

The atomic motions involved with the characteristic amide bands have been studied extensively by Miyazawa and co-workers (12), with the result that the amide I–III bands have been shown to arise from "in-plane" vibrations of the peptide group, as shown in Fig. 5.5. It can be seen that each band

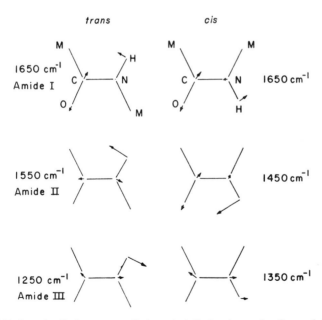

Fig. 5.5. Atomic displacements of characteristic in-plane vibrations of the CONH group (*N*-methylacetamide). [Reprinted from Miyazawa (2) p. 69, courtesy of Marcel Dekker, Inc.]

arises from a combination of vibrations; the relative contribution of each vibration is given in Table 5.2. The amide I appears as a very strong band in the infrared and is also very strong in the Raman effect, whereas the amide II, which is very strong in the infrared, is usually weak or missing in the Raman. The amide III vibration generally gives rise to a strong infrared band or a moderate Raman line. The subtleties of band position and dichroism and their relation to conformation are presented later. Although the amide IV band, which is also categorized as resulting from "in-plane" vibrations, may be grouped with amide I, II, and III bands, it is not localized to the peptide group and appears in the infrared (of *N*-methylacetamide) at 647 cm^{-1}. It is rarely identifiable in either the infrared or Raman spectrum of polypeptides.

TABLE 5.2

Vibrational Components of Amide Bands (Infrared)

Approx. wavelength (cm^{-1})	Amide band	Composition (%)		
		N-Methylacetamide	Polyglycine I	Polyglycine II
1630–1650	I	C=O Stretch, 78	C=O Stretch, 78	C=O Stretch 78
1520–1560	II	N—H Bend, 60	N—H Bend, 70	N—H Bend, 70
		C—N Stretch, 40	C—N Stretch, 30	
1250–1310	III	C—N Stretch, 35	C—N, 44	C—N, 30
		N—H Bend, 29	N—H, 18	N—H, 24
			C—C, 22	C—C, 20–25

Amide V, VI and VII Bands

The infrared bands classified as V, VI, and VII occur at lower frequencies (below 800 cm^{-1}) and involve vibrations that are "out of (peptide) plane." Displacements and band assignments as presented by Miyazawa for *N*-methylacetamide are shown in Fig. 5.6. Recently considerable attention has been paid to the amide V band since it appears to be particularly sensitive to conformation and local environment. It appears in both the infrared and Raman but is weak in the latter. Amides VI and VII are rarely discernible in the infrared spectra of polypeptides, although a number of Raman lines appear in the frequency range expected for skeletal modes. Complete assignments have not generally been made.

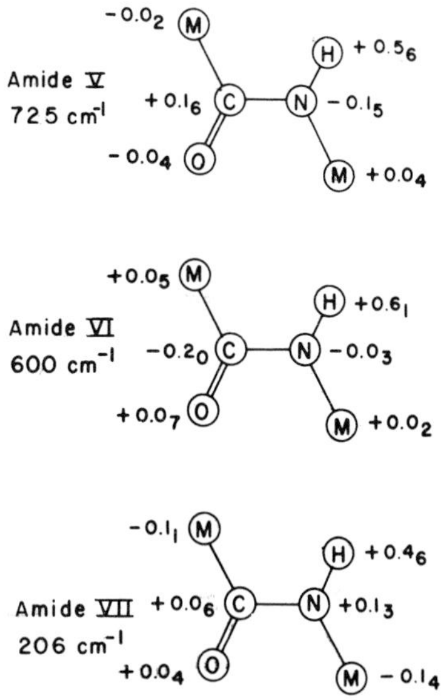

Fig. 5.6. Atomic displacements of the out-of-plane vibrations of N-methylacetamide. [Reprinted from Miyazawa (2) p. 69, courtesy of Marcel Dekker, Inc.]

Vibrational Analysis

For a polypeptide chain of infinite length, i.e., where end effects are negligible, which is in some regular conformation, the vibrations of the various atoms in the peptide group may be treated in terms of a natural frequency v_0, which is modified by vibrational coupling with neighboring groups. This coupling can be intrachain through the α carbon atom or interchain across peptide hydrogen bonds. As will be seen below, the perturbation depends on the phase differences δ, δ', etc., between the peptide group motions within the unit cell. The actual values of δ for which coupling occurs depend on the molecular and crystal symmetry. Figure 5.7 shows how vibrational inter-actions among various peptide groups of a parallel two-strand β structure can occur. The algebraic expression (13) for the general case of the type de-picted in Fig. 5.7 is

$$v(\delta, \delta') = v_0 + \sum_s D_s \cos(s\delta) + \sum_{s'} D'_{s'} \cos(s'\delta') \tag{5.1}$$

where the primed entities refer to interchain and unprimed to intrachain parameters, D and D' are the respective coupling constants, and s and s' are the numbers of successive peptide groups along the same and neighboring chains. In this approach the vibrations are treated as localized in the peptide

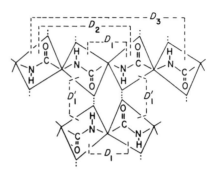

Fig. 5.7. Vibrational interactions among peptide groups. [Reprinted from Miyazawa (2) p.69, courtesy of Marcel Dekker, Inc.]

groups, which is, for certain purposes, a useful approximation. A more generalized mathematical treatment for the vibrations of helical chains has been presented (3,14–16) and applied to polyamino acid spectra (17).

For present purposes, the more limited approach represented by Eq. (5.1) will be used. The solution of Eq. (5.1) may readily be achieved if the conformational geometry is known. For an extended peptide chain (twofold screw axis) the allowed transitions are those in which the transition moments are in phase or π radians out of phase. These two possibilities correspond to the vector sum of the moments, being parallel or perpendicular to the chain axis (respectively). If two antiparallel chains are considered with coupling to only the nearest-neighbor peptide groups, then Eq. (5.1) reduces to

$$v(\delta, \delta') = v_0 \pm D_1 \pm D_1' \tag{5.2}$$

having four solutions,*

$$v_{\parallel}(0, 0) = v_0 + D_1 + D_1' \tag{5.2a}$$

$$v_{\perp}(\pi, 0) = v_0 - D_1 + D_1' \tag{5.2b}$$

$$v_{\parallel}(0, \pi) = v_0 + D_1 - D_1' \tag{5.2c}$$

$$v_{\perp}(\pi, \pi) = v_0 - D_1 - D_1' \tag{5.2d}$$

These equations then represent the frequencies at which vibrational modes should appear for the antiparallel β-sheet arrangement in polypeptides. However, of these four frequencies, one, the symmetrical $v_{\parallel}(0, 0)$ is infrared inactive since it does not lead to a net change in dipole moment. Although all four bands should be Raman active, in practice the strong symmetrical $v_{\parallel}(0, 0)$ mode is usually the only observable band. Of the three infrared bands, two are usually clearly observable and the third, $v_{\perp}(\pi, \pi)$, is weak. It should

---

* Note that four peptide units appear in the unit cell of the antiparallel β arrangement.

be stressed that this approach is applicable only if the vibrational motion is localized in the peptide group, and in interpreting spectra it is usual to concentrate on the amide I and II bands for this reason. Unfortunately, the amide II band is very weak in most Raman spectra and thus complete experimental resolution of the parameters in Eq. (5.2) is most readily achieved by examination of the amide I band.

The parallel extended-chain or β-sheet arrangement of peptide chains leads to a somewhat different situation. Only two peptide residues appear in the unit cell, both in the same chain. Out-of-phase vibrations between unit cells do not contribute to the infrared spectrum. A summary of some available experimental and calculated data for the amide I band of β-structures is presented in Table 5.3.

TABLE 5.3

Assignment of Amide I Bands in β-Sheet Conformation

| Type | Designation | Observed | | Calculated parameters | |
		IR[a]	Raman[a]	Coupling constants (cm^{-1})	ν_0 (cm^{-1})
Parallel chain	$\nu_\parallel(0, 0)$	1648(w)		$D_1 = 8$	$\nu_0 = 1658$
(keratin)[b]	$\nu_\perp(\pi, 0)$	1632(s)		$D_1' = -18$	
Antiparallel chain	$\nu_\parallel(0, 0)$		1674(m)		
(polyglycine I)[c]	$\nu_\parallel(0, \pi)$	1684(w)		$D_1 = 19$	
	$\nu_\perp(\pi, 0)$	1696(s)		$D_1' = -5.5$	$\nu_0 = 1660.5$
	$\nu_\perp(\pi, \pi)$	(1647)[d]			

[a] In this and following tables of this chapter, observed intensities of infrared bands or Raman lines are characterized as strong (s), medium (m), weak (w), very strong (vs), and very weak (vw).

[b] Data from Ref. 31.

[c] Data from W. B. Rippon, J. L. Koenig, and A. G. Walton, *J. Agric. & Food Chem.* **19**, 692 (1971).

[d] Calculated band.

For the α helix there are three infrared-active vibrational transitions. The first is the in-phase coupling along the helix axis, and the second, which is doubly degenerate, occurs at a phase angle of $\delta = 2\pi/n$, where n is the number of residues per turn. The dichroism in this latter case is perpendicular to the helix axis. The two allowed infrared transitions are given by

$$\nu_\parallel(0) = \nu_0 + D_1 + D_3 \tag{5.3}$$

$$\nu_\perp(\delta) = \nu_0 + D_1 \cos \delta + D_3 \cos 3\delta \tag{5.4}$$

There are no intermolecular contributions considered in these equations; the coupling arises from nearest-neighbor D_1 and third-neighbor D_3 interactions. The latter arises because each third residue is coupled by hydrogen bonding in the helix (see Chapter 2). The amide vibrations for the random-coil forms are believed to be essentially uncoupled, and therefore only one mode, $v = v_0$, is active. The "in-phase" and "out-of-phase" vibrations may be strictly represented by the [A] ($\delta = 0$) and [E] ($\delta = 2\pi/n$) terminology. Although splitting of the amide I and II infrared bands has been observed for helical molecules, only one strong amide I line has been noted in the Raman spectrum and thus, in this region at least, little additional information emerges from the combined use of IR and Raman spectroscopy. However, the symmetric [A] modes are expected to be strongest in the Raman, the [E] modes in the infrared.

Some data for α-helical structures are presented in Table 5.4. It can be seen that there is only a very small wavelength separation between bands.

TABLE 5.4

Infrared and Raman Data for Amide I Band of α-Helical Structures

Material	Designation	Raman (cm^{-1})	Infrared (cm^{-1})
Poly-L-alanine[a]	$v_{\parallel}(0)$[A]	1654	
	$v_{\perp}(2\pi/3.6)$[E]		1651
Poly-L-lysine[b]	[A]	1650	
	[E]		1645
Poly-L-lysine HCl[c]	[A]	1652	
	[E]		1650

[a] From J. L. Koenig and P. L. Sutton, *Biopolymers* **8**, 167 (1969).
[b] From J. L. Koenig and P. L. Sutton, *Biopolymers* **9**, 1229 (1970).
[c] From text Ref. 18.

The infrared weak [A] modes may sometimes be observed as weak shoulders on main amide bands. The D_1 and D_3 parameters for the α helix have been reported (18) as $D_1 = 10$ cm^{-1} and $D_3 = -18$ cm^{-1} on the basis of a random chain, uncoupled fundamental frequency $v_0 = 1658$ cm^{-1}.

The polyglycine II structure has also been studied by the localized vibration method. This molecule is helical and has the two active infrared modes, $v(0)$, $v(\delta)$. In this case, $\delta = 2\pi/3$ for the threefold helix; however the form of Eq. (5.1) is modified to remove the intramolecular hydrogen bonding and introduce intermolecular hydrogen bonding. Again, splitting of the main amide bands is expected, and the strong Raman band at 1654 cm^{-1} is to be compared with the strong infrared band at 1644 cm^{-1} probably corresponding to $v(0)$ and $v(\delta)$, respectively (19).

Other conformations that have been examined by infrared and Raman methods, but perhaps in less detail, are the poly-L-proline and charged coil forms. Table 5.5a contains a summary of Raman and infrared experimental

TABLE 5.5a

Data for Peptide Bands of the Antiparallel β Form and PGII

Conformation band	β (Antiparallel)[a]		pGII	
	IR (cm^{-1})	Raman (cm^{-1})	IR (cm^{-1})	Raman (cm^{-1})
Amide A			3303(s)	3305(w)
	3308(s)	3301(m)		3278(m)
Amide B	3088(m)		3086(m)	3085(m)
Amide I	1690(w) $\nu(0, \pi)$		[A], 1644(s)	[E], 1654(vs)
	$\nu(0, 0)$	1663(vs)		
	1650(w) $\nu_\perp(\pi, \pi)$			
	1623(vs) $\nu_\perp(\pi, 0)$			
Amide II		[A], 1554(s)		[E], 1560(w)
	1517(s)			
Amide III	1236(m)	[E], 1283(m)		1283(m)
	1214(w)		[A] 1249	
Amide IV				673(m)
			573(s)	566(m)
			363(s)	353(vw)
Amide V			751(w)	752(vw)
Amide VI			573(s)	566(m)

[a] The PGI form (19).

data for the various peptide bands, including the localized vibrational assignments for the amide I–III bands. It can be seen that there are clear differences in band position, dichroism, and intensity for the different conformations, which can be particularly useful when vibrational spectroscopy is applied to conformational analysis.

Although the preceding method of analyzing the vibrational spectrum is convenient and straightforward it is certainly an oversimplification and does not allow the *ab initio* calculation of vibrational frequencies. The more complex mathematical methods (14,17) involve matrix algebra based on the symmetry and displacement of atomic elements in the helix in question. This approach has been applied to the polyglycine II helix (19) and to poly-L-alanine (20). Although there is fairly good overall agreement, detailed coincidence with experimental data is lacking and it is sometimes difficult to resolve assignments. Presumably, refinements in the calculations will eventually allow close analysis.

Conformational Analysis

As can be surmised from Table 5.5a, the conformation of regular poly-peptides can be determined relatively easily from infrared data alone. Much of the early work in infrared spectroscopy of polypeptides centered around defining the position of the amide I and II bands. The β form generally gives rise to strong bands at lower frequency than the α conformation. On the other hand, the closeness of the bands between the random and α forms prevents a definite conformational assignment. A major step forward was taken when polarized infrared measurements were made with oriented samples, since the dichroic nature could be determined. The dichroism measurements allow the vector of the vibrational transition moment to be determined and thus many possible ambiguities are removed. It is to be noted, for example, that the major component of the amide I band arises from C=O vibration which occurs parallel to the chain axis for the α helix and perpendicular to the chain axis for β forms. A large number of polyamino acids have been investigated by this method. Table 5.5b contains examples of these data and, on the basis

TABLE 5.5b

Experimental Data for the Principal Amide Bands of
α-Helices (Solid State)

Band	Raman (cm^{-1})	IR	Dichroism
Amide I	1650–5(vs)	1648–58(vs)	$\parallel$
Amide II	vw or absent	1540–60(s)	$\perp$
Amide III	1310–35(s)	1320–30(s)	$\parallel$
Amide V	vw or absent	610–20(m)	$\perp$

of such observations, Blout and co-workers (21) have been able to classify poly-α-amino acids into conformational categories (Table 5.6). It is generally believed that poly-α-L-amino acids with unbranched C_β atoms tend to be helix formers unless there is an active side group available for intermolecular bonding. Substituents on the C_β atom invariably cause the β-sheet conforma-tion to be most stable. These results are in good agreement with the theoretical conformational aspects explored in Chapter 2.

Most infrared dichroism studies are for materials whose x-ray diffraction patterns indicate that the molecular axis and fiber axis coincide (except for cross-β). We have seen in Chapter 4 that truly crystalline polyamino acids generally crystallize with chain and fiber axes mutually perpendicular, so that it can only be assumed that the orientational procedures used for dichroic measurements (brushing, stroking, stretching) change the three-dimensional

TABLE 5.6

Infrared Dichroic Bands of Some Poly-α-Amino Acids and Derivatives[a]

Amino acid or derivative	Amide I (cm^{-1})	Dichroism	Amide II (cm^{-1})	Dichroism	Conformation
Poly-L-alanine	1651	‖	1555,1536		α
Poly-L-valine	1638	⊥	1545	‖	β
Poly-L-leucine	1650	‖	1543	⊥	α
Poly-L-serine[b]	1621		1512		β
Poly-O-acetylserine	1637	⊥	1520	‖	β
Poly-S-methylcysteine	1632	⊥	1540,1525	‖	β
Poly-S-benzylcysteine	1632	⊥	1524	‖	β
Poly-L-methionine	1648	‖	1540	⊥	α
Poly-β-benzylaspartate	1668	‖	1563	⊥	α
Poly-γ-benzylglutamate	1658	‖	1542	⊥	α
Poly-L-lysine HCl[b]	1652		1540		α
Poly-L-lysine-ε-carbobenzoxy[b]	1650		1560		α

[a] Assembled data of Blout and co-workers (text Ref. 21).
[b] P. L. Sutton, M.S. thesis, Case Western Reserve University, Cleveland, Ohio, 1968.

order. Since it is usually quite difficult to produce biopolymers with crystalline morphology, there is generally little danger of misinterpreting the infrared dichroism. However, a clear distinction should be drawn between crystalline, amorphous, and random biopolymers.

Although the laser beam is inherently polarized, dichroic studies of polypeptides by Raman spectroscopy have so far been strictly limited. One exception is the study of poly-L-alanine (20). In this work the detailed group theoretical tensor analysis for the Raman spectra of helical polymers is given and compared with the polarized Raman spectrum. Table 5.7 contains some of the salient features.

It has been suggested that the far infrared region is useful for conformational diagnosis (though some of the following evidence refutes such a contention). In this case, information is obtained from the amide V band. Poly-γ-methyl-L-glutamate, the equivalent benzyl ester, and poly-L-alanine, all in the α form, show a strong band with perpendicular dichroism at 620 or 610 cm^{-1}. Poly-DL-alanine, polyserine, and sodium polyglutamate films, all supposedly in the random state, show bands in the 650–665 cm^{-1} region, and β forms appear to have the strongest amide V band at 700 cm^{-1}. However, α-poly-L-alanine often shows an intense band at 655 cm^{-1}, which may or may not result from a partially random character (22). The position of the three bands in the amide V region has been calculated for poly-L-alanine by Miyazawa and co-workers (23). The bands in the infrared that arise from

TABLE 5.7

Polarized IR and Raman Data for Poly-L-Alanine Bands Characteristic of
α Helix of Polyalanine[a]

Infrared			Raman	
$\Delta\nu$ (cm^{-1})	Dichroism	Assignment	$\Delta\nu$ (cm^{-1})	Assignment[b]
1657(vs)	‖	[A]	1659(vs)	[A]
1537(vs)	⊥	[E]	1543(vw)	[E]
1274(m)	⊥	[E]		
1107(s)	⊥	[E]	1106(vs)	[E]
906(m)	‖	[A]	909(vs)	[E]
893(s)	⊥	[E]	889(w)	[A]
528(s)		[A]	531(vs)	[E]
375(s)		[E]	878(s)	[A]

[a] From Ref. (20).
[b] Determined from line polarizations.

TABLE 5.8

Some Characteristic Amide V Bands

Compound	Conformation	Raman (cm^{-1})	IR (cm^{-1})	Reference[a]
Poly-L-alanine	α (Some random and β?)	698, m	690, w	1
		662, w	655, m	
			615, w	
			610, s	2
			650, m	
Poly-L-leucine	α	702, w?	694, m	3
			657, vw	
			614, m	
PBLG	α	627, s (C—C)	697 (C—C)	3
			614, s (am V)	
PCLL	α	627, s (C—C)	606, m	3
PCLL HCl	α		610, w	3
Hexa-L-alanine	β		690, s	
Poly-L-valine	β		709, s	3
			606, w	

[a] Key to references:
1. J. L. Koenig and P. L. Sutton, Biopolymers **8**, 167 (1969).
2. T. Miyazawa, Y. Masuda, and K. Fukushima, J. Polym. Sci. **62**, 562 (1962).
3. P. L. Sutton, M.S. thesis, Case Western Reserve University, Cleveland, Ohio, 1969.

out-of-plane N—H bending modes are mainly absent in the Raman. Unfortunately, the situation is not always straightforward since, for example, the strongly α-helical polybenzyl-L-glutamate (PBLG) shows a strong IR band at 697 cm^{-1}, which is assigned to a sidechain vibration. Poly-L-leucine, which is also believed to be α-helical, shows medium-intensity bands at 694 and 614 cm^{-1} and a weak band at 657 cm^{-1} in the infrared, but no lines in the Raman (see Table 5.8). Also poly-ε-carbobenzoxy-L-lysine (PCLL) shows only a medium IR band at 606 cm^{-1}, whereas the hydrochloride gives a weak 690 cm^{-1} IR band and a very weak Raman line at 690 cm^{-1}.

These differences may arise from inhomogeneous samples. A similar situation exists with β structures; poly-L-valine gives the expected strong 709 cm^{-1} band (no corresponding Raman line), but poly-L-serine in a form showing a strong Raman line at 1698 cm^{-1} indicative of the antiparallel β-conformation has no defined IR or Raman bands in the 500–800 cm^{-1} region.

The amide V band for the polyglycine II structure occurs in the infrared at 740 cm^{-1}, which is seen to be at much higher frequency, the effect arising from strong intermolecular coupling.

Polyimino Acids

As pointed out in Chapters 2, 4, and 8, poly-L-proline exists in at least two conformational forms as determined by x-ray diffraction, electron diffraction, ORD–CD, and theoretical analysis. Since there is no amide available for intra- or intermolecular hydrogen bonding, the polyimino acids possess unique infrared and Raman characteristics. Although detailed theoretical treatment of the spectra is not available in the literature, considerable attention has been applied to the question of the sensitivity of infrared and Raman spectroscopy in determining spectral differences between the various forms. The infrared spectra of the right-hand 10_3 helix of *cis* poly-L-proline I and the left-hand 3_1 helix of *trans* poly-L-proline II have been reported (24). Although infrared measurements for form II have, as suggested, shown that the band assigned to a mode consisting predominantly of the C=O stretching vibration exhibits perpendicular dichroism in accord with motion perpendicular to the chain axis, unexpected perpendicular dichroism was observed for this mode in form I, where the dichroism should be parallel. In the infrared spectrum of forms I and II (both in the solid state) there are apparently only two bands unique to each conformation, at 1355 and 960 cm^{-1} for form I and at 670 and 400 cm^{-1} for form II. On the other hand, there are a large number of unique lines in the Raman spectra (25). Table 5.9 summarizes these characteristic frequencies. Although assignments have not been made for several of the Raman lines, it is probable that most arise from skeletal vibrations, and thus,

TABLE 5.9

Summary of the Raman Scattering Lines and Infrared Absorption Bands Used to Distinguish the Two Forms of Poly-L-Proline[a]

Raman		Infrared	
Poly-L-proline I (cm^{-1})	Poly-L-proline II (cm^{-1})	Poly-L-proline I (cm^{-1})	Poly-L-proline II (cm^{-1})
		1355	
	1198		
1187	1187		
	1176		
957	1000	960	
781			
	722		670
662			
540	530		
363	400		400

[a] From Ref. (25).

TABLE 5.10

Tentative Assignments of the Raman Lines Common to the Monomer and Both Polymeric Forms of L-Proline

Proline (cm^{-1})	Poly-L-proline I (cm^{-1})	Poly-L-proline II (cm^{-1})	Lit.[a] (cm^{-1})	Tentative assignment
1446	1446	1447	1454[b]	CH_2 Bend
1261	1261	1266	1287[b]	
1237	1237	1241	1221[b]	
		1198		CH_2 Wagging–twisting
	1187	1188		
1172		1176	1161[c]	
1080	1080	1093	1082[b]	CH_2 Rocking
1055	1044	1046	1025[b]	CH_2 Wagging
983	957	1000	980[b]	Ring stretching
914	917	919		Ring breathing
896			902[b]	
840	838	839	838[d]	CH_2 Out-of-plane bend

[a] Values reported for pyrrolidine, 2-pyrrolidone, and pyrrole.
[b] Text Ref. 31.
[c] M. A. Smith, A. G. Walton, and J. L. Koenig, *Biopolymers* **8**, 29 (1969).
[d] A. R. Kalvetsky, *Quart. Rev. Chem. Soc.* **4**, 353 (1959).

at least on this evidence, the Raman technique is somewhat more sensitive to conformational subtleties than is the infrared.

Apart from the peptide bands extensively explored above, the bands assigned to vibrations of the side chains can be sensitive to conformation. It is usual to compare the monomer with the polymer; for poly-L-proline, differences resulting from environmental changes of the pyrrolidine residue are clearly evident (Table 5.10). A third form of poly-L-proline, the "collapsed" or random form produced by precipitation from concentrated salt (calcium chloride) solutions, has also been reported (26) and, again, has a distinctly different Raman spectrum, as shown in Fig. 5.8 (25). The infrared and Raman spectra of poly-L-hydroxyproline (27) differ from those of poly-L-proline mainly in that the band at 1660 cm^{-1} for the latter appears at 1640 cm^{-1} in poly-L-hydroxyproline, presumably due to intermolecular hydrogen bonding from the hydroxyl group to the peptide C=O.

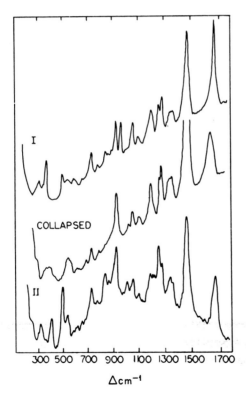

Fig. 5.8. Raman spectra of poly-L-proline form I, collapsed, and form II, each in the solid state. [From Ref. (25), with permission of the American Chemical Society.]

Aqueous Solutions

The natural medium for solution studies of ionic polyamino acids and proteins, i.e., water, is opaque in several regions of the infrared spectrum, particularly in the 1400–1800 cm^{-1} range where the amide I and II bands are observed. The well-known maneuver of using heavy water (D_2O) to circumvent this difficulty has certain disadvantages (exchange with labile hydrogens, opacity in other regions) but has been used with some success. In this respect the Raman technique has the advantage that water is only a very weak scatterer and is ideal for solution work.

A number of polyamino acids have been examined by the Raman method in solution, including poly-L-proline (28), poly-L-hydroxyproline (27), poly-L-lysine (29) and its hydrochloride (30), poly-L-glutamic acid (29) and various oligomers of glycine (17), L-alanine (31), and L-proline (21). Generally there are small, but discernible differences between the Raman spectra in the solid state and in aqueous solution. The peptide bands are usually in a similar position for both states but specfic side-chain interactions are modified. For example, the 1216 cm^{-1} line assigned to the CH_2 twisting mode observed for solid samples of poly-L-lysine and the hydrochloride salt, moves to 1243 cm^{-1} in aqueous solution, the significance of which is not entirely clear. Poly-L-proline II, on the other hand, shows changes in the 826-836 cm^{-1} region associated with the pyrrolidine ring carbon modes, as well as a shift in the 1660 cm^{-1} line to the 1645 cm^{-1} region. This latter effect is apparently caused by a transition from the non-hydrogen-bonded solid state to the aqueous solution in which water molecules are hydrogen bonded to the peptide carbonyl. A similar change is not noted with poly-L-hydroxyproline, which is hydrogen bonded in both solid and aqueous states.

The conclusion drawn from a comparison of the solid and solution spectra of poly-L-proline II is that they have essentially the same conformation in both states. However, both on theoretical grounds (Chapter 2) and from light-scattering measurements (32) it seems likely that poly-L-proline II is more highly contracted in aqueous solution. It is not possible to discern readily the actual conformation of biomolecules in any state by infrared and Raman methods; only changes can be detected. Slight changes such as that for polyproline may be reflected in relative intensities of the Raman bands but such an analysis has yet to be performed.

Variations in line intensity have been used to good effect in the study of proline oligomers (25), and they offer a subsidiary method of conformational analysis. Band intensities are known to be first-order perturbation effects, whereas band frequencies are second-order perturbation effects likely to be less sensitive to minor conformational changes.

Conformational Transitions of Polyamino Acids in Solution

The acidic and basic polyamino acids are soluble in water and undergo conformational transitions induced by changes in pH, which may be followed by Raman spectroscopy. However, most polyamino acids are not water soluble and the majority of conformational transitions, particularly helix–coil transitions, have been followed in mixed nonaqueous solvents by various methods, of which CD/ORD have been the most popular. These solvent mixtures are invariably Raman scatterers, and also often absorb infrared radiation: of the two, the infrared technique is usually preferred. A region of the spectrum that has received particular attention in this respect is the *near* infrared, particularly 7000–5800 cm^{-1}. No bands of any significant intensity other than those due to NH and OH modes occur in this range. The overtone N—H bands, which occur between 5800 and 6250 cm^{-1}, are quite sensitive to the chemical state of the peptide bond (33), and thus studies of this region of the spectrum in conjunction with conformational transformations should provide information concerning the mechanism of the transition (1). Figure 5.9 (33) shows band assignments for three of the commonly observed bands. For the helix–coil transition we might expect that the normal intramolecular hydrogen bonds of the α helix would disrupt as a function of

CHEMICAL STATE OF −NH IN POLYPEPTIDES	POSITION OF −NH OVERTONES (cm^{-1})
PEPTIDE HYDROGEN BOND 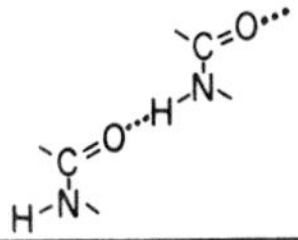	649, 6330
PROTONATED PEPTIDE 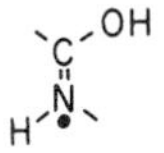	6620
PEPTIDE − ACID HYDROGEN BOND 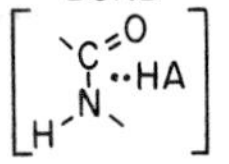	6710

Fig. 5.9. Band assignments for the first overtone of the −NH stretching vibration in polyamino acids in the near infrared. [From Ref. (33), with permission of the American Chemical Society.]

solvent interaction to give protonated or solvated peptide bonds. Accompanying the interaction there is a shift of infrared absorbance from 6490 or 6330 cm^{-1} to 6710 or 6620 cm^{-1}. Such changes are observed in the spectra of poly-L-alanine, poly-L-methionine and poly-γ-benzyl-L-glutamate in dichloroacetic acid/ethylene dichloride (DCA/EDC) mixtures, where the hydrogen-bonded —NH is converted to the protonated form (33). Figure 5.10 (33)

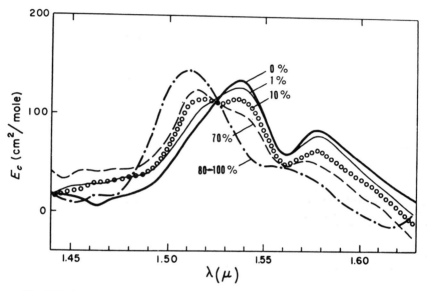

Fig. 5.10. Spectra of poly-γ-benzyl-L-glutamate corrected for benzyl ester side-chain effects in DCA–EDC mixtures. Percentages indicate amount of DCA in solution for each curve. [From Ref. (33), with permission of the American Chemical Society.]

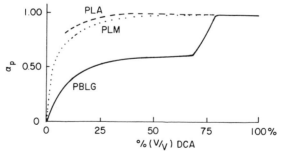

Fig. 5.11. Protonation of the peptide residues of polyamino acids in dichloroacetic acid–ethylene dichloride mixtures. The fraction, α_p, of protonated residues at 25°C is plotted as a function of solvent composition in percent (v/v) acid for poly-γ-benzyl-L-glutamate (PBLG), poly-L-methionine (PLM), and poly-L-alanine (PLA). (From Ref. 1 with permission.)

shows the change in spectrum of PBLG with solvent composition. Quantitative estimates of the protonated and hydrogen-bonded forms may be derived from peak areas, and the fraction of protonated form as a function of solvent composition is shown in Fig. 5.11. It is tempting to identify the protonated form with the random coil, and the hydrogen-bonded form with the α helix. However, estimates of helix content by this method lie substantially below ORD values for the same systems. The situation is further complicated by the fact that neither poly-L-alanine nor poly-L-leucine show evidence of any hydrogen-bonded form in $CHCl_3$–CF_3COOH mixtures.

Sequential Polypeptides and Proteins

Applications of the Raman and infrared methods to polyamino acids and their esters seem to hold some promise for more complicated systems, particularly the Raman technique, since the richness in backbone vibrational modes along with the sensitive response of band intensities seem to afford a discriminating conformational tool. In practice, Raman spectroscopy is seriously hampered by the so-called "fluorescence response" of samples. Even simple polyamino acids often fluoresce under the influence of the exciting laser beam. Although the fluorescence problem may often be overcome by intensive purification or particularly by exposure of the sample for long periods in the beam, this latter method does not lend itself to an investigation of sequential polypeptides or proteins that are thermally unstable or readily denatured. Nevertheless, some progress has been made in this area, particularly with globular proteins.

For the cross-β (antiparallel) polydipeptides (Ala-Gly)$_n$, [Glu(OEt)Gly]$_n$ and polytripeptide [Glu(OEt)Glu(OEt)Gly]$_n$, the Raman spectrum provides the infrared-inactive $v_{\parallel}(0, 0)$ band, as with the homopolymers (34). A more subtle Raman and infrared study has been carried out with poly (Gly-Gly-Ala), which has three stable conformations: cross-β, polyglycine II, and true random. As noted in Chapter 6, the charged coil forms of poly-L-glutamic acid and poly-L-lysine are now believed to have a polyglycine II conformation and, since the charged coil form has been identified traditionally as random, it is evidently of some interest to identify differences between PGII and random forms by infrared spectroscopy. In fact conformational differences between the PGII and true random forms of poly (Gly-Gly-Ala) show up in the infrared spectrum in the amide II and III bands and in the weak $-CH_2$ modes. These latter modes are strong Raman scatterers. A comparison of the three conformations is thus possible by a combination of infrared and Raman data (see Table 5.11).

TABLE 5.11

Some Distinctive IR and Raman Bands observed for specimens of Poly(Gly-Gly-Ala) with different chain conformations

Cross-β, IR (cm^{-1})	PGII IR (cm^{-1})	PGII Raman (cm^{-1})	Random, IR (cm^{-1})	Assignment
3070	3082		3072	Amide B
2929	2935	2936	2931	CH$_2$ Antisymmetric stretch
1697				
1621	1656	1660	1655	Amide I
1515	1550		1535	Amide II
1435	1435	1436	1420	CH$_2$ Twist
1225	1245	1247(w)	1240	Amide III/CH$_2$ bend
			(Broad)	
690	735		730	Amide V
610	670		650	Amide IV

These data suggest that it is feasible to obtain Raman spectra from fibrous proteins, and indeed there is some evidence that fibrous proteins in the native state will yield such data (35). Collagen is the only such material for which Raman data have been reported, and the quality of the spectrum is rather poor.

Infrared spectroscopy of proteins has received considerable attention over the past two decades. The approach to fibrous proteins is much the same as those expounded above for the synthetic polyamino acids. Apart from the position and dichroism of peptide bands, which are useful in identifying prominent conformations and peptide orientation, the amide A bands in the 3300 cm^{-1} region are often used as a quantitative measure of NH $\cdots$ O=C hydrogen bond distances. For example, the normal amide A at 3300 cm^{-1} corresponds to a N $\cdots$ O distance of approximately 2.7 Å, whereas in collagen the band at 3330 cm^{-1} is believed to be indicative of longer hydrogen bonds ($\sim$3.1 Å) formed in the triple-helical conformation (36). The situation with globular proteins is somewhat more complicated since they invariably contain a mixture of conformations as well as bound water molecules. Popular studies with globular proteins include exchange techniques, which can indicate the proportion of amide groups that are buried or readily accessible, and denaturation studies which reveal various conformations. In both cases severe limitations are imposed by the infrared absorbance of water, as previously noted. The application of infrared spectroscopy to proteins has been reviewed extensively (37).

The conformational transitions of bovine serum albumin (BSA) as a function of pH may be followed in D$_2$O or in films cast from aqueous solution.

Figure 5.12 shows some spectra of various proteins. In general it might be said that the infrared spectra are too rich in information in the sense that broad unresolved bands are constituted from local vibrations which subtly reflect molecular structure, but the information cannot generally be delineated.

The first reported Raman spectrum of a globular protein in aqueous solution was that of lysozyme by Garfinkel and Edsall (38) in 1958. The spectrum was obtained using mercury-arc excitation and photographic recording, and showed fourteen faint lines. More recently Raman spectra of crystalline lysozyme, pepsin, and α-chymotrypsin have been obtained (39) using (He–Ne) laser excitation. These spectra also showed rather high background noise and poor quality. Only recently have high-quality spectra been obtained for a few globular proteins (40). Figure 5.13 shows the Raman spectrum of lysozyme in aqueous solution at pH 5.2 and, for comparison, the addition spectrum of the constituent amino acids. The assignment of many of the bands is straightforward in terms of the characteristic side-chain modes. Apart from the amide bands, the protein spectrum differs at several important points from the component amino acid spectrum including the 500–725 cm^{-1} region, which contains frequencies assigned to C—S and S—S vibrations. It

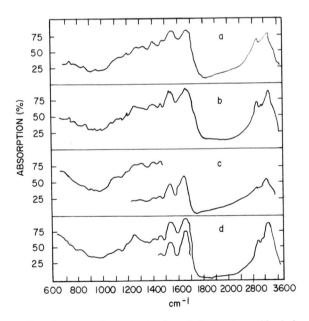

Fig. 5.12. Infrared spectra of proteins having a folded polypeptide chain configuration: (a) human serum albumin, (b) zinc insulin, (c) elephant hair, (d) porcupine quill. [From M. Beer, G. B. B. M. Sutherland, K. N. Tanner, and D. L. Woods, *Proc. Roy. Soc. London* **A249**, 147 (1959), with permission.]

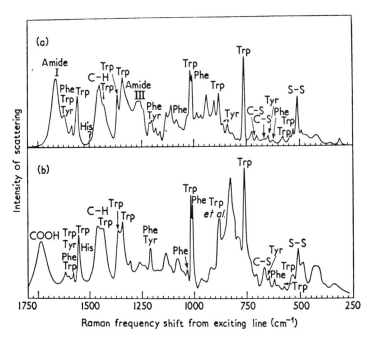

Fig. 5. 13. (a) Raman spectrum of lysozyme in water, pH 5.2. (b) Superposition of Raman spectra. Sum of the constituent amino acids, pH 1.0. [From Ref. 40 with permission.]

has been suggested that the relative intensities in this region reflect the configuration of the C—S—S group. If so, it may be that future work with Raman spectroscopy will show that specific molecular geometry of side-chain and cross-linking groups can be detected from the line intensities.

The two major amide bands (I and III) indicate that the peptide backbone is probably not strongly hydrated, and it appears that the component intensities of the amide III band correspond to different conformations; it is thus possible to calculate the relative proportions of these conformations in the protein. Evidence of the ionization state of carboxyl groups was also extracted from the lysozyme spectrum.

Similar analyses of ribonuclease and α-chymotrypsin (40) have been presented. For the former, the C—S—S angles are found to be smaller than for lysozyme on the basis of relative band intensities. Another interesting Raman study has been concerned with the denaturation of chymotrypsinogen in aqueous solution (41), which is reversible providing that the solution is not taken above 65°C. A detailed analysis of these spectra should throw considerable light on the unfolding and denaturation mechanisms.

Spectroscopy of Polynucleotides

The vibrational spectroscopy of nucleic acids and polynucleotides is an order of magnitude more complicated than that of polypeptides. Apart from vibrations of the various entities in the nucleotide, various conformational inhomogeneities and ionization states must be delineated. So far no success has been achieved in a fundamental theoretical approach to prediction of the spectra, but a number of published reports concerned with qualitative aspects of the infrared and Raman spectra have now enabled certain fundamental assignments to be made. However, a major part of the difficulties of spectral analysis arises from the complex vibrational transitions. The logical approach to a spectroscopic study of the nucleic acids is to examine separately the base, sugar, and phosphate moieties, their combination in nucleosides and nucleotides, polymerization into homopolynucleotides, and comparison with nucleic acids. The approximate vibrational frequency ranges for the components of nucleic acids are shown in Table 5.12. A review of the status of infrared studies of polynucleotides to 1964 has been published (42), but the work on synthetic homopolymers and Raman studies have developed predominantly since that time.

TABLE 5.12

Approximate Frequency Ranges for Characteristic Vibrations of Nucleic Acids and Related Compounds (Infrared)

Frequency (cm^{-1})	Mode[a]	Origin of Vibration
2800–3500	νOH	Water, sugar
	νNH	Base residues
	νCH	Sugar and base residues
1800–1500	$\nu C{=}O$	
	$\nu C{=}N$	Base residues, mixed bodies
	$\nu C{=}C$	
	δNH	
	δHOH	Water
1250–950	$\nu PO_2{}^-$	Antisymmetric stretch
	$\nu PO_2{}^-$	Symmetric stretch
	νCO	Sugar
1000–700	νPO	Phosphate
	νCO	Sugar
	τNH	Base residues
600–300		Skeletal deformations, all groups

[a] Modes: ν, stretching; δ, in-plane bending; τ, out-of-plane bending.

Bases and Nucleosides

The fundamental modes associated with base residues occur throughout the infrared and far-infrared spectrum and thus overlap bands from other entities in nucleic acids. However, between 1500 and 1750 cm^{-1} several strong bands occur which are practically free from interference by phosphate and sugar groups, and much attention has been focused on these bands in relation to the structure of the purine and pyrimidine rings in nucleosides, nucleotides, and nucleic acids. These bands arise from strongly coupled vibrations of the C=O, C=N, C=C types with in-plane NH deformation. The Raman and infrared spectra often show remarkable similarity in this region. Results for cytidine crystallized from neutral solution (42,43) are shown in Table 5.13. Band positions of the bases are modified to some

TABLE 5.13

Vibrational Spectra of Cytidine in 1500–1700 cm^{-1} Region

Infrared (cm^{-1})		Raman (cm^{-1})		
Neutral	Acidic	Neutral	Acidic	Assignment[a]
	1725		1730	C=O Stretch
1670	1680	1665	1684	NH Bend?
1647		1635⎱		
1603		1600⎰	1652	C=O and N=C Stretch
1529	1535	1532	1545	
1500		1498		

[a] Based on IR assignments (37). A somewhat different set of assignments has been made on the basis of Raman measurements (43).

extent both by the addition of ribose or deoxyribose (at position 3 in the pyrimidines and 9 in the purines, and by the state of protonation.*) The pK of cytidine is 4.2, and from more acidic solution a protonated form is produced in which the cluster of five IR bands in Table 5.13 changes drastically to give three bands: 1725 cm^{-1} (C=O stretch), 1680 cm^{-1} (NH bend), and 1535 cm^{-1} (unassigned). Similar changes are observed in the other bases. On the basis of deuteration and protonation studies, predominately of the nucleosides (base–sugar), the following has been concluded (42):

1. The acidic form of adenosine has a proton attached to one of the ring nitrogens (44), not to —NH$_2$.

* The ring numbering system for bases is shown in Fig. 1.10.

2. The deprotonated form (ionic) of uridine produces an electron delocalization lowering all double-bond stretching frequencies (44).

3. Guanosine exists in acidic, neutral, and basic forms with two possible neutral forms (44). Each form has a different IR spectrum. The basic form is believed to be deprotonated in position 1, and the acidic form to be protonated in position 7.

4. Deprotonation of inosine also causes π-electron delocalization (42) with concomitant decrease in C=O stretching frequency from 1675 to 1595 cm^{-1}.

The attachment of the ribose or deoxyribose component produces spectral changes that can vary in extent. For example the differences between the Raman spectra of cytidine and its derivatives are centered around two of the five bands (1532 and 1600 cm^{-1}), whereas the weak adenine band at 1530 cm^{-1} disappears, and a new band appears at 1575 cm^{-1} for both nucleosides (see Table 5.14.).

TABLE 5.14

Raman Spectra of Bases and Nucleosides in 1500–1700 cm^{-1} Region^a

Cytosine	Cytidine	Deoxycytidine	Adenine	Adenosine	Deoxyadenosine
1500	1498				
1540	1532	1524	1530		
1575	1600	1592		1575	1570
			1600	1600	1613
1635	1635	1631			
1665	1665	1670			

^{a} All values in reciprocal centimeters.

Phosphate Derivatives

The phosphate groups linking the sugars in the polynucleotide backbone (phosphodiester linkage) can exist in two different states, protonated and charged:

$$\text{RO}-\overset{\overset{\textstyle OR'}{|}}{\underset{\underset{\textstyle O}{\|}}{P}}-\text{OH} \;\rightleftharpoons\; \left[\text{RO}-\overset{\overset{\textstyle OR'}{|}}{\underset{\underset{\textstyle O}{|}}{P}}\raisebox{0.3ex}{$\rightharpoonup$}\text{O}\right]^{-} + \text{H}^{+}$$

In the monomeric nucleotides, where the phosphate group occupies a terminal position, a further deprotonated RPO_3^{2-} state is feasible. The phosphate ions have been studied in some detail in small molecules and the spectral features associated with them are fairly well understood (45). In the infrared spectrum the P—O modes are considerably stronger than those for

TABLE 5.15

Infrared Phosphate Bands (cm^{-1})

Solid	Solution	Assignment	
1230		PO_2^-	Antisymmetric stretch
1100		PO_5^{2-}	Degenerate stretch
1080		PO_2^-	Symmetric stretch
980		PO_3^{2-}	Symmetric stretch
	530–570	PO_3^{2-}	Symmetric stretch
	450–470	PO_3^{2-}	Degenerate stretch
	460–470	PO_2^-	Deformation stretch

the P—O—R linkage (expected frequencies are listed in Table 5.15). The antisymmetric modes are generally the stronger in the infrared, and the symmetric are stronger in the Raman spectrum. In a Raman study of the adenosine phosphates (46) it has been shown that lines at 980 and 1082 cm^{-1} are good indications of the degree of ionization. These are the symmetric PO_3^{2-} and PO_2^- modes, respectively, and the relative intensity of the Raman bands seems to reflect, fairly accurately, the relative concentrations. Figure 5.14 shows the change in intensity of these two Raman bands with pH.

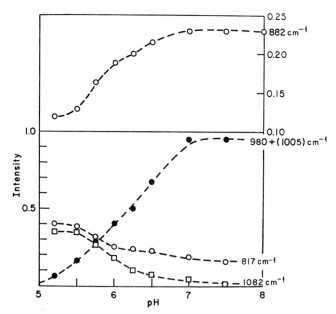

Fig. 5.14. Relative Raman line intensities in arbitrary units for the transitions at 817, 882, 980, and 1082 cm^{-1} in AMP around the pK_2 for terminal phosphate proton ionization. [Taken from Ref. 46, with permission.]

Sugar Moiety

The contributions of saccharide components to polynucleotide spectra are not well understood. In the infrared spectra, the most characteristic group of bands occurs in a broad distribution in the 1000–1100 cm^{-1} region and arises from C—O—C and C—C(OH)—C modes. These modes are, however, rather weak Raman scatterers and it is likely that several of the Raman bands in the 100–500 cm^{-1} arise from skeletal vibrations in the sugar ring. They have not yet been identified as such, and indeed the Raman spectra of polynucleotides do not seem to be greatly influenced by the sugar moiety. In the ribose poly-nucleotides there is generally a band in the 1120 cm^{-1} region which is absent in the corresponding deoxy compounds. Alternatively, the deoxy derivatives often exhibit a sharp band at 975 cm^{-1} which is absent in the ribonucleates. There is therefore some indication that DNA's and RNA's may be distin-guished in this manner.

Polynucleotides

From the above considerations, it appears that the 1500–1750 cm^{-1} region of the polynucleotide spectrum would provide the most readily accessible information since the sugar and phosphate bands might overlap the 1000–1100 cm^{-1} region. Indeed most attention in the infrared spectroscopy of homo-polynucleotides has centered on this region. The infrared spectra of poly-cytidylic acid (poly C) and cytidine are very similar in this region, implying a similar chemical structure. The Raman lines for solid poly C occur at 1524, 1606, and 1645 cm^{-1} compared with 1532, 1600, 1635, and 1665 cm^{-1} for cytidine, showing subtle but real differences that are as yet uninterpreted. Actually the 1500–1750 cm^{-1} lines are by no means the strongest originating from the base in the Raman effect. Figure 5.15 shows the Raman spectrum of the potassium salt of poly C in the solid state. From various investigations of the derivatives of cytidine it is possible to infer that the prominent lines at 780, 987, 1243, 1287, and 1364 cm^{-1} originate in vibrations of the ring system. It seems quite likely that the 1097 cm^{-1} line originates in the PO$_2$ symmetrical vibration, as noted previously. It is of some interest to note changes in this spectrum as the material is taken into aqueous solution, particularly at low pH, where it forms a double helix hydrogen bonded through the bases. (For the base-pairing scheme see Fig. 12.13.)

The solution spectrum of poly C at neutral pH is shown in Fig. 5.16. Comparison with Fig. 5.15 shows considerable similarity, perhaps the major change being the strongly decreased intensity of the 987 cm^{-1} line, possibly due to hydrogen bonding of the base to the aqueous environment. At low

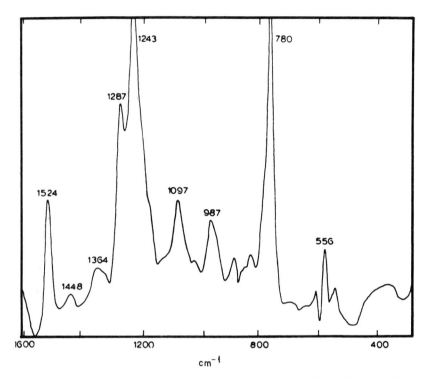

Fig. 5.15. Raman spectrum of the potassium salt of polycytidylic acid in the solid state. [From Ref. 43, with permission of the American Chemical Society.]

pH (<3.3) poly C in solution, and precipitated from solution, shows a considerable decrease in the number of definable bands, as shown in Table 5.16. The main changes appear to involve a coalescence of the 1243 and 1287 cm^{-1} bands to form one band at 1257 cm^{-1}, the shift from 1524 to 1545 cm^{-1}, and the appearance of a new line for the solution spectrum at 1477 cm^{-1}. Whereas these changes have not been analyzed in detail they would seem to be consistent with the formation of a hydrogen-bonded, base-paired, ordered conformation arsing from the double-helical arrangement.

Polyadenylic acid also may be converted from a single strand above pH 6 to a double-stranded form in aqueous solution at low pH. Infrared spectra of the random form in D_2O are similar to adenosine spectra. At low pH (pD) the spectral indications are that partial protonation of the adenine (probably at position 1) occurs, initiating double-helix formation.

A comparison of the infrared and Raman spectra for poly A shows a particularly interesting difference in the 1000–1100 cm^{-1} region. In the infrared spectrum a strong broad band is observed in this region which is

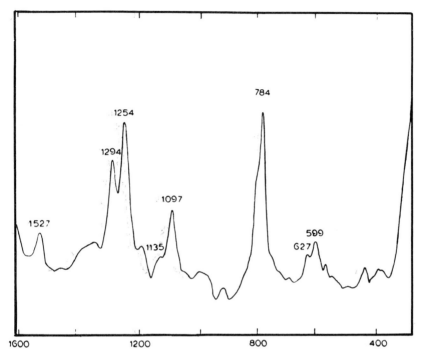

Fig. 5.16. Raman spectrum of aqueous polycytidylic acid, potassium salt, pH 7.0 (20% by weight). [From Ref. 43, with permission of the American Chemical Society.]

TABLE 5.16

Raman Bands (cm⁻¹) of Poly C in Solid and Solution
(Strong Bands Arising from the Base)

Solid K⁺	H⁺	Neutral soln	Acid soln
1645 (w)	1659 (m)		1659 (s)
1606 (w)		1613 (m)	
1524 (m)	1538 (m)	1527 (m)	1545 (w)
1364 (w)		1368 (vw)	
1287 (s)		1294 (m)	
1243 (vs)	1257 (vs)	1254 (vs)	1257 (vs)
987 (m)		991 (w)	
780 (vs)	787 (m)	787 (vs)	791 (s)

thought to be due mainly to the sugar residue, but it probably includes other bands. In the Raman spectrum the most intense band is a medium band at 1093 cm^{-1} assigned to the symmetrical phosphate vibration.

As can be seen from Table 5.17, the Raman scattering bands originating in the adenine base are essentially the same for solid salt and for aqueous solution, but they change considerably for the double helix, once again reflecting the hydrogen-bonded and stacked base pairs. An analysis of the bands

TABLE 5.17

Raman Spectra (cm^{-1}) of Poly A in Various Forms
(Bands Originating in the Adenine Base)

Solid salt	Neutral soln	Solid double helix
1578 (m)	1570 (m)	1595 (vw)
1480 (w)	1473 (m)	1552 (w)
1371 (m)	1371 (m)	1408 (m)
1335 (vs)	1335 (vs)	1360 (m)
1309 (s)	1302 (s)	
1254 (m)	1246 (m)	1254 (m)
725 (vs)	721 (s)	

originating in the ribose is more difficult because of their low relative intensity. A listing of tentatively assigned bands in Table 5.18 seems to indicate that there are few, if any, significant changes in position or relative intensity of bands with the environment or conformation of the polynucleotide.

The relative independence of the 1473 cm^{-1} sugar band from environmental or conformational effects has made it useful as an internal standard for comparison with the Raman scattering intensity from the base (47). Figure 5.17 shows the effect of temperature on the relative intensity of the 1473

TABLE 5.18

Raman Bands (cm^{-1}) of Poly A Originating in the
Ribose Moiety

Solid Salt	Neutral soln	Solid double helix
1506 (w)	1502 (m)	1502 (w)
1415 (w)	1419 (w)	
1213 (w)		1231 (vw?)
		1025 (m)
1006 (w)	1044 (vw)	907 (w)
635 (w)	635 (vw)	635 (vw)
		599 (vw)

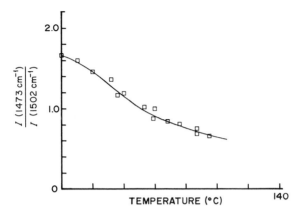

Fig. 5.17. Ratio of the intensities (I) of the 1502 and 1473 cm^{-1} lines plotted against temperature, for Poly A in neutral solution. [From Ref. 47, with permission of the American Chemical Society.]

and 1502 cm^{-1} bands for poly A in neutral solution. As the temperature is increased, the intensity of the 1502 cm^{-1} line increases and probably corresponds to an unstacking of the adenine bases (48).

The infrared spectra of the polyribonucleotides polyguanylic acid (49), polyuridylic acid (50–52) and polyinosinic acid (53,54) have been studied fairly extensively, with the conclusion that the spectra closely resemble those of the respective bases or nucleosides. No Raman data are available as yet.

Mixed Polynucleotides

It is well known that the polyribonucleotides not only undergo self-association to form double and sometimes triple helices, but also undergo the

TABLE 5.19

Infrared Spectra (cm^{-1}) of Combined Polynucleotides

Polynucleotide	I	II	III	IV	V
Poly A				1628	
Poly U	1692		1657		
Poly (A + U)	1691	1672		1631	
Poly (A + 2U)	1696	1677	1657		
Poly I		1677			
Poly C			1653		1617
Poly (I + C)	1696		1648	1630 (sh)^a	

^{a} sh = shoulder.

TABLE 5.20a[a]

Principal Raman Lines (cm^{-1}) of Poly A and Poly U Complexes

Poly A (soln, pH 7)	Poly 2A (soln, pH < 4)	Poly U (soln, pH 7)	Poly (A + U)		Poly (A + 2U)		Comments
			Soln	Solid	Soln	Solid	
		1681 (vs)	1688 (s)	1663 (s)	1674 (vs)	1669 (s)	Uracil C=O
		1630 (vs)	1620 (m)	1622 (vs)	1635 (vs)	1627 (vs)	Uracil double bond
1570 (m)							Adenine ring vibration
1502 (m)							Adenine ring vibration
1473 (m)		1397					Uracil single bond
1371 (m)	1360 (m)		1338 (vs)				Adenine single bond
1335 (vs)	1324 (vs)						Adenine single bond
1302 (s)				1302 (m)			Adenine $C-NH_2$
1246 (m)	1025 (m)	1231 (vs)	1234 (s)		1257 (vs)		Antisymetric $O-P-O^{-2}$
		784 (s)	784 (m)		784 (m)		Antisymetric $O-P-O^{-}$
811 (m)	815 (s)						Base ring
721 (s)	721 (s)						Base ring

[a] Unpublished data of N. N. Aylward and J. L. Koenig.

same association with complementary polynucleotides (see Chapter 12). Thus poly A and poly U form poly (A + U) and poly (A + 2U), and poly I and poly C form poly (I + C). Similarly poly G and poly C form poly (G + C) (their principal Raman lines are shown in Table 20b). The infrared absorption spectra of these complexes in the 1500–1750 cm^{-1} region reflect base–base interactions which are not found in the one-component systems (50–55). A tabulation of these bands is made in Tables 5.19 and 5.20. Studies of the Raman spectra of mixed polynucleotides have also been carried out; data for the same spectral region are reported in Table 5.20.

TABLE 5.20b[a]

Principal Raman Lines (cm^{-1}) of Poly G and Poly C and Their Complex

| Poly G (soln, pH 7) | Poly C (soln, pH 7) | Poly (G + C) | | Comments |
		Soln	Solid	
		1688 (s)		
	1657 (w)	1649 (vs)	1652	C=O Cytosine
		1635 (vs)		
1578 (s)				Guanine ring
1480 (vs)				
1364 (s)				
1331 (s)		1338 (w)		C—NH$_2$ Stretch
	1294 (m)	1298 (m)		Single-bond vibration
1257	1254 (vs)	1254 (m)	1261	Single-bond vibration
1093 (m)	1101 (m)	1101 (w)		O—P—O$^-$ Symmetric
	311 (s)			
	787 (vs)			
682 (m)				
	496 (m)			

[a] Unpublished data of N. N. Aylward and J. L. Koenig.

Nucleic Acids

As expected, the infrared spectra of DNA and RNA samples show features in common with their simpler analogs. Figure 5.18 shows calf thymus DNA and rat liver RNA spectra (42) under conditions of high and low humidity. In the 1500–1750 cm^{-1} region there are broad, poorly resolved bands corresponding to the now-familiar complex ring modes. The strong phosphate (antisymmetric) mode is in the 1240 cm^{-1} region, and the generally broad and complex band in the 1000–1100 cm^{-1} region contains the sugar vibrations. The strong band in the 3400 cm^{-1} region, which contains the NH stretching

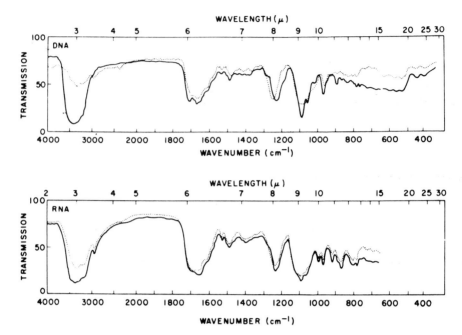

Fig. 5.18. Infrared spectra of films of sodium deoxyribonucleate from calf thymus and sodium ribonucleate from rat liver observed at 75% relative humidity (——) and at 0% relative band intensity (......). (From Ref. 42 with permission.)

mode, is greatly influenced by the presence of water. In fact one of the major difficulties of studying even solid-state polynucleotides is that water almost invariably crystallizes with the nucleotides. On the other hand, water molecules do not interfere with the Raman spectra, in which the bands are generally quite sharp. Although fluorescent impurities have hampered advances in the Raman study of proteins and synthetic protein models, the difficulties have not, so far, been so severe with synthetic polynucleotides, and it seems likely that Raman spectra of native materials will prove most helpful in future work.

Polysaccharides

Infrared spectroscopy and, more recently, Raman spectroscopy has added considerably to our knowledge of the solid-state structure of polysaccharides. Extensive study has been made of the polymorphic forms of cellulose and chitin and the other common homopolysaccharides. Although many permutations of monomers and glycosidic linkages are found in natural polysaccharides, most of the abundant polysaccharides are homopolymers of a few monomers, such as D-glucose, D-xylose, D-mannose, etc., or their derivatives. For example, cellulose and amylose are both polymers of 1,4-linked D-glucose. As will be seen below, the spectra of these two polysaccharides are extremely similar, and the small differences are due to the different anomeric configurations (α or β), chain conformations (2_1 or 6_1 helices), and chain packing. In some cases, additional complexities such as branching and chain folding have to be considered. Before describing the infrared and Raman work on the individual polysaccharides, it is valuable to look at studies of D-glucose and the band assignments common to many of the more abundant polymers.

D-Glucose

The Raman spectrum of crystalline α-D-glucose is shown in Fig. 5.19 (56), and the observed band frequencies and intensities for both the infrared and Raman spectra (56) are listed in Tables 5.21a and b. For the $C_6H_{12}O_6$ molecule we predict $(3N-6=)66$ fundamental modes. α-D-Glucose has four molecules in the unit cell and additional bands due to coupling may be observed, although the interactions should be negligible in most cases. Some overtone and combination bands may also occur, but these are expected to be very weak.

For the α-D-glucose molecule we would expect to observe five O—H stretching bands. The O—H stretching region for many sugars is poorly resolved as a result of absorption of water by the specimen. Six O—H stretching bands have been observed for α-D-glucose, indicating that some type of coupling must take place between these motions in the crystalline

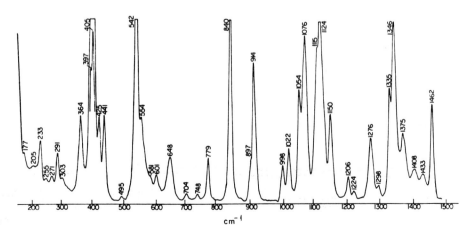

Fig. 5.19. Raman spectrum of crystalline α-D-glucose in the 1500–150 cm^{-1} region. [Taken from Ref. (56), with permission.]

TABLE 5.21a

Observed Frequencies (cm^{-1}) in the O—H and C—H Stretching Region of the Infrared Spectrum of α-D-Glucose[a]

O—H Stretching	C—H Stretching
3405	2964
3390	2946
3350	2913
3320	2893
3255	2878
3200	2850

[a] Data from A. J. Michell, *Aust. J. Chem.* **21**, 1259 (1968),

TABLE 5.21b

Vibrational Assignments for α-D-Glucose below 2000 cm^{-1}[a]

α-D-Glucose		D-Glucose solution	Deuteration assignment	Band assignment
Raman (cm^{-1})	IR (cm^{-1})	Raman (cm^{-1})		
1462	1457	1461	CH$_2$	CH$_2$
	1442			
1433	1427			
1408	1402	1405	C-2—H	C—H Bending with some O—H
1375	1378	1373		bending contribution
	1369			
	1360		C-1—H	
1346		1349	C—O—H	C—O—H Bending
1335	1337	1335	CH$_2$	CH$_2$
	1328	1328		
1298	1293	1298		
1272	1270	1278	C-6—O—H or	C-6—O—H and/or
			C-1—O—H	C-1—O—H related vibration
	1250		C-1—H	C—H Deformation
1224	1219	1222	CH$_2$	CH$_2$
1206	1197	1206		
	1189			C—O and
1153		1152		C—C Stretchings with
	1142			
1124		1130		C—H, C—O—H deformations
1115	1116			
	1164			

[a] Taken from Ref. 56, with permission.

TABLE 5.21b—continued

α-D-Glucose		D-Glucose solution		
Raman (cm⁻¹)	I.R. (cm⁻¹)	Raman (cm⁻¹)	Deuteration Assignment	Band Assignment
1076	1076	1071	C-1—H and C—O—H	C-1—H and C—O—H Deformation
1054	1047	1041	C-1—H	C-1—H Deformation
1022	1026	1020	C—O—H	C—O—H Deformation
998	1011		CH_2	CH_2
	988			
914	911	913	C-1—H and C—O—H	C-1—H and C—O—H deformation
897	890	898		C-H Vibration for β-form
		859		
840	836	847	C-1—H	C-1—H Deformation
779	768	771		
748		747		
	721			C—C, C—O Stretchings
704		705		
648	645	633		
	622			
601	603			
581				
554	555			
542		541		
	522	514		
495		498		Mainly skeletal modes
441		443		
425		423		
405		409		
397				
		381		
364				
		341		
303				Torsional modes
294		291		
271				
255				
233				
205				
177				

TABLE 5.21c[a]

Calculated and Observed Frequencies and Potential Energy Distribution for α-D-*Glucose*

Frequency observed (cm^{-1})		Frequency calculated (cm^{-1})	Approximate potential energy distribution (%)
IR	Raman		
		3397	OH (99)
		3397	OH (93)
	3550	3397	OH (97)
	broad	3397	OH (98)
		3397	OH (94)
	2961	2985	CH (96)
	2947	2944	CH (94)
		2939	CH (90)
		2937	CH (98)
	2911	2933	CH (97)
	2891	2929	CH (89)
	2877		
	2850	2883	CH (95)
1460	1462	1469	OCH (50), CH$_2$ (43)
1442			
1427	1433	1434	OCH (85), CCH (14)
1405	1408	1375	CCH (28), OCH (27), COH (24)
1378	1375	1360	CCH (50), OCH (29), COH (13)
		1358	CCH (50), OCH (25), COH (13)
		1356	CCH (44), OCH (18), COH (17)
	1346	1347	CCH (48), OCH (24), COH (14)
1337	1335	1337	CCH (59), CO (15), OCH (12)
1328		1335	CCH (47), COH (25)
		1320	CCH (59), OCH (37)
1293	1298	1295	OCH (61), CCH (37)
		1290	COH (43), CCH (30)
		1284	OCH (30), COH (29), CCH (22)
	1272	1256	OCH (59), CCH (36)
	1224	1226	COH (44), OCH (24), CCH (18)
1219		1216	CCH (28), OCH (21), COH (21), CO (20)
1197	1206	1201	CO (39), CCH (19), COH (16)
		1194	OCH (33), COH (26), CCH (18)
		1185	CO (29), COH (24), CCH (22), OCH (20)
1152	1153	1149	CO (50), CC (27), COH (18)
		1123	CO (50), CCH (14), COH (13)
	1124	1120	CO (47), CC (47)
	1115		
1104		1108	CC (45), CO (34), COH (17)
	1022	1083	CO (55), COH (24), CC (17)
1076	1076	1075	CO (57), CC (25), COH (23)

[a] Taken from Ref. (56), with permission

TABLE 5.21c—continued

Frequency observed (cm⁻¹)		Frequency calculated (cm⁻¹)	Approximate potential energy distribution (%)
IR	Raman		
1047	1054	1051	CO (79), CC (19)
1026			
1011		1014	CC (45), OCH (15), CCH (13), CCO (13)
998	998	996	CC (43), CO (30), CCH (15)
937		946	CC (42), CO (39), CCH (19)
911	914		
890	897	902	CC (35), CO (25), CCO (20)
836	840	863	CO (43), CCH (26), CC (18)
768	779		
	748	736	CO (16), CCO (15), CC (14), CCH (14)
721			
	704		
645	648	625	CCO (57), CCH (20)
622		616	CCO (41), CCH (20), CO (11), CC (11)
603	601		
	581		
555	554	537	CCO (32), CC (18), CCH (16), CO (14)
	542	531	OCC (35), OCO (18), CO (17)
	495		
	441	443	OCC (38), CCC (15)
	425	428	CCO (34), CCC (30), CCH (26)
	405		
	397	390	CCO (46), OCO (16), CCC (11)
	364	359	CCO (41), CCH (16), CC (16)
		316	CCO (59), CC (13)
	303	307	CCO (59), CC (11)
	291	294	CCO (63), CCH (17)
	271		
	255	252	CO (τ)[b] (49), CCO (20), CCH (16)
		249	CO (τ) (59), CCO (14)
	233	235	CO (τ) (92)
		229	CO (τ) (85)
		221	CO (τ) (86)
	219		CO (τ) (55), CCO (24)
	205	200	CCO (35), CO (τ) (31)
	177		
		127	CCC (25), COC (21), CC (τ) (10)

[b] τ indicates torsional mode.

form. Similarly, seven bands due to C—H (five) and CH$_2$ (two) stretching are predicted, and this number is indeed observed. In fact the O—H stretching region of the spectra is frequently where the greatest differences among the various carbohydrates are seen. A non-hydrogen-bonded O—H group would be expected to give rise to a band at 3650 cm^{-1}. All the O—H bands occur in the range 3200–3500 indicating that each one is involved in hydrogen bonding. The actual lowering of the frequency due to hydrogen bonding depends on the O $\cdots$ H—O bond length and the O $\cdots$ Ĥ—O bond angle. Prediction of the frequency shift for a particular hydrogen bond is a problem akin to prediction of the potential energy of such a structure. Correlation of frequency shifts with bond lengths have been made by Nakamoto *et al.* (57), and the predictions based on this work have sometimes been useful in assignment of bands to individual O—H groups in polyhydroxy structures. Nevertheless, more than for any other bands, the frequencies of the O—H stretching bands depend on the crystal structure and hydrogen-bonding network. As will be seen below, one of the simplest and most reliable methods of distinguishing between the polymorphic forms of polysaccharides is by the characteristic O—H stretching bands. However, it is one thing to identify bands resulting from structural or conformational differences. It is quite another to determine which aspects of the structure give rise to the observed effects of the spectrum. Nevertheless, for polysaccharide structures the O—H and C—H stretching regions contain much useful information when additional dichroic data are available for oriented speciments.

The lower regions of the spectra, below 2000 cm^{-1}, contain the O—C—H, C—C—H, and C—O—H deformation and the C—C and C—O stretching modes, bands due to skeletal vibrations, etc. The region between 1500 and 700 cm^{-1} for α-D-glucose and for the oligomers and polymers of D-glucose are very similar, both in the frequencies and in the intensities of the observed bands. Identification of the CH$_2$ and C—O—H deformation frequencies for D-glucose means that the same assignments can probably be made for cellulose, amylose, etc. Some of these bands have been identified by deuteration methods (58). The C—O—H bands were identified by examination of the spectra of D-glucose, cellobiose, maltose, and dextran in separate solutions in both H$_2$O and D$_2$O (58). These solutions were examined using the Raman technique, which is ideal for aqueous-solution work, as discussed above. The solid-state and solution Raman spectra of α-D-glucose are almost identical, and the few differences are probably due mainly to the fact that the solution contains a mixture of α and β anomers (see below). Raman lines at 1349, 1071, 1020, and 913 cm^{-1} (figures for D-glucose) are absent from the spectra of all four carbohydrates in D$_2$O and these lines are assigned to C—O—H related modes. Some of the C—H modes have been identified (58) by comparison of the infrared spectra of α-D-glucose and three different C-deuterated-D-glucoses. Comparison for the -6, $6-d_2$ derivative, in which the CH_2 group is replaced by CD_2, allows assignment of the lines at 1457, 1337, 1219, and 1011 cm^{-1} to CH_2 deformation modes. Some of the bands due to C-1—H and C-2—H modes were identified by comparison of the spectra of α-D-glucose and derivatives with these particular carbons deuterated; the resulting assignments are listed in Table 5.21b. It must be noted that the above bands should be listed as C—O—H, C—C—H, CH_2, etc., *related modes*. For a complex molecule such as glucose, many of the bands in this region probably arise from complex molecular modes involving vibrations of a number of groups and thus cannot be assigned to simple individual motions.

The theoretical vibrations of the α-D-glucose molecule have been studied by normal coordinate analysis (56). The atomic coordinates were those determined by neutron diffraction, which gives accurate positioning of the hydrogens as well as the carbon and oxygen atoms. The force constants were taken from previous studies of paraffins, aliphatic ethers, alcohols, and carboxylic acids. The computed vibrational frequencies and potential energy distribution are listed in Table 5.21c. There is fairly good agreement (in most cases, within 10 cm^{-1}) between the observed and calculated frequencies, and the match is considered very satisfactory for such a large molecule. The potential energy distribution shows that most of the modes are highly coupled. It must be remembered that for any mode containing a significant contribution from, e.g., C—O—H bending, replacement of the hydrogen by a deuterium will change the coupling and cause that band to be absent from the spectrum of the deuterated compound. Almost all the assignments are in accord with the potential energy distribution. However, the analysis predicts rather more C—O—H related modes than are indicated by the small effects observed in deuteration. Further refinement of the model, e.g., by variation of the force constants, will be necessary to improve the agreement between the frequencies and potential energy distribution.

Study of the infrared spectra of a number of sugar monomers and oligomers led Barker *et al.* (59) to propose that pyranose ring structures with an axial C-1—O-1 bond as in α-D-glucose possess a typical band at ~845 cm^{-1}. Structures with an equatorial C-1—O-1 bond as in β-D-glucose have a band at approximately 893 cm^{-1}. For polysaccharide structures this effect is generally observed and is indicative of the type of linkage, although for the α-1,4-linked polymers of D-glucose, i.e., amylose, amylopectin, and glycogen, both bands are observed (56). The deuteration studies have shown that the band at 845 cm^{-1} in the α-D-glucose spectrum declines on deuteration of the OH groups and of C-1—H, and this band is probably the result of a complex mode dependent on the anomeric configuration (58).

Cellulose

The infrared spectra of cellulose were studied by a number of groups in the 1950's and early 1960's, when it was recognized that the different polymorphic forms could be differentiated quite easily by the bands in the O—H and C—H stretching regions. The polarized infrared spectra in the range 4000–2500 cm^{-1} for native and mercerized cellulose are shown in Fig. 5.20a (60) and b (61), and the band frequencies, intensities, and dichroisms are listed in Table 5.22. These data are taken from the work of Liang and Marchessault (60–62). It can be seen that cellulose II gives two strong O—H bands at

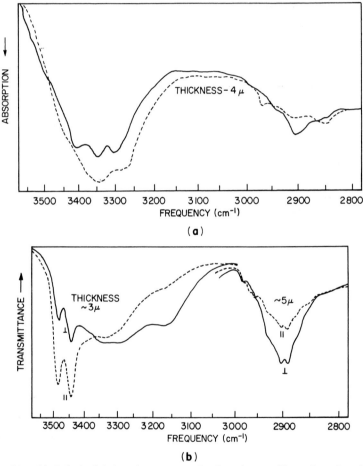

Fig. 5.20. (a) Polarized infrared spectrum of oriented crystallites of ramie cellulose (cellulose I) in the O—H and C—H stretching regions. Solid line: electric vector perpendicular to the fiber axis; broken line: electric vector parallel to the fiber axis. (b) Polarized infrared spectrum of oriented mercerized crystallites of ramie cellulose (cellulose II) as in (a). [Taken from Refs. 60 and 61, with permission.]

TABLE 5.22a

Band Assignments for the 2000–4000 cm^{-1} Region of the Infrared Spectrum of Cellulose I[a]

Frequency (cm^{-1})			
Valonia and bacterial cellulose	Ramie cellulose	Polarization	Assignment
2853	2853	‖	CH$_2$ Symmetric stretching
2870	2870	⊥	
2897		⊥	
	2910	⊥	
2914		⊥	
2945	2945	⊥	CH$_2$ Antisymmetric stretching
2970	2970	‖	
3245		‖	
3275	3275	‖	
3305	3305	⊥	O—H Stretching (intermolecular hydrogen bonding in the 10$\bar{1}$ plane)
3350	3350	‖	O—H Stretching (intramolecular O-3—H $\cdots$ O-5 hydrogen bonding)
3375	3375	‖	
3405	3405	⊥	O—H Stretching (intermolecular hydrogen bonding in the 101 plane)
—	3450	‖	Cellulose II impurity in ramie?

[a] Taken from Ref. 60 with permission.

TABLE 5.22b

Band Assignments for the 2000–4000 cm^{-1} Region of the Infrared Spectrum of Mercerized Ramie Cellulose (Cellulose II)[a]

Frequency (cm^{-1})	Polarization	Assignment
2850	‖	CH$_2$ Symmetric stretching
2874	⊥	
2891	⊥	
2904	⊥	
2933	⊥	CH$_2$ Antisymmetric stretching
2955	‖	
2968	‖	
2981	⊥	
3175	⊥ ⎫	O—H Stretching (intermolecular hydrogen bonding)
3305	⊥ ⎬	
3350	⊥ ⎭	
3447	‖ ⎫	O—H Stretching (intramolecular hydrogen bonding)
3488	‖ ⎭	

[a] Taken from Ref. 61 with permission.

3480 and 3447 cm^{-1} with parallel dichroism. No such bands are observed for cellulose I, where the dominant O—H band is at 3350 cm^{-1} with parallel dichroism. The O—H stretching regions for cellulose III and IV are also distinctive; different spectra are obtained depending on whether the polymorph (III or IV) was prepared from cellulose I or II (63).

The most detailed spectra for cellulose are those obtained for the highly crystalline specimens from *Valonia ventricosa* (64). The polarized infrared spectrum obtained from a mat of oriented fibers is shown in Fig. 5.21 (64) together with the Raman spectrum for an unoriented specimen. The observed band frequencies, intensities, and IR dichroisms are listed in Table 5.23. The bands in the frequency range 1500–800 cm^{-1} are very similar to those of α-D-glucose and probably have similar origins. The situation in the O—H and C—H stretching region is more complex. The glucose residue in cellulose has three O—H groups. However, nine O—H stretching bands are observed in the Raman for native cellulose, and the most likely cause of this is that the glucose residues are not arranged in an identical manner. The Meyer and Misch unit cell for cellulose I (65) contains two residues, each of two separate cellulose chains. Each chain is believed to have a twofold screw axis along its length. Thus the environments of successive residues should be identical. However, the two chains in the cell cannot be arranged in an identical manner. Thus the O—H groups of the glucose residues of neighboring chains are involved in different hydrogen-bonding networks and give rise to different O—H stretching bands. Such considerations would account for at most six O—H bands, but probably the intramolecular hydrogen band would be the same for each chain, so that we would expect only five bands. However, for *Valonia*, the detailed x-ray photograph shows that the unit cell possesses eight cellulose chains (66) and the most likely differences between these chains will be in the nature of the hydrogen-bonding network. The microfibrils of *Valonia* cellulose have a cross-sectional width of 210 Å and are believed to consist of a crystalline array of smaller units, known as elementary fibrils, with widths of 35 Å (67,68). The additional O—H stretching frequencies may be due to the hydrogen bonding between the elementary fibrils. A further possibility is that there is a coupling mechanism between the O—H modes giving rise to additional bands.

In native cellulose (I), the strong band at 3350 cm^{-1} with parallel dichroism has been assigned to the O—H group involved in the O-3—H $\cdots$ O-5 intramolecular hydrogen bond (60), as shown in Fig. 5.22a. This band is absent from the cellulose II spectrum, in which there are two new bands at 3447 and 3480 cm^{-1}. The hydrogen bonding for the O—H groups giving rise to these bands must be weaker than that for the band assigned to the intramolecular bond in cellulose I. One proposed explanation is that there are two intramolecular bonds in cellulose (II), as shown in Fig. 5.22b. One is the O-3—H

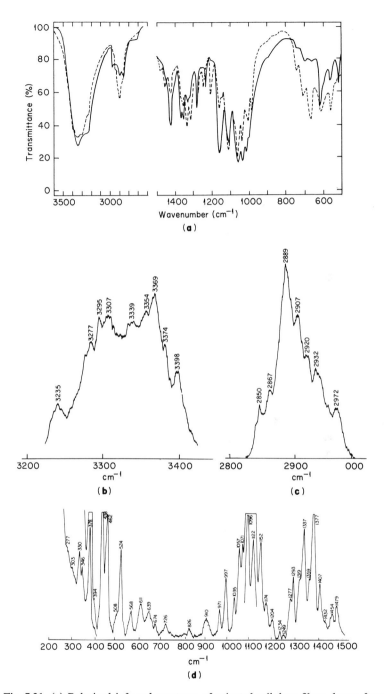

Fig. 5.21. (a) Polarized infrared spectrum of oriented cellulose fibers drawn from the cell walls of the algae *Valonia ventricosa*. Solid line: electric vector parallel to the fiber axis; broken line: electric vector perpendicular to the chain direction. (b) Raman spectrum of an unoriented specimen of *Valonia* cellulose in the O—H stretching region, 3500–3200 cm⁻¹; (c) in the C—H stretching region 3000–2800 cm⁻¹; (d) in the 1500–200 cm⁻¹region. [From Ref. 64, with permission.]

TABLE 5.23

Infrared and Raman Band Assignments for Valonia Ventricosa Cellulose[a]

Raman (cm⁻¹)	Infrared (cm⁻¹)	IR Dichroism	Assignment
3398 w	3408 w	⊥	
3374 w	3376 w	=	
3369 m		=	
3354 m	3347 vs	=	O—H Stretching
3339 m			
3307 m	3306 w	⊥	
3295 m		=	
3277 m	3271 m	=	C—H Stretching
3235 w	3238 m	=	CH₂ Antisymmetric
2972 w	2966 w	⊥	stretching
2932 m	2942 w		
2920 w	2919 vw	?	
2907 m	2911 vw	⊥	C—H Stretching
2889 m	2894 m	⊥	
2867 w	2866 w	⊥	CH₂ Symmetric
2850 w	2853 w	=	stretching

Raman (cm⁻¹)	Infrared (cm⁻¹)	IR Dichroism	Assignment
1204 vw	1206 w	⊥	Antisymmetric
1174 w	1161 s	=	Ring mode
1152 m	1143 w	⊥	
1122 vs	1125 w	?	
1090 vs	1083 w	⊥	C—O Stretchings
1071 w			ring modes, etc.
10575 m	1058 vs	=	
1035 w	1033 vs	=	
	1011 m	=	
997 m	1003 w	⊥	
971 w	984 w	?	
910 w	893 vw	?	C-1—H-β Deformation
	858 vw	⊥	
	745 w	⊥	
	708 w	⊥	
	668 m	⊥	CH₂ Rocking
	628 w	?	
	619 m		
	600 w	=	

Raman (cm⁻¹)	Infrared (cm⁻¹)	IR Dichroism	Assignment	Raman (cm⁻¹)	Infrared (cm⁻¹)	IR Dichroism	Assignment
1479 m	1482 w	⊥	O—H Deformation	568 w	560 w	⊥	
1454 w	1455 w	⊥		524 m	530 vw	⊥	
1432 vw	1426 m	∥	CH₂ Bending		518 w	∥	
1407 m	1405 w	∥	O—H Deformation	462 s	455 w	∥	
	1376 vw	∥	C—H Deformation	438 s	445 w	⊥	
1377 s	1372 m	∥			431 w	∥	
1359 vw	1359 m	⊥	C—H Deformation		420 w	⊥	
1337 s	1337 m	⊥	O—H Deformation	394 vw	395 w	?	
1319 vw	1316 m	⊥	CH₂ Wagging	378 s			
1293 m	1293 w	⊥	CH₂ Twisting	366 vw			
1277 w	1279 m	∥		346 w	339 m	?	
	1271 w	⊥		330 m			
1249 vw	1249 w	⊥		303 w			
1234 w	1233 w	∥	O—H Deformation	277 vw			
				211 w			

ᵃ Taken from Ref. 64, with permission.

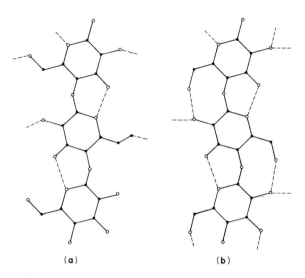

(a)   (b)

Fig. 5.22. Possible sidechain conformations for the cellulose chain: (a) CH_2OH groups forming intermolecular hydrogen bonds; (b) CH_2OH group forming $O-6-H \cdots O-2'$ intramolecular hydrogen bonds. The CH_2OH side-chain conformations in (a) and (b) are described respectively as *gt* and *tg* (see Fig. 5.23).

$\cdots O-5'$ bond as in cellulose I; the other involves the CH_2OH side chain in the $O-6-H \cdots O-2'$, bond. Both of these intramolecular bonds would have to be weaker than the single $O-3-H \cdots O-5'$ bond in cellulose I, where the CH_2OH group is probably involved in intermolecular hydrogen bonding. Set against this is the fact that both these bands at 3447 and 3480 cm^{-1} occur in the spectrum of α-chitin (69), which has the same basic chain conformation. It seems unlikely that there are two intramolecular hydrogen bonds in chitin, since it requires an $O-6-H \cdots N_2$ bond to the side of the amide group.

The basic conformation of the cellulose chain is defined by x-ray measurements and model building. Spectroscopy is useful for deciding between possible hydrogen-bonding networks, as demonstrated above. Identification of the bands due to CH_2 modes allows some questions to be resolved on the orientation of the CH_2OH groups. In pyranose sugars, the CH_2OH group is always in one of the three staggered rotational conformations about the $C-5-C-6$ bond. These staggered positions, shown in Fig. 5.23, are labeled *gg, gt*, and *tg* (69). Normal coordinate analysis (56) for α-D-glucose predicts an essentially pure CH_2 bending mode, and Liang and Marchessault (60) assigned the band at 1430 cm^{-1} with parallel dichroism to this vibration. The only other observed bands in this region are assigned to $O-H$ modes by deuteration work on ramie cellulose. The predicted dichroisms, from the

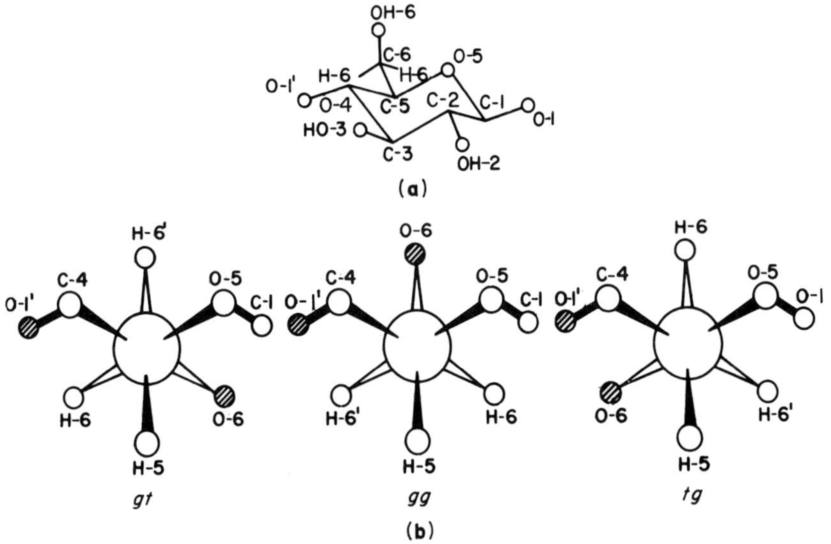

Fig. 5.23. The possible conformations of the —CH₂OH side group in cellulose. (a) The pyranose ring. (b) Newman projections of the three side-group conformations looking along C-5—C-6. (Taken from Blackwell and Marchessault, in "Cellulose and Cellulose Derivatives," N. Bikalis and L. E. Segal, eds., vol IV, Interscience, New York, 1971, by permission of John Wiley and Sons, Inc.)

model of a cellulose chain, are parallel bands for the *gt* and *tg* conformation and perpendicular bands for *gg*. Thus, the *gg* conformation can be ruled out, but it is not possible in this case to distinguish between *gt* and *tg*. The dichroisms for the CH_2 bands on cellulose II are the same as in cellulose I, and a similar dilemma arises. Cellulose I is converted to cellulose II by swelling in reagents such as caustic soda solution, and it has been argued by Warwicker and Wright (70) that the cellulose structure swells in sheets rather like the swelling of chitin (see Chapter 11). It is suggested that the CH_2OH groups are involved in intermolecular hydrogen bonds in cellulose I which are broken when the structure is swollen. On removal of the swelling agents, the CH_2OH groups form intramolecular hydrogen bonds of the type O-6—H ⋯ O-2, as shown in Fig. 5.22. Such a mechanism would most likely involve a change from *gt* CH_2OH groups in cellulose I to *tg* groups in cellulose II.

Conversion of cellulose I to cellulose II produces changes in intensity of some of the infrared bands. Several research groups have plotted the intensity of these bands for specimens of cellulose I treated with different concentrations of sodium hydroxide. The bands at 1430 and 1111 cm⁻¹ for cotton cellulose show a sharp decrease in intensity over a range of sodium hydroxide concentration from 9 to 13%, while those at 990 and 893 cm⁻¹ increase over the same range (71). The most significant change is in the band at 1430 cm^{-1}. Between 9 and 13% concentration the specimen is probably a mixture of forms I and II. However, the relative proportions cannot be determined since the crystallinity also decreases during the mercerization process. The band at 1430 cm^{-1} has been shown to decrease in intensity significantly when the specimen is ground in a ball mill (72).

The band at $\sim$893 cm^{-1}, which is considered typical of β-D-glucans, is very weak for the highly crystalline *Valonia* cellulose, but it becomes more intense for the less crystalline forms, such as ramie and flax; it is also more intense in the mercerized and regenerated celluloses, which are less crystalline than native ramie. The relation to crystallinity has not been examined in detail.

Chitin

Chitin has a structure similar to cellulose, except that a NHCOCH$_3$ group replaces the hydroxyl on C-2. The glucan backbone has the same chain conformation as cellulose, with the result that the infrared spectra are very similar (73). The chitin of the spines of the diatoms *Cyclotella cryptica* and *Thalasiosira fluviatilis* has high crystallinity comparable with that of the cellulose from *Valonia*. This is evidenced by comparison of the spectrum of *Cyclotella* chitin in Fig. 5.24 (74) with that of *Valonia* cellulose in Fig. 5.21a. The presence of the amide group means that the numerous amide bands observed in polypeptides will be present and we can expect additional bands due to the CH$_3$ group frequencies.

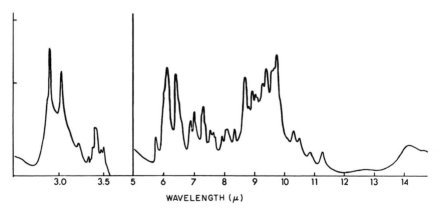

Fig. 5.24. Infrared spectrum of the chitin spines of the diatom *Cyclotella cryptica*. [Taken from Ref. 74, with permission.]

Chitin occurs in nature in at least three polymorphic forms (α-, β-, and γ-) and the arrangement of the chains in these structures is shown in Figs. 11.12, 11.14, and 11.16. The three polymorphic forms of chitin are distinguished spectroscopically by the frequency of the amide bands. The frequencies of these bands for α-, β-, and γ-chitins are listed in Table 5.24 (73,75). The

TABLE 5.24

Frequencies (cm^{-1}) of Amide Bands Observed in Infrared Spectra of α-, β-, and γ-Chitins

Amide band	α	β	γ
A	3264 s	3292 s	3263 s
B	3106 m	3102 m	3100 m
I	1656 m	1656 s	1656 s
	1631 sh	1631 s	1626 s
	1621 s		
II	1555 s	1556 s	1550 s
III	1310 m	1309 m	
V	730 w	703 w	

infrared spectra for the three forms of chitin in the 4000–2500 cm^{-1} range are shown in Fig. 5.25 (75). The α- and β-chitins give rise to amide A bands at 3160 and 3190 cm^{-1} respectively. This frequency difference has been correlated with a longer $C=O \cdots H-N$ hydrogen bond in the β-form (76). x-Ray measurements show that the unit cell dimension in the hydrogen bond direction is 4.76 Å for α-chitin and 4.85 Å for β-chitin, and since the chains have the same conformation, most of this increase for the β form must be due to a longer hydrogen bond (77). The frequency difference of 30 cm^{-1} probably corresponds to approximately 40 cm^{-1} for the unperturbed $N-H$ stretch, and the data of Nakamoto *et al.* (57) suggest this kind of shift for an increase in hydrogen bond length of 0.1 Å for amide groups. For γ-chitin, the amide A band is broad and covers the range between both α- and β-form frequencies. The proposed structure for γ-chitin has two parallel chains followed by one antiparallel chain. Such an arrangement has two sets of α-chain contacts followed by one β-chain contact. This structure might act like a mixture of α and β forms, and the broad band may be a doublet at 3160 and 3190 cm^{-1}. Deuteration removes the high-frequency side leaving a sharper band at 3160 cm^{-1} (76). This is consistent with the above hypothesis since the amide A is removed on deuteration of β-chitin, but α-chitin cannot be deuterated in the crystalline regions, and the 3160 cm^{-1} band is largely unaffected.

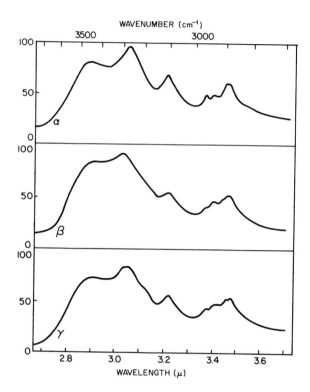

Fig. 5.25. Infrared spectra of α-, β-, and γ-chitins in the region 3600–2800 cm^{-1}. [Taken from Ref. 75.]

There has been much discussion earlier in this chapter of the complex nature of the amide I band in polypeptides. The regions of the polarized I.R. spectra of α-chitin containing the amide I and II bands are shown in Fig. 5.26. The amide I band can be resolved into three components, and is at least a doublet in β- and γ-chitins (Table 5.24). In chitin, as in nylon 66 (78), the amide groups are well separated along the chain and coupling between amide groups along the main chain should be negligible. Coupling between hydrogen-bonded amide groups leads to two modes, only one of which is infrared active. Consequently, for chitin structures we would expect to observe a single band at about 1630 cm^{-1}, 18 cm^{-1} below the unperturbed value of approximately 1648 cm^{-1}. In β-chitin the lower amide I band is at 1631 cm^{-1} and is probably this perturbed mode. In α-chitin there is a much shorter hydrogen bond and the perturbed mode is probably that at 1621 cm^{-1}. The equivalent band in γ-chitin occurs at ~1625 cm^{-1}. What, then, are the higher-frequency bands? β-Chitin is known to form a series of hydrates (77), with water molecules held between the chains. Water gives rise to a deformation band at about 1640 cm^{-1}. The 1650 cm^{-1} band in β-chitin is almost completely removed on deuteration (76), and it is likely that this is due to water present in hydrated structures. In α-chitin, the 1656 cm^{-1} band is slightly diminished when attempts are made to exchange with D_2O. It has been argued that water of crystallization is present in α-chitin which cannot be reached by the D_2O since the structure does not swell (76). Set against the above argument, the structure proposed by Carlstrom (79) on the basis of x-ray work contains no space large enough to hold a water molecule. Furthermore, most sugar hydrates give only a weak band in this region.

An alternative explanation is that the bands at higher frequencies are amide I modes that result from some other mechanism. The model for coupling proposed by Miyazawa may not be the entire story for amide structures. Crystalline anhydrous β-N-acetyl-D-glucosamine, the monomer for chitin, gives rise to a doublet amide I at 1624 and 1612 cm^{-1} (80), while the Miyazawa model predicts a single peak. The unit cell for this structure contains two molecules stacked together in the same way as in α-chitin, with the amide groups on neighboring molecules arranged side by side. It may be that coupling occurs between these neighboring groups, which would account for the double amide peak. This effect may be responsible for the 1631 cm^{-1} peak in α-chitin, although an alternative explanation could be the presence of β-chitin as a contaminant. Raman spectra of chitin have not yet been recorded. When these data become available they may help to resolve some of the above questions.

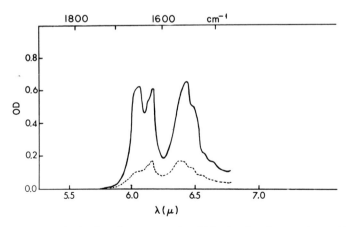

Fig. 5.26. Polarized infrared spectrum of α-chitin showing the complex amide I and II bands. Solid line: electric vector perpendicular to the fiber axis; broken line: electric vector parallel to the fiber axis. [Taken from Ref. (75).]

Amylose, Amylopectin, and Starch

Specimens of the starch polysaccharides are never as highly crystalline as those of cellulose and chitin, and consequently the spectra are less detailed. In addition, these polysaccharides crystallize as hydrates which are unstable in dry atmospheres. Consequently the O—H stretching region is indistinct (81,82) due to the intense broad band originating in the absorption by water. This water cannot be removed by deuteration without converting all the O—H groups to O—D, and thus information concerning the hydrogen-bonding network is not available from the stretching region. The CH stretching region is not well resolved, and again no useful information has been obtained. The lower regions of the spectrum are almost identical for starch amylopectin and the polymorphic forms of amylose. This is not unexpected since the chains have very similar 6_1 helical conformations (83). It is argued that the conversion of the V-amylose helix to the extended B (and A) form is achieved by breaking the O-6—H $\cdots$ O-2' intramolecular bond and inserting a water molecule to give

$$O\text{-}6\text{—}H \cdots \underset{\underset{H}{|}}{O}\text{—}H \cdots O\text{-}2$$

bonding (see Fig. 11.30). This can be achieved without altering the rotational conformation of the CH_2OH group.

The infrared and Raman spectra of the starch polysaccharides appear to be closely similar in the range below 2000 cm^{-1}, although a few small differences can be detected. Casu and Reggiani (81) have noted that the weak band at 1112 cm^{-1} in the infrared spectrum of β-amylose is absent from the spectra of the less extended V form and amorphous specimens. Similarly, the V form exhibits a weak infrared band at ~ 1300 cm^{-1} which is absent for the other forms. Vasko and Koenig (82) have followed the intensity changes of the 1300 cm^{-1} band, which they found sensitive to annealing and humidification. They proposed that this band is indicative of the extent of chain folding in the V structure.

β-1,3-Xylan

The β-1,3-xylan from seaweeds was shown by Atkins *et al.* (84) to have a triple-helical structure with three 6_1 helices coiled together around a common threefold axis (Fig. 11.25b). Polarized infrared spectra allowed a choice between possible models involving inter- or intramolecular hydrogen bonding. Experience with other polysaccharides indicated that the intramolecular bond might be more likely. The model for this structure predicted an O—H stretching band with parallel dichroism. However, the O—H stretching region of the infrared spectrum [Fig. 5.27 (84)] showed an intense O—H band with strong perpendicular dichroism at 3300 cm^{-1}. The band at 3330 cm^{-1} is resistant to deuteration, indicating that it is probably caused by OH groups in the center of the triple-helical structure. This led to the proposed inter-molecular bonding by which three O—H groups, in a plane perpendicular to the helix axis, form the hydrogen-bonded triplet shown in Fig. 11.25c.

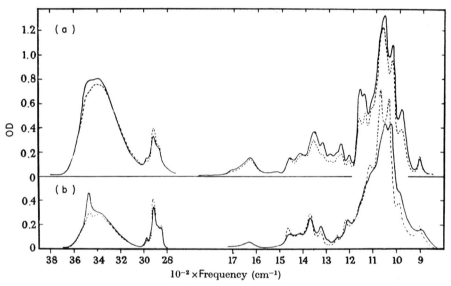

Fig. 5.27. Polarized infrared spectrum of β-1,3-xylan (a) before deuteration, (b) after deuteration. Solid line: electric vector perpendicular to the fiber axis; broken line: electric vector parallel to the fiber axis. [Taken from Ref. (84), with permission.]

β-1, 4-*Xylan*

This polysaccharide, prepared by deacetylation of *O*-acetyl xylan from hardwood, was shown to have a 3_1 helical conformation by Settineri and Marchessault (85). As with the 1,3-xylan above, the absence of a CH_2OH side chain makes for a simpler spectrum in the O—H stretching region. Figure 5.28 (86) shows the polarized infrared spectra from oriented films of this 1,4-xylan (86), which has an intense band in the O—H stretching region with parallel dichroism. This is considered to be a confirmation of the intra-molecular hydrogen bond between contiguous residues proposed from a study of molecular models (85).

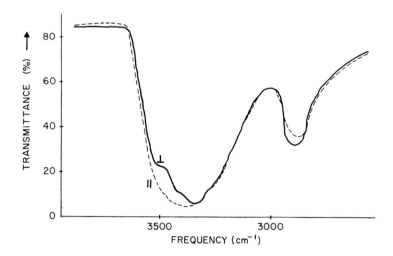

Fig. 5.28. Polarized infrared spectrum of β-1,4-xylan in the O—H and C—H stretching regions. Solid line: electric vector perpendicular to the fiber axis; broken line: electric vector parallel to the fiber axis. [Taken from Ref. 86, with permission.]

κ- and ι-Carrageenans

Structure determination for these seaweed polysaccharides has shown them to have a double-helical conformation (87), in which two chains, each with three disaccharide units per turn, are coiled together with the same sense. Study of molecular models showed that the most likely structure had an intermolecular hydrogen bond (between the two chains of the double helix) between O-6 of one galactose and the O-2 of the galactose on the second strand. Oriented films of both κ and ι structures show little detail in the O—H stretching region. However, after deuteration a sharp band remains with strong perpendicular dichroism at 3330 cm^{-1}. This observation confirms the presence of an intermolecular hydrogen bond in the interior of the double helix.

REFERENCES

1. S. Hanlon, *in* "Spectroscopic Approaches to Biomolecular Conformation" (D. W. Urry, ed.), p. 161. Amer. Med. Ass. Press, Chicago, 1970.
2. T. Miyazawa, *in* "Poly-α-Amino Acids" (G. D. Fasman, ed.), p. 69. Dekker, New York, 1967.
3. T. Miyazawa, *J. Polym. Sci.* **C7**, 3–137 (1964).

4. P. J. Hendra and C. J. Vear, *Analyst* (*London*) **95**, 322 (1970).
5. J. L. Koenig, *Appl. Spectrosc, Rev.* **4**, 233 (1971).
6. G. M. Barrow, "Structure of Molecules." Benjamin, New York, 1964.
7. H. A. Szymanski, "Theory and Practice of Infrared Spectroscopy." Plenum, New York, 1964.
8. J. R. Nielsen and A. H. Woollett, *J. Chem. Phys.* **26**, 1391 (1957).
9. M. C. Tobin, *J. Chem. Phys.* **23**, 891 (1955).
10. I. Sandeman and A. Keller, *J. Polym. Sci.* **19**, 401 (1956).
11. T. Miyazawa, *J. Mol. Spectrosc.* **4**, 155 (1960).
12. T. Miyazawa, T. Shimanouchi, and S. Mizushima, *J. Chem. Phys.* **29**, 611 (1958).
13. T. Miyazawa, *J. Chem. Phys.* **32**, 1647 (1960).
14. P. Higgs, *Proc. Roy. Soc. London* **A220**, 472 (1953).
15. T. Miyazawa, Y. Ideguchi, and K. Fukushima, *J. Chem. Phys.* **35**, 691 (1961).
16. T. Miyazawa, Y. Ideguchi, and K. Fukushima, *J. Chem. Phys.* **38**, 2709 (1963).
17. K. Fukushima, Y. Ideguchi and T. Miyazawa, *Bull. Chem. Soc. Japan* **36**, 1301 (1963).
18. T. Miyazawa and E. R. Blout, *J. Amer. Chem. Soc.* **83**, 712 (1961).
19. E. W. Small, B. Fanconi, and W. L. Peticolas, *J. Chem. Phys.* **52**, 4369 (1970).
20. B. Fanconi, B. Tomlinson, L. A. Nafie, W. Small, and W. L. Peticolas, *J. Chem. Phys.* **51**, 3993 (1969).
21. E. R. Blout, *in* "Polyamino Acids, Polypeptides and Proteins" (M. A. Stahmann, ed.), p. 275. Univ. of Wisconsin Press, Madison, 1962.
22. P. L. Sutton and J. L. Koenig, *Biopolymers* **9**, 615 (1970).
23. T. Miyazawa, T. Shimanouchi, and S. Mizushima, *in* "Conformations of Biopolymers" (G. N. Ramachandran, ed.), Vol. 2, p. 537. Academic Press, New York, 1967.
24. E. R. Blout and G. D. Fasman, *Recent Advan. Gelatin Glue Res. Proc. Conf.* **1**, 122 (1957).
25. W. B. Rippon, J. L. Koenig, and A. G. Walton, *J. Amer. Chem. Soc.* **92**, 7455 (1970).
26. M. L. Tiffany and S. Krimm, *Biopolymers* **6**, 1767 (1968); **8**, 347 (1969).
27. M. J. Deveney, A. G. Walton, and J. L. Koenig, *Biopolymers*, **10**, 615 (1971).
28. M. Smith, A. G. Walton, and J. L, Koenig, *Biopolymers* **8**, 173 (1969).
29. D. H. Wallach, J. M. Graham, and A. R. Oseroff, *FEBS Lett.* **7**, 330 (1970).
30. J. L. Koenig and P. L. Sutton, *Biopolymers* **9**, 1229 (1970).
31. S. Krimm, *J. Mol. Biol.* **4**, 528 (1962).
32. S. H. Carr, E. Baer, and A. G. Walton, *J. Macromol. Sci. Phys.*, **B5**, 789 (1971).
33. S. Hanlon, *Biochemistry* **5**, 2049 (1966).
34. W. B. Rippon and A. G. Walton, unpublished data.
35. A. G. Walton, M. J. Deveney, and J. L. Koenig, *Calcif. Tissue Res.* **6**, 162 (1970).
36. G. N. Ramachandran, ed., *in* "Collagen." Academic Press, New York, 1967.
37. H. Susi, *in* "Structure and Stability of Biological Macromolecules" (S. N. Timasheff and G. D. Fasman, eds.), Chapter 7. Dekker, New York, 1969.
38. D. Garfinkel and J. T. Edsall, *J. Amer. Chem. Soc.* **80**, 3818 (1958).
39. M. Tobin, *Science* **161**, 68 (1968).
40. R. C. Lord and N.-T. Yu, *J. Mol. Biol.* **50**, 509 (1970).
41. B. Frushour and J. L. Koenig, unpublished data.
42. T. Shimanouchi, M. Tsuboi, and Y. Kyogoku, *Advan. Chem. Phys.* **7**, 435 (1964).
43. N. N. Aylward and J. L. Koenig, *Macromolecules* **3**, 583 (1970).
44. M. Tsuboi, Y. Kyoguku, and T. Shimanouchi, *Biochim. Biophys. Acta* **55**, 1 (1962).
45. J. S. Ziomek, J. R. Ferraro, and D. F. Peppard, *J. Mol. Spectrosc.* **8**, 212 (1962).
46. L. Rimai, T. Cole, J. L. Parsons, J. T. Hickmott, and E. B. Carew, *Biophys. J.* **9**, 320 (1969).

47. N. N. Aylward and J. L. Koenig, *Macromolecules* **3**, 590 (1970).
48. D. Poland, J. N. Vournakis, and H. A. Scheraga, *Biopolymers* **4**, 223 (1966).
49. H. T. Miles and J. Frazier, *Biochim. Biophys. Acta* **79**, 216 (1964).
50. H. T. Miles, *Chem. Ind. (London)*, p. 591 (1958).
51. H. T. Miles, *Biochim. Biophys. Acta* **30**, 324 (1958).
52. H. T. Miles, *Biochim. Biophys. Acta* **45**, 196 (1960).
53. H. T. Miles, *Biochim. Biophys. Acta* **35**, 274 (1959).
54. H. T. Miles, *Proc. Nat. Acad. Sci. U.S.* **47**, 791 (1961).
55. H. T. Miles and J. Frazier, *Biochem. Biophys. Res. Commun.* **14**, 21, 129 (1964).
56. P. D. Vasko, J. Blackwell, and J. L. Koenig, *Carbohydr. Res.*, **23**, 407 (1972).
57. K. Nakamoto, M. Margoshes, and R. E. Rundle, *J. Amer. Chem. Soc.* **77**, 6480 (1955).
58. P. D. Vasko, J. Blackwell, and J. L. Koenig, *Carbohyd. Res.*, **19**, 297 (1971).
59. S. A. Barker, E. J. Bourne, M. Stacey, and D. H. Wiffen, *J. Chem. Soc. London*, p. 171 (1954).
60. C. Y. Liang and R. H. Marchessault, *J. Polym. Sci.* **37**, 385 (1959).
61. R. H. Marchessault and C. Y. Liang, *J. Polym. Sci.* **43**, 31 (1960).
62. C. Y. Liang and R. H. Marchessault, *J. Polym. Sci.* **39**, 269 (1959).
63. J. Mann and H. J. Marrinan, *J. Polym. Sci.* **32**, 357 (1958).
64. J. Blackwell, P. D. Vasko, and J. L. Koenig, *J. Appl. Phys.* **41**, 4375 (1970).
65. K. H. Meyer and L. Misch, *Helv. Chim. Acta* **20**, 232 (1937).
66. G. Honjo and H. Watanabe, *Nature* **181**, 326 (1958).
67. A. Frey-Wyssling, K. Mühlethaler, and R. Muggli, *Holz. Roh Werkst.* **24**, 443 (1966).
68. K. H. Gardner and J. Blackwell, *J. Ultrastruct. Res.*, **36**, 725 (1971).
69. M. Sundaralingam, *Biopolymers* **6**, 189 (1968).
70. J. O. Warwicker and A. C. Wright, *J. Appl. Polym. Sci.* **11**, 659 (1967).
71. A. W. McKenzie and H. G. Higgins, *Sv. Papperstidn.* **61**, 893 (1958).
72. R. Jeffries, *J. Appl. Polym. Sci.* **12**, 425 (1968).
73. F. G. Pearson, R. H. Marchessault, and C. Y. Liang, *J. Polym. Sci.* **43**, 101 (1960).
74. J. Blackwell, K. D. Parker, and K. M. Rudall, *J. Mol. Biol.* **28**, 383 (1967).
75. J. Blackwell, Ph.D. thesis. University of Leeds, England, 1967.
76. K. D. Parker, cited by K. M. Rudall, *Advan. Insect Physiol.* **1**, 257 (1963).
77. J. Blackwell, *Biopolymers* **7**, 281 (1969).
78. T. Miyazawa and E. R. Blout, *J. Amer. Chem. Soc.* **83**, 712 (1961).
79. D. Carlström, *J. Biophys. Biochem. Cytol.* **3**, 669 (1957).
80. J. Blackwell and K. D. Parker, in preparation.
81. B. Casu and M. Reggiani, *J. Polym. Sci.* **C7**, 171 (1964); *Die Stärke* **7**, 218 (1966).
82. P. D. Vasko and J. L. Koenig, *J. Macromol. Sci. (Phys)* **B4**, 347 (1970).
83. J. Blackwell, A. Sarko, and R. H. Marchessault, *J. Mol. Biol.* **42**, 379 (1969).
84. E. D. T. Atkins, K. D. Parker, and R. D. Preston, *Proc. Roy. Soc. (London)*, **B173**, 209 (1969); E. D. T. Atkins and K. D. Parker, *Nature* **220**, 784 (1968).
85. W. Settineri and R. H. Marchessault, *J. Polym. Sci.* **C11**, 142 (1965).
86. R. H. Marchessault and C. Y. Liang, *J. Polym. Sci.* **59**, 357 (1962).
87. N. S. Anderson, J. W. Campbell, M. M. Harding, D. A. Rees, and J. W. B. Samuel, *J. Mol. Biol.* **45**, 85 (1969).

6

ELECTRONIC SPECTROSCOPY

Introduction

We have seen in the previous chapter that light whose wavelength is longer than that pertaining to the visible spectrum may be absorbed by biological molecules to cause changes in the vibrational energy of the different components. Such processes form the basis of infrared and Raman spectroscopy. Light of visible or shorter wavelength may also be absorbed by biological molecules. In these cases, an electronic energy change accompanies the absorption, and the theoretical formalism relies heavily on quantum mechanics. In some cases, the molecules may emit radiation after exposure to an exciting source. This luminescence, of which there are two forms, fluorescence and phosphorescence, has been turned to good account in recent years in conformational studies of biological macromolecules. However, most work in electronic spectroscopy has been in the area of absorption or the related dispersion process. Essentially, one can measure the *absorbance* of monochromatic light by a sample or its *refractive index*. The incident radiation may be polarized either linearly or in such a manner that the electric vector precesses with distance such that it is said to be circularly polarized. This latter type of polarization is particularly useful when one examines helical structures of optically active materials, as may be intuitively obvious. The combination of these absorptive and dispersive methods with the polarization of the incident radiation gives us six possibilities, each of which has provided the basis for an important research technique; these are included in Table 6.1.

Before proceeding to a discussion of the application of these various techniques to biopolymers, it is perhaps worthwhile to review the basis for the electronic transitions that can occur. The calculation and prediction of electronic transition energies and absorption wavelengths involve some fairly advanced quantum mechanical formulation. An effort will be made here to extract some of the basic concepts and present several fundamental equations without encompassing detailed derivations.

TABLE 6.1

Relation between Absorption and Dispersion Processes[a]

Polarization	Absorption	Dispersion
None	UV–Visible absorption, ε	Refractive index, n
Linear	UV Dichroism, $\varepsilon_\parallel - \varepsilon_\perp$	Birefringence, $n_\parallel - n_\perp$
Circular	Circular dichroism, $\varepsilon_L - \varepsilon_R$	Optical rotary dispersion, $n_L - n_R$

[a] ε is the extinction coefficient and $\parallel$, $\perp$, L, and R refer, respectively, to the property measured with parallel, perpendicular, left- or right-hand polarized radiation.

The Chromophores

Since the electronic transitions occurring in the near-ultraviolet and visible regions generally involve π electrons, the presence of double bonds or portions of the molecule containing delocalized π electrons is of central importance. For this reason, polypeptides and polynucleotides yield UV–visible spectra, whereas the saturated polysaccharides do not. The origin of electronic transitions in polypeptides may be associated basically with the peptide bond and certain absorbant side groups. In the nucleic acids and polynucleotides, the origin is associated with the bases and phosphate groups. For a more detailed description we must resort to the language of molecular orbital theory; a very brief and elementary description follows.

The Peptide Bond

It is usual to suppose that the wave motion of electrons in the individual atoms of a molecule may be combined in a linear manner to build a molecular orbital. Thus, for two atoms that have wavefunctions ψ_1 and ψ_2, the square of each being equal to the electron density at any point (ψ_1 and ψ_2 might refer to 2s and 2p electrons respectively on different atoms, for example), we can combine these to express a molecular orbital, $\phi = c_1\psi_1 + c_2\psi_2$.

Coefficients c_1 and c_2 are initially unknown but can be related to the charge distribution between the atoms. The wavefunctions ψ_1 and ψ_2 result from a mathematical solution of the Schrödinger wave equation and may be positive or negative in sign. Thus, in a two-atom system, there is a combination of atomic orbitals that can be additive or subtractive, and these define two molecular orbitals which we call bonding or antibonding. The schematic diagram in Fig. 6.1 illustrates that the low-energy, and thus stabilizing, bond-

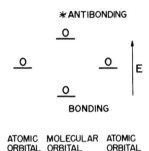

Fig. 6.1

ing orbital is filled with electrons first, and the high-energy or destabilizing orbital is filled last. The energy required to promote electrons from the lower to the upper orbital may be provided by radiation, as given by the Planck relation,

$$\Delta E = h\nu \tag{6.1}$$

In the peptide bond, as we have seen in Chapter 2, there are four π electrons provided by the three atoms O, C, and N. One is from the 2p orbital of oxygen, one from the 2p orbital of carbon, and two (the "lone pair") from the nitrogen 2p orbital.

In this case we can define three energy levels corresponding to solutions of equations based on the three-center molecular orbital,

$$\phi = c_1\psi_O + c_2\psi_C + c_3\psi_N \tag{6.2}$$

The three molecular orbitals can be classified as bonding, nonbonding, or antibonding, as shown in Fig. 6.2. Actually, if the oxygen, carbon, and nitrogen atoms were equally electronegative, the orbitals would be symmetrical in energy, as shown. However, in practice the "nonbonding" orbital is slightly "bonding." Each of the three orbitals can contain two electrons, so that the four delocalized π electrons fill the two lowest molecular orbitals. The

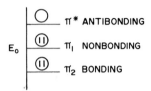

Fig. 6.2

minimum energy required for electronic excitation is that which promotes an electron from the "nonbonding" to antibonding orbital, thus producing a $\pi_1 - \pi^*$ transition. Often the subscript is omitted and the transition is denoted by $\pi \to \pi^*$. In addition to the delocalized π electrons, the oxygen atom also possesses two nonbonding electrons in the atomic p_y orbital. One of these electrons can also be promoted to the π^* orbital, the so-called $n - \pi^*$ transition. The relative energies (wavelengths) of these and two other related transitions to the antibonding σ orbital are shown in Fig. 6.3. (The data are based on experimental values for formamide, dimethylformamide, and myristamide.) It can be seen that the lowest energy (highest wavelength) transition is the $n - \pi^*$ followed by $\pi_1 - \pi^*$, $\pi_2 - \pi^*$ and $n - \sigma^*$. The last two transitions occur in the vacuum ultraviolet range and are not observable with conventional visible–UV spectrophotometers. Two further factors must be considered when the ultraviolet spectrum of polypeptides is observed. (a) The intensity of the lowest-energy transitions may be such that they are not observed. (b) The conformation of the polymer and electronic environment may shift the absorption wavelength away from the expected value. In order to identify the correct transitions, it is helpful to consider intensity and directional properties of spectral transitions.

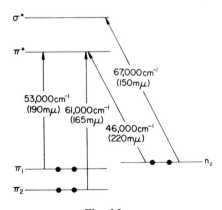

Fig. 6.3

First, it is useful to recognize that, because of charge redistribution on the oxygen, carbon, and nitrogen and attached atoms, the peptide group can be represented by a dipole. Figure 6.4 shows the results of a very approximate Hückel quantum mechanical calculation for charge distribution on the O, C, N atoms and the representative dipole, all for the ground electronic state. The dipole moment is defined by

$$\mu = \sum e\mathbf{d} \tag{6.3}$$

and is thus a product of charge and charge separation. (The charge can be calculated from the coefficients of the molecular orbitals; for details of the method, see Ref. 1.) As electrons are promoted from the ground to one or the other of the excited states, a redistribution of charge occurs and, hence, the magnitude of the dipole changes, and as a result an oscillating dipole is produced. This oscillating dipole can be regarded as the result of resonance between various charged forms in the peptide group. We shall see

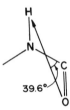

Fig. 6.4. Permanent dipole orientation originating from charge distribution (ground state) for the peptide unit.

later that, by use of oriented samples and polarized light, the orientation of this oscillating dipole can be established experimentally. Some dipole moments, associated electronic transitions, and directional properties calculated theoretically are shown in Fig. 6.5 (2).

The intensity of ultraviolet absorption can be established quantum mechanically by calculating the so-called oscillator strength. In essence, this relates the probability of transition between ground and excited states. The oscillator strength f is related to the extinction coefficient by

$$f = 8.66 \times 10^{-9} \int_{\sigma}^{\nu} \varepsilon(\bar{\nu})\, d\bar{\nu} \tag{6.4}$$

where $\bar{\nu}$ is the reciprocal wavelength in question. The extinction coefficient is as usual

$$\varepsilon(\bar{\nu}) = \frac{1}{cx} \log_{10} I_0(\bar{\nu})/I_x(\bar{\nu}) \tag{6.5}$$

where c is the molar concentration and x the path length.

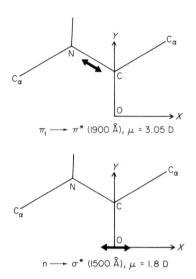

$\pi_1 \longrightarrow \pi^*$ (1900 Å), $\mu = 3.05$ D

$n \longrightarrow \sigma^*$ (1500 Å), $\mu = 1.8$ D

Fig. 6.5. The direction and magnitudes of two of the transition moments in an amide group. [From I. Tinoco, A. Halpern and W. T. Simpson, "Conformation and Light Absorption" in "Polyamino Acids, Polypeptides and Proteins," Ed., M. A. Stahmann with permission.]

The calculation of oscillator strengths theoretically (ab initio) is quite complex (3, 4); however, calculated values of f for the $\pi - \pi^*$ and $n - \pi^*$ transitions for the peptide group are approximately 0.24 and 0.004, respectively, and it is expected that the $n - \pi^*$ will be, therefore, only $\frac{1}{60} - \frac{1}{70}$ as intense as the $\pi - \pi^*$ transition. Consequently, the major band in the UV absorption spectrum of polypeptides is expected to be that of the $\pi - \pi^*$ transition. The electronic environment of the peptide group is clearly dependent upon conformation, and it might be expected that changes in conformation could be detected by changes in the intensity, wavelength, and band shape of the absorption curves.

Certain amino acids have side chains with delocalized electrons, which can contribute strongly to the ultraviolet spectra of polypeptides and proteins into which they are incorporated. The most important of these are arginine, histidine, phenylalanine, tyrosine, and tryptophan. There is also some indication that proline may contain partially delocalized electrons in its (saturated) ring structure, although it is not normally included in the above category. The sulfur-containing amino acids show some side-chain absorptivity. (The absorption characteristics of these amino acids and their chromophoric side chains are shown in Table 6.2 and Fig. 6.6.)

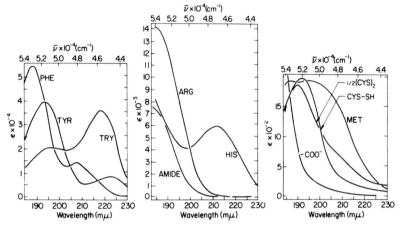

Fig. 6.6. Ultraviolet absorption spectra of the chromophoric amino acid side chains, corrected for α-carboxyl and amino absorption. (Data of R. McDiarmid, Ph.D. thesis, Harvard University, Cambridge, Mass., 1965.)

TABLE 6.2

Principal UV Absorption Bands of
Peptide Chromophores

Chromophore	λ_{max} (mμ)	ε
Try	288	4,400
	280	5,500
	266	4,400
	220	32,000
	196	21,000
Tyr	282	1,100
	275	1,200
	222.5	8,000
	192	47,000
Phe	206	9,000
	187.5	58,000
His	211.5	5,800
$\frac{1}{2}$(Cys)$_2$	190	2,510
Cys—SH	190	2,070
Arg	185	14,200

Nucleotides

The origin of electronic transitions in nucleotides may be treated by similar manner by molecular orbital theory. However, because of the large number of delocalized electrons and possibilities of different ionic and tautomeric states, each nucleotide must be considered separately. Furthermore, the possible additional delocalization and conjugation of the π electrons through the hydrogen bonds of base pairs (see Chapter 12) is worthy of consideration. Evidently, theoretical considerations become extremely complex, but the origin of the longest-wavelength absorption bands in the visible–UV spectrum can be expected to arise from promotion of π electrons (ten or more per base). A general correlation between the number of available π electrons, the size of the system, and the position of the absorption maximum has long been recognized on both theoretical (free-electron theory) and experimental grounds.

The experimental data for the UV absorption of several bases and their derivatives are shown in Table 6.3. Usually two main maxima are observable in the 200–300 mμ range. In some cases another major band appears in the far-ultraviolet region (280–300 mμ). From Table 6.3 it may be deduced that introduction of sugar and phosphate moieties does not radically alter the spectrum of the base, although certain minor changes are evident. The effect of pH on the spectra (via protonation of the bases) certainly seems to be as great if not greater than the effect of the sugar and phosphate groups.

The detailed identification of these absorption bands in terms of the electron transitions and the direction of the transition dipole moment is complicated by the fact that combinations of the basic transitions are involved. Ts'o has reviewed the fairly extensive experimental and theoretical information (5). Probably the most fundamental experimental work relating to polarized absorption spectra of base derivatives has been accomplished by Stewart and co-workers (6, 7), who have studied thin sections of single crystals of 1-methylthymine and 9-methyladenine.

For the purine derivative (9-methyladenine), the first transition band ($\lambda_{max} \sim$ 275 mμ) contains two components, the major component being polarized at 3° from the C-4–C-5 axis, which is inclined toward N-7, as shown in Fig. 6.7a (5–7). The minor component ($\lambda_{max} \sim 255$ mμ) is approximately perpendicular to the major component (see Fig. 6.7a), i.e., along the major molecular axis. A second transition band ($\lambda_{max} < 230$ mμ) appears also to be polarized along the long axis. The pyrimidine derivative (1-methylthymine) shows the first absorption band ($\lambda_{max} \sim 275$ mμ) to be polarized at 19° to the N-1–C-4 axis inclined towards the $-CH_3$ group, as shown in Fig. 6.7b. The second UV band ($\lambda_{max} < 230$ mμ) is polarized approximately perpendicular to the first.

The major components of the high-wavelength bands for both purine and

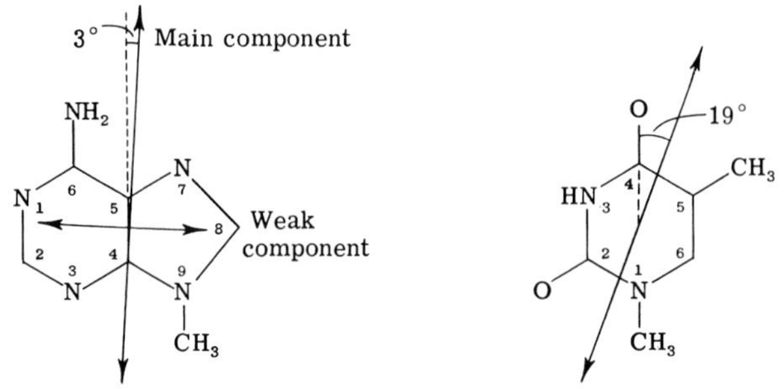

Fig. 6.7. Direction of polarization of the first transition moment in (a) 9-methyladenine and]b) 1-methylthymine. [Reproduced from Ref. 5, p. 90 based on Refs. 6 and 7 by permission of Marcel Dekker, Inc., publishers.]

pyrimidine bases arise from $\pi - \pi^*$ transitions; however, an additional weak shoulder occurs, arising from an $n - \pi^*$ transition. Electrons involved in the $n - \pi^*$ transition originate in the nonbonding atomic orbital of nitrogen. Consequently, factors affecting these N atom centers, such as solvent binding or protonation, will affect the $n - \pi^*$ transition and also, in the latter case, cause an electronic rearrangement that affects the $\pi - \pi^*$ transitions. Such shifts caused by pH in aqueous solution are apparent in the data of Table 6.3 and have been the subject of intensive investigation in a variety of solvent media (5).

Apart from the specific interaction of solvents and the nature of the electronic transitions themselves, much effort has been expended in investigation of the effect of base pairing and base stacking on the spectroscopic properties. The subject is particularly relevant to the absorption spectroscopy of polynucleotides and nucleic acids. It is generally found that in the crystalline state, or in solutions where aggregation and base stacking occur, the high-wavelength transitions show a fairly substantial decrease in integrated intensity. This effect is known as hypochromicity (see below). This effect is believed to originate in the dipole–dipole interaction between transition moments in neighboring oscillators. Theoretical studies by De Voe and Tinoco (8) indicate that the interactions involve the transition moment for the lowest-energy $\pi - \pi^*$ transition with those of higher energy. On this basis hypochromic effects are predicted in stacked bases, such as in DNA or certain crystals, and the intensity lost by the longest-wavelength band should be acquired by transitions further into the ultraviolet. Experimental studies to date show hypochromic effects for bands above 185 mμ, so that verification of the above concepts cannot at present be achieved. There is also some indication (9, 10) that hypochromicity of the long-wavelength band also occurs when base pairs such as A · U, G · G, G · C, and I · C are formed. (The dichroic properties of nucleotides will be presented in a later section.)

TABLE 6.3

The Extinction Coefficients of the Maxima and Minima of the Ultraviolet Spectra of Nucleic Acid Components[a]

Substance	pH	λ_{max}(mμ)	$\varepsilon \times 10^3$	λ_{min}(mμ)	$\varepsilon \times 10^3$	λ_{max}(mμ)	$\varepsilon \times 10^3$	λ_{min}(mμ)	$\varepsilon \times 10^3$	λ_{max}(mμ)	$\varepsilon \times 10^3$
Adenine	7.0	260.5	13.4	225	2.6	207	23.2	190	12.9	190	19.8
	2.5	262.5	13.2	229	2.5	200.5	19.9	190.5	15.5	190	18.7
Adenosine	7.9	259.5	14.9	227	2.3	206.3	21.2	195.3	19.3	188.5	20.6
	1.5	257	14.6	230	3.4	205	21.2	192.5	18.5	187.5	19.1
Deoxyadenosine	7.9	259.5	14.9	226.3	2.2	207.5	21.0	195	18.1	188.8	20.1
	1.5	257	14.6	228.5	2.8	203.5	22.4	192	18.3	189.5	19.7
Adenylic acid	7.9	260	15.0	227	2.4	206.5	20.9	195	19.5	190	18.3
	4.5	260	15.0	228	2.7	205.8	21.2	195	19.1	185	20.8
	2.3	257	14.6	229	3.2	205	21.2	192.5	18.2	187.5	22.3
Deoxyadenylic acid	7.9	260	15.0	226.5	2.3	207.5	20.8	196	18.7	187.5	19.0
	6.8	260	15.0	227	2.3	207	20.4	195.5	18.9		
	2.2	257.5	14.3	228	2.6	205	21.9	193.5	17.9		
Cytosine	8.8	267	6.1	248	4.3	196.5	22.5	192	5.1		
	2.5	276	10.0	237.3	1.2	209	9.7				
Cytidine	8.2	271	9.1	250	6.6	230	8.2	225	8.2	198	22.3
	2.2	280	13.4	240	1.7	212	10.2	193	14.6		
Deoxycytidine	7.8	272	9.1	250	6.2	197.5	22.4				
	2.2	280	13.2	241.5	1.6	212.5	10.1	195	5.4		
Cytidylic acid	7.9	270	8.9	250	6.9	232.5	8.1	225	7.9	197.5	22.3
	4.6	274	9.7	245	4.7	198.5	15.1				
	2.5	277	13.5	241	1.8	212	10.4	193	5.5		
Deoxycytidylic acid	7.8	272	9.3	250	6.0	197	22.3	188.5	11.3		
	4.6	275	10.9	245	4.2	198.5	15.1	193	5.4		
	2.4	280	13.8	242	1.5	212	10.6				
Guanine	6.2	275	8.1	263	7.0	246	10.7	224	4.5	196	22.1
	1.6	273	7.4	266	7.2	248.5	11.4	224	3.6	192.5	22.5

Guanosine	5.5	252.5	13.7	223	2.8	188.3	26.8	
Deoxyguanosine	5.4	252.5	13.7	223	2.6	188.5	27.4	
Guanylic acid	7.7	252.5	13.5	223	3.8	188	26.6	
	4.7	252.5	13.5	223	2.9	189	26.2	
Deoxyguanylic acid	1.4	256.5	12.2	227	2.9	190	19.9	
	7.7	252.5	13.5	223	2.7	187.8	26.2	
	4.7	252.5	13.5	222.5	2.6	188	26.5	
	1.4	249	14.1	224.5	4.4	192.5	27.7	
Thymine	7.0	264.5	7.9	233.5	1.9	205	9.5	187	5.6
Thymidine	7.2	267	9.7	234.5	2.3	206.5	9.8	189	6.1
Thymidylic acid	7.1	268	9.5	235	2.1	207	9.4	188.5	5.3
	4.6	267.5	9.5	235	2.2	206.5	9.6	188.5	5.9
Uracil	7.0	259.5	8.2	228.5	1.6	202.5	9.2	188.8	6.5
Uridine	7.3	261	10.1	231	2.1	205	9.8	191	7.2
Uridylic acid	7.6	262	10.1	230	2.1	205	9.1	190	6.5
	3.2	261.5	10.1	230.5	2.9	205	9.2	190	6.9
Pyrimidine	3.7	243	3.2	239.8	0.28	237.5	3.0	217.5	6.46
	0.4	243	2.4	212.5	1.5	189.5	2.4	188	
Purine	5.8	263	8.0	220	1.4	188	23.6		5.2
	0.4	260	6.2	227.5		201	23.0		

ᵃ No extinction coefficients could be found in the literature for the deoxynucleosides and deoxynucleotides of adenine and guanine, and for deoxycytidylic acid. Their extinction coefficients were therefore set equal to that of the corresponding nucleoside or nucleotide at 260 mμ.

ᵇ Taken from D. Voet, W. B. Gratzer, R. A. Cox, and P. Doty, Biopolymers 1, 193 (1963), with permission.

Line Broadening

Ultraviolet absorption curves are generally broad and often overlapping. The sources of this broadening are many; some are instrumental and some are due to the sample itself. In the latter case, intermolecular interactions cause broadening as do intramolecular interactions. Even if the peptide chromophore could be completely isolated from intra- and intermolecular interactions, there are still two basic causes of line broadening. The first is the Doppler effect, i.e., the motion of the absorbing unit relative to the radiation source. This effect produces a line, Gaussian in shape and conforming to the relation

$$I(v) = \alpha \exp\left[- \frac{Mc^2(\Delta v)^2}{2v_{max}^2 kT} \right] \tag{6.6}$$

where M is the mass of the molecular assembly, Δv is the frequency displacement from v_{max} and c is the velocity of light. It can be seen that the Doppler broadening is strongly reduced if measurements are performed at low temperatures.

The second effect, which is of minor importance experimentally, is the so-called "natural linebreadth." This inherent property of the molecule has been explored theoretically (11) and has the form of a Lorentzian curve decreasing linearly in Δv^2.

Oligomers

If the ultraviolet spectra of peptides or nucleotides were independent of their neighboring environment, including the spatial arrangement of neighboring entities, the field would not have received the attention that it has. There are perhaps three aspects to the problem which are concerned with: (a) the effect of adjacent groups (b) the effect of the total conformation; (c) the effect of solvent. Solvent effects are examined in more detail in Chapter 9. In an attempt to understand the first two points, studies of very short chain polyamino acids (oligomers) and their derivatives have been conducted. If the absorptivity of a series of oligomers, i.e., di-, tri-, tetramers, etc., is proportional to the number of residues and the effect of the peptide bond is additive, then a linear increase with the number of peptide bonds should result, provided that there are no chemical or conformational effects. In practice, it is found that the terminal residues contribute differently than the internal ones, and consequently the change of extinction coefficient with degree of polymerization may be written

$$\Delta\varepsilon = (n - 2)\,\Delta\varepsilon_{int} + 2\,\Delta\varepsilon_{term} \tag{6.7}$$

where n is the number of peptide residues. However, when n becomes greater than about 8–10, the oligomer often starts to take on a definite secondary structure (in some cases, e.g., proline oligomers, a helical form begins with about four residues), and this secondary structure influences the absorbance. Goodman and co-workers (12) have, for example, shown that the onset of helix formation, as detected by a strong deviation from the linear relation expressed above, begins at approximately eight residues for γ-methyl-L-glutamate in 2,2,2-trifluoroethanol, whereas the onset of helix formation occurs at approximately twelve residues for β-methyl-L-aspartate in the same solvent. This information in itself is quite interesting since the glutamate has a strong tendency to form right-hand α helices, whereas the aspartate has a much weaker tendency to form left-hand α helices. Apparently, the co-operative effect of more residues is required for helix formation in the latter case. These results are reproduced in Fig. 6.8 (12).

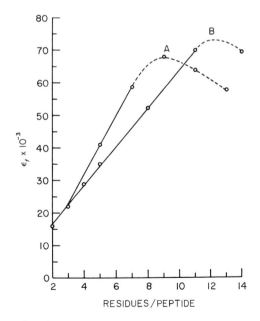

Fig. 6.8. Total molar absorptivity as a function of chain length for oligomers of (A) γ-methyl-L-glutamate and (B) β-methyl-L-aspartate in 2,2,2-trifluoroethanol. [From Ref. 12, by permission of John Wiley and Sons, Inc.]

It can be seen that the absorptivity per residue decreases markedly after the appearance of the secondary structure and actually drops to about one-third of the dipeptide value in the high polymer. This decreased absorption (hypochromicity) is an important property of ordered structures in solution. Concomitant with the decreased absorption per residue, the position of the maximum-absorption wavelength changes, usually to higher wavelength. This effect is called a "bathochromic shift." The nucleotides show particularly strong hypochromicity; the nature of this effect will be discussed again later.

Ultraviolet Absorption by Biopolymers

Polypeptides—Theory and Experiment

Since the ultraviolet spectrum of a polypeptide in a helical array is significantly different in both form and absorption intensity from that of the basic unit, it follows that specific interactions within the biopolymer modify the nature of the electronic transitions. Although we have so far considered the peptide bond from an essentially static point of view, with a charge distribution defining an effective dipole, it is useful to remember that in fact the peptide chromophore may also be represented as resonating between several charged forms and that the quantum mechanical picture represents an average over these combinations. The atoms over which the charge is distributed also vibrate with respect to one another, so that electrons in the ground state may be promoted to a state that is simultaneously excited, both electronically and vibrationally, the so-called *vibronic transition*. Furthermore, and most important to this discussion, the radiation-excited states in the polymer can undergo resonance coupling to give what is known as exciton splitting of the energy level.

There are essentially two models for the interaction of adjacent chromophores. The first, proposed by Förster (13), assumes a very weak coupling (interaction), which is transferred by vibrational relaxation resonance. The second model, proposed by Davydov (14) and called the *exciton* model, supposes that if the coupling is weak, then localized regions, e.g., the peptide chromophore, are covered by the excited-electron distribution, whereas if there is strong coupling, the complete system can be treated (quantum mechanically) as collective excited states, and traveling waves of excitons are produced. A result of the Davydov theory is that the absorption (exciton) band has a width dependent upon the oscillator strength in the chromophore, the structural geometry, and the mutual arrangement of transition dipoles.

When coupling is strong, the wavefunctions for electronic and vibrational states may be separated and a complete theoretical solution achieved. Extinction coefficients $> 10^4$ are usually involved. Very weak coupling involves the vibronic transitions in which the vibrational and electronic states cannot be separated. Nevertheless, theoretical solutions can be achieved which show only slight broadening of the chromophore spectrum and $\varepsilon(v) < 10^2$. Unfortunately, the intermediate coupling process is the one most commonly experienced in biopolymers, and a complete theoretical solution has not yet been achieved. An excellent review of the exciton theory is available in Ref. 15.

The case of strong coupling may be taken as an illustration of some of the effects to be found in biopolymers (although we note that detailed predictions may not be accurate). For a dimer, three possibilities for the coupling of transitions may be demonstrated, as in Fig. 6.9. The transition dipoles can

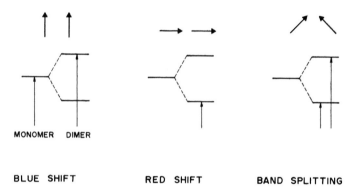

MONOMER   DIMER

BLUE SHIFT   RED SHIFT   BAND SPLITTING

Fig. 6.9. Diagrams of exciton band structure in molecular dimers with various arrangements of transition dipoles. [From M. Kasha, M. A. El-Bayoumic, and W. Rhodes, *J. Chim. Phys. Physicochim. Biol.* **58**, 916 (1961).]

either reinforce or oppose each other, leading to a decreased or increased energy gap and a corresponding shift in the absorption spectrum. Although the exciton resonance splitting is shown for a dimer in which two excited levels are shown, in general an n-mer will have n very closely spaced levels causing a continuous band.

We are now in a position to examine the transition dipoles of helical structures. In the α helix the peptide chromophores lie in planes that almost describe a square. There are four combinations of the peptide transition dipoles, two of which have net transition dipoles perpendicular to the helix axis as shown in Fig. 6.10a (15). This arrangement of transition dipoles causes a degenerate polarization. On the other hand, a further, parallel polarization occurs as the result of the translational arrangement of dipoles along the helix,

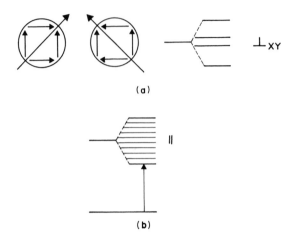

(a)

(b)

Fig. 6.10. Diagrams for exciton band structure in helical molecular polymers with four nonconjugated molecular chromophore units per turn. (a) Transition dipoles perpendicular to helix axis; (b) parallel polarization.

as shown in Fig. 6.10b (15), in which the energy of transition corresponds to a red shift.

The experimentally observed ultraviolet absorption spectrum of biopolymers may be subdivided into the visible to far-ultraviolet range (1 $\mu \rightarrow$ 180 mμ) and the vacuum ultraviolet region (180–100 mμ), based on current instrumental capabilities. Most work has been accomplished in the near- and far-UV regions, where many biopolymers show interesting electronic transitions and many nonabsorptive solvents may be found; α helices in the 180–250 mμ region usually show a strong peak at 185 mμ, a shoulder in the 205 mμ region, and a tail, which appears to contain an additional component. A typical curve (actually for α-helical poly-γ-methyl-L-glutamate) is shown in Fig. 6.11 (16) and demonstrates the assignments based on the polarization of the bands (i.e., the use of plane-polarized UV radiation shows absorbance enhancement when it is in phase with the oriented dipoles). It can be seen that, as expected from theory, there are two bands corresponding to $\pi - \pi^*$ transitions of perpendicular and parallel polarization, the latter occurring with a red (bathochromic) shift. In addition, a weak contribution from the $n - \pi^*$ transition (intensity approximately one-tenth of the $\pi - \pi^*$ transition) has been identified.

The theoretical ultraviolet spectrum corresponding to the β conformation of polypeptides has been considered only recently by two different authors (17a, b). Although there is some measure of agreement there are significant differences between results of the two treatments. Energy-level shifts arising from Davydov splitting were calculated, and the levels corresponding to

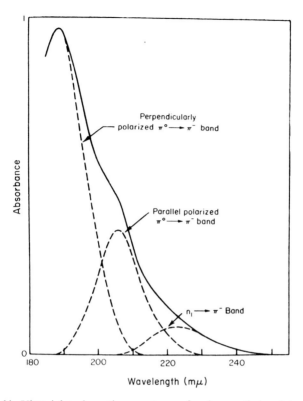

Fig. 6.11. Ultraviolet absorption spectrum of poly-γ-methyl-L-glutamate in the α-helical form, in 2,2,2-trifluoroethanol, with schematic breakdown into three components. [From Ref. (16), reprinted by permission of the American Chemical Society.]

parallel and antiparallel sheet arrangements are shown in Fig. 6.12. Also included are comparative oscillator strengths (17a), which are indicative of the relative intensities. It can be seen that in the parallel arrangements, the $\pi - \pi^*$ transition is split into two bands; one is polarized parallel to the chain axis, and the second is polarized mainly in plane but perpendicular to the chain axis (17b). In the antiparallel arrangement, three components are produced, two perpendicular to the molecular axis and one parallel. It is noticeable that the theoretical absorption spectrum of the antiparallel conformation is shifted to the red compared with the parallel arrangement. Poly-L-lysine in the β conformation in aqueous (saline) solution (18) shows a maximum absorbance at 195–196 mμ, but the conflicting calculated values do not allow for discrimination between parallel or antiparallel conformation. The $n - \pi^*$ is expected to occur at 218 mμ for both parallel and antiparallel structures, but its contribution is probably even smaller than for the α helix.

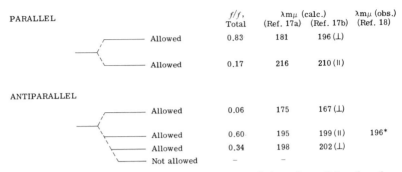

		f/f, Total	λmμ (calc.) (Ref. 17a)	λmμ (calc.) (Ref. 17b)	λmμ (obs.) (Ref. 18)
PARALLEL	Allowed	0.83	181	196 (⊥)	
	Allowed	0.17	216	210 (‖)	
ANTIPARALLEL	Allowed	0.06	175	167 (⊥)	
	Allowed	0.60	195	199 (‖)	196*
	Allowed	0.34	198	202 (⊥)	
	Not allowed	–	–		

Fig. 6.12. Energy levels arising from Davydov splitting of parallel and antiparallel β-sheet forms. Diagram also shows oscillator strengths and band positions. Asterisk value is for poly-L-lysine in aqueous solution and films.

The so-called "disordered" form of poly-L-lysine has an absorption maximum at 192 mμ, so that the formation of antiparallel or parallel structures in solution would involve a red or blue shift, respectively. It is not expected that a splitting of the 195–198 mμ band for the antiparallel arrangement could be detected experimentally, and indeed it has not been observed.

Other polypeptide conformations which have been studied, both theoretically and experimentally, are those of poly-L-proline (19). For both *cis*-poly-L-proline I and the *trans* form, poly-L-proline II, the $\pi - \pi^*$ amide absorption band is expected to be split into two components, as shown in Fig. 6.13, with the high-energy and low-energy branches polarized perpendicular and parallel to the helices, respectively. It can be seen that there are considerable expected differences in the ultraviolet spectra of the two forms. The $n - \pi^*$ transition has been calculated (19) to be two orders of magnitude smaller in intensity than the $\pi - \pi^*$ bands of form I, and even less for form II, and is not expected to contribute to the ultraviolet spectrum. If a random form exists it is expected to have an absorption maximum at approximately 206 mμ. Figure 6.14 shows

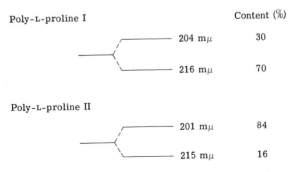

Poly-L-proline I		Content (%)
	204 mμ	30
	216 mμ	70
Poly-L-proline II		
	201 mμ	84
	215 mμ	16

Fig. 6.13. Calculated band splitting and relative intensities for poly-L-proline.

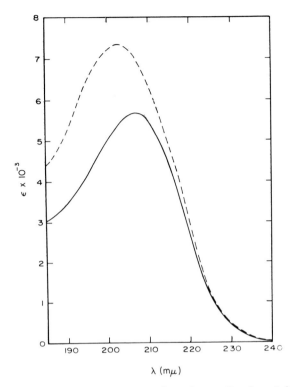

Fig. 6.14. Ultraviolet absorption curves for poly-L-proline form I (solid line) and II (dashed line). [From G. D. Fasman and E. R. Blout, *Biopolymers* **1**, 3 (1963).]

experimental UV spectra of forms I and II. It can be seen that form II has an absorption maximum shifted slightly to lower wavelength, in qualitative accord with theory.

Vacuum Ultraviolet Spectra

To search for the predicted transitions of the chromophore in the vacuum ultraviolet region, sophisticated equipment must be used and only solid films can be examined. A recent study (20) of a number of polyamino acids and sequential polypeptides has shown that generally three to five bands occur in the 100–200 mμ range, with two of them lying in a general band of increasing intensity below 150 mμ. This tail may result from a combination of side-chain transitions and ionization, and definite assignments cannot be made at present. A typical spectrum (for poly-L-valine) is shown in Fig. 6.15.

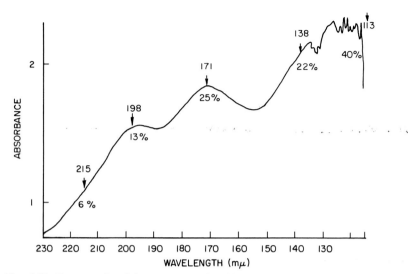

Fig. 6.15. Vacuum ultraviolet spectrum of a film of poly-L-valine cast from TFA; percentages refer to the proportion of the total integrated intensity. (Taken from Ref. 20.)

The main new feature is a band at about 170 mμ, which probably originates in the $\pi_2 - \pi^*$ transition. This band appears to be sensitive to specific residues and their conformation. The effect of the side chains on peak position appears most clearly in the (aliphatic) amino acids themselves. Increasing bulk causes a hypsochromic shift (Table 6.4a). Similarly, there is a blue shift with decreasing proline content of polypeptides and proteins in the poly-L-proline II (or similar) conformation, as shown in Table 6.4b. The band at ~170 mμ is generally sensitive to conformation via its intensity (see Table 6.5). If aromatic groups are present in the side chain, additional bands are observed; for example, α-helical poly-γ-benzyl-L-glutamate has the usual n $- \pi^*$ (219 mμ), $\pi - \pi^*$ (196 mμ), and $\pi_2 - \pi^*$ (164 mμ), with additional

TABLE 6.4a

Vacuum Ultraviolet Band Position for Aliphatic Amino Acids

Amino acid	Side-chain mol wt	$\pi_2 - \pi^*$ Position (mμ)	Intensity (%)
Glycine	1	175	11
L-Alanine	15	170	14
L-Valine	43	166	14
L-Leucine	57	163	15

TABLE 6.4b

Vacuum Ultraviolet Band Position for Polymers in the Polyproline II Conformation

Polymer	Conformation	Band position, $\pi_2 \to \pi^*$ (mμ)	Pro (%)
Poly-L-proline	PPII	172	100
Poly-L-hydroxyproline	PPII	171	100
Poly (Gly-Pro-Pro)	PPII triple helix	170	66
Collagen	PPII triple helix	165	~ 20
Poly (Gly-Gly-Ala)	PPII	162	0

side-chain bands at 209 and 186 mμ. There are also, as usual, two bands that appear to underlie the low-wavelength tail at 136 and 119 mμ.

Brahms and co-workers (21) have attempted a further refinement of vacuum ultraviolet measurements by using polarized radiation. Figure 6.16 (21) shows the spectra of oriented poly-γ-ethyl-L-glutamate in polarized and nonpolarized radiation. The former clearly established the $n - \pi^*$, two $\pi_1 - \pi^*$ transitions, and the additional parallel band at 155 mμ. The low wavelength band is probably due to the $\pi_2 - \pi^*$ transition.

TABLE 6.5

Conformation/Intensity Relations

Conformation	Ratio $\dfrac{190 \text{ m}\mu}{165 \text{ m}\mu}$
Beta or extended	
Poly-L-valine	0.50
Poly [Gly-Ala-Glu(OEt)]	0.44
Poly-L-alanine	0.65
Polyglycine I	0.75
Poly-L-serine	0.57
Helical polypeptides	
Poly-L-proline I	1.2
Polyhydroxy-L-proline II	2.1
Poly (Gly-Pro-Pro)	1.2
Poly (Ala-Gly-Gly)	2.0
Poly-L-proline II	1.8
Poly-γ-benzyl-L-glutamate	1.4

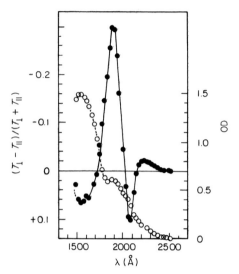

Fig. 6.16. The spectra of oriented poly-γ-ethyl-L-glutamate: solid line, linear dichroism transmittance expressed as $(T_\perp - T_\parallel)/(T_\perp + T_\parallel)$ $p = (T_\perp - T_\parallel)/2T_m$; dashed line, conventional absorbance (optical density). (From Ref. 21, with permission of the National Academy of Sciences.)

Conformational Transitions in Polypeptides

The helix–coil transition in proteins and nucleic acids has received a great deal of attention over the years. Popular models for studying such conversions have been poly-L-glutamic acid (PLGA) and poly-L-lysine (PLL).

The conformation of PLGA changes from an α helix at low pH, where the side chains are protonated $-COOH$, to a "coil" form at high pH, where the side chains become charged $-COO^-$. The conformational change is induced by the strong electrostatic repulsion between charged side chains and has been followed by a large number of techniques, including most forms of spectroscopy. The transition is generally found to occur at about pH 5 in the absence of additional electrolyte.

Similarly, PLL undergoes a transition from the charged "coil" form at neutral pH ($-\varepsilon NH_3^+$) to the helix form above pH 10 (neutral side group). At elevated temperature and high pH, an additional conformational change is observed for PLL, the polypeptide taking on a β conformation, which may be stabilized by superfolded chains or by molecular aggregation. The ultraviolet spectra of PLL are shown in Fig. 6.17.

Although there is fairly good evidence that the neutral forms of PLGA

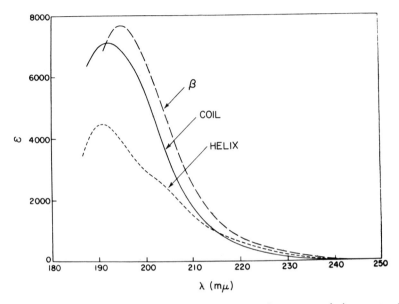

Fig. 6.17. Ultraviolet absorption spectra of poly-L-lysine in aqueous solution: extended coil, pH 6.0, 25°C; α helix, pH 10.8, 25°C; β conformation, pH 10.8, 52°C. [From K. Rosenheck and P. Doty, *Proc. Nat. Acad. Sci. U.S.* **47**, 1775 (1961) with permission.]

and PLL are indeed α helical (for example, the polarization of the UV bands is in agreement with theory), there has arisen some controversy over whether the coil form is truly random in nature. We shall see that when circular dichroism measurements are used, three types of structure are discernible for PLGA and PLL solutions at room temperature. The first is the α helix, the second is a coil, (or charged extended helix), and the third is a " collapsed " form produced in concentrated salt (e.g., calcium chloride) solutions, which is thought to be truly random (22). The coil form is believed to resemble a threefold left-hand (PGII) helix with approximately 2.5 peptides/turn maintained by electrostatic repulsion between side chains (23), whereas the " collapsed " form resembles denatured proteins, particularly gelatin. For the present purposes, the terminology helix–coil transition will be maintained in line with the vast majority of the literature.

From Fig. 6.17 it can be seen that considerable hyperchromism, i.e., increase in absorption, accompanies the helix–coil transition of PLL in the $\pi - \pi^*$ region, although there is apparently slight hypochromism in the $n - \pi^*$ tail region of the curve. A similar effect is generally seen on denaturation of proteins.

Ultraviolet Spectra of Polypeptides with Chromophoric Side Chains

If an ultraviolet absorption analysis is being performed with a protein or polypeptide containing absorbing side groups, steps have to be taken to assess this contribution before a conformational analysis can be carried through. The molar absorptivity of the most significant of these amino acids is given in Table 6.2. Clearly the largest absorption arises from the three aromatic amino acids: tryptophan, tyrosine, and phenylalanine, and in principle it should be readily possible to assess the contribution of the side chains to the total molecular absorbance from the overall primary structure of the material.

In practice we may expect that the contributions to the spectrum of these side chains will change with the environment. For example, the absorptivity of a tyrosine residue in the interior of a protein is very different from that when the residue is freely exposed to the surrounding solvent. In an interesting demonstration of this effect, it has been shown (24) that the spectra of tyrosine and histidine are modified when the amino acids are enclosed in micellar structures, the interior of which might be taken as a model protein environment. In any case, the total changes are rather small if there is only a small content of such residues in the polypeptide or protein in question.

The helix content of a polypeptide or protein can now be obtained from the ultraviolet spectrum by use of the relation

$$H_c = \frac{N\varepsilon_c + \Sigma_i \varepsilon_i n_i - 10A_\lambda M/cl}{N(\varepsilon_c - \varepsilon_h) + \Delta \Sigma_i \varepsilon_i n_i} \tag{6.8}$$

where A_λ is the measured absorbance at wavelength λ, M is the molecular weight, N is the total number of amino acid residues, ε_h and ε_c are the molar absorptivities in the helical and "random" states, ε_i is the molar absorptivity of the ith side chain, and n_i is the number of such side chains per molecule; c and l are as usual the percent weight concentration and path length (in centimeters) of the light in the absorption cell, respectively. The summation term in the denominator which accounts for environmental effects on the side chains has generally been treated as negligible; however, in practice the above equation gives only a very rough estimate of helicity and is now rarely used. In fact, the assumption is made in the derivation that only true random and helical forms are present in proteins, an assumption that is not generally valid.

Polynucleotides

As shown in the previous section, on transformation of polypeptides from helix to random form in solution, there are changes in intensity, frequency, polarization, and line shape in the ultraviolet absorption spectrum. On the other hand, polynucleotides show little apparent change in frequency and band shape on transformation of double-stranded helix to random-coil form. There is, however, a pronounced decrease in absorption intensity (hypochromism) in the $\pi - \pi^*$ transition near 260 mμ and a slight increase in the n $- \pi^*$ high-wavelength component of the tail. The hypochromicity has been used extensively as a measure of the ordered conformation in nucleic acids.

The usual method of data presentation is in terms of hyperchromicity, i.e.,

$$h_{(\lambda)} = [H_\lambda^{-1} - 1] = \left(\frac{\varepsilon(\lambda) \text{ [disordered form]}}{\varepsilon(\lambda) \text{ [ordered form]}}\right) - 1$$

$$= \left(\frac{A(\lambda) \text{ [disordered form]}}{A(\lambda) \text{ [ordered form]}}\right) - 1 \qquad (6.9)$$

where the A's are absorbancies. However, it is to be noted that the absorbance of the disordered (e.g., heat-denatured) forms of polynucleotides need not necessarily coincide with the sum of the component monomers. In fact heat-denatured DNA and synthetic polymers each show residual hypochromicity. Following the notation of h representing the hyperchromicity of the denatured form and h^1 being the value based on monomer absorption, data in Table 6.6 shows that the two values do not coincide (24). The hyperchromicity values for various RNA's and DNA's generally fall in the range $0.18 < h < 0.56$.

Although the exciton theory of polynucleotides has not yielded accurately predictable values, the relation between the polarization of various absorption

TABLE 6.6

Hyperchromism of Synthetic
Polynucleotides at $\lambda_{\max}$

Polynucleotide[a]	h	h^1
Poly A		0.54
Poly C		0.40
Poly G		0.39
Poly U	0.06	0.10
Poly T		0.40
Poly I		0.62
Poly (A + U)	0.24	
Poly (dC + dC)		0.21
Poly (dG + dC)	0.21	0.42
Poly (dI + dC)	0.53	0.95
Poly (dBC + dBC)		0.25
Poly (dI + dBC)	0.45	0.63
Poly (dAdT + dAdT)	0.42	0.84

[a] BC is 5-bromocytidylic acid.

bands and the orientation of the biopolymer are of some interest. We have seen that the high-wavelength bands have their origin in the bases, and further information has been derived from a consideration of the vacuum ultraviolet spectra (21). Only two spectra seem to be currently available, one for poly (A + U) and the other for DNA (calf thymus). In Fig. 6.18 the absorbance

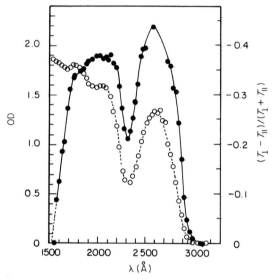

Fig. 6.18. The spectra of oriented poly (A + U) double-stranded helical complex: solid line, degree of polarization expressed as a ratio $(T_\perp - T_{\parallel})/2T_m$; dashed line, absorbance (optical density). (From Ref. 21 with permission of the U.S. National Academy of Sciences.)

and polarization $(T_\perp - T_{\parallel}/T_\perp + T_{\parallel})$ of an oriented film of poly (A + U) are compared in the 150–300 mμ spectral region. (Note that the polarization scale on the right-hand side is negative.) A large negative contribution arises from electronic transitions perpendicular to the helix axis (the fiber and helix axes are assumed to coincide), and such is true for the familiar 260 mμ band, as well as for bands at 206 and 177 mμ. Since the 260 mμ band arises from $\pi - \pi^*$ transitions in the plane of the purine and pyrimidine bases, the spectrum (Fig. 6.18) provides supporting evidence for the perpendicular arrangement of these bases. The n $- \pi^*$ transition, which should be perpendicular to the base plane, could not be detected. The two other negative bands have also been correlated with transitions in the bases which occur at 210 and 190 mμ in uridine and 206 and 190 mμ in adenosine.

The vacuum ultraviolet spectrum of thymus DNA is shown in Fig. 6.19 and here, apart from the two main bands at 255.5 and 189 mμ, there are

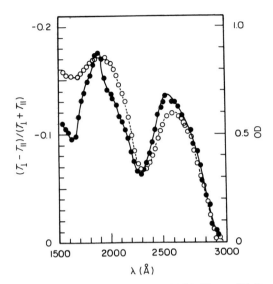

Fig. 6.19. The spectra of oriented calf thymus DNA film: solid line, degree of polarization; dashed line, absorbance (optical density). (From Ref. 21 with permission.)

several shoulders indicating components at 213, 200, and 177 mμ, details that do not readily emerge from the absorbance curve. A rather interesting aspect of the long-wavelength band in the polarized spectrum is its displacement from 260 mμ observed in the regular transmission curve. This appears to indicate that there are additional weak transitions involved which lie parallel to the helix axis.

Optical Rotation

In the previous section, attention has been focused on the absorption of unpolarized or linearly polarized light and the accompanying forced oscillations of the charge distribution in the absorbing molecule. The linear displacement of charge results in oscillations of the electric dipole moment; the directional properties of such oscillations often lead to identification of electronic transitions in the ultraviolet and visible spectrum. In the optical rotation of biological macromolecules, it is the product of the transition electric and transition magnetic dipole moments that controls the magnitude of the effect. Although many of the early studies of the optical properties of biological molecules were made by measuring the optical rotation of light at a single wavelength (usually the sodium D line) it is now well recognized that much more information can be obtained by studying the absorption or dispersion of radiation as a function of wavelength. Some excellent reviews of these topics have appeared recently (5, 25, 26), and the reader is referred to these sources for detailed information.

Optical Rotatory Dispersion

In general, there is a straightforward mathematical relationship between absorptive and dispersive properties, which is given by the Kronig–Kramers transform (27). This relation is shown schematically in Fig. 6.20. It can be

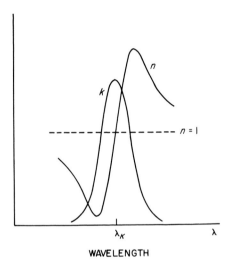

WAVELENGTH

Fig. 6.20. Refractive index (n) and absorption coefficient (k) in the vicinity of an isolated band. [From A. Abu Shumays and Jack J. Duffield, *Anal. Chem.* **38**, 29A (1966) by the American Chemical Society. Reprinted by permission of the copyright owner.]

seen that the dispersive effect remains substantial in the wavelength region far removed from the specific absorption band, and it is this property that enables optical rotation measurements to be made in the visible spectral region.

However, the preceding relations apply near or in an absorption band. In an optically transparent region of the spectrum, there is some contribution to the dispersive effect from any distant absorption bands; a single band dominates the dispersion near its absorption maximum. On this basis various equations have been proposed as quantifying the optical rotatory dispersion of biopolymers. It is now recognized (28) that the splitting of initially degenerate electronic states (exciton splitting) can also cause contributions that

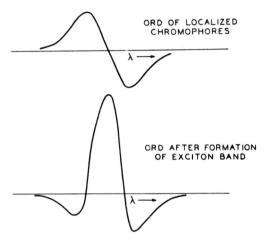

ORD OF LOCALIZED
CHROMOPHORES

ORD AFTER FORMATION
OF EXCITON BAND

Fig. 6.21. Schematic representation of the changes in optical rotatory dispersion on formation of an exciton band. [From Ref. 21, by permission of John Wiley and Sons Inc.]

considerably modify the earlier interpretations of the dispersion effect [see Fig. 6.21 (28)].

The quantum theory of optical rotation was presented in 1928 by Rosenfeld (29), who noted that the mean residue rotation at a wavelength far removed from an absorption band was given by

$$[m']_\lambda = \left[\frac{3}{n^2 + 2}\right] \frac{M}{100N} [\alpha]_\lambda$$

$$= \frac{96\pi N_0}{hc} \sum \text{Im} \langle \mu_i \rangle \langle m_i \rangle \frac{\lambda_i^2}{\lambda^2 - \lambda_i^2} \tag{6.10}$$

where n is the refractive index of the solvent at wavelength λ, the first bracketed term on the right being the Lorentz correction for the effect of the solvent medium; M is the molecular weight, N the number of peptide residues per molecule, $[\alpha]_\lambda$ the specific rotation at wavelength λ; $[\alpha]_\lambda = \alpha/l'c$ where l' is the light path in centimeters and c is the concentration in grams per milliliter; N_0 is Avogadro's number and the summation term includes the imaginary part of the (vector) contribution of the electric (μ_i) and magnetic (m_i) dipole moments; λ_i is the absorption wavelength of the ith band.

In interpreting the above equation in terms of a physical model, it can be seen that the rotation effect originates in a coupling of the electric and magnetic moments. Since the vector product is zero for perpendicular electric and magnetic moments, which occur in symmetric, i.e., nonoptically active, molecules, only molecules with optically asymmetric centers can cause rotation of incident radiation.

Application to Polypeptides

In terms of the elementary molecular orbital theory used previously, it can be noted that, for example, in an $n - \pi^*$ transition, a charge rotation occurs along the O–C axis of the peptide bond which is associated with a large magnetic moment. Thus, in an optically active molecule, the $n - \pi^*$ transition is likely to contribute more strongly than it does in the absorption situation, where only a small electric moment is involved.

Experimental ORD curves often show more than one maximum or minimum. These ORD "bands" or Cotton effects are termed negative if the rotational band is negative as the wavelength decreases and positive if the peak is positive with *decreasing* wavelength. Positive and negative Cotton effects are shown schematically in Fig. 6.22.

For polypeptides in the wavelength region $> 300 \, m\mu$, Eq. 6.10 reduces approximately to

$$[m']_\lambda = a_0 \lambda_0^2 / (\lambda^2 - \lambda_0^2) \tag{6.11}$$

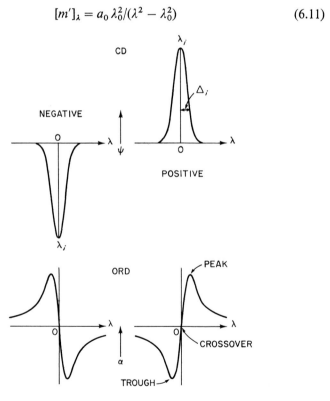

Fig. 6.22. Idealized Cotton effect of an isolated, optically active absorption band with its maximum at λ_i. The left-half is a negative and the right-half a positive Cotton effect.

known as the Drude equation. Both a_0 and λ_0 are empirical constants, which, for random polypeptide chains in aqueous solution at $\sim 25°C$, are approximated by $a_0 = -600$ and $\lambda_0 = 220$ mμ. The value of a_0 changes with temperature and solvent. Proteins also obey the Drude equation for $\lambda > 350$ mμ with considerable variations in a_0 and λ_0 due to specific structural and conformational effects.

An extension of the Drude equation to shorter wavelength requires the introduction of a second term and has been found applicable to helical polypeptides (i.e., polypeptides known to exhibit the α helix in the solid state). The equation introduces one further empirical constant b_0:

$$[m']_\lambda = \frac{a_0 \lambda^2}{\lambda^2 - \lambda_0^2} + \frac{b_0 \lambda_0^4}{(\lambda^2 - \lambda_0^2)^2} \qquad (6.12)$$

In a theoretical approach, Moffitt and Yang (30) derived this equation on the basis of no exciton splitting. The term b_0 has been found especially to be characteristic of the helical nature of the polypeptide and its sign was expected to reflect the left- (positive) or right- (negative) handedness of the polypeptide helix. Although certain aspects of this theory have been found unsatisfactory, it is now known that essentially all α-helical molecules do obey this equation, with $b_0 \sim -630$, $\lambda_0 \sim 212$ mμ, and a_0, a solvent-dependent constant. The Moffitt–Yang equation is still used extensively for measuring the helical content of polypeptides and proteins in solution. Most α helices of L-amino acids are believed to be right-handed, both in the solid state and in solution, and give b_0 values close to -630. Exceptions are poly-L-tyrosine and poly-L-tryptophan, which are right-handed but which give positive b_0 values and certain esters of poly-L-aspartic acid, which are left-handed and also give b_0 of approximately $+600$. An analysis of this effect is presented later.

If it is assumed that for random structures that $b_0 = 0$, then the " effective " helical content of proteins and polypeptides may be estimated by

$$f_{\text{helix}} = |b_0(\text{measured})/630| \qquad (6.13)$$

It should be noted that although each amino acid in a polypeptide is optically active, the contributions to the ORD and CD curves originate almost entirely in the conformationally sensitive peptide transition moments, and thus these techniques are well suited for conformational measurements in solution.

The relationship between structures other than the α helix and random form are relevant to a discussion of ORD. The technique is often first used as a qualitative guide to the conformation of a polypeptide in solution or in a solid film. In such cases the wavelength of the maxima and minima of the Cotton effect and the crossover points are noted and compared with standard

TABLE 6.7

Optical Rotatory Dispersion Data for Polypeptides[a]

Spectral Characteristics	β Sheet		Helices					True random	Gelati
	I	II	α	PPI	PPII	Charged coil[b]	PGII[c]		
Max (+)	205	210–215	198 215 (sh)	194	189 230[d]	226	228		
Min (−)	230 190	240 194	233	203	216	204	201	206	204
Crossover	220 196	230 205	224 190	211	203	198			200

[a] All wavelengths in millimicrons.
[b] Charged PGA, PLL; see Ref. 19.
[c] For poly (Gly-Gly-Ala); see Ref. 26.
[d] Very weak peak—with negative rotation.

conformations. Table 6.7 contains a summary of data for different conformations. It is immediately evident that each conformation gives a distinctive dispersion curve. The two β-sheet forms (31) are based on films of (a) poly-L-lysine precipitated at pH 11 and heated to 52°C for 10 min and (b) poly-CBZ-methyl-L-cysteine. Although the specific conformations have not been identified for both of these forms, poly [Gly-Ala-Glu(OEt)] gives form βI, and is known to be a superfolded cross-β sheet; form βII has not been thoroughly characterized in any known counterpart. (It appears possible that the βII form is identical with form I, but the spectral curves are shifted by the presence of the blocking group, CBZ.) Another interesting feature is the similarity between the coil form and the polyglycine II helix as recently identified for poly (Gly-Ala-Ala) (32). Such a relationship appears to give some weight to the concept that the coil form is not as random as was previously thought. Furthermore, the disordered form of poly (Gly-Gly-Ala) resembles that of gelatin and polypeptides treated with calcium chloride, supporting the concept that a truly random form has finally been attained.

It is evident that with these basic conformations contributing in varying degrees to the total dispersion curves for proteins, some difficulty may be encountered in separating them. Attempts to characterize the percentage helicity of protein molecules might therefore appear to be of limited validity. Table 6.8 contains b_0 values for various uniform conformations. Although all the forms of Table 6.7 are not available for comparison it is rather remarkable that three of the conformations give b_0 values of close to zero, i.e., they obey

TABLE 6.8

ORD (b_0 Parameter) for
Various Conformations

Conformation	b_0
α Helix[a]	−630
β Sheet	≤ −200
Coil	0
PPII	0
Collagen	0

[a] Poly-L-lysine.

the Drude equation down to a region close to the Cotton effect. No data seem to be available for the collapsed or disordered form, but here again values close to zero might be expected. Estimates of the α-helix content of proteins have tended to ignore the β contribution, which renders the equation unsolvable. With this reservation, however, the apparent helix content of a number of proteins has now been examined, and wherever it has been possible to check the data against other methods (x-ray crystallography and nmr methods) the numbers generally appear to be in fairly good agreement. Table 6.9 contains some of these data.

TABLE 6.9

Estimated Helical Contents of Several
Proteins in Aqueous Solutions

Protein	Fraction α helix
Tropomyosin	0.90
Hemoglobin	0.70–0.80
Myoglobin	0.70–0.80
Serum albumin	0.45
Insulin	0.40
Fibrinogen	0.35
Ovalbumin	0.30
Lysozyme	0.30
Histone	0.20
Ribonuclease	0.15
Chymotrypsinogen	0.10
β-Lactoglobulin	0–0.10

Circular Dichroism

Plane-polarized light may be resolved into two circularly polarized components by passage through a quarter-wave plate. A schematic representation of the plane-polarized wave and its relation to the circularly polarized components is shown in Fig. 6.23; OL and OR are the rotating components of the left- and right-hand polarized light and OA is the vector sum, which represents the amplitude of the plane-polarized light. As OL and OR rotate, OA describes a sine wave motion along the axis of the light path. In essence the quarter-wave plate causes a phase difference between the left- and right-hand polarized light, which may then be used in separating the components.

The absorption of circularly polarized light by optically active materials occurs only in the regions of electronic transitions. In ORD terminology, these are the regions of the Cotton effects. The advantage of the CD technique over ORD is that these Cotton effects may be more clearly resolved and identified with the underlying electronic transition because of the " sharpness " associated with the absorption phenomena compared with dispersive phenomena. Put another way, the dispersion effect is concerned with the sum total of electronic transitions whereas, at least in principle, a circular dichroism band arises from a single transition.

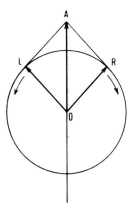

Fig. 6.23. Resolution of electric vector of plane-polarized light into electric vectors of right and left circularly polarized light.

Interaction with Optically Active Materials—Ellipticity

If an optically active compound is exposed to alternating left- and right-hand polarized light in the absorption region of its spectrum, then one of the components will be absorbed to a greater extent than the other. As a result,

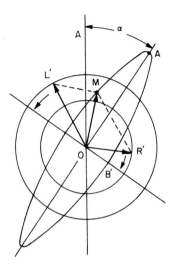

Fig. 6.24. Action of optically active medium on plane-polarized light. Right circularly polarized light absorbed to a greater extent than left, and index of refraction for left greater than that for right, as depicted. (From L. Velluz, M. Legrand, and M. Grosjean, "Optical Circular Dichroism," Academic Press, New York, 1965.)

the emergent light will be elliptically polarized. A diagrammatic represen-tation of this situation is shown in Fig. 6.24 where, as before, the left- and right-hand components precess and the resultant is vector OM, which rep-resents the intensity of transmitted light. Optically active materials by defini-tion have different refractive indices for left- and right-hand polarized light, which is equivalent to saying that the velocity of left- and right-hand polarized light is different. Hence OL' and OR' precess at different rates and their vector sum, OM, traces an elliptical path. The ellipticity per unit length of sample, θ, is then defined in terms of the minor and major axes of this ellipse, $\tan \theta = OB'/OA'$. The angle α is the optical rotation, to which we have already referred, and is related to the refractive index of the medium by the Fresnel equation,

$$\alpha(\text{radians}) = \pi l(n_L - n_R)/\lambda \tag{6.14}$$

l, is again the path length of the light in the medium.

The molar ellipticity† $[\theta]_\lambda = 3300 \, (\varepsilon_L - \varepsilon_R)$, and evidently can be positive or negative in sign depending upon the relative magnitude of the molar

---

† It is to be noted that all work with macromolecules is on the basis of moles of residues, thus $[\theta]$ and $[m]$ are subsequently in this chapter defined as mean ellipticity or mean rotation per mole residue. Solvent corrected parameters are designated $[\theta']$ and $[m']$ respectively.

extinction coefficients ε_L and ε_R. The relationship between the CD and ORD of an isolated optically active absorption band is once again given by the Kronig–Kramers transform. The correlation between positive and negative Cotton effects in ORD curves and CD bands is shown schematically in Fig. 6.22. The algebraic expression of these relationships is given by

$$[m_i]_\lambda = 2/\pi \int_0^\infty [\theta_i]_\lambda [\lambda/(\lambda^2 - \lambda'^2)] \, d\lambda \qquad (6.15)$$

or alternatively

$$[\theta_i]_\lambda = -(2/\pi\lambda) \int^\infty [m_i]_\lambda [\lambda'^2/(\lambda^2 - \lambda'^2)] \, d\lambda \qquad (6.16)$$

Generally the ellipticity curves are displayed on current instrumentation and form the basis for conformational analysis. However, for theoretical purposes, a parameter R_i, the rotational strength, analogous to the oscillator strength in absorbance work, is often quoted. Its relationship to the molar ellipticity is given by

$$R_i = 0.696 \times 10^{-42} \int^\infty ([\theta_i]_\lambda/\lambda) \, d\lambda \qquad (6.17)$$

Although CD bands for simple homobiopolymers are usually fairly well resolved, in more complex spectra resulting from proteins or polynucleotides it is sometimes necessary to resort to curve-fitting procedures. In these cases, the form of individual bands is taken as Gaussian and use of the band half-width Δ_i is made, i.e.,

$$[\theta_i]_\lambda = [\theta_i]^0 \exp[-(\lambda - \lambda_i^0)^2/\Delta_i^2] \qquad (6.18)$$

(Actually Δ_i is the wavelength interval over which θ falls to e^{-1} of its maximum value $[\theta_i]^0$ at wavelength λ_i^0).

Circular Dichroism of Polypeptides

As with ORD measurements, CD may be used both qualitatively and quantitatively. In the first place the familiar $n - \pi^*$ and twin $\pi - \pi^*$ transitions of parallel and perpendicular polarization for the α helix are readily delineated. In fact the CD curves for several of the more common polypeptide conformations are shown in Fig. 6.25 and are summarized with tentative assignments in Table 6.10. It is again noteworthy that the coil form of polyglutamic acid and poly-L-lysine are remarkably similar to the polyglycine II CD spectrum as established for poly (Gly-Gly-Ala). Figure 6.26. shows the CD spectrum of poly-N-methylalanine, which is believed to be in the right-

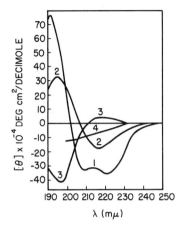

Fig. 6.25. Circular dichroism curves showing (1) α, (2) β, (3) charged coil, and (4) random (CaCl₂ solutions) of poly-L-lysine.

handed PGII conformation. Also included is poly (Gly-Gly-Ala) which, by inference from its inverted form, is a left-hand PGII helix in aqueous solution.

It should be noted that apart from the normal instrumental difficulties associated with the absorbance of oxygen in the far-UV region, many ions also absorb below 200 mμ. There has been an increasing interest in the circular dichroism of polypeptide films, and in many cases spectra identical with those obtained from solutions can be produced. However, if the material is strongly birefringent and/or oriented, additional complexities can result (31, 33).

TABLE 6.10

Circular Dichroism Data for Polypeptides[a]

Spectral characteristics	β Sheet		Helices				Random[c]	Gly-Gly-Ala	
	I	II	α	PPI	PPII	Extended[b]		H₂O	Heated
Max (+)	196	203	191	215	226	216		213	
Min (−)	219[d]	228[d]	221[e] 209	200	207	201	205	192	199
Crossover	207	217	207	205	220	210		202	

[a] All wavelengths in millimicrons.
[b] Charged PGA, PL; see Ref. 19.
[c] Polypeptide in 4 M CaCl₂ or denatured collagen.
[d,e] All maxima and minima probably correspond to $\pi - \pi^*$ transitions, with the exception of [d] mixed $\pi - \pi^*$ and $n - \pi^*$ and [e] $n - \pi^*$.

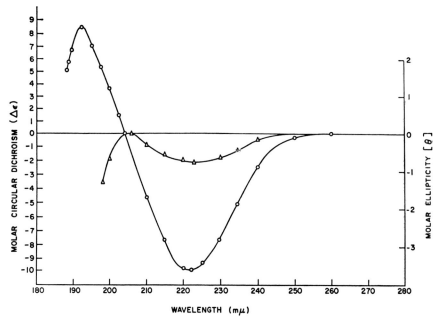

Fig. 6.26. Circular dichroism of poly-*N*-methyl-L-alanine (○) and *N*-acetyl-*N*-methyl-L-alanine methyl ester (△) in trifluoroethanol [From Goodman *et al.*, *J. Amer. Chem. Soc.* **89**, 1265 (1967), by permission of copyright owner.]

Absorbing Side Chains

Poly-α-amino acids with side chains that absorb in the UV/visible region have provided particularly challenging conformational problems. Since these bands usually occur at wavelengths > 220 mμ they tend to confuse the evaluation of ORD data and the misassignment of helix sense has not been uncommon. The CD analysis may be divided into three parts: that of the water-soluble aromatic biopolymers poly-L-tyrosine, poly-L-histidine, and poly-L-tryptophan; the insoluble biopolymers, such as poly-L-phenylalanine; and solid-state films.

In the case of the aromatic homopolypeptides, the conformations of all of these are pH sensitive. The most extensive work has been performed with poly-L-tyrosine. Its CD spectrum at high pH, where it is a right-hand α helix, is shown in Fig. 6.27. The strong negative ellipticity at 224 mμ affirms the α-helical assignment, but the ellipticity is smaller than would be expected for the n − π* peptide transition alone. Two positive ellipticity bands occur at 270 and 248 mμ due to the side chains; the former shows up in L-tyrosine at neutral pH, the latter at high pH. The "random" form of poly-L-tyrosine shows positive bands at 245 and 220 mμ.

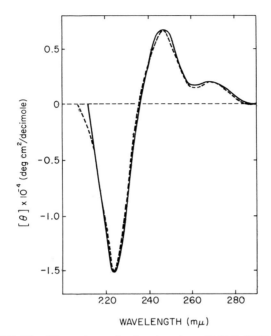

Fig. 6.27. The CD spectrum of poly-L-tyrosine, pH 11.2, 24°C, 0.2 *M* NaCl. [From Ref. 34, reprinted by permission of the American Chemical Society.]

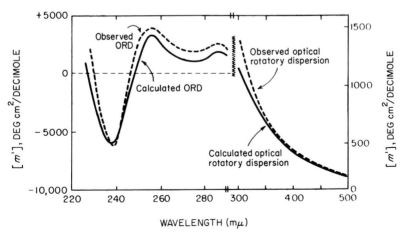

Fig. 6.28. Comparison of experimental and calculated ORD curves for helical poly-L-tyrosine at pH 11.2 in 0.2 *M* NaCl; [M¹] is the residue rotation. [From Ref. 34, with permission.]

It has proven possible to perform the transform of Eq. 6.16 on the CD data of poly-L-tyrosine to produce a fairly good representation of the ORD curve, as illustrated in Fig. 6.28 (34). In this manner it has been possible to show that the incorrect assignment of helix sense based on a Moffit–Yang positive b_0 value arises mainly because of the complicating side bands.

Poly-L-histidine and poly-L-tryptophan show similar but much less marked side-chain effects in the high-wavelength region, although they show strong pH effects, as described in Chapter 9, due to conformation changes. The CD bands of the amino acids themselves are shown in Table 6.11 and all

TABLE 6.11

Molar Ellipticity of Aromatic Amino Acids

	pH Interval					
	1–2		6–8		13	
Amino Acid	λ_{max} (mμ)	$[\theta]$	λ_{max} (mμ)	$[\theta]$	λ_{max} (mμ)	$[\theta]$
L-Tyrosine	200	+			210	7,260
	226	8,315			230	−2,310
	274	1,320			293	1,110
L-Tyrosine	270	960			289–290	1,325
ethyl ester	225	7,600			No discernible band	
					between 210 and	
					260 mμ	
N-Acetyl-L-	270–280	Negative band[a]			285[b]	890
tyrosine		~ −500				
ethyl ester	226	17,300			235–238[c]	13,000
L-Tryptophan	207	−13,860	195	−30,360		
	225	18,810	223	26,070	275	19,470
	265	1,980	269	1,155	270	1,155
	222	17,400	222	8,300	227	19,900
	265	2,600	265	1,900	265–270	2,000

[a] Negative band also observed with N-acetyl-L-tyrosine amide in acid; positive band in alkali.
[b] Spectral maximum at 295 mμ.
[c] Spectral maximum at 245 mμ in all alkaline tyrosine derivatives.

show high-wavelength positive bands. Poly-L-phenylalanine also shows weak, high-wavelength bands in nonaqueous solvents, although much of the recent work on it and the other polymers with active side chains has been performed by transmission through films.

It is of some interest to note that in addition to the polypeptides with aromatic side chains, the cystine residue is also a potential source of CD bands in the high-wavelength region. The free amino acid, for example, gives a fairly strong negative band at 256 mμ with $[\theta] \sim 3 \times 10^{-3}$. Although it is not the intention here to proceed to a detailed study of the CD spectra of proteins, many proteins have been examined by this method, and high-wavelength bands, due presumably to specific residues, are often observed (35).

Measurement of Helix Content by CD

If the CD bands can be assigned specifically to the peptide group without interference from side bands, then the ratio of the oscillator strengths should define the helicity. It is, however, necessary to define oscillator strengths for all conformations, and such data are not currently available in a particularly accurate form. Table 6.12 has been compiled from the literature. In principle,

TABLE 6.12

CD Rotation Strengths[a] for Various Bands Associated with Polypeptide Conformations

Conformation	Band	λ (mμ)	$R_k \times 10^{40}$
α	$n - \pi^*$	222	-19
	$\pi - \pi^*$ ($\parallel$)	205	-19
	$\pi - \pi^*$ ($\perp$)	191	
β(I)	$n - \pi^*$	217	-11
(antiparallel)	$\pi - \pi^*$	195	$+14$
		226	$+6$
	$\pi - \pi^*$ ($\parallel$)	215[b]	
PPII (H$_2$0)		206	-33
	$\pi - \pi^*$ ($\perp$)	201[b]	
	$n - \pi^{*\,b}$	236	
PPI	$\pi - \pi^*$ ($\parallel$)[b]	215	28
	$\pi - \pi^*$ ($\perp$)[b]	203	-18
Extended[c]		217	$+2$
		202	-15
Random		198	-4

[a] Averaged literature values cited.
[b] Theoretical (see Ref. 19).
[c] Charged form of poly-L-glutamic acid.

it should be possible, by working with different bands, to give a quantitative description of the fractions of various conformations present in the solid state (films), which is not possible by the ORD approach. Little progress has been made in this direction.

Nucleosides and Nucleotides

As we have seen, polynucleotides and their constituent subunits absorb radiation in the near-UV region, primarily as a result of electronic transitions in the bases. These base chromophores are, by themselves, optically inactive but when they are attached to optically active sugars they produce Cotton effects in the region of the absorption bands. Thus specific effects that are likely to show up in CD–ORD measurements are concerned with sugar–base configuration, orientation, and localized interactions.

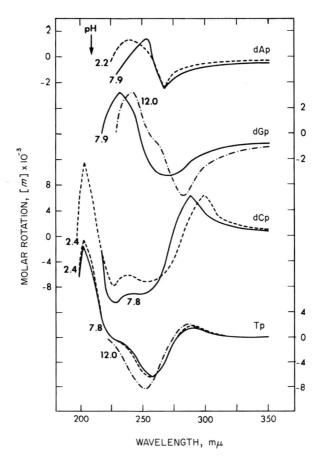

Fig. 6.29. The ORD curves of 5′-deoxyribomononucleotides. [From Ref. 36, by permission of John Wiley and Sons Inc.]

Mononucleosides and Mononucleotides

The ORD curves for the mononucleotides and mononucleosides all show Cotton effects in the 260–290 mμ range. The data in Table 6.13 refer to the various entities at neutral pH, and the crossover point, as expected, correlates with the UV absorption data of Table 6.3. Perhaps the most interesting general feature of the data in Table 6.13 is that the purine derivatives always have a negative Cotton effect and the pyrimidine derivatives a positive effect in the high-wavelength region independent of linkage to ribose or deoxyribose, or of linkage to phosphate. In the shorter-wavelength range, the nucleotides that have been studied (36) all show a strong positive Cotton effect, particularly the pyrimidine derivatives. It is clear, however, from the shape of the curves, that several electronic transitions are involved [see Fig. 6.29 (36)].

Apart from the objective of providing basic ORD data for the study of polynucleotides and nucleic acids, there are two pieces of configurational information which may readily be obtained from ORD spectra, the first being concerned with distinguishing between anomeric forms. If, instead of the β linkage at the C-1 atom of a purine nucleoside, the α anomeric form is synthesized, the sign of the Cotton effect is reversed, i.e., high-wavelength positive effects are observed. On the other hand, the pyrimidine α nucleosides give negative Cotton effects and the β anomers give positive Cotton effects (37).

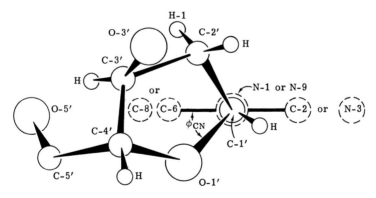

Fig. 6.30. Schematic illustration of the torsion angle in a pyrimidine or purine nucleoside. The plane of the base is viewed end-on with the glycosyl bond between C-1′ and N-1 (or N-9) of the base, perpendicular to the paper. The torsion angle ϕ_{CN} is the dihedral angle between the plane of the base and the plane formed by the C-1′—O-1′ bond of the furanose ring, and the C-1′—N-1′ bond. The angle is zero when O-1′ lies directly in front of C-6 (or C-8 for a purine), and positive angles are measured when C-1′—O-1′ is rotated in a clockwise direction when viewing from C-1′ to N. [Drawing from A. E. V. Hachemeyer and A. Rich, *J. Mol. Biol.* **27**, 369 (1967) and Ref. 5.]

TABLE 6.13

The Cotton Effects of Mononucleosides and 5'-Mononucleotides[a,b]

Substance	Peak λ(mμ)	Peak [m]×10⁻³	Crossover λ(mμ)	Trough λ(mμ)	Trough [m]×10⁻³	Amplitude, [m]×10⁻³	Absorption. λ$_{max}$(mμ)
dA	260	−1.55	258	245	+2.25	−3.80	259.5
A	275	−3.30	256	245	+1.60	−4.90	260
dAMP	268	−2.49	258	253	+1.30	−3.79	260
AMP	272	−3.00	257	246	+1.80	−4.80	260
dG	258	−2.40	250	231	+3.65	−6.05	252.5
G	258	(−2.90)[c]	249	230	(+1.95)	(−4.85)	253
dGMP	270	−3.35	248	233	+3.10	−6.45	252.5
GMP	281	(−1.61)	247	237	(+0.60)	(−2.21)	253
dC	288	+5.85	273	250	−7.20	+13.1	272
C	288	+6.80	273	242	−13.6	+20.4	271
dCMP	288	+6.30	273	245	−9.10	+15.4	272
CMP	288	+7.70	272	240	−13.0	+20.7	270
dT	290	+2.20	277	260	−4.30	+6.50	267
dTMP	290	+1.50	278	257	−6.20	+7.70	268
U	283	+4.00	271	253	−9.80	+13.8	261
UMP	283	+3.50	272	255	−9.70	+13.2	262

[a] For solutions in 0.1 M phosphate buffer (pH 7.5–7.9).
[b] Table from Yang and Samejima (46) p. 233.
[c] Numbers in parentheses are rough estimates.

Secondly, ORD data provide information on the *syn* and *anti* configurations with respect to the sugar–base torsion angle ϕ_{CN} about the glycosyl bond (38), as shown in Fig. 6.30.† There are essentially two favored rotational regions centering around $-30°$ (*anti*) and $+150°$ (*syn*). An extensive consideration of the relation between the electronic transition moment and the geometry of the asymmetric center (5, 37) has led to the conclusion that the ORD spectra of naturally occurring uridine, thymidine, and cytidine and the purine nucleosides all have the *anti* configuration.

Dinucleoside Phosphates

The sixteen dinucleoside phosphate pairs of adenine (A), uracil, (U), guanine (G), and cytosine (C) have been examined by absorption and ORD methods at various pH (39). Although the absorption curves differ only slightly from the sum of the components the ORD curves are significantly different. The low-wavelength hypochromicity and longer-wavelength hyperchromicity are taken as an indication of the tendency for base stacking. Although it is possible to take only an arbitrary (3% hypochromism) "cut off" for stacking, it is possible to establish which pairs prefer to stack, as shown in Table 6.14.

Although the absorption curves seem to show little dependence on the sequence, the same is not true of ORD curves. For example, ApG and GpA

TABLE 6.14

Base Stacking of Dinucleoside Phosphate Pairs[a]

Stacked		Unstacked
GpG	UpG	ApU
CpG	CpC	UpA
GpC	CpU	UpC
ApA	GpA	GpU
ApC	ApG[b]	UpU

[a] For solutions in 0.01 M phosphate buffer $+$ 0.08 M KClO$_4$ (pH 7.0).

[b] Borderline.

† Different conventions have been used for defining the torsion angle. The convention used here is that of the quoted references whereas in Chapter 2, the angle x is taken as a rotation in the opposite sense and differs by $180°$.

at pH 7 have very different ORD curves [see Fig. 6.31 (39)]. Thus, potentially, ORD can be used for analysis of sequential isomers. A theoretical calculation of the exciton splitting for the dinucleoside phosphates has been carried out (40) with the conclusion that the base pairs tend to form part of a right-hand helix.

It is also concluded from the ORD studies of the dinucleoside phosphates (and polynucleotides) that the multiple Cotton effects seen in the near-UV region reflect base stacking.

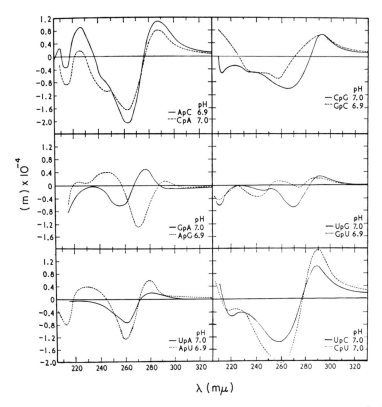

Fig. 6.31. The optical rotary dispersion of the 12 sequence isomers at neutral pH, an ionic strength of 0.1, and 25°C. [From Ref. 39, with permission of Academic Press Inc.]

Trinucleoside Diphosphates and Higher Oligomers

Several trinucleoside diphosphates have been prepared and their ORD spectra examined. Nine combinations are shown in Fig. 6.32. In addition several trinucleoside triphosphates (trinucleotides) have been examined (ApGpCp, GpApCp, ApApGp, ApApCp, ApCpGp, CpApGp, UpApGp,

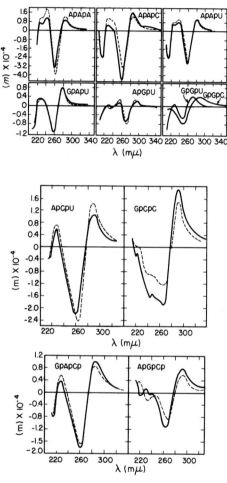

Fig. 6.32. Molar rotation per base of nine trinucleoside diphosphates and two trinucleotides. Solid line, experimental; broken line, calculated using the nearest-neighbor approximation. [From Cantor and Tinoco (48) and *Biopolymers* **5**, 821 (1967).]

ApUpGp, and UpUpGp) (41, 42). The additional phosphate does not seem to affect the optical properties of the trimers greatly and the homotrinucleosides resemble the dinucleotide of the same base. Again, however, the ORD spectrum is quite dependent on sequence; see, for example, GpApU and ApGpU in Fig. 6.32.

Studies of the higher oligomers have concentrated on the homopolymers, mainly with the objective of establishing whether additional secondary or tertiary structure can be detected. The oligomers of adenylic acid show increasing ellipticity per residue of the CD bands as a function of degree of

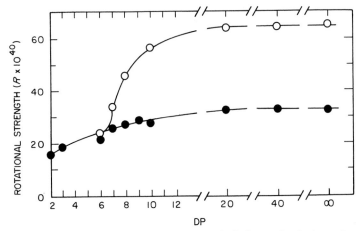

Fig. 6.33. Rotational strength of the positive band of oligo and poly A as a function of the degree of polymerization, DP. Open circles, pH 4.5; filled circles, pH 7.4; temperature, about 0°C. [From Ref. 43, with permission of The Royal Society (London).]

polymerization and also show a marked pH dependence when DP > 6. Figure 6.33 (43) shows the rotational strength of the 270 mμ band of adenylic acid oligomers as a function of DP and neutral and acidic pH. At pH 7.4 the oligomers (and polymer) are believed to be in the single-strand arrangement, whereas at low pH the polymer forms the double-stranded arrangement as do oligomers with DP > 15. A similar situation exists with oligocytidylic acids

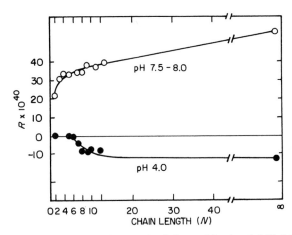

Fig. 6.34. Changes in rotational strength of the positive band (pH 7.5) and negative band (pH 4.0) of 3'→5'-cytidylate oligomers at 0°C (in 0.1 M NaCl, 0.01 M Tris (pH 7.5), and 0.1 M NaCl, 0.05 M acetate (pH 4.0) as a function of chain length. [From Brahms et al. (44) with permission.]

and poly C, as can be seen in Fig. 6.34 (44). Both of the preceding nucleotides have a stacked-base arrangement in the single-strand form, as indicated by the decrease in ellipticity with increasing temperature of solution. This is not apparently the case with the oligomers and polymer of uridylic acid (45) as might be expected from the reticence of di-U (UpU) to form base stacking (Table 6.14). This leads one to expect, in fact, that there might be some fairly straightforward relation between pair interactions and the properties of the polymer. Such a prospect has indeed attracted considerable attention. The approach is discussed below (46, 47).

ROLE OF NEAREST-NEIGHBOR INTERACTIONS

The molar rotation (M) at any wavelength of a dinucleoside phosphate XpY might be written as

$$[M_{xy}] = [M_x] + [M_y] + I_{xy} \qquad (6.19)$$

where I_{xy} is the contribution arising from nearest-neighbor interactions. The molar rotations of nucleosides and nucleotides are usually of comparable magnitude and are small compared with their coupled oligomers. Thus the molar rotation of pY $\sim$ Y $\equiv [M_y]$, etc. For a trinucleoside phosphate XpYpZ, a similar relationship may be written

$$[M_{xyz}] = [M_x] + [M_y] + [M_z] + I_{xy} + I_{yz} \qquad (6.20)$$

If the dinucleoside and trinucleoside are in the same conformation, Eqs. 6.19 and 6.20 may be combined to give

$$[M_{xyz}] = [M_{xy}] + [M_{yz}] - [M_y] \qquad (6.21)$$

If we denote a mean residue rotation $[m] = [M_{ij}]/2$ (for ith and jth residues), then

$$[M_{xyz}] = \{2[m_{xy}] + 2[m_{yz}] - m_y\}/3 \qquad (6.22)$$

For a homopolymer, Eq. 6.22 becomes

$$[m_{poly}] = 2[m_{xx}] - [m_x] \qquad (6.23)$$

But for a random-sequence polynucleotide, the equation becomes much more cumbersome, i.e.

$$[m_{polymer}] = \sum_{x=1}^{4} \chi_x \left\{ 2 \sum_{y=1}^{4} \chi_y [m_{xy}] - [m_x] \right\} \qquad (6.24)$$

where χ_x and χ_y are the mole fractions of X and Y, respectively.

The calculated ORD curves for several trinucleoside diphosphates are in quite good accord with experiment (48), as shown in Fig. 6.32. However the equations have not yet been used for higher oligomers or polymers. Some refinements may be necessary to account for differences in nucleotide and nucleoside spectra (36).

Polynucleotides

The general ORD profile of all polynucleotides (except poly I) consists of two peaks and one trough centered around the 260 mμ absorption band; additional Cotton effects lie at shorter wavelength ($<$230 mμ). Some tabulated data are presented in Table 6.15.

The ORD/CD methods have been used to study the melting characteristics of the homopolymers (at neutral and low pH, and in salt solutions) and complexes of complementary polynucleotides and random copolynucleotides (48). Figure 6.35 shows the ORD curves for the poly (A + U) complex

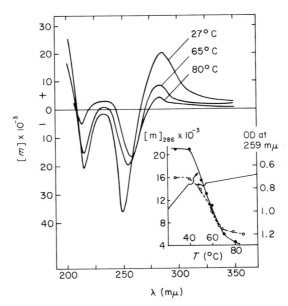

Fig. 6.35. The ORD of poly (A + U) at pH 7.5 in 0.15 M KF. Inset: variation of rotation and optical density with temperature. [From P. K. Sarkar and J. T. Yang, *Biochemistry* **4**, 1238 (1965), with permission.]

at various temperatures and the melting profile based on the 286 mμ Cotton effect. The melting " temperature " is clearly around 60°C, as confirmed by hyperchromicity measurements and relative viscosity data. [The melting refers to the disruption of the base-paired double helix.] Homopolynucleotides generally undergo very broad " melting " in which the base pairs unstack

TABLE 6.15

The Cotton Effects of Polynucleotides

Polymer and solvent	Temperature (°C)	Peak 1		Trough 1		Peak 2		Trough 2	
		λ (mμ)	$[m] \times 10^{-3}$	λ (mμ)	$[m] \times 10^{-3}$	λ (mμ)	$[m] \times 10^{-3}$	λ (mμ)	$[m] \times 10^{-3}$
Poly A (0.15 M KF, pH 7.5)	~1–2	283	+33	257	−101	238	~15	212	−90
	27	283	+26	257	−76	238	~15	212	−66
	80	283	+6	257	−21	238	~11	212	−16
Poly U (0.15 M KF, pH 7.5)	~1–2	284	+25	257	−46	230	~0	213	−15
	27	284	+10	260	−18	220			
	80	284	+6	260	−15	220			
Poly (A + U) (0.15 M KF, pH 7.5)	27	286	+21	250	−36	232	−2	218	−21
	80	284	+5	257	−17	235	+3	212	−5
Poly C (0.1 M NaCl plus 0.01 M cacodylate, pH 7.0)	27	293	+35	268	−45				
	80	293	+17	268	−19				
Poly G^a (0.1 M NaClO$_4$ + 0.001 M cacodylate, pH 7.0)	25	271	+15.5	249	−29	224	+1.0		
Poly (G + C) (0.1 M tris, pH 7.5)	~27–85	276	+33	245	−37	~220–230	−23		
	95	294	+9	270	−26	~240–245	−1.3		
Poly d(AT) (0.15 M NaCl + 0.015 M sodium citrate)	27	283	+7.5	255	−17	236	+16.5		
	85	287	+2.7	263	−11	235	+8.3		

^{a} With another small Cotton effect near 290 mμ, $[m]_{296}$ (peak) $= +9 \times 10^3$ and $[m]_{283}$ (trough) $= +8 \times 10^3$.

in a weakly cooperative manner, i.e., local ordering is believed to occur over relatively short sequences in the single-strand polymer. Poly (G + C) complex melts at a considerably higher temperature ($\sim 90°$) indicating the stronger base pairing for this combination (see Chapter 12).

As mentioned previously, poly A and also poly C both form double helices at low pH. However, the ORD spectrum of AMP† is virtually the same in neutral and acid solution, whereas the CMP† spectrum varies considerably with pH. A comparison of the poly C spectrum in neutral and acid solutions suggests that protonation accompanies the conformation change and development of the double helix (46).

An interesting comparison may be made between the polyribonucleotides and their deoxy counterparts. The difference, for example, between the ORD spectra of poly dA and poly A at neutral pH is quite striking (see Fig. 6.36).

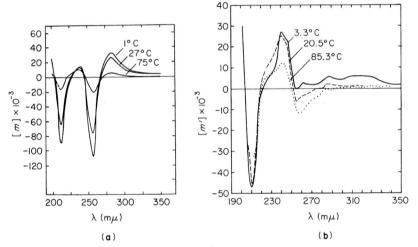

Fig. 6.36. The ORD of (a) poly A (from Sarkar and Yang, see Fig. 6.35) and (b) poly dA [from P. O. P. Ts'o *et al.* (49) with permission.]

This difference becomes minimal at low pH where both occur in the double-helix conformation. The deoxynucleotide poly dA undergoes the transformation at lower pH than poly A, indicating that the 2'-hydroxyl group plays a part in double-helix formation, probably through intramolecular hydrogen bonding, which enhances base stacking (49).

The opposite effect is observed with poly C and poly dC since the former is found to form the less stable double helix (49). This effect is thought to arise from a combination of two effects, the first originating in the reduced rota-

† Monophosphates of bases A and C.

RNA and DNA

tional freedom around the glycosyl bond caused by hydrogen bonding of the 2′-hydroxyl to the 2-carbonyl group, the second originating in the modified dissociation of the ribosyl cytosine group.

Based upon the previous considerations of polyribonucleotides and polydeoxyribonucleotides, it might be expected that the nucleic acids RNA and DNA would have certain similarities and certain specific differences in their ORD/CD spectra, and this has indeed proven to be the case. Figure 6.37 shows typical curves in the 200–320 mμ range. If the peak in the 195–200 mμ region is taken into account, there are three peaks in the general ORD

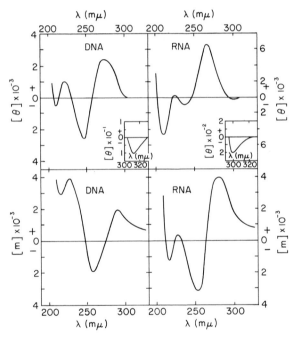

Fig. 6.37. The CD (upper half) and ORD (lower half) of Salmon DNA and *Escherichia coli* ribosomal RNA (16 S and 23 S). All solvents are made up of 0.005 M tris, 0.005 M Mg^{2+}, and 0.1 M KCl. χ And α are the specific ellipticity and optical rotation respectively, but may be correlated, via the mole residue concentration to θ and m. [From P. K. Sarkar, B. Wells, and J. T. Yang, *J. Mol. Biol.* **25**, 563 (1967), with permission.]

spectra of RNA and DNA, the other two lying at 225–235 mμ; and 290–295 mμ. The two major troughs are at 210–220 mμ and 255–265 mμ, a third very minor trough occurs at 300–315 mμ. A compendium of data for various DNA's and RNA's is given in Table 6.16. Perhaps the most striking difference

TABLE 6.16

The Cotton Effects of Nucleic Acids[a,b,c,d]

Substance or source	Temperature (°C)	Peak 1 λ (mμ)	Peak 1 [m]	Trough 1 λ (mμ)	Trough 1 [m]	Peak 2 λ (mμ)	Peak 2 [m]	Trough 2 λ (mμ)	Trough 2 [m]	Peak 3 λ (mμ)	Peak 3 [m]	T_m From 290 mμ peak (°C)	T_m Literature value (°C)
DNA													
1. Aerobacter aerogenes	27	290	+2310	257	−2920	225	+4650	217	+2980			91	93.5
	96	292	+1200	260	−2770	225	+615						
2. Bacillus megaterium	27	200	+1820	258	−860	225	+4480	215	+3550	200	+8900	82	85
	87	292	+700	260	−1180	225	+2070						
3. Escherichia coli	27	290	+2200	257	−2230	228	+4200	215	+3090			86	90.5
	90	292	+1080	260	−2590	225	+1640						
4. Mycobacterium tuberculosis	27	290	+2610	256	−3590	225	+4360	215	+3390			92	97
	96	292	+1260	260	−1380	228	+870	215	−1160				
5. Serratia marcescens	27	290	+2410	258	−2540	228	+4130	217	+2680			90	93.5
	96	292	+1170	260	−1420	228	+1450						
6. T2 Phage	27	290	+1380	258	−2410	235	+3900	218	+3150			80	83
	90	292	+770	262	−2120	230	+1060						
7. Calf thymus	27	290	+1910	257	−1920	228	+3730	216	+2830			84	87
	96	292	+850	260	−1980	230	+660						
8. Salmon sperm	27	290	+1930	256	−2000	228	+3920	215	+2940	200	+10700	85	87.5
	96	292	+950	262	−1900	228	+950						
9. Poly d(AT)	85	283	+2450	255	−5780	236	+5480	247	−5490			63[c]	65
	27	287	+900	263	−3730	235	+2770						
10. Poly dG · poly dC	27	310	+250	285	−830	264	+2520	242	−2190	225	−206		
	90	320	−130	295	−600	270	+1610						
RNA													
11. Yeast tRNA	27	280	+3140	251	−2770	227	−160	217	−1360	196	+10300		
	90	292	+800	258	−2040	227	−750	215	−1030				
12. Tetrahymena rRNA	27	283	+3060	253	−3220	225	+240	215	−1430				
	90	290	+730	258	−1790			210	−1380				
13. Rat liver	27	280	+3400	252	−3500	228	+120	217	−1150	195	+11600		
	90	290	+980	260	−2580	228	−750	210	−3100				

[a] Data taken from T. Samejima and J. T. Yang, *J. Biol. Chem.* **240**, 2094 (1965).

[b] Solvents used: 0.15 *M* KF (pH 7.4), except substance 1, 0.01 *M* NaCl; substance 3, 0.13 *M* KF; substance 9, 0.15 *M* NaCl + 0.015 sodium citrate; substance 10, 0.02 *M* NaCl + 0.01 *M* phosphate buffer (pH 7.1); substance 12, 0.15 *M* KF + 0.01 *M* phosphate buffer (pH 7.0).

[c] From 283 mμ peak.

[d] Taken from Ref. 46, p. 264, with permission.

in the ORD spectra lies in the fact that the long-wavelength band (290–295 mμ) is considerably smaller (about one-half) than the second peak (225–235 mμ) for DNA, whereas the reverse is true for RNA; in the latter, the second peak is very small and exhibits a rotation near zero. A second minor difference is that the first peak in the RNA's is generally at slightly shorter wavelength (280 mμ) than that of the DNA's ($\sim$290 mμ).

Table 6.16 also contains data for the copoly d(AT) and the complex poly d(G + C). It is noticeable that the former shows a slight blue shift of the band near 290 mμ but in general has an ORD parameter very similar to those of the DNA's. The homopolymer complex poly d(G + C) on the other hand, shows a strong displacement to longer wavelength for the Cotton effects, and the intensity of the high-wavelength peak is much reduced.

It has been found empirically, that a simple linear relationship exists between the rotation at the 290 mμ peak and the guanine and cytosine content of the various DNA's:

$$[\alpha]_{290} = 26.5(G + C) + 850 \qquad (6.25)$$

for native DNA's, where G and C are expressed in mole percent, and

$$[\alpha]_{290} = 15.4(G + C) + 220 \qquad (6.26)$$

for heat-denatured DNA's.

Figure 6.38 demonstrates these relationships for a number of naturally occurring DNA's. It can be seen that with the exception of native T2 phage DNA, the relationship works quite well. The discrepancy for T2 phage probably originates in the fact that it contains a hydroxymethyl derivative of cytosine, which may not contribute the same rotation as that of cytosine itself.

There is also a relation between the melting temperature of the DNA, as listed in Table 6.17, and guanine plus cytosine content, reflecting the strong base pairing of G–C. Figure 6.38 shows data based on the ORD 290 mμ peak.

The origin of the differences between DNA and RNA spectra is not fully understood but offers some intriguing possibilities. Since native DNA molecules are generally double helical and the RNA's are single stranded (cloverleaf, hairpinlike, etc.) the discrepancies could arise from this different chain arrangement. However, certain double-stranded RNA's (rice dwarf virus, cytoplasmic polyhedral virus, etc.) show ORD curves essentially similar to those of single-stranded RNA's. Thus the double-strand–single-strand arrangement cannot be the origin of the variation in ORD curves. The second possibility, that the hydroxyl group is the primary cause, can be eliminated because the Cotton effects of the ribose and deoxyribose derivatives of the bases are very similar. It appears, therefore, that the hydroxyl group has some secondary influence on conformation (particularly the arrangement of bases). Yang and Samejima have considered this possibility in

TABLE 6.17

Position and Magnitude of Ultraviolet Cotton Effects in Mucopolysaccharides[a]

Compound	ORD Cotton effects				CD Bands	
	λ, mμ (1)	$[m]\,\Phi \times 10^{-2}$	λ, mμ (2)	$[m]\,\Phi \times 10^{-2}$	λ, mμ	$[\theta] \times 10^{-2}$
N-acetyl-D-galactosamine	225 Trough	7	200 Peak	136	(−) 212	48
					(+) ~189	122 (Inflection)[b]
N-acetyl-D-glucosamine	222.5 Trough	−1.7	196 Peak	70	(−) 211	65
Chondroitin sulfate A	218 Trough	−40			(−) 208.5	73
	205 Peak	1				
Chondroitin sulfate C	216.5 Trough	−59				
	202 Peak	2.5				
Sulfated chondroitin sulfate C	216.5 Trough	−63			(−) 208	113
	202 Peak	4				
Chondroitin sulfate B2	222 Trough	−80	199 Trough	−148	(−)~212	33
					(−) 188	165

Chondroitin sulfate B1					
Hyaluronic acid	220 Trough	−118	200 Trough	~130	(−) 208	107
	199 Peak	−7.7			
Heparin	~230 Trough	25	200 Peak	89	(−) 210	13
					(+) 192	26
Keratan sulfate	220 Trough	−47	198.5 Peak	186	(−) 210	93
					(+) 190	79
Sulfated keratan sulfate	220 Trough	−48	198 Peak	79	(−) 207.5	95
					(+) 189	59
Heparitin sulfate 1					(−) 209	46
					(+)~190	115
Heparitin sulfate 2	212.5 Trough	−55	196 Peak	94	(+) 240	~24
					(−) 202.5	128
					(+) 188	?c

[a] Taken from Ref. 50, with permission.

[b] This peak is largely unresolved with 189 mμ a rough approximation of its center based upon the inflection seen on the low-energy side of a large positive Cotton effect centered below 185 mμ.

[c] Experimental error at 188 mμ for this sample is large relative to the possible small positive CD; values obtained for $\phi \times 10^{-2}$ vary from 0 to +13.

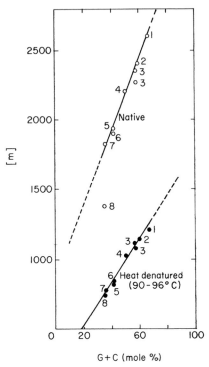

Fig. 6.38. The relationship between rotation at the 290 mµ peak and the (G + C) content of DNA. Symbols: 1, *Mycobacterium tuberculosis*; 2, *Serratia marcescens*; 3, *Aerobacter aerogenes*; 4, *Escherichia coli*; 5, salmon sperm; 6, calf thymus; 7, *Bacillus megaterium*; 8, T2 phage. [From T. Samejima and J. T. Yang, *J. Biol. Chem.* **240**, 2098 (1965) with permission of the American Society of Biological Chemists Inc.]

some detail (46) and hypothesize that the base tilting, which in turn reflects the presence of the ribosyl hydroxyl group, is primarily responsible for the relative magnitude of the Cotton effects. Their conclusions are summarized as follows:

1. The relative magnitudes of the two peaks of ORD, at 290 and 230 mµ, and of the corresponding positive and negative CD bands of polynucleotides are primarily determined by the geometry of the stacked bases, which in turn is influenced by the presence or absence of the 2′-hydroxyl group on the pentose.

2. The stacking of bases perpendicular to the helical axis would lead to a higher ORD peak near 225–230 mµ and to a larger negative CD band, e.g., DNA. Tilting of the bases would reduce the peak near 230 and increase that

at approximately 290 and increase the magnitude of the CD extreme, e.g., RNA.

3. The deoxyribosyl polymers can have stacked bases perpendicular to the helical axis or tilted, whereas the stacked bases of the ribosyl polymers are always tilted.

4. For any polynucleotide or nucleic acid, the magnitude of the Cotton effects decreases with unstacking of the bases (e.g., at elevated temperature).

5. Protonation or deprotonation of the bases can change the magnitude of the Cotton effects, even though the stacking of bases may remain unchanged.

6. For RNA and ribosyl polymers, the formation of base pairs leads to a significant blue shift of the ORD and CD. In contrast, only a very small blue shift could be detected for the formation of a double helix of DNA.

Optical Properties of Polysaccharides

Polysaccharides do not, in general, have high-probability electronic transitions in the visible/near-ultraviolet region of the spectrum, and consequently detailed studies of the type outlined for polypeptides have not been forthcoming. There are, however, electronic transitions in the far-ultraviolet region that have enabled some recent advances in conformational studies to be made. Probably the most extensive ORD measurements have been made on polysaccharide complexes with iodine and various dyes (50). By such methods the absorption spectrum is shifted to the visible region of the spectrum, and Cotton effects and their change with conformation, as reflected through the dye spectrum, may be followed. The mucopolysaccharides, however, can readily be examined without recourse to the dye- or iodine-binding procedure. These A–B polysaccharides are characterized by the presence of amide groups in which electronic transitions may be identified. At the present time only fairly qualitative data for ORD/CD spectra are available. Chondroitin sulfate, heparin, hyaluronic acid, and keratosulfate all show Cotton effects in the 180–220 mμ region which arise from the amide group on the C-2 atoms. Figure 6.39 (50) shows the ORD spectrum of the monosaccharide *N*-acetylgalactosamine and the polysaccharide chondroitin sulfate C. It would seem that the curves arise from several electronic transitions, but the CD curves indicate the presence of two main transitions centered at 212 mμ (negative) and about 189 mμ (positive) for the galactosamine. It is tentatively assumed that these bands correspond with the $n - \pi^*$ and $\pi - \pi^*$ transitions in the *N*-acetyl group, though other possibilities exist. Although CD data have not been presented for chondroitin sulfate C, data are available for a number of other mucopolysaccharides (see Table 6.17). It is evident that most yield two

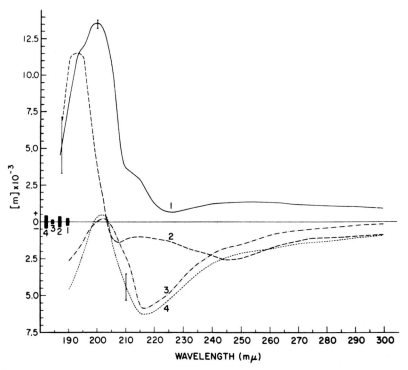

Fig. 6.39. Ultraviolet optical rotatory dispersion of mucopolysaccharides containing *N*-acetylgalactosamine: 1, sample *N*-acetylgalactosamine; 2, beef ganglioside; 3, chondroitin sulfate C; 4, sulfated chondroitin sulfate C. The bars along the zero baseline indicate sensitivity of the measurements. The lines through the experimental curves at the low wavelengths show the experimental error due to noise at those wavelengths. Values of ϕ are not corrected for the refractive index of the water solvent. [From Stone (50), by courtesy of Marcel Dekker Inc.]

bands in the 210 and 190 mμ region and that these transitions do not change position significantly on passing from the monomers *N*-acetylgalactosamine and *N*-acetylglucosamine to the polymers. There are, however, some significant differences in the molar ellipticities in many cases, indicating interaction between the active chromophores. The detailed nature of the solution conformation of these polysaccharides is not known at present and although certain helical models and bonding schemes have been proposed (51), further supporting evidence is still being sought.

As stated previously, most sugars and polysaccharides do not display significant absorption bands above 200 mμ. Below 200 mμ there is the tail of a band that centers below the accessible limit of current ORD/CD instrumentation and that has not yet been studied by vacuum ultraviolet spectro-

scopy. Figure 6.40 (52) shows the circular dichroism spectra for D-glucose
and D-galactose. The CD band, of which only the tail is evident, probably
arises from an electronic transition in the ring oxygen (53). It would seem,
therefore, that the stereochemical alignment of groups about the ring oxygen
would determine the characteristics of the first active absorption band (54).
A second electronic transition occurring at somewhat lower wavelength is
believed to originate in the hydroxyl group (55). Axial hydroxyl groups
introduce a new axis of polarizability, equatorial groups giving positive
bands and axial groups negative bands. The opposite signs of the CD bands
for sugars in the D-galactose and D-glucose series may be attributed to these
factors. It appears, then, that the contributions of individual stereochemical
alignments of atoms or groups in polysaccharides may be assessable by
analysis of the CD spectrum below 200 mμ. (52).

The CD data for some disaccharides shown in Table 6.18 are in agreement
with the general proposal that for sugars in the C1 conformation, the presence
of an axial hydroxyl group at C-4 determines a negative CD band, while an
equatorial hydroxyl group at C-4 gives a positive band. Similarly, sugars
containing axial hydroxyl groups at C-2 should exhibit positive curves. Thus

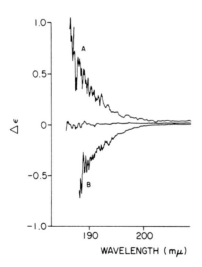

Fig. 6.40. Circular dichroism spectra for D-glucose (curve A) and D-galactose (curve B)
in water. The original spectra from the recorder charts of the Durrum–Jasco CD spectro-
photometer are reproduced. The measurements were made on 0.11 M sugar solutions after
mutarotational equilibrium was attained. A cell of 1 mm optical path length was used. The
CD absorption, $\Delta\varepsilon$, was calculated from the relationship $\Delta\varepsilon = \Delta D/cd$ where ΔD is the
circular dichroic optical density, c is the concentration in moles per liter, and d is the light
path in centimeters. [From Ref. 52, with permission of Academic Press Inc.]

TABLE 6.18

Ultraviolet Circular Dichroic and Optical Rotatory Dispersion Properties of Selected Mono- and Disaccharides

Compound	$[\theta] \times 10^{-2}$ at 190 mμ (deg cm²/decimole)	Rotational direction at 190 mμ
Methyl-α-D-glucopyranoside	+2.0	(+)
Methyl-β-D-glucopyranoside	+2.0	Level
Methyl-α-D-galactopyranoside	−0.7	(+)
Methyl-β-D-galactopyranoside	−6.6	(−)
α-D-Glucopyranosyl-α-D-glucopyranoside	+3.3	(+)
4-O-α-D-Glucopyranosyl-D-glucose	+0.7	(+)
4-O-β-D-Glucopyranosyl-D-glucose	+2.6	(+)
4-O-β-D-Galactopyranosyl-D-glucose	−7.2	(−)
4-O-β-D-galactopyranosyl-D-fructose	−7.0	(−)
6-Deoxy-D-galatose	−5.2	(−)
D-Mannose	+12.0	(+)
Methyl-α-D-mannopyranoside	+5.6	(+)
D-Talose	+0.9	(−)

D-mannose and methyl (α-D-mannopyranoside) exhibit positive bands of greater intensity than those of the corresponding glucose derivatives. Furthermore, D-talose, containing C-2 and C-4 axial substituents that contribute in opposite directions, exhibits a very weak positive CD curve.

Conformational Transitions of Polysaccharides

Evidence from ORD/CD for conformational transitions in polysaccharides has been obtained. Heparin (56) shows considerable change in both molar refraction and position of the high-wavelength Cotton effect as a function of pH. Figure 6.41 (56) demonstrates the ORD curves for heparin. It has been deduced, partially from these data, that the low pH form is more "ordered," although the detail of this order is still missing.

A rather interesting conformational change has been observed for segments of (desulfated) plant polysaccharide ι-carrageenan (57). In this case the solid-state structure of the intact polymer is known to be a double helix (58), and it is believed that at normal temperatures the double helix is preserved in solution (for the modified partially desulfated form.) Carrageenans do not contain amino sugars and consequently do not show Cotton effects above 200 mμ. However, optical rotation at high wavelength (546 mμ) is observed to undergo

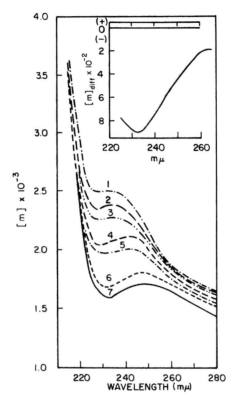

Fig. 6.41. Changes with pH of heparin ORD in the spectral range 280–220 mμ. Curves 1–7 are those at pH 7, 6.1, 5.5, 5.0, 4.4, 3.1, and 2.7, respectively, for a given heparin solution; 1.52 mg/ml at a path length of 0.5 cm. The insert is the difference curve between curves 1 and 7; ϕ values are uncorrected for the refractive index of the water solvent. [From Ref. 56, with permission of MacMillan and Co. Ltd.]

changes as a function of temperature [Fig. 6.42 (57)]. These thermal effects are thought to arise from the dissociation of the helix into random-coil components. The transformation curve is remarkably similar to that for polypeptides and polynucleotides, intramolecular cohesion between sugar residues being the probable driving force for helix stabilization.

Optical rotatory curves for oligo- and polysaccharides may generally be fitted to Drude-type equations with $\lambda_c = 145 - 156$ mμ, although it is not clear whether information on the helix content can be drawn from such an approach, and no conformational transitions appear to have been treated in this manner (59).

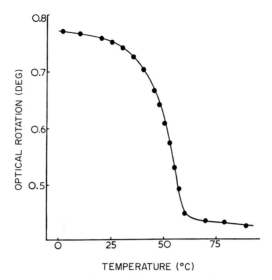

TEMPERATURE (°C)

Fig. 6.42. Optical rotation changes with temperature for ι-carrageenan segments in distilled water (5.6% w/v at 546 $m\mu$). [From Ref. 57, with permission of the Chemical Society.]

Spectra of Complexes

Probably the most fruitful approach to the spectroscopy of polysaccharides has been through the analysis of dye complexes. It is not the intention here to review this area, which is being applied to polynucleotides and proteins as well as to the polysaccharides. Basically, the principles are that if dye molecules absorb or chemically complex to the polymer molecule in such a manner that they interact electronically, then the study of the electronic spectra should reveal conformational characteristics of the underlying polymer substrate. For example, the appearance of Cotton effects with dye complexes is usually taken to indicate the presence of helix content of the polymer, imposing an asymmetry on the bound dye molecules. Furthermore, the displacement of the absorption band from that of the individual dye molecules themselves may be taken as an indication of electronic conjugation. The complex of amylose with iodine, demonstrated by x-ray diffraction to have a helical conformation in the solid state, has an absorption maximum at 600 $m\mu$. From "free-electron theory" it is deduced that long segments of conjugated electron systems give rise to a lower absorption frequency, i.e., higher wavelength. Although calculations based on this model are likely to be unreliable, the absorption wavelength probably corresponds to a conjugated system of the order of 100 Å in length; if the complexing atoms are associated with helical segments then some estimate of segment length may be obtained. The iodine complex of carboxymethyl amylose absorbs with a broad maximum centered at 550 mμ, indicating rather shorter regions of regular ordered conformation in aqueous solution.

Probably the best examples of relevant polysaccharide–dye complexes relate to the mucopolysaccharides, where the staining of such materials in intact tissue represent an important part of histochemistry. Generally, dye molecules such as methylene blue, acridine orange, etc., complex in a manner such that there is one molecule per anionic saccharide site. It is thought that stacking of dye molecules occurs which leads to electronic interaction through the coupling of transition moments to form an exciton band. Credible attempts to calculate the separation of stacked dye molecules in solution have been made (60) and an extension to mucopolysaccharide/dye complexes has also been reported (61). In essence the calculation is carried out as follows: the transition moment $\mu_1 = eQ$ interacts with the excited states of neighboring dye molecules. Exciton theory predicts (62, 63) that for a "card pack" arrangement of neighboring molecules the spectral shift $\Delta\sigma$ (in ergs per mole) is related to the maximum separation by

$$r_{max} = (4e^2 Q^2/\Delta\sigma)^{1/3} \qquad (6.27)$$

Now Q may be calculated from the observed absorption band using the relation

$$Q^2 = -1.09 \times 10^{-19} \int \varepsilon \, d \ln \lambda \qquad (6.28)$$

where ε is the molar extinction coefficient.

Application of this method to sodium carrageenates complexed with methylene blue yields $r_{max} \sim 9$ Å. Similarly the separation of toluidine blue molecules complexed with hyaluronic acid indicates a 7 Å separation of carboxyl groups. Although in the solid state the separation of carboxyls, is found by x-ray diffraction to be 9.5 Å, the rotation of sugar rings in solution could produce a 7 Å separation (64).

Thus the use of dye complexes, in conjunction with the methods of absorption and optical rotatory dispersion spectroscopy, seems to afford quantitative tools for estimation of polysaccharide conformation and structure in solution.

REFERENCES

1. A. Streitweiser, "Molecular Orbital Theory." Wiley, New York, 1961.
2. A. Halpern and W. T. Simpson, *in* "Polyamino Aids, Polypeptides and Proteins" (M. Stahmann, ed.), p. 147. University of Wisconsin Press, Madison, Wisconsin, 1962.
3. E. U. Condon and G. H. Shortley, "The Theory of Atomic Spectra." Cambridge Univ. Press, London and New York, 1959.

4. S. Yomosa, *in* "Quantum Aspects of Polypeptides and Polynucleotides" (M. Weissbluth, ed.). Wiley (Interscience), 1964.

5. P. O. P. Ts'o, *in* "Fine Structure of Proteins and Nucleic Acids" (G. D. Fasman and S. N. Timasheff, eds.), Chapter 2. Dekker, New York, 1970.

6. R. F. Stewart and N. Davidson, *Biopolym. Symp.* **1**, 465 (1964).

7. R. F. Stewart and L. H. Jensen, *J. Chem. Phys.* **40**, 2071 (1964).

8. H. De Voe and I. Tinoco, *J. Mol. Biol.* **4**, 518 (1962).

9. G. J. Thomas and Y. Kyogoku, *J. Amer. Chem. Soc.* **89**, 4170 (1967).

10. W. B. Gratzer and C. W. F. McClare, *J. Amer. Chem. Soc.* **89**, 4224 (1967).

11. V. Weisskopf and E. Wigner, *Z. Phys.* **63**, 54 (1930); **65**, 18 (1930).

12. M. Goodman, I. Listowsky, Y. Masuda, and F. Boardman, *Biopolymers* **1**, 33 (1963).

13. Th. Förster, *in* "Comparative Effects of Radiation" (M. Burton, J. S. Kirby-Smith, and J. L. Magee, eds.), p. 300. Wiley, New York, 1960.

14. A. S. Davydov, "Theory of Molecular Excitons" (translated by M. Kasha and M. Oppenheimer, Jr.). McGraw-Hill, New York, 1962.

15. M. Kasha, *Radiat. Res.* **20**, 55 (1963).

16. G. Holzwarth and P. Doty, *J. Amer. Chem. Soc.* **87**, 218 (1965).

17a. E. S. Pysh, *Proc. Nat. Acad. Sci. U.S.* **56**, 825 (1966).

17b. K. Rosenheck and B. Sommer, *J. Chem. Phys.* **46**, 532 (1967).

18. J. Applequist and J. L. Breslow, *J. Amer. Chem. Soc.* **85**, 2869 (1963).

19. E. S. Pysh, *J. Mol. Biol.* **23**, 587 (1967).

20. C. McMillin, M.S. thesis. Case Western Reserve University, Cleveland, Ohio, 1971.

21. J. Brahms, J. Pilet, H. Damany, and V. Chandrasekharan, *Proc. Nat. Acad. Sc.i U.S.* **60**, 1130 (1968).

22. M. L. Tiffany and S. Krimm, *Biopolymers* **8**, 347 (1969).

23. S. Krimm and J. E. Mark, *Proc. Nat. Acad. Sci. U.S.* **60**, 1122 (1968).

24. H. R. Mahler and E. H. Cordes, "Basic Chemistry," p. 159. Harper, New York, 1964.

25. H. Eyring, H.-C. Lin, and D. Caldwell, *Chem. Rev.* **68**, 525 (1968).

26. I. Tinoco and C. R. Cantor, *in* "Methods of Biochemical Analysis" (D. Glick, ed.), Vol. 18. Wiley (Interscience), New York, 1970.

27. See, for example, W. Moffitt and A. Moscowitz, *J. Chem. Phys.* **30**, 648 (1959).

28. Y. Pao, R. Longworth, and R. L. Kornegay, *Biopolymers* **3**, 519 (1965).

29. L. Rosenfeld, *Z. Phys.* **52**, 161 (1928).

30. W. Moffitt and J. T. Yang, *Proc. Nat. Acad. Sci. U.S.* **42**, 596 (1956).

31. L. Stevens, R. Townend, S. N. Timasheff, G. D. Fasman, and J. Potter, *Biochemistry* **7**, 3717 (1968).

32. W. B. Rippon and A. G. Walton, *Biopolymers,* **10**, 1207 (1971).

33. D. W. Urry, *in* "Spectroscopic Approaches to Biomolecular Conformation" Ed. D. W. Urry, American Medical Assoc. Pub., Chicago, Illinois (1970) p. 105.

34. S. Beychok and G. D. Fasman, *Biochemistry* **3**, 1675 (1964).

35. S. Beychok, *in* "Poly-α-Amino Acids" (G. D. Fasman, ed.), p. 314. Dekker, New York, 1967.

36. J. T. Yang, T. Samejima, and P. K. Sarkar, *Biopolymers* **4**, 623 (1966).

37. T. R. Emerson, R. J. Swan, and T. L. V. Ulbricht, *Biochemistry* **6**, 843 (1967).

38. A. E. V. Haschemeyer and A. Rich, *J. Mol. Biol.* **27**, 369 (1967).

39. M. M. Warshaw and I. Tinoco, *J. Mol. Biol.* **20**, 29 (1966).

40. C. A. Bush and I. Tinoco, *J. Mol. Biol.* **23**, 601 (1967).

41. J. N. Vournakis, H. A. Scheraga, G. W. Rushizky, and H. A. Sober, *Biopolymers* **4**, 33 (1966).

42. Y. Inoue, S. Aoyagi, and K. Nakanishi, *J. Amer. Chem. Soc.* **89**, 5701 (1967).

43. J. Brahms, *Proc. Roy. Soc. London* **A297**, 150 (1967).
44. J. Brahms, J. C. Maurizot, and A. M. Michelson, *J. Mol. Biol.* **25**, 465 (1967).
45. H. Simpkins and E. G. Richards, *J. Mol. Biol.* **29**, 349 (1967).
46. J. T. Yang and T. Samejima, *Progr. Nucl. Acid Res. and Mol. Biol.* **9**, 223 (1969).
47. J. Brahms and S. Brahms, *in* "Fine Structure of Proteins and Nucleic Acids" (G. D. Fasman and S. N. Timasheff, eds.), Chapter 3. Dekker, New York, 1970.
48. C. R. Cantor and I. Tinoco, *J. Mol. Biol.* **13**, 65 (1965).
49. P. O. P. Ts'o, S. A. Rapaport, and F. J. Bollum, *Biochemistry* **5**, 4153 (1966).
50. A. L. Stone, *in* "Structure and Stability of Biological Molecules" (S. N. Timasheff and G. D. Fasman, eds.), Chapter 5. Dekker, New York, 1969.
51. A. L. Stone and H. Moss, *Biochim. Biophys. Acta* **136**, 56 (1967).
52. I. Listowsky and S. Englard, *Biochem. Biophys Res. Commun.* **30**, 329 (1968).
53. L. W. Pickett, N. J. Hoeflick, and T. C. Liu, *J. Amer. Chem. Soc.* **73**, 4865 (1951).
54. I. Listowsky, G. Avigad, and S. Englard, *J. Amer. Chem. Soc.* **87**, 1765 (1965).
55. A. J. Harrison, B. J. Cederholm, and M. A. Terwilliger, *J. Chem. Phys.* **30**, 355 (1959).
56. A. L. Stone, *Nature* **216**, 551 (1967).
57. A. A. McKinnon, D. A. Rees, and F. B. Williamson, *Chem. Commun.*, p. 701 (1969).
58. N. S. Anderson, J. W. Campbell, M. M. Harding, D. A. Rees, and J. W. B. Samuel, *J. Mol. Biol.* **45**, 85 (1969).
59. S. Beychok and E. A. Kabat, *Biochemistry*, **4**, 2625 (1965).
60. G. S. Levinson, W. T. Simpson, and W. Curtis, *J. Amer. Chem. Soc.* **79**, 4314 (1957).
61. A. L. Stone, L. G. Childers, and D. F. Bradley, *Biopolymers* **1**, 111 (1963).
62. I. Tinoco, *J. Chem. Phys.* **33**, 1332 (1960).
63. D. F. Bradley and M. K. Wolf, *Proc. Nat. Acad. Sci. U.S.* **45**, 944 (1959).
64. M. D. Schoenberg and R. D. Moore, *Biochim. Biophys. Acta* **83**, 42 (1964).

7

PHYSICAL PROPERTIES OF BIOPOLYMERS IN DILUTE SOLUTIONS*

Symbols

a Radius of capillary; major axis of an ellipsoid; exponent for viscosity–molecular weight relationship

b Minor axis of an ellipsoid

B Second virial coefficient

c_2 Concentration of solute, weight basis

c_3 Concentration of salt, MX, in saline solvent, weight basis

C Third virial coefficient

d Solution cell path length; diameter

D Translational diffusion coefficient

f Translational friction factor of any dissolved particle

f_0 Translational friction factor for a spherical particle

F Force; ratio of fraction factors

g Gravitational constant, 980 cm/sec^2

G Gibbs free energy, ergs per mole

h Height of liquid column

i Index for summing individual components

$i_s(\theta)$ Intensity scattered in the direction of θ

I_0 Intensity incident upon a scatterer

* By Stephen H. Carr, Department of Materials Science, The Technological Institute, Northwestern University, Evanston, Illinois.

J Flux; ratio of friction factors

k' Flow constant in Huggins equation

k'' Flow constant in Kraemer equation

K Optical constant

K' Proportionality constant for viscosity–molecular weight relationship

l Length of one segmental unit in a chain, angstroms

$\ln(x)$ Natural logarithm of x, to the base e as opposed to the base 10

L Length of capillary; length of a rodlike particle; ratio of friction factors

M_0 Molecular weight of one chemical repeat unit in a chain

M_1 Molecular weight of solvent

M_2 Molecular weight of solute

M_3 Molecular weight of salt, MX, in saline solvent

M_i Molecular weight of the ith component

M_r Molecular weight of one repeat unit (residue) in a chain

$\overline{M}_n$ Number-average molecular weight of solute

$\overline{M}_w$ Weight-average molecular weight of solute

$\overline{M}_z$ z-Average molecular weight of solute

n Index of refraction of medium or solution

n_0 Index of refraction of pure solvent

n_e Number of electrons in a molecule

n_i Number of moles of the ith component

n_p Number of moles of polyelectrolyte

n_+ Number of moles of positive salt ion

n_- Number of moles of negative salt ion

N Avogadro's number, 6.023×10^{23} particles per mole

p Ratio of axes in an ellipsoid, b/a

P Any pressure

P_0 Atmospheric pressure

q Number of residues in a chain

$\dot{Q}$ Flow rate, volume per unit time

r Distance between point of observation and light or x-ray emission source; radius

R The gas constant, 8.314×10^7 ergs/°K mole

$R_{\theta u}$ Rayleigh ratio as a function of θ, for unpolarized incident light

R_G Radius of gyration

s_2 Sedimentation coefficient

$s_2{}^0$ Sedimentation coefficient at infinite dilution

$s_{20,w}^0$ Sedimentation coefficient at infinite dilution and corrected to water at 20°C

S Entropy

t Time

T Absolute temperature, degrees Kelvin

u Relative velocity

v Velocity

v_1 Specific volume of solvent, cubic centimeters per gram

v_2 Specific volume of solute, cubic centimeters per gram

v_x Excluded volume of a particle in solution

V Volume of solution

V_e Volume of one particle

$\overline{V}$ Molar volume

$V_1{}^0$ Molar volume of pure solvent, cubic centimeters per mole

X_1 Mole fraction of component 1

z Electron-mole per gram-molecule
Z Net electric charge on one polyelectrolyte molecule (it may be positive or negative)
α Polarizability of electrons in a particle
β Parameter from which sedimentation data may be related to ellipsoid axial ratio
$\dot{\gamma}$ Shear rate (or velocity gradient), reciprocal seconds
δ Parameter from which flow-birefringence data may be related to ellipsoid axial ratio
ΔP Pressure difference across the capillary length
ζ Rotational frictional factor
ζ_0 Rotational frictional factor for a spherical particle
η Viscosity of a solution
η_0 Viscosity of medium (solvent)
η_{rel} Relative viscosity
η_{sp} Specific viscosity
$[\eta]$ Limiting viscosity number, cubic centimeters per gram
θ Angle between scattered light and direction of incident beam (see Fig. 7-10)
Θ Rotational diffusion coefficient
λ Wavelength of incident light in medium
μ_1 Chemical potential of component 1
μ_1^0 Chemical potential of pure 1 in its reference state
ν Simha's factor for ellipsoidal particle
π Osmotic pressure
ρ Density of solvent, weight per unit volume
τ Turbidity
ϕ Volume fraction
Φ Constant, a combination of parameters
χ Angle between cross of isocline and $\pi/4$ with regard to polars
ω Angular velocity, radians per second
Ω Torque

Introduction

As studies on the physical chemistry of biological systems have progressed, an increasingly large number of experimental techniques have emerged as useful tools for probing physical properties of biopolymers in solution. They range from such relatively well-used techniques as osmometry to less commonly practiced procedures such as small-angle x-ray scattering. In each case, one or more bulk property of the solution is measured and related to physical properties of the dissolved biopolymers. Data that can be obtained are usually related to the size and shape of the biopolymers being studied, and such information must frequently be correlated with other findings in order to learn about details which approach the atomic-size scale.

Physical chemical techniques have, for quite some time, been applied to solutions of native biopolymers, but a reasonably quantitative description of many of their characteristics has been the result of more recent studies on model biopolymers (see Refs. 1 and 2). Most of the original work was on homopolymers, but the demand for more realistic models has been met with sequenced polypeptides and polynucleotides. Most of these were polydisperse in molecular weight so that the data obtained are the result of an average over many different chain lengths. Synthetic polynucleotides have yielded information relating to the coiling and uncoiling of deoxyribonucleic acid (see Ref. 3). Polysaccharides have also been studied extensively in solution in an effort to determine whether they have extended or random conformations. The techniques and the information that may be derived from them are explored in the following pages.

Determination of Particle Size and Shape

A number of the solution techniques considered in this chapter enable values of molecular weight to be obtained experimentally, and if the effective density of the dissolved particles is known, it is then possible to calculate molecular dimensions. Most procedures can also provide some measure of the asymmetry in particle shape. Several physical chemical techniques described in the following pages measure colligative properties of macromolecular solutions. These properties result from the fact that there is a difference between the chemical potential of pure solvent and that of solvent in solution with a macromolecule. Colligative solution properties which are experimentally accessible include vapor pressure, freezing point depression, boiling point elevation, and osmotic pressure. A technique that is very versatile and widely used measures osmotic pressure.

Osmometry

Although the experimental apparatus varies widely, the device pictured in Fig. 7.1 serves to illustrate the principles of osmometry. It has two chambers separated by a rigidly fixed membrane. This membrane is commonly cut from a sheet of cellulose acetate and is thought to possess a network of pores whose size is in the range of 30 Å. Solvent molecules, which are relatively small, diffuse readily through these channels and are thus able to pass rather freely between the two chambers of the osmometer. When one chamber contains a solute solution and the other contains pure solvent, molecules move under the influence of a concentration gradient through the membrane pores and pass into the chamber containing pure solvent. However, if the solute molecules are nearly as large or larger than the pores of the membrane,

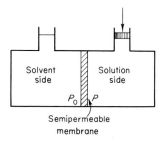

Fig. 7.1. A schematic representation of an osmometer in which increased hydrostatic pressure in the solution side opposes the net movement of solvent through the membrane into the solution.

diffusion of the solute from one chamber to the other is very unlikely, if not impossible. Such a membrane is said to be semipermeable because it provides for passage of solvent but not macromolecular solute. As seen in the figure, the solvent side is open to atmospheric pressure P_0, and the solution side is confined by a movable piston with which the pressure P in that chamber can be adjusted.

General Background

Solvent molecules tend to move through the membrane and enter the solution chamber, and thus it is necessary to impose a hydrostatic driving force to counterbalance that due to the solute concentration differential. Equilibrium is attained when the pressure in the solution chamber of the osmometer has been raised by some amount, which causes the net flux of solvent to become zero. This pressure increment π is equal to the osmotic pressure of the solution under study. Experimentally, there are at least three good ways to obtain the value of π:

1. Use the preceding method, i.e., determine how much the pressure must be raised on the solution side in order to prevent the flow of solvent through the membrane (Fig. 7.1).

2. Measure the amount by which the pressure in the solvent chamber must be reduced so that the net passage of solvent through the membrane is zero (Fig. 7.2).

3. Allow solvent to pass through the membrane into the solution side until equilibrium is reached. At this point, the flow of solvent has ceased because (a) a hydrostatic head has developed which exerts a pressure on the solution and (b) the concentration of the solution falls as solvent enters its

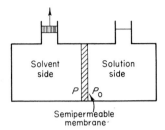

Fig. 7.2. A schematic representation of an osmometer in which reduced hydrostatic pressure in the solvent side opposes the net movement of solvent through the membrane into the solution.

chamber, thus reducing the amount of pressure needed to reach equilibrium (Fig. 7.3).

It is possible to start with the definition of chemical potential for component number 1 in an ideal binary solution and show how osmometry may be used to determine molecular weight.

$$\mu_1 = \mu_1{}^0 + RT \ln X_1 \qquad (7.1)$$

In Eq. (7.1), μ_1 is chemical potential at any value of mole fraction X_1, μ_1^0 is chemical potential of pure component 1, R is the gas constant, and T is absolute temperature. If subscript 1 denotes solvent and subscript 2 denotes solute

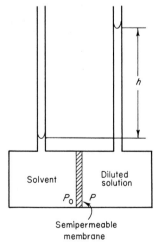

Fig. 7.3. A schematic representation of an osmometer in which the movement of solvent into the solution occurs until the polymer concentration falls to a point where the hydrostatic pressure opposes further passage of solvent.

(in this case macromolecular), it can be shown that $\mu_1 - \mu_1^0$ is related to polymer concentration and molecular weight by

$$\mu_1 - \mu_1^0 = -RT\left[\frac{V_1^0}{M_2}c_2 + \frac{1}{2}\left(\frac{V_1^0}{M_2}\right)^2 c_2^2 + \frac{1}{3}\left(\frac{V_1^0}{M_2}\right)^3 c_2^3 + \cdots\right] \qquad (7.2)$$

Here, V_1^0 is molar volume of pure solvent, M_2 is solute molecular weight, and c_2 is solute concentration. Also, the osmotic pressure may be related to the chemical potential of the solvent:

$$-(\mu_1 - \mu_1^0)/V_1^0 = \pi \qquad (7.3)$$

One can then write Eq. (7.4).

$$\frac{\pi}{c_2} = RT\left[\frac{1}{M_2} + \frac{V_1^0}{2M_2^2}c_2 + \frac{(V_1^0)^2}{3M_2^3}c_2^2 + \cdots\right] \qquad (7.4)$$

Equation (7.4) can also be written as a general virial expansion:

$$\pi/c_2 = RT[(1/M_2) + Bc_2 + Cc_2^2 + \cdots] \qquad (7.5)$$

Here, B and C represent the second and third virial coefficients, respectively. A plot of reduced osmotic pressure, π/c_2, against solute concentration will have an intercept equal to RT/M_2 and a slope at $c_2 = 0$ from which the second virial coefficient B may be obtained. In many cases these plots are essentially linear, so one can conclude that virial coefficients higher than the second are negligible. On some occasions measurements are made on solutions that contain only one kind of macromolecular species, for example a pure protein, but at other times the solute is comprised of a number of different species, as happens with a mixture of proteins or a polydisperse sample of some synthetic polypeptide. The latter case represents a multicomponent system in which one component, namely the solvent, is present in great excess. It is the other components, then, that contribute to the measured osmotic pressure. Equation (7.5) can be written in the following manner:

$$\lim_{c_2 \to 0}\frac{\pi}{c_2} = RT \lim_{c_2 \to 0}\frac{\sum_i(c_i/M_i)}{\sum_i c_i} = \frac{RT}{\overline{M}_n} \qquad (7.6)$$

That is, the molecular weight obtained is the number average $\overline{M}_n$ (see Appendix). A schematic representation of one kind of commercial osmometer is shown in Fig. 7.4. In general, osmometry should not be used on polymer samples containing appreciable fractions of molecules smaller than 10^4 daltons*, because diffusion of this solute through commercial membranes

---

* A dalton is the weight of N particles, expressed in atomic mass units, for example, grams per mole.

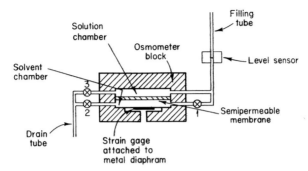

Fig. 7.4. Schematic representation of one kind of automated osmometer. When determining π for a solution, the system is first filled entirely with solvent, and then valves 1 and 2 are closed. Next, solution is run into its chamber, and valve 3 is closed when the meniscus in the filling tube has fallen to its predetermined level. The tendency of solvent to pass into the solution chamber gives rise to a signal from the strain gage, and at equilibrium it can be related to π through a calibration factor.

may introduce errors in the data. The upper limit for osmometry is determined by the sensitivity with which pressures can be measured, and presently this restriction becomes significant for molecular weights above 7×10^5 daltons.

Figure 7.5 (4) shows a plot of π/c_2 versus c_2 for poly-L-proline (in the PP II conformation) in propionic acid. The number-average molecular weight in this case is 19,500 daltons and the second virial coefficient B is 2.1×10^{-4} cm^3 mole/gm^2.

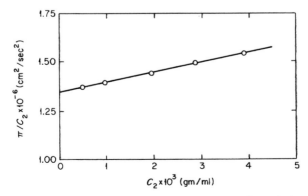

Fig. 7.5. Osmometry data from poly-L-proline in propionic acid plotted as reduced osmotic pressure versus concentration. From the intercept at infinite dilution one may compute the number-average molecular weight. (Taken from Ref. 4, with permission.)

Interpretation of the Second Virial Coefficient

Three major factors can be shown to affect the magnitude of the second virial coefficient observed in osmometry: excluded volume effects, the Donnan effect (in the case of polyelectrolytes) (5), and aggregation of solute molecules.

The first factor to be discussed is the excluded volume effect. Every particle in solution occupies a certain small fraction of the total volume, but each has access to a volume smaller than that calculated simply as the difference between total solution and volume occupied by all solute particles. As illustrated in Fig. 7.6 for the case of solid spheres, this is because the center of gravity

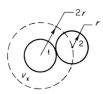

Fig. 7.6. Contact between two spherical particles. The center of sphere 2 will never be able to occupy the region of radius $2r$ indicated by the dotted circle.

of particle number 2 will never be allowed inside a spherical region of diameter $4r$ centered upon particle number 1. This is the region that corresponds to the excluded volume v_x. A different value for v_x is found for rodlike particles, such as that represented in Fig. 7.7 by elongated cylinders of length L and radius r. It is clear from the figure that a different v_x would correspond to each mode of approach. In a suspension of rods, virtually all angles of approach would exist because there is a large population of particles. Thus it is necessary to describe v_x statistically. The net result is that the volume accessible to solute particles is reduced by the presence of other solute particles, and some orientations of asymmetric particles are more predominant than others; these

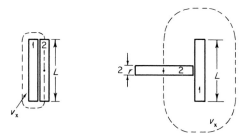

Fig. 7.7. Two possible modes of contact between a pair of cylindrical particles. The smallest excluded volume would occur when one particle approached the other with its long side; the largest excluded volume would occur when one particle approached the other with an end.

factors affect colligative properties because they alter the character of solute–solvent interactions. Thus the quantity $\mu_1 - \mu_1^0$ in Eq. (7.3) is expressed differently (5) and the value of B changes from $V_1^0/2M_2^2$ to

$$B = Nv_x/2M_2^2 \tag{7.7}$$

where N is Avogadro's number. For spherical solute particles, the excluded volume is expressed in Eq. (7.8).

$$v_x(\text{sphere}) = 8M_2 v_2/N \tag{7.8}$$

and, thus, the second virial coefficient would be given by the following equation:

$$B_{\text{sphere}} = 4v_2/M_2 \tag{7.9}$$

For rods like those shown in Fig. 7.7, the excluded volume may be expressed as in Eq. (7.10).

$$v_x(\text{rod}) = LM_2 v_2/rN \tag{7.10}$$

Here, L is rod length, r is its radius, and v_2 is specific volume of the solute rods. This results in a different expression for the second virial coefficient.

$$B_{\text{rod}} = Lv_2/2rM_2 \tag{7.11}$$

A simple calculation for, say, poly-L-lysine of molecular weight 50,000 daltons, in the α-helix form (assumed to be a rigid rod with radius r of 6Å), would give $B = 8.0 \times 10^{-4}$ cm^3 mole/gm^2, whereas B for the same material in a true random form would be 5.9×10^{-5} cm^3 mole/gm^2.

Rods and spheres are extreme cases in terms of excluded volume, but for a coiled macromolecule, which might be considered as a long series of short, rodlike segments, it becomes substantially more difficult to derive an expression for the excluded volume. Regardless of how such polymer chains might be described statistically, it is clear that they would not be expected to have nearly such a large v_x as a rod, and, on the other hand, they would in all likelihood exclude more volume than impenetrable spheres. All factors affecting the expansiveness or compactness of randomly coiled polymers would thus affect significantly the value of the second virial coefficient.

In this regard the nature of the solvent appears to be of considerable importance. If the interaction between solvent molecules and segments of polymer molecules is energetically unfavorable, the chain would be expected to coil more tightly in order to minimize the frequency of its contacts with solvent. This smaller particle size should lead to a smaller value for B than would be expected for the same polymer if it were dissolved in a solvent with which

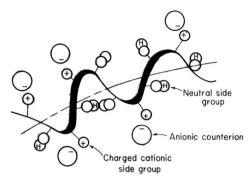

Fig. 7.8. Schematic representation of a chain segment of a polyelectrolyte. Some of the side groups are ionized, while others are not. Each ionized side group is associated with one counterion of opposite charge.

these contacts are more favorable (lower free energy). The nature of the solution composition (see Fig. 7.8 and the next section) can also be shown to have an effect on the degree of chain coiling for polyelectrolytes. As ionic strength is increased, each charged side group becomes more effectively shielded by its counterion cloud, and thus electrostatic forces between other charged side groups become diminished. This means that in the case of poly-L-lysine, with a substantial fraction of ε-amino groups ionized, the chains will be rather extended due to some repulsion from neighboring side groups. Raising the ionic strength of the solvent should decrease B because side-group repulsion would decrease, and these macromolecules would then contract into smaller, more nearly random coils.

In the case of polyelectrolyte solutions, another distinct factor contributing to the second virial coefficient measured from osmometry is the Donnan effect. Its basis is related to the fact that ionized groups on the polymer molecules are present on one side of the membrane only. For example, assume the sample is a polyelectrolyte of some net electric charge $-Z$, dissolved in water made saline with some salt MX (which ionizes into M^+ ions and X^- ions). For the sake of clarity, it is necessary to identify components in such a system as follows.

Component number	Species identification
1	Solvent, water
2	Polyelectrolyte, P^{-Z}
3	Salt, MX

At the moment the osmotic pressure experiment is started, the situation on each side of the membrane could be considered as follows.

Solvent side	Solute side
c_3/M_3 moles of free M^+ ion	$n_p = c_2/M_2$ moles of polyelectrolyte of charge $-Z$
c_3/M_3 moles of free X^- ion	$n_+ = \|Z\| c_2/M_2$ moles of bound counterion M^+
	$(c_3/M_3 - \|Z\| c_2/M_2)$ moles of unbound positive ions M^+
	$n_- = (c_3/M_3 - \|Z\| c_2/M_2)$ moles of free negative ions X^-

Here, M_3 is molecular weight of the salt. Equation (7.12), which shows that the number of negative charges and the number of positive charges are equal on the solution side, is now written to satisfy electroneutrality.

$$|Z| n_p + n_- = n_+ \tag{7.12}$$

As opposed to osmotic pressure experiments (where the chemical potential difference of the solvent is due only to macromolecules existing on the side of the membrane), here there is also a molar imbalance of polyelectrolyte across the membrane. Thus, as equilibrium is approached, both solvent and some ions of positive charge tend to move into the solution side, while some ions of negative charge shift into the solvent side. These adjustments occur because the polyelectrolyte, with its negative charge, cannot move across the membrane. As a consequence, the second virial coefficient is given (see Ref. 5) by the expression in Eq. (7.13).

$$B_{\text{Donnan}} = M_3 Z^2/4c_3 M_2{}^2 \tag{7.13}$$

The Donnan effect does not alter the limiting value of reduced osmotic pressure, and thus the value obtained for polyelectrolyte molecular weights does not need to be modified. Using the previous example of poly-L-lysine (M_n 50,000 daltons) and a 0.01 M NaBr solution, it is found that B_{Donnan} is 1.89 cm^3 mole/gm^2.

The final factor to be mentioned which contributes to the second virial coefficient is solute aggregation. Although it may be possible to write an analytical expression for B corresponding to any given scheme of aggregation, practically every real situation requires a different one. This would be a sizable task, especially in view of the fact that samples of most model biopolymers have the added complication of a molecular weight distribution. However, regardless of the energetics or statistics, it is evident that aggregation of macromolecules in a solution will reduce the effective number of thermodynamically independent particles. As a consequence, a lower reduced osmotic pressure will be measured. In the case of many native or model biopolymers, approach to infinite dilution is accompanied by dissociation of aggregated species into single molecules. This will result in a B that is negative and possibly quite large in magnitude, although the value of the molecular weight calculated from the limiting value of π/c_2 will still correspond to that for individual polymer chains. Table 7.1 is a summary of data for some biological macromolecules (4,6–10).

TABLE 7.1

Values of Second Virial Coefficient B for Some Biopolymer Systems That May Undergo Concentration-Promoted Aggregation

Polymer	Solvent	T (°C)	B (cm^3 mole/gm^2)	Reference
Insulin	pH 1.9	25	-1.5×10^{-5}	6
	pH 2.6	25	-2.5×10^{-5}	6
Collagen (probably nonassociating)	2 M KCNS	25	$+3.0 \times 10^{-4}$	7
Gelatin	pH 5.1	40	-3.0×10^{-3}	8
Gelatin (probably nonassociating)	pH 5.1 $+2\,M$ KCNS	25	$+3.9 \times 10^{-4}$	8
Fibrinogen	pH 6.66	20	-5.4×10^{-6}	9
3-Phospho-D-glyceraldehyde dehydrogenase	glycine–NaOH–NaCl	20	-7.1×10^{-5}	10
Poly-L-proline I	Propionic acid	23	-2.9×10^{-3}	4

Light Scattering

Application of this physical method to solutions of macromolecules can produce values of particle mass as well as values of particle size. The basic phenomenon underlying light-scattering techniques is turbidity resulting from the difference in polarizability of the solute particles and that of their suspending medium (solvent). Chains scatter light in direct proportion to their mass, and it happens that the angular dependence of scattering can be related directly to the radius of gyration of a z-average particle (see Appendix).

General Background

Work published by Lord Rayleigh (11) in 1871, which was derived for the scattering of light by particles quite small relative to the wavelength of light, λ, is the foundation for this technique. The physical situation involves an isolated molecule being encountered by a photon whose energy becomes stored, momentarily, in the form of an oscillating dipole. This excitation is considered to be parallel in direction to that of the electric vector in the incident photon. Following this momentary interaction, another photon is reemitted in any direction except that which is parallel to the oscillation of the dipole. The probability of it going in some particular direction θ is proportional to the length of a vector between the scatterer and the surface shown in Fig. 7.9. If a ray of vertically polarized light is incident on a group of these molecules in a unit volume, the intensity of light scattered in any particular direction will be directly proportional to this probability distribution function. Thus, if the intensity of the incident beam (vertically polarized) is I_0 and the

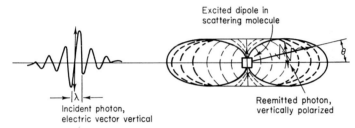

Excited dipole in
scattering molecule

Incident photon,
electric vector vertical

Reemitted photon,
vertically polarized

Fig. 7.9. Schematic representation of the interaction of one photon, illustrated here as a wave packet, with a molecule. In this case, where the electric vector lies in the plane of this page, a photon would be reemitted in any direction except that parallel to the electric vector. The relative probability of a photon being scattered in some particular direction is directly proportional to the length of vector between the scatterer and the surface shown.

intensity of light scattered at θ with respect to the horizontal plane is $i_s(\theta)$ then the ratio of these quantities can be computed by the expression

$$\frac{i_s(\theta)}{I_0} = \frac{16\pi^4\alpha^2\cos^2\theta}{\lambda^4 r^2} \tag{7.14}$$

where r is the distance between observer and scattering point. Equation (7.14) is commonly known as the Rayleigh equation.

The molecular properties that give rise to scattering are combined into the polarizability factor α. For dilute solutions of macromolecules illuminated with unpolarized light, the Rayleigh equation becomes

$$\frac{i_s(\theta)}{I_0} = \frac{2\pi^2(1 + \cos^2\theta)n_0^2(dn/dc_2)^2 c_2}{N\lambda^4 r^2[(1/M_2) + 2Bc_2 + 3Cc_2^2 + \cdots]} \tag{7.15}$$

Here n is index of refraction of a solution, and n_0 is that for pure solvent. Figure 7.10 shows the distribution of scattered light predicted by Eq. (7.15).

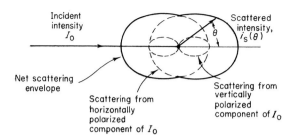

Fig. 7.10. Cross section of "scattering envelope" produced when unpolarized incident light is reemitted from a collection of molecules. The diagram indicates that scattering of horizontally polarized photons will be equal in all directions within the plane of this page, while scattering of vertically polarized photons will make an additional contribution at all angles except θ equal to $\pi/2$ and $3\pi/2$.

It is customary to rearrange this equation in terms of the following quantities:

Rayleigh ratio for unpolarized light

$$R_{\theta,u} = \left(\frac{i_s(\theta)}{I_0}\right)\left(\frac{r^2}{1 + \cos^2 \theta}\right) \qquad (7.16)$$

Optical constant

$$K = \frac{2\pi^2 n_0^2 (dn/dc_2)^2}{N\lambda^4} \qquad (7.17)$$

Thus Eq. (7.15) becomes the Debye equation,

$$Kc_2/R_{\theta,u} = (1/M_2) + 2Bc_2 + 3Cc_2^2 \cdots \qquad (7.18)$$

It is applicable to small polymer molecules, but for larger particles, either bigger macromolecules or species made by aggregation of chains, destructive interference between light reemitted from individual portions of the same particle occurs. Such would be the case for particles of size larger than $\lambda/20$. This requires a modification of Eq. (7.18) by a particle scattering factor $P(\theta)$, and when the analytical expression for $P(\theta)$ is inserted into the Debye equation, it becomes the working equation for the reciprocal intensity plots described by Zimm (12). Such a method for plotting light-scattering data can be seen to have an intercept which yields M_2 (as weight average, $\overline{M}_w$).

$$\lim_{\substack{\theta \to 0 \\ c_2 \to 0}} \frac{Kc_2}{R_{\theta,u}} = \frac{1}{M_2} \quad \text{or} \quad \frac{1}{\overline{M}_w} \qquad (7.19)$$

From the extrapolation line with $c_2 = 0$, a value of particle radius of gyration R_G can be obtained.

$$R_G^2 = \frac{\text{limiting slope}}{\text{limiting intercept}} \cdot \frac{3\lambda^2}{16\pi^2} \qquad (7.20)$$

Equation (7.20) gives a direct value of R_G which corresponds to a particle with a z-average molecular weight (see Appendix). Because the slope of the line for $\theta = 0°$ is twice the second virial coefficient, it is possible to consider B in terms of properties discussed previously, namely excluded volume and aggregation.

Experimental

A schematic diagram of typical light-scattering equipment is shown in Fig. 7.11. Scattered light is collected by a photomultiplier tube, the voltage generated being directly proportional to the intensity of light incident on the tube.

Solutions used are, typically, of concentrations less than 10^{-2} gm/ml and must be exhaustively clarified to avoid scattering from solid particulate matter. Data must be gathered on the angular dependence of light scattered from the stock solution and from its successive dilutions. In this way it is possible to extrapolate to both $\theta = 0°$ and $c_2 = 0$. One should note that the equations are derived for the excess scattering of the solutions over that of pure solvent, so that it is necessary to determine the scattering envelope of solvent in the sample cell and subtract this from scattering values obtained from every solution studied. Also, it is frequently necessary to obtain the refractive index increment, dn/dc_2, using a differential refractometer, although values of this quantity may in some cases be found among those tabulated in the literature. (13).

One technique for treating the data obtained, often called the dissymmetry method, relies on precalculated curves relating $i_s(\theta = 45°)/i_s(\theta = 135°)$ to $P(\theta)$, but this method requires that the specific shape of the scattering particles

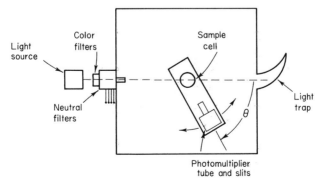

Fig. 7.11. Schematic representation of a commonly used light-scattering photometer (see text for details).

be known. A more convenient way to handle light-scattering data involves the construction of plots as described by Zimm (12). The advantages of a Zimm reciprocal intensity plot are as follows: one need not measure depolarization (a property of polymer solutions related to optical anisotropy); nor is it necessary to assume a particular shape for the particle, since an absolute determination of the root mean square radius of gyration can be made. Although a slightly larger amount of data must be taken when using the Zimm method, its advantages justify the effort.

Typical data for poly-L-proline are shown in Fig. 7.12 (the system is the same as that in Fig. 7.5). The extracted information is $\overline{M}_w = 58{,}500$ daltons, $R_G = 540$ Å, and $B = 2.2 \times 10^{-4}$ cm^3 mole/gm^2. It is interesting to note that this radius of gyration is considerably lower than might have been expected if poly-L-proline II were a rigid rodlike helix. With care, light-scattering techniques may be used to determine molecular weight values between 10^3

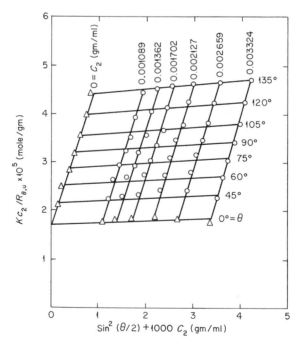

Fig. 7.12. Reciprocal intensity plot of light-scattering data of poly-L-proline in propionic acid. The stock solution and subsequent dilutions are put in the photometer, and the angular dependence of scattered light is measured in each case. Actual data are indicated by circles, while points derived by extrapolation are indicated by triangles. [Data of Carr, *et al.* (4).]

and 10^7. If very large macromolecules such as native deoxyribonucleic acid are under investigation, experimental and computational refinements such as those described by Harpst *et al.* (14) should be used.

Small-Angle x-Ray Scattering by Solutions of Macromolecules

In a manner somewhat analogous to the scattering of visible light, the scattering of x-rays can also provide values of molecular weight and radius of gyration. In the previously described technique, the scattering of light arises from interactions between photons and index of refraction fluctuations. Polymer solutions also scatter photons of small wavelength, such as x-rays, because the individual dissolved particles represent fluctuations in the electron density. The part of the scattering envelope from which molecular parameters can be obtained is confined to a few degrees on either side of the main beam, hence the term small-angle x-ray scattering. Although this technique is not used as frequently as light scattering, it has several distinct advantages. Most significantly, it is capable of yielding meaningful data on a greater range of particle sizes, namely 20–20,000 Å. The shape of the scattering curve may be related to parameters giving specific information on the three-dimensional shape of the particles, and some inferences may be drawn on hydration effects of the dissolved particles. Finally, some experimental advantage lies in the fact that relatively small amounts of sample (about 10–50 mg) are required for a complete experiment.

General Background

It has been noted that a highly collimated x-ray beam becomes broadened when it passes through a medium comprised of dispersed particles. To infer properties of the particles from the characteristics of the broadened beam, the following considerations are made (15). If the particles are spherical, the scattering envelope may be assumed Gaussian; this is called the Guinier approximation. Thus the angular dependence of scattering may be described by Eq. (7.21).

$$i_s(\theta) = I_0 \exp\left(- \frac{16\pi^2}{3\lambda^2} R_G{}^2 \theta^2 \right) \tag{7.21}$$

From Eq. (7.21) it is clear that a plot of $\ln i_s(\theta)$ versus θ^2 should be linear and have a slope from which the particle radius of gyration may be calculated. It should be noted from Fig. 7.13 that the scattering is observed at an angle 2θ, in accordance with other x-ray diffraction concepts. In addition to this central maximum, which is centered about $\theta = 0°$, much smaller side maxima exist at slightly larger angles. However, if elongated particles such as ellipsoids are being investigated, the scattering curves are broader in the main peak and have less pronounced side maxima. Analytical equations have been written (16) from which scattering envelope shapes for these nonspherical particles may be calculated. Thus it is possible for one to compare an experimentally derived curve with a calculated curve by assuming a certain shape. If the calculated and observed curves coincide, then the shape of the unknown particles may be identical to that used for the calculated curve. Having determined R_G from the plot of $\ln i_s(\theta)$ versus θ^2, one may then compute actual dimensions of the particle. The following table lists formulas for $R_G{}^2$ which are useful when making such calculations (16).

Particle	$R_G{}^2$
Sphere of radius a	$\tfrac{3}{5}a^2$
Ellipsoid with semiaxes a, b, c	$\dfrac{a^2 + b^2 + c^2}{5}$
Elliptic cylinder of height h and semiaxes a and b	$\tfrac{1}{4}\left(a^2 + b^2 + \dfrac{h}{3}\right)$
Cylindrical rod of length h and radius a	$\dfrac{a^2}{2} + \dfrac{h^2}{12}$

Because x-ray techniques can quantitatively measure the scattered energy (intensity or number of photons per unit area per unit time), it is possible to obtain values of molecular weight through use of the following relationships. For the case of incident photons encountering a single electron, the angular dependence of the scattered beam is given by

$$i_s(\theta) = \left(\frac{7.9 \times 10^{-26}}{r^2}\right)\left(\frac{1 + \cos^2 2\theta}{2}\right)I_0 \qquad (7.22)$$

The molecular weight is given by

$$M_2 = (21r^2/z_2{}^2c_2d)(i_s/I_0) \qquad (7.23)$$

In this equation z_2 is derived from the difference in electron density between the solute and solvent, d is path length of beam through the solution, and i_s is, in this case, the total energy scattered in all directions. If the sample contains a distribution of macromolecular species, the computed molecular weight is a weight average, $\overline{M}_w$.

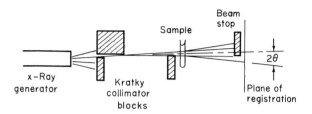

Fig. 7.13. Schematic diagram for the determination of small-angle x-ray scattering from a dilute macromolecular solution. The Kratky collimator produces a very narrow beam of nearly parallel x-rays and minimizes the scattering generated by the slits. (Divergence of the beam is exaggerated in the figure for the sake of clarity.)

Experimental Details in Small-Angle x-Ray Scattering

The apparatus for making x-ray scattering measurements is illustrated in Fig. 7.13. Although lenses can be used to focus visible light, collimators must be used in the case of x-rays. Figure 7.13 shows a schematic representation of the collimator geometry attributed to Kratky (see Ref. 15). The scattering envelope is recorded either on photographic film and then analyzed with an optical densitometer or plotted directly from the output of a scintillation counter which scans the scattered x-rays. [It should be noted that correction for any instrumentally induced line broadening must be made when interpreting results (17).]

Some examples of this approach have been reported for bovine serum albumin (18,19) and poly-γ-benzyl-L-glutamate (20). Figure 7.14 (18) shows $\ln i_s(\theta)$ versus θ^2 for bovine serum albumin; R_G at infinite dilution is determined (by extrapolation) to be 29.8 Å.

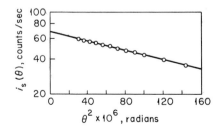

Fig. 7.14. Small-angle x-ray scattering from a solution of bovine serum albumin (BSA); 4.73% BSA, pH 4.61, 0.05 M NaCl. The linearity of the data shown by this semilog plot supports the use of the Guinier approximation for this biopolymer. [Data of Anderegg *et. al.* (18) with permission.]

Viscometry

The hydrodynamic properties of dissolved particles may be considered to result from rotational friction and translational friction. Three experimental techniques (21) can be used to characterize these molecular parameters. Flow birefringence is a means of measuring the rotational hydrodynamics of dissolved particles. Sedimentation rate studies can evaluate translational friction properties. Both of these techniques will be discussed in the following sections. This section will deal with viscometry and the way it can provide useful information from the combination of both rotational and translational friction. By measuring momentum transport in solutions being studied, inferences may be drawn on both shape and size of the dissolved particles.

Briefly, these molecular parameters may be described as follows (22). For particles moving, without rotating, at a velocity u relative to the medium in which they are suspended, a force F will be required to sustain the velocity, where:

$$F = fu \tag{7.24}$$

The value f, which amounts to a proportionality constant, is called the translational friction factor. By analogy, for stationary particles rotating with some angular velocity ω relative to that of the surrounding medium (equal to half the local shear rate), a torque Ω must be continually applied, where:

$$\Omega = \zeta\omega \tag{7.25}$$

The value ζ, then, is the rotational friction factor. Expressions have been derived for f and ζ corresponding to either spherical or ellipsoidal particles, but a randomly coiling polymer presents a more difficult case. A chain macromolecule may be considered, if it is flexible, to be a linear ensemble of much smaller particles. If it is reasonably rigid (perhaps as is the case with many proteins), the entire macromolecule might effectively behave as a rather solid particle (23).

General Background

Each of these friction properties influences the viscous properties of polymer solutions (24). A very useful way to characterize how the dissolved polymer molecules contribute to the viscous dissipation of energy is by computing the limiting viscosity number $[\eta]$. This is done by obtaining the relative viscosity η_{rel}, as the ratio of solution viscosity $\eta_{solution}$ to solvent viscosity $\eta_{solvent}$:

$$\eta_{rel} = \eta_{solution}/\eta_{solvent} \tag{7.26}$$

This may also be expressed as specific viscosity.

$$\eta_{sp} = \eta_{rel} - 1 = \frac{\eta_{solution} - \eta_{solvent}}{\eta_{solvent}} \tag{7.27}$$

The two following equations may be used to relate η_{sp} and η_{rel} to concentration of solute.

Huggins (25) $\quad\quad \dfrac{\eta_{sp}}{c_2} = [\eta] + k'[\eta]^2 c_2 + \cdots \tag{7.28}$

Kraemer (26) $\quad\quad \dfrac{\ln(\eta_{rel})}{c_2} = [\eta] - k''[\eta]^2 c_2 + \cdots \tag{7.29}$

In these equations, k' and k'' are flow constants. It can be seen that a plot of either η_{sp}/c_2 or $\ln(\eta_{rel})/c_2$ versus c_2 will be nearly linear and have a common intercept at $[\eta]$. The proper units for $[\eta]$ are milliliters per gram, so it is necessary to use c_2 values in grams per milliliter. If c_2 units are in grams per deciliter (that is, grams per 100 ml), the intercept will be 100 times smaller, and in such cases $[\eta]$ will be called intrinsic viscosity. It is possible to show that k' and k'' should be related such that their sum equals one-half, but this is not usually found to be exactly true in practice. Nevertheless, k' (or k'') is a parameter that will vary when different solvents are used with a given polymer sample. The physical significance of different values of k' with various solvents is not easily established, although Eirich and Riseman (27) have offered a number of reasons, such as rigidity of particle or penetration of solvent, for certain relative trends in k'.

One approach to calculating the effect that dissolved particles will have on viscosity has been described by Simha and co-workers (28). The limiting viscosity number may be related to a factor v by

$$\lim_{c_2 \to 0} \frac{\eta_{sp}}{c_2} = [\eta] = v_2 v \tag{7.30}$$

Values of v have been calculated for ellipsoidal particles and are plotted in Fig. 7.15 as a function of axial ratio, a/b. Thus for each experimentally determined limiting viscosity number, one can obtain two values of axial ratio which correspond to rigid prolate or rigid oblate ellipsoids. If it can be assumed that the particles being studied are both rigid and ellipsoidal, then their axial ratio can successfully be determined by this technique, but because rigid particles of such simple shapes contribute to solution viscosity solely through their asymmetry, no information may be obtained on their size. Thus it can be seen that all suspensions of uniformly sized rigid spheres should have the same limiting viscosity number (that is, 2.5) if $v_2 = 1$.

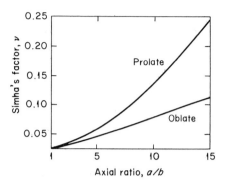

Fig. 7.15. Relationship between axial ratio of ellipsoidal particle and Simha's factor ν, for interpretation to their intrinsic viscosity, which is 10^{-2} times the limiting viscosity number. [From Tanford (22, p. 335), with permission.]

However, a dependence of $[\eta]$ on particle size, or molecular weight, is usually encountered in practice. This is due partly to the fact that chain macromolecules permit some flow of solvent through them when they are subjected to a shear field, and it is also due to excluded volume effects. These factors enter Eq. (7.30) when the specific volume term is modified by quantities related to chain segmental size, coiling effects, and especially molecular weight. The results can be expressed (29) by Eq. (7.31).

$$[\eta] = \Phi R_G{}^3/M_2 \qquad (7.31)$$

where

$$\Phi = 19.94 \times 10^{24} \quad \text{Debye–Beuche}$$
$$= 6.13 \times 10^{24} \quad \text{Kirkwood–Riseman–Zimm}$$
$$= 5.32 \times 10^{24} \quad \text{Flory}$$

Radius of gyration R_G is also a function of molecular weight (see Ref. 29). For example, if Gaussian chain statistics may be assumed, R_G may be written

$$R_G = \left(\frac{l^2}{6}\frac{M_2}{M_0}\right)^{1/2} \qquad (7.32)$$

Therefore, the net result of Eqs. (7.31) and (7.32) is that $[\eta]$ should be a function of $M_2{}^{1/2}$. A semiempirical relation (30) is often used which takes into account the combined contribution of various effects on Φ and R_G, i.e.,

$$[\eta] = K'M_2{}^a \qquad (7.33)$$

For example, poly-γ-benzyl-L-glutamate exhibits $K' = 1.4 \times 10^{-7}$ and $a = 1.75$ in dimethylformamide solutions, while dichloroacetic acid solutions (31) of the same synthetic polypeptide yield values of $K' = 2.78 \times 10^{-3}$ and $a = 0.87$. The rather large difference between these two pairs of parameters strongly suggests that the degree of expansion or contraction of the biopolymer differs considerably in the two solvents. Equation (7.33) is of great practical significance because M_2 values may be computed directly from the easily obtained limiting viscosity numbers. Typical data for a variety of biopolymers are presented in Table 7.2 (31–38).

TABLE 7.2

Values of K' and a for Some Biopolymers

Polymer	Solvent	K'(ml/gm)	a	T (°C)	Reference
Poly-γ-benzyl-L-glutamate	Dimethylformamide	1.4×10^{-7}	1.75	25	31
Poly-γ-benzyl-DL-glutamate	Dichloroacetic acid	2.78×10^{-3}	0.87	25	31
	Dimethylformamide	3.77×10^{-2}	0.55	25	32
Poly-ε-N-carbobenzoxy-L-lysine	Dichloroacetic acid	2.85×10^{-3}	0.85	25	32
	Dimethylformamide	2.83×10^{-3}	1.26		33
Poly-DL-phenylalanine	Chloroform	3.46×10^{-6}	1.48	25	34
Poly-N-benzyl-β-alanine	Dichloroacetic acid	1.2×10^{-1}	0.525	25	35
Polysarcosine	Water	5.6×10^{-2}	0.88	20	36
Amylose	Dimethyl sulfoxide	3.97×10^{-3}	0.82	20	37
	Water	1.32×10^{-2}	0.68	20	37
Cellulose	Cuprammonium	1.05×10^{-1}	0.66	20	38

Equipment

Probably the most convenient apparatus for measuring limiting viscosity number is the capillary viscometer. More sophisticated equipment is also available, as perhaps best represented by the Weissenberg rheogoniometer, which is capable of making exceedingly precise measurements on viscosity, shear rate dependence, and normal stress effects of solutions being tested. Yet for the information desired, glass capillary viscometers are often completely satisfactory. The chief advantages are good temperature control, inert surfaces, low shear rates, internal dilution capability, and small polymer sample requirements. Viscometry may be used, in conjunction with both Eq. (7.33) and a good calibration, to measure molecular weights over a range that is at last as wide as light scattering.

From Poiseuelle's law for capillary flow, it can be shown that the time t necessary for a given quantity of solution to pass through a capillary is directly proportional to the viscosity of the liquid:

$$t = \frac{\eta}{\rho} \frac{8L}{\pi g a^4} \int_{h_1}^{h_2} \frac{1}{h} d(V) \qquad (7.34)$$

where, η is solution viscosity, ρ is solution density, L is capillary length, g is the gravitational constant, a is capillary radius, h is height of solution above capillary exit, and V is volume of solution. The term η/ρ is the kinematic viscosity. Because the limiting viscosity number is calculated by taking ratios of solution to solvent viscosities, and because the density of solution converges to that of the solvent at infinite dilution, the ratio of flow times becomes equivalent to the ratio of actual viscosities in the limit as concentration approaches zero. A good type of capillary viscometer is due to Ubbelohde, which is pictured in Fig. 7.16. Data obtained for poly-L-proline in propionic

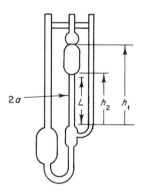

Fig. 7.16. Schematic diagram of an internal dilution capillary viscometer. The flow time is obtained by measuring the time required for solution in the center tube to drain from h_1 to h_2 by passing through the capillary section of diameter $2a$ and length L. The tube on the right always maintains the bottom of the capillary at atmospheric pressure, independent of liquid inventories above or below that point.

acid (same sample as for Figs. 7.5 and 7.12) are shown in Fig. 7.17. From the intercept, the limiting viscosity number is $[\eta] = 91.4$ ml/gm. The upper set of data was computed using the Huggins relationship, Eq. (7.28), while the lower set was generated from the Kraemer equation (7.29). Within reason, a common intercept can be derived using these data.

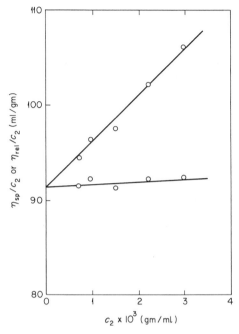

Fig. 7.17. Viscometry data for poly-L-proline in propionic acid. When c_2 is expressed as grams per milliliter, the intercept on the ordinate is called limiting viscosity number. [Data of Carr *et al.* (4).]

Sedimentation

As stated in the previous section, sedimentation analyses measure solution properties related to the friction acting on macromolecules moving through a stationary medium. By placing the solution in a cell rotating about an axis, each dissolved particle is subjected to a force equal to the difference between centrifugal and buoyancy forces. This statement is expressed mathematically in Eq. (7.35).

$$F = (\omega^2 r M_2/N) - (\omega^2 r \rho_1 v_2 M_2/N) \qquad (7.35)$$

This force F is opposed by the drag force fu when the velocity u rises to a sufficiently high level. It can be seen from Eqs. (7.24) and (7.35) that knowledge of the velocity u will give information on the translational friction factor f. In order to obtain appreciable particle velocities, sedimentation experiments are usually performed in a high-speed ultracentrifuge.

General Background

The size and shape of the dissolved macromolecules may be ascertained from the experimental data through the translational friction factor. Stokes derived the following expression for f_0 corresponding to impenetrable spheres.

$$f_0 = 6\pi\eta_0 r \qquad (7.36)$$

Here, η_0 is viscosity of the medium through which spherical particles of radius r are moving. Later Herzog *et al.* (39) obtained analogous equations for solid ellipsoidal particles. In order to adapt the concept embodied in this relationship to coiling polymers, the particle radius r is represented by $R_G A$, where A is some compensating factor. The expression for A [from the work of Kirkwood and Riseman (40)] takes into account such factors as chain contour length and solvent–polymer friction per unit chain length, and the result is a value (22) in the neighborhood of 0.665–0.875 (provided the chains contain more than 1000 effective segments).

The Sedimentation Coefficient

The velocity with which particles move under the influence of the gravitational field, $u = dr/dt$, can be related to the sedimentation coefficient s_2 by

$$s_2 = u/\omega^2 r = M_2(1 - v_2 \rho_1)/Nf \qquad (7.37)$$

Here, ω is the angular velocity of the solution and ρ is solvent density. At infinite dilution Eq. (7.37) can be rearranged such that the molecular weight of the solute may be computed:

$$\lim_{c_2 \to 0} \left| \frac{s_2{}^0 RT}{D_2^0(1 - v_2 \rho_1)} \right| = M_2 \qquad (7.38)$$

Since $u = dr/dt$ can be obtained directly as a function of r from ultracentrifuge photographs taken at various times, it is possible to compute s_2 directly from the middle part of Eq. (7.37). One can get $s_2{}^0$ by extrapolation of s_2 to zero concentration. The easily accessible values of v_2 and ρ_1 must also be obtained. Another parameter that must be obtained is the limiting value of the translational diffusion coefficient D_2^0. In practice the measurement of D_2^0 is fairly difficult, and the results obtained are frequently regarded with some uncertainty. Consequently, Eq. (7.37) is not generally used to determine M_2. Instead, semiempirical correlations are often constructed to relate M_2 directly with $s_2{}^0$.

Equipment

One experimental situation is illustrated in Fig. 7.18. It involves an aluminum rotor, approximately 20 cm in diameter, which contains near its periphery one (or more) cells. The sides of a cell are made to be exactly radial, so that spurious effects, which might result from sedimenting particles impinging on these walls, are eliminated. A photographic system is aligned so that each cell is viewed every revolution of the rotor. Included in this system are schlieren optics for the detection of concentration gradients in the cell. When

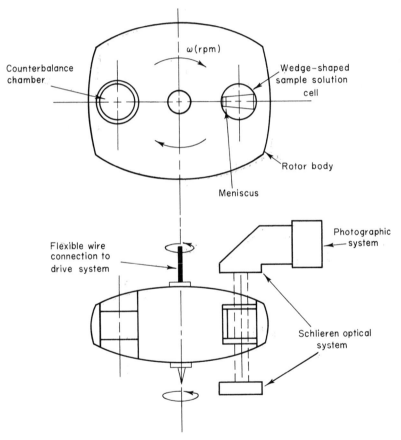

Fig. 7.18. Schematic diagram of an ultracentrifuge. The rotor, which is suspended from the drive system by a flexible wire, is located such that the wedged-shaped solution cell is exposed to the schlieren optical system every revolution. When the image of a line is viewed through the sedimenting solution, it is recorded by the photographic system as a curve that gives the concentration gradient directly.

a straight line is viewed through the rotating cells, its image is distorted into a curve that is directly related to the refractive index gradient in the solution, and, in turn, the index of refraction can be converted directly into solute concentration as a function of r. At the moment the cell starts rotating, the concentration of solute molecules is uniform throughout, but as time elapses after the centrifugal field develops, the average location of each particle is displaced outward (see Fig. 7.19). The result is that the top of the cell, which is closest to the center of rotation, becomes virtually depleted of solute and the bottom of the cell begins to accumulate solute particles. It can be seen in Fig. 7.20 that this gives a radial distribution of concentration that is zero at the top of the cell, is the same as the initial concentration in the middle of the cell, and is a very high value at the bottom. The derivative of this concentration distribution is also shown in Fig. 7.20. It can be obtained directly from photographs of the schlieren pattern. The peak which appears in the gradient curve corresponds to the boundary between solvent in the top and the solute that has been spun downward. These peaks become increasingly broad as time elapses, due to the tendency of solute particles in the vicinity of this boundary to diffuse into the region where concentration is zero.

If more than one solute component is present, it is likely that each component will move at a different rate, depending on its particular friction factor and on any interaction effects between the moving particles. In the case of a polydisperse polymer sample, the number of components may be enormous if one considers each different chain length as representing a distinct component. A relationship may be derived which indicates that the sedimentation coefficient s_2 depends on $M_2^{2/3}$. Other factors to be mentioned below, however, make the experimental exponent lie somewhat closer to a direct proportionality between s_2 and M_2. In any event, peaks corresponding to the solute boundary in a sedimentation experiment will represent, to a first approximation, the distribution of molecular sizes (molecular weights).

The interfering factors previously alluded to are as follows. First, the friction factor is a function of shape; for example, s_2 for rodlike particles

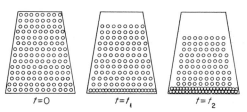

Fig. 7.19. Schematic representation of a polymer solution undergoing sedimentation. As time elapses a boundary forms between pure solvent at the top and a zone in the middle which contains polymer at its original concentration; particles also begin to pile up at the bottom of the cell. [From Tanford (22, p. 366) with permission.]

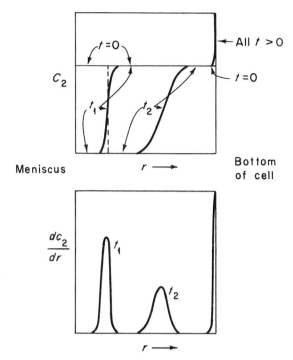

Fig. 7.20. The upper graph shows the concentration profile in the cell at times corresponding to those indicated in Fig. 7.19. The lower graph shows the corresponding concentration gradients for t_1 and t_2; these peaks are essentially what would appear in schlieren photographs of this sedimenting solution. [From Tanford (22, p. 366) with permission.]

depends on the first power of M_2, rather than the two-thirds power. Friction factors of certain macromolecules rise with concentration of solute, and because the viscosity of the medium is raised by the presence of other solute molecules, an effect called hypersharpening may occur. Finally, the Johnston–Ogston effect (41) may also interfere with the independent sedimentation of solutes having different values of s_2. This may best be explained for the case of two macromolecular components, one with large and one with small s_2. The result is a partial exclusion of the slower component from the bottom region of the cell when the faster component has undergone appreciable sedimentation. Thus there is a "builddown" of the slower species near the bottom of the cell with the concomitant "buildup" of this component in the zone between its own boundary and the boundary made by the faster component. Whether the reasons for this are related to the interaction energies between the two kinds of solutes or whether they are due to a physical effect like viscosity remains unclear, but if the Johnston–Ogston effect exists in some biopolymer system, it will certainly interfere with quantitative measurements.

An alternative technique, called zone sedimentation, is now as commonly used as boundary sedimentation. In zone sedimentation, the centrifuge cell is filled with the solvent rather than the solution to be studied. Next, a small aliquot of solution is layered on the top (42). As the rotor speed is increased the macromolecules begin to move through the solvent zone. By measuring their velocity $u = dr/dt$, s_2 can be computed. In these experiments it is frequently found that the movement of sedimentating particles is more sharply resolved. Also a smaller amount of biopolymer is needed to conduct the experiment. Furthermore it is possible to use zone sedimentation to separate different macromolecular components on the basis of sedimentation velocity. Two significant disadvantages, however, should be noted. Firstly, the concentration dependence of the sedimentation coefficient cannot be studied, and secondly, special requirements for solvents restrict the possible choices while increasing chances of inadvertently altering chain conformation. One of the many studies in which this technique has been employed is that of Wetmur and Davidson (43) on DNA denaturation kinetics.

As an example of the application of the sedimentation boundary method, the poly-L-proline system may again be quoted. The observed sedimentation coefficient is calculated from

$$s_{obs} = \omega^{-2}\, d(\ln r)/dt \tag{7.39}$$

Figure 7.21 shows a plot of $\ln r$ versus t for poly-L-proline II. The slope gives $s_{obs} = 0.518 \times 10^{-13}$ sec. After obtaining data of this type it is conventional to adjust the value to standard conditions (water, 20°C) and to obtain

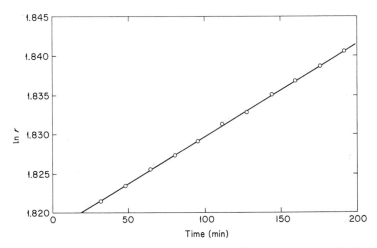

Fig. 7.21. Plot of sedimentation data for poly-L-proline in propionic acid. The movement of the boundary is plotted against ellapsed time. From the slope of this line one can compute the sedimentation coefficient for this particular concentration.

a series of values at different concentration such that extrapolation to zero concentration may be achieved as required by Eq. (7.38). For the data in Fig. 7.21 the value is $s_{20,w}^0 = 0.85$.

The Archibald Approach to Equilibrium Techniques

Operating the ultracentrifuge at appropriately slow speeds can also produce a concentration gradient from which solute molecular weight M_2 can be computed. The chief requirement is to establish conditions within the cell such that the net flux of solute is zero. This can be achieved experimentally by finding some rotor speed at which back-diffusion stabilizes the concentration gradient developed as a result of sedimentation effects. However, a simpler method for producing conditions that meet the criterion of zero net solute flux has been described by Archibald (44). He noted that at the upper and lower confines of the solution in the centrifuge, namely, at the top (or meniscus) and at the bottom of the cell, no flux of solute can exist. Thus, the expressions for flux due to diffusion and flux due to sedimentation can be combined to yield solute molecular weight:

$$M_2 = \frac{RT}{(1 - v_2 \rho_1)\omega^2} \left(\frac{dc_2/dr}{c_2 r} \right) \tag{7.40}$$

Equation (7.40) is generally valid only for conditions which exist at the top and bottom of the cell, provided some solute is present at each location.

Experiments are conducted in a manner similar to those designed to yield values of s_2, except that the rotor speed is significantly slower. It is necessary to know the refractive index increment for the biopolymer–solvent system being studied, so that the index of refraction gradient, which is obtained directly from a calibrated schlieren optical system, can be converted into the actual concentration gradient. Values of molecular weight must then be obtained from a number of solutions with different concentrations, and extrapolation of experimental values to the case of infinite dilution is required. The technique commonly employs a synthetic boundary at the bottom; this is achieved by layering the biopolymer solution on top of some higher-density liquid in which no constituents of the test solution are soluble. By doing this, one moves the lower boundary into the view of the optical system and also provides a way for contaminating particles to pass into the more dense liquid. This eliminates the accumulation of material at the lower boundary where the critically important concentration gradient is to be obtained.

An example of the application of the Archibald technique may be drawn from the work of Smith *et al.* (45) on lysozyme. Their data are summarized in Fig. 7.22, where $(1/rc_2)(dc_2/dr)$ is plotted versus r. The molecular weight computed from Eq. (7.40) is found to be 14,100 on average for the top of the

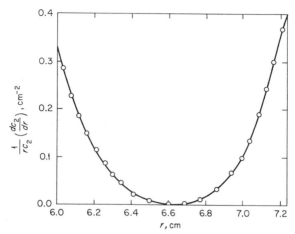

Fig. 7.22. Plot of slow-sedimentation data for lysozyme. Extrapolation of the curve to its value at cell top or cell bottom facilitates use of the Archibald equation. [Data of Smith *et al.* (45).]

cell (14,500 from the specific data in Fig. 7.22) and 18,700 for the bottom of the cell (18,700 in the figure). The values should be the same, but in this case it was felt that high molecular weight impurities were present, and the larger value was consequently rejected.

It is, however, quite possible to observe a concentration dependence of molecular weight, and it is also possible to encounter some biopolymer that exhibits time-dependent dc_2/dr values at the boundaries. Such interfering effects are eliminated by plotting the data in a fashion similar to the Zimm reciprocal intensity method for light scattering. This means that boundary values of $(1/rc_2)(dc_2/dr)$ are obtained at a number of times and at a number of concentrations and then are plotted on the same graph. Data at corresponding times are extrapolated to infinite dilution, and these limiting values are extrapolated to zero time. By performing this operation for data from both cell top and cell bottom, the extrapolation can sometimes be made more accurately, but contamination (as was evident in the example with lysozyme) will generally cause boundary values of $(1/rc_2)(dc_2/dr)$ at the cell bottom to be higher.

Two distinct advantages in using the Archibald technique are worth mentioning here. The first is that a minimum of time is required to obtain the experimental data with the ultracentrifuge, although without modern computational equipment a fair amount of time would be required to process these data. The second advantage is that large-scale separation of components, as occurs in sedimentation velocity experiments, is not involved. Thus effects such as Johnston–Ogston, large concentration dependence on sedimentation, or aggregation–dissociation behavior are largely avoided.

A Correlative Procedure Based on Sedimentation

Scheraga and Mandelkern (46) have reported a technique from which the experimentally determined quantities: sedimentation coefficient, limiting viscosity number, and molecular weight, can be correlated with the axial ratio of a hydrodynamically equivalent ellipsoid. In their work, a parameter F was defined as the ratio of friction factors for a spherical and a nonspherical (ellipsoidal) particle, i.e., f_0/f. Two rather complicated analytical expressions can be written relating F to the axial ratio a/b; one corresponds to the case of prolate ellipsoids (cigar-shaped), while the other applies to oblate ellipsoids (discus-shaped). Equations (7.30) and (7.37) can be combined with the preceding definition to give the following expression:

$$ F\left(\frac{vN}{162\pi^2}\right)^{1/3} = \frac{Ns_2^{\,0}[\eta]^{1/3}\eta_0}{M_2^{\,2/3}(1 - v_2\rho_1)} \equiv \beta \qquad (7.41) $$

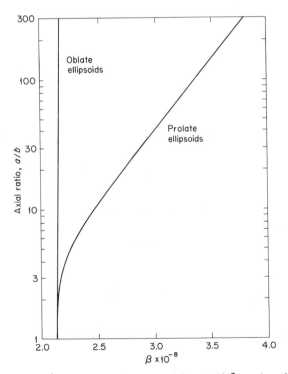

Fig. 7.23. Plot of β [in Eq. (7.41)] versus axial ratio. If β can be calculated, it is, in principle, possible to obtain a value of a/b for ellipsoidal particles which would behave equivalently.

Either combination on the left side of this equation is defined as equal to a parameter, β. The middle term is composed largely of material quantities and the other contains terms that can be related to an assumed ellipsoidal particle shape. Figure 7.15 shows this relationship for the parameter v, and Fig. 7.23 contains a plot of computed values of β corresponding to ellipsoids with different axial ratios.

In practice one must obtain values of M_2, $s^0_{20,w}$, and $[\eta]$ in order to use Fig. 7.23. It provides the axial ratio of an ellipsoid which would behave hydrodynamically the same as the biopolymer being studied. However, one must first make the assumption that the macromolecular particles in question are ellipsoidal, and then it is necessary to decide whether these particles are prolate or oblate. One cannot, on the other hand, declare that the particles in the sample solution are ellipsoidal simply because a plausible value of β (and therefore a/b) is obtained. Noting that the magnitude of β falls in a rather narrow range of values (especially if oblate ellipsoids are being investigated), Eq. (7.41) could, in principle, be used as a check on the accuracy of the experimental values obtained. Because this correlation was derived for a monodisperse polymer system, its applicability to polydisperse samples is questionable.

Flow Birefringence

As mentioned earlier, the frictional properties of dissolved particles can be characterized both by a friction factor f, related to resistance to flow through its surrounding medium, and by another friction factor ζ related to resistance to rotation about its volume centroid. Viscometry measures properties of the macromolecules that arise from a combination of resistance to translation and rotation, while ultracentrifugation provides values of s_2, which is inversely proportional to the translation friction factor. Flow birefringence, on the other hand, is a technique that permits an assessment of the other kind of hydrodynamic friction, namely, rotational.

In these experiments, a velocity gradient $\dot{\gamma}$ is applied to the sample solution. The result is that the domain of each macromolecule tends to become elongated and oriented into the flow direction. The degree to which the dissolved chains become elongated depends on their flexibility, but if their unperturbed shape is elongated (perhaps ellipsoidal), these particles are subjected to a smaller amount of deformation. Torque arising from the imposed shear field is opposed by a torque due to rotational diffusion. Thus, if any velocity gradient existed in a polymer solution, nonspherical particles would undergo some rotation, but for a larger part of the time their long axis would be parallel to the fluid flow. Thermal energy would cause the particles to diffuse away from this preferred direction. Such a diffusion process is characterized by a rotary diffusion coefficient Θ, which is described as follows:

$$\Theta = RT/N\zeta \tag{7.42}$$

The term ζ for spherical particles is given by the Stokes equation:

$$\zeta = 8\pi\eta_0 r^3 \tag{7.43}$$

Perrin (47) has derived expressions for ζ which correspond to ellipsoids, but Scheraga and Mandelkern (46) have shown how knowledge of Θ can lead to the determination of the ellipsoid axial ratio a/b. They defined a quantity L as the ratio of rotary friction factors for spherical and ellipsoidal particles:

$$L = \zeta_0/\zeta \tag{7.44}$$

It is possible to show that Eqs. (7.42), (7.43), and (7.44) can be combined to produce the following expression:

$$Lv = 6\eta_0 \Theta[\eta]M_2/RT \equiv \delta \tag{7.45}$$

Either term on the left is thus equated to a parameter δ, and from Fig. 7.24 it can be seen that axial ratios of hydrodynamically equivalent prolate and oblate ellipsoids can be found once the quantities Θ, $[\eta]$, and M_2 have been obtained experimentally.

The extent to which the elongated particles tend to become oriented in the shear field can be determined from the birefringence that develops in the sample solution. In order for this to occur, the dissolved macromolecules must possess the following two properties: they must be either inherently nonspherical or capable of deforming under shear into some elongated shape, and they must be optically anisotropic. Shape asymmetry is necessary for the shear field to have any orienting effect on the initially random alignment of particles, while the optical anisotropy is necessary to provide some means to determine the extent to which particles have become oriented by the shear field. Thus it would be possible to have spherical particles which could be very optically active but which would not produce any flow-induced birefringence because the shear field could not cause preferential alignment of individual particles. Likewise, highly nonspherical particles, such as discs or rods, would not give rise to birefringence if they lacked optical activity. In reality all chain molecules possess some degree of optical anisotropy, but in a few cases it cannot be observed because of its weakness.

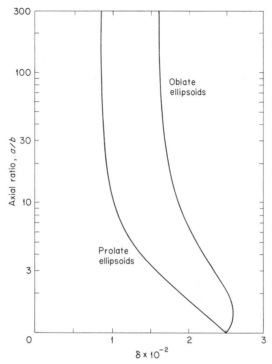

Fig. 7.24. Plot of δ [in Eq. (7.45)] versus axial ratio. This plot, which is applicable to flow-birefringence measurements, can be used in a manner that is analogous to Fig. 7.23.

Experimental Apparatus

As mentioned earlier, information on axial ratio can be calculated once the rotary diffusion coefficient Θ has been obtained. Further, the ability of dissolved particles to deform under shear can also be inferred from knowledge of Θ at different shear rates. The most commonly used technique for obtaining the rotary diffusion coefficient of particles that meet the two previously mentioned requirements employs couette flow produced between concentric cylinders (48). An alternative technique utilizing flow between two stationary plates has also been described (49). As shown in Fig. 7.25, particles in a solution placed between two concentric cylinders tend to become aligned when either cylinder (in this case the outer one) begins to rotate.

This orientation is analyzed by viewing the solution in the annular space with crossed polars, as shown in Fig. 7.25. As the outer cylinder starts to rotate, a very small shear field is created in the solution, and the particles gradually become oriented. This means that the polarized light entering the

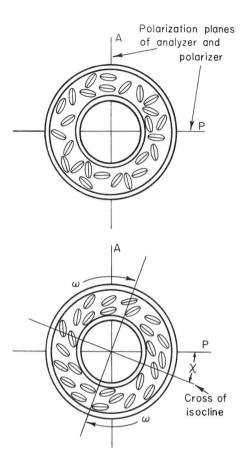

Fig. 7.25. Schematic representation of the measurement of flow birefringence with a concentric cylinder device. The dissolved particles are represented as prolate ellipsoids with their optic axis parallel to the major axis of the particle. No preferred orientation exists when there is no hydrodynamic shear in the solution, and thus light passing through the annular space will be completely blocked by the analyzer. When the outer cylinder is rotated, the particles tend to assume a tangential orientation, and the resulting birefringence permits some light to pass through the analyzer, except where the optic axes of the particles are oriented parallel to either the analyzer or the polarizer. These regions make what is called the "cross of isocline."

solution will be slightly rotated as it passes through the annulus, and consequently some light will be able to pass through the analyzer. However, at four locations, which are χ degrees from the plane of either the polarizer or analyzer, the net optic axis of the solution will be parallel to polarizer or analyzer. Thus no rotation of the plane of the incoming light will occur, and as a result no light will pass through the analyzer. These four points of extinction make what is called the "cross of isocline." As the shear field is increased, the particles tend to acquire a more nearly tangential orientation, and thus χ will correspondingly decrease toward zero.

Experimentally, one applies increasing levels of shear rate to the solutions being studied and measures the angle between the cross of isocline and the coordinates of the crossed polars. This can be done either visually or with greater sensitivity using a photoelectric scanning technique described by Harrington (50). It can be shown (22,51) that χ is related to Θ, and for small values of $\dot{\gamma}$ this relationship may be expressed by the following series expansion:

$$\chi = \frac{\pi}{4} - \frac{1}{12\,\Theta}\,\dot{\gamma}\left[1 + \frac{1}{108\,\Theta^2}\,\dot{\gamma}^2\left(1 + \frac{24}{35}\frac{a^2 - b^2}{a^2 + b^2} + \cdots\right)\right] \qquad (7.46)$$

It should be noted that χ is a function of the particle axial parameters a and b, but it should also be noted that as $\dot{\gamma}$ approaches zero, the expression tends

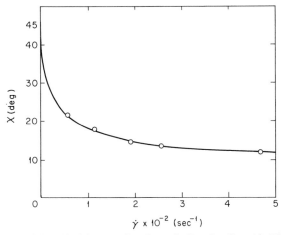

Fig. 7.26. Plot of flow-birefringence data for polyriboadenylic acid. The angle between the cross of isocline and the plane of the polarizer (or analyzer if rotation is in the opposite direction) is plotted against shear rate produced by rotating the outer cylinder.

toward linearity, with a slope of $-1/(12\,\Theta)$ and a vanishing dependence on particle shape parameters. From the slope of plots of χ versus $\dot{\gamma}$, one can then compute Θ. For a more complete treatment of these relationships the reader is also referred to the work of Scheraga *et al.* (52). Finally, parallel measurements to zero concentration (where solution nonidealities disappear) can be achieved.

An example of the application of flow birefringence may be quoted from the work of Takashima (53) on polyriboadenylic acid. A plot of $\dot{\gamma}$ versus χ is shown in Fig. 7.26. From the limiting slope Θ is calculated, via Eq. (7.46), to be 4.98×10^{-2} sec^{-1}. This is an unusually small value which may reflect a very extended shape for the polynucleotide chains. The preceding value of Θ was obtained for a 0.333 mg/ml solution of the potassium salt. "True" values of Θ are generally obtained by extrapolation to infinite dilution (54). If values of molecular weight and limiting viscosity are available, it is possible to proceed from Θ values via Eq. 7.45 to an estimation of ζ and thence the axial ratio of biopolymers in the unperturbed state.

Characterization of Conformation, Structure, and Conformational Transitions in Solution

There are, of course, many techniques for characterizing conformation of biopolymers in solution. The spectroscopic techniques, ORD/CD, NMR, infrared, Raman, etc., have in common the assumption that biopolymers with a common geometry give rise to a common spectral feature. The basic geometry of the initial member has generally been determined by other techniques, particularly x-ray diffraction. However, the other techniques spelled out in the first portion of this chapter also have the potential of yielding features related to the dimensions of biomolecules. Thus "rigid" rods of coiled coils and superhelices often reflect the underlying monomer repeat and molecular weight; helices may give rise to extended or "worm-like" solution conformations; and random structures appear compact with small radii of gyration. There are other techniques which give useful information concerning conformation, structure, and conformational transitions in solution; one such method is dielectric dispersion which is mentioned in the following chapter.

Hydrogen Exchange and Titration

The conformation of biopolymers which contain ionizable groups or exchangeable hydrogens may be investigated experimentally for conformation-dependent changes in the chemical properties of these groups. If part of the side group on a polypeptide can ionize, its interconversions between charged and noncharged states can be followed by titration. Studies on the rate of this conversion can be related to helical or coil conformations of backbone, and in work done with poly-L-glutamic acid (55,56), investigators were able to deduce thermodynamic differences between its α-helical and coil states. Studies on polyriboadenylic acid by Holcomb and Timasheff (57) have successfully demonstrated the ability of titration techniques to identify both its single and double stranded helices. Results from other investigations can be found in Table 7.3 (55–58).

TABLE 7.3

Solution-State Transformations as Obtained by Acid/Base Titration

Macromolecular species	Temperature (°C)	Solvent	Transformation	pK	Reference
Poly-L-glutamic acid	30	N-Methylacetamide	Charged coil → α helix	4.5[a]	55
	15	0.1 M KCl	Charged coil → α helix	5.2[a]	56
Poly-L-lysine	25	0.1 M KCl	Charged coil → α helix	9.9[a]	56
Polyriboadenylic acid	0	0.001 M KCl	Double helix → single helices	7.5	57
	40	0.15 M KCl	Double helix → single helices	5.5	57
Polyribocytidylic acid	25	0.05 M NaCl	Single chains → double helix	5.8	58
	25	0.05 M NaCl	Double helix → protonated aggregates	3.2	58
Polyribocytidylic acid + polyriboinosinic acid	25	0.05 M NaCl	C : I Double helix → half protonated double helix	4.9	58
	25	0.05 M NaCl	Half protonated double helix → fully protonated double helix	3.9	58

[a] Values are apparent pK, computed according to B. H. Zimm and S. A. Rice, *Mol. Phys.* **3**, 391 (1960).

Hydrogen exchange experiments can also give a measure of the accessibility of chemical groups which contain "labile" hydrogens. Such groups would include peptide linkages, different amino acid side groups, and all purine and pyrimidine moieties in nucleic acids. The rate at which the hydrogens in question exchange can be determined in either of two basic ways. One technique starts by equilibrating the polymer in deuterated (or tritiated) water. Subsequently, it is put in normal water for a specified length of time. Next it is separated from the exchange medium, possibly by using a gel permeation column, and the amount of isotope lost by the polymer is measured. Conversely, experiments can also be designed whereby the sample macromolecules are exposed to deuterated (or tritiated) water for specific time intervals, and the uptake of isotope is then measured. Because deuterium is heavier than hydrogen and tritium is radioactive, their presence can be readily monitored. In the case of polypeptides it is necessary to establish the exchange rate for amide hydrogens and also for labile hydrogens in side groups. If the α-helix were present, the amide hydrogens would be bound by hydrogen bonds and, to a certain extent, inaccessible to solvent. Thus their exchange rate is found to be slow. On the other hand, if the polypeptide were in a randomly coiled state, these same hydrogens would be much more readily exchanged. Side group hydrogens show differences in exchange rate usually as a result of pH changes. The reader is referred to work by Ikegami *et al.* on poly-L-glutamic acid (59) for an example of these effects. A novel approach to characterizing these rate differences has been reported by Englander and Poulsen (60) in which hydrogen exchange in poly-DL-alanine and poly-DL-lysine was studied. Because neither polymer can form any kind of ordered conformation, they were thus able to establish values of exchange rate for nonhelical polypeptides. Other work, measuring tritium exchange-out rates are summarized in Table 7.4 (61).

TABLE 7.4

Tritium Exchange-Out Rates for Some Peptide Linkages (61)

Material	pH	Conformation	First-order rate constant (min^{-1})
Triglycine	4.7	Random coil	0.795
Poly-DL-alanine	4.7	Random coil	0.795
Poly(D-glutamic acid/L-alanine)	4.7	Random coil	0.398
Poly-L-glutamic acid	4.7	α Helix	0.025
Poly(L-glutamic acid/L-alanine)	4.7	α Helix	0.016

Dilatometry and Calorimetry

In addition to the technique mentioned in the preceding section, calorimetry and dilatometry are two other physical techniques which are well suited to investigating the process whereby one conformation transforms into another. Examples of these transformation processes may be of the helix-to-helix type, such as occurs with poly-L-proline (62), or of the more commonly studied helix-to-coil type. It may be possible to follow a transformation by establishing physicochemical properties before and after changing conditions which stabilize one form or another. The two techniques to be described in this section are not so well suited to determining which form is present in a solution as they are capable of yielding quantitative information about the interconversion of conformations.

Most dilatometric equipment can measure the volume of a biopolymer solution as functions of such variables as temperature, pressure, time, pH, or ionic strength. Commonly, it is found that solution density will change linearly with the variable in question until some point is reached where a transition is observed. By use of other experimental techniques, the conformation above and below the transition can be obtained. The advantage of dilatometry is its high sensitivity to small changes in volume expansion coefficient. Consequently this technique can pinpoint exactly when, and under what conditions, conformational changes occur. Furthermore, it is then possible to compute thermodynamic quantities once the transformation has thus been characterized. Studies reported by Noguchi and Yang (63) can serve as one example of how dilatometry has been used to follow the helix-to-coil transformation in poly-L-glutamic acid.

The most direct means for obtaining thermodynamic parameters associated with conformational transformations is represented by calorimetry. Heat effects that accompany changes in solution conformation can be measured by differential thermal analysis or differential scanning calorimetry (64–66). The output of both techniques is an electrical signal which may be related to exothermic or endothermic processes within the solution. Hydrogen bond formation, such as that accompanying the formation of an α-helix in poly-α-amino acids or a double helix in polynucleotides, may result in the reduction of enthalpy in the system, but the net change in entropy (from those molecules which had previously occupied the hydrogen bonding sites), may offset any heat effects. Conversion of polyriboadenylic acid from a double-stranded helix to a single-stranded helix can be followed calorimetrically (67). Complementary double helices made from polyriboadenylic acid and polyribouridylic acid have also been studied with this technique (68), and Krakauer and Sturtevant (69) were able to demonstrate with calorimetry the existence of a triple

TABLE 7.5

Transformations in the Solution State as Observed by Calorimetry

Initial macromolecular species	Solvent	Transition	Temperature (°C)	Enthalpy (kcal/mole residue)	Reference
Polyriboadenylic acid + polyribouridylic acid	pH 6.5; 0.06 M cations	A: U Double helix → two single chains	51	6.9	68
Polyriboadenylic acid	pH 4.89; SSC	Double helix → two single chains	57	5.13	67
Polyriboadenylic acid + 2 polyribouridylic acid	Na⁺ Counterions	A: 2U Triple helix → A single chain + U double helix	68	12.7	69
Poly-γ-benzyl-L-glutamate	18% Ethylene dichloride in dichloroacetic acid	Random coil → α helix	40 (Concentration dependent)	0.95	70

helix containing two chains of polyribouridylic acid and one of polyribo-adenylic acid. Table 7.5 contains the principal points of these and other calorimetric investigations (67–70).

Appendix: Statistical Ways of Averaging a Distribution of Molecular Weights

For polymer samples which contain macromolecules of differing sizes, it is necessary to speak of a characteristic molecular weight average. Some chains contain q residues; larger chains contain $q + 1, q + 2, \ldots$ residues. Thus in a polymer sample there would be n_q chains having q residues, and there would be $n_{q+1}, n_{q+2}, \ldots$ molecules, each of chains having $q + 1, q + 2, \ldots$ residues, respectively. One way to calculate an average molecular weight would be according to the following equation:

$$\overline{M}_n = \frac{\sum_i n_i M_i}{\sum_i n_i} \tag{A-1}$$

The result is a number-averaging procedure, and, relative to the other molecular weight averages, this value is weighed heavily by the smaller molecules present. Another way of calculating a molecular weight average is

$$\overline{M}_w = \frac{\sum_i n_i M_i^2}{\sum_i n_i M_i} \tag{A-2}$$

This produces the weight-average molecular weight. As opposed to Eq. (A-1), where each molecular weight is represented by the number of molecules present, Eq. (A-2) multiplies each molecular weight by the weight of each kind of molecule in the sample. The so-called z-average molecular weight is computed by the following equation:

$$\overline{M}_z = \frac{\sum_i n_i M_i^3}{\sum_i n_i M_i^2} \tag{A-3}$$

Higher-order averages may also be calculated from analogous equations, but none of them would have any practical significance with regard to the material presented in this chapter.

REFERENCES

1. C. H. Bamford, A. Elliott, and W. E. Hanby, "Synthetic Polypeptides." Academic Press, New York, 1956.
2. G. D. Fasman, ed., "Poly-α-Amino Acids." Dekker, New York, 1967.
3. G. Felsenfeld and H. T. Miles, *Annu. Rev. Biochem.* **36**, 407 (1967).

4. S. H. Carr, E. Baer, and A. G. Walton, *J. Macromol. Sci., Phys.* B **6**, 15 (1972).
5. C. Tanford, "Physical Chemistry of Macromolecules," Chapter 4. Wiley, New York, 1961.
6. P. Doty and G. E. Myers, *Discuss. Faraday Soc.* **13**, 51 (1953).
7. H. Boedtker and P. Doty, *J. Amer. Chem. Soc.* **78**, 4267 (1956).
8. H. Boedtker and P. Doty, *J. Phys. Chem.* **58**, 968 (1954).
9. H. Ende, G. Meyerhoff, and G. V. Schulz, *Z. Naturforsch.* B **13**, 713 (1958).
10. H. G. Elias, *Angew. Chem.* **73**, 209 (1961).
11. J. W. Strutt (Lord Rayleigh), *Phil. Mag.* [4] **41**, 107 (1871).
12. B. H. Zimm, *J. Chem. Phys.* **16**, 1099 (1948).
13. J. Brandrup and E. H. Immergut, eds., "Polymer Handbook," pp. IV-279. Wiley (Interscience), New York, 1966.
14. J. A. Harpst, A. I. Krasna, and B. H. Zimm, *Biopolymers* **6**, 585 (1968).
15. O. Kratky, *Progr. Biophys.* **13**, 105 (1963).
16. P. Mittelbach and G. Porod, *Acta Phys. Austr.* **15**, 122 (1962).
17. R. Hosemann, *Z. Phys.* **113**, 751 (1939).
18. J. W. Anderegg, W. W. Beeman, S. Shulman, and P. Kaesberg, *J. Amer. Chem. Soc.* **77**, 2927 (1955).
19. V. Luzzati *et al.*, *J. Mol. Biol.* **3**, 566 (1961).
20. V. Luzzati, J. Witz, and A. Nicolaieff, *J. Mol. Biol.* **3**, 379 (1961).
21. H. Morawetz, "Macromolecules in Solution," Chapter VI. Wiley (Interscience), New York, 1965.
22. C. Tanford, "Physical Chemistry of Macromolecules," Chapter 6. Wiley, New York, 1961.
23. H. Benoit, L. Freund, and G. Spach, *in* "Poly-α-Amino Acids" (G. D. Fasman, ed.), Chapter 3, p. 105. Dekker, New York, 1967.
24. J. T. Yang, *Advan. Protein Chem.* **16**, 323 (1961).
25. M. L. Huggins, *J. Amer. Chem. Soc.* **64**, 2716 (1942).
26. E. O. Kraemer, *Ind. Eng. Chem.* **30**, 1200 (1938).
27. F. Eirich and J. Riseman, *J. Polym. Sci.* **4**, 417 (1949).
28. J. W. Mehl, J. L. Oncley, and R. Simha, *Science* **92**, 132 (1940).
29. P. J. Flory, "Principles of Polymer Chemistry," Chapter XIV. Cornell Univ. Press, Ithaca, New York, 1953.
30. H. Mark, "Physical Chemistry of High Polymeric Systems," Vol. 2, p. 289. Wiley (Interscience), New York, 1940.
31. P. Doty, J. H. Bradbury, and A. M. Holtzer, *J. Amer. Chem. Soc.* **78**, 947 (1956).
32. G. Spach, *C.R. Akad. Sci.* **249**, 543 (1959).
33. E. Daniel and E. Katchalski, *in* "Polyamino Acids, Polypeptides, and Proteins" (M. A. Stahmann, ed.), p. 183. Univ. of Wisconsin Press, Madison, 1962.
34. J. Marchal and C. Lapp, *J. Chim. Phys.* **61**, 999 (1964).
35. A. Zilkha and Y. Burstein, *Biopolymers* **2**, 147 (1964).
36. J. H. Fessler and A. G. Ogston, *Trans. Faraday Soc.* **47**, 667 (1951).
37. W. Burchard, *Makromol. Chem.* **64**, 110 (1963).
38. W. G. Harland, *in* "Recent Advances in Chemistry of Cellulose and Starch" (O. Honeyman, ed.), p. 265. Wiley (Interscience), New York, 1959.
39. R. O. Herzog, R. Illig, and H. Kudar, *Z. Phys. Chem., Abt.* A **167**, 329 (1934).
40. J. G. Kirkwood and J. Riseman, *J. Chem. Phys.* **16**, 565 (1948).
41. H. Fujita, "The Mathematical Theory of Sedimentation Analysis," Chapter III, Academic Press, New York, 1962.
42. J. Vinograd *et al.*, *Proc. Nat. Acad. Sci. U.S.* **49**, 902 (1963).

43. J. G. Wetmur and N. Davidson, *J. Mol. Biol.* **31**, 349 (1968).
44. W. J. Archibald, *J. Phys. Colloid Chem.* **51**, 1204 (1947).
45. D. B. Smith, G. C. Wood, and P. A. Charlwood, *Can. J. Chem.* **34**, 364 (1956).
46. H. A. Scheraga and L. Mandelkern, *J. Amer. Chem. Soc.* **75**, 179 (1953).
47. F. Perrin, *J. Phys. Radium* [7] **5**, 497 (1934).
48. V. N. Tsvetkov, *in* "Newer Methods of Polymer Characterization" (B. Ke, ed.), Chapter XIV, p. 567. Wiley (Interscience), New York, 1964.
49. P. R. Callis and N. Davidson, *Biopolymers* **7**, 335 (1969).
50. R. E. Harrington, *Biopolymers* **9**, 141 (1970).
51. A. Peterlin and H. A. Stuart, *Z. Phys.* **112**, 129 (1939).
52. H. A. Scheraga, J. T. Edsall, and J. O. Gadd, Jr., *J. Chem. Phys.* **19**, 1101 (1951).
53. S. Takashima, *Biopolymers* **6**, 1437 (1968).
54. J. T. Yang, *J. Amer. Chem. Soc.* **80**, 5139 (1958).
55. J. B. Harry and J. S. Franzen, *Biopolymers* **8**, 433 (1969).
56. A. Ciferri *et al.*, *Biopolymers* **6**, 1019 (1968).
57. D. N. Holcomb and S. N. Timasheff, *Biopolymers* **6**, 513 (1968).
58. G. Giannoni and A. Rich, *Biopolymers* **2**, 399 (1964).
59. A. Ikegami, S. Yamamoto, and F. Oosawa, *Biopolymers* **3**, 555 (1965).
60. S. W. Englander and A. Poulsen, *Biopolymers* **7**, 379 (1969).
61. A. Ikegami and N. Kono, *J. Mol. Biol.* **29**, 251 (1967).
62. J. M. Rifkind and J. Applequist, *J. Amer. Chem. Soc.* **90**, 3650 (1968).
63. H. Noguchi and J. T. Yang, *Biopolymers* **1**, 359 (1963).
64. Bacon Ke, *in* "Newer Methods of Polymer Characterization" (B. Ke, ed.), Chapter IX, p. 347. Wiley (Interscience), New York, 1964.
65. R. F. Schwenker and P. D. Garn, eds., "Thermal Analysis," Vol. 1. Academic Press, New York, 1969.
66. M. J. O'Neill, *Anal. Chem.* **36**, 1238 (1964).
67. H. Klump, TH. Ackermann, and E. Neumann, *Biopolymers* **7**, 423 (1969).
68. H. J. Hinz, O. J. Schmitz, and TH. Ackermann, *Biopolymers* **7**, 611 (1969).
69. H. Krakauer and J. M. Sturtevant, *Biopolymers* **6**, 491 (1968).
70. TH. Ackermann and E. Neumann, *Biopolymers* **5**, 649 (1967).

8

ELECTRICAL AND MAGNETIC FIELD EFFECTS

Introduction

Several techniques involving the interaction between biopolymers and applied electrical or magnetic fields (as opposed to electromagnetic radiation) have evolved over the past few years as useful tools for studying conformation. Two of these are nuclear magnetic resonance (particularly proton magnetic resonance) and dielectric relaxation. Both methods have, of course, become standard tools in physical organic chemistry but the application to biological macromolecules has encountered certain difficulties. Most developments have occurred recently with the advent of more sophisticated equipment.

Proton Magnetic Resonance

It is not appropriate to record here details of the proton magnetic resonance technique, many excellent reviews being available (see Refs. 1 and 2). However, in essence, many nuclei have spin, the angular momentum of which may be expressed in terms of the nuclear spin quantum number I:

Spin angular momentum $= [I(I + 1)]^{1/2}(h/2\pi)$ $\qquad I = 1/2, 3/2 \ldots$ (8.1)

We can imagine these spinning nuclei as small magnets that can align with or against a constant, homogeneous external field, and consequently a splitting

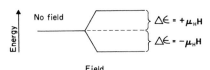

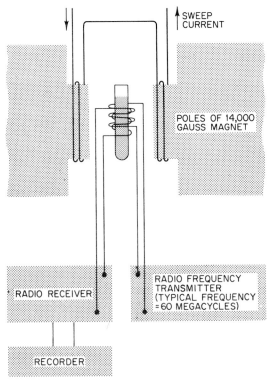

Fig. 8.1. Diagram showing the energetics of uncoupling proton spin states in a magnetic field.

of the nuclear energy level occurs as shown schematically in Fig. 8.1. The external field is denoted by **H**, and μ_H is the nuclear magnetic moment for the atom in question, in this case, hydrogen. The total energy difference between nuclei in the upper and lower energy levels $\Delta\varepsilon_T$ is known to be approximately 4×10^{-19} erg. From Boltzmann's equation, the relative population of the lower and upper states at room temperature is approximately

$$N_L/N_U = \exp(\Delta\varepsilon_T/kT) = 1,00001 \qquad (8.2)$$

showing that essentially the same population in the two energy states is to be expected.

Fig. 8.2. Schematic diagram of an nmr spectrometer (60 megacycle). (Taken from G. M. Barrow, "Physical Chemistry," McGraw-Hill, New York, 1966, with permission.)

Proton magnetic resonance (pmr) spectroscopy is based upon the detection of energy absorbed when the proton spins of the system come "into resonance" with the frequency of a superimposed alternating field. For such resonance,

$$\Delta\varepsilon_T = 2\mu_H H = h\nu \tag{8.3}$$

where ν is the frequency of transition between high and low spin states. Transitions between the energy levels are stimulated by pulses of energy and occur only if the quanta of energy match the nuclear energy-level spacing. The diagrammatic outline of equipment designed for such a purpose is shown in Fig. 8.2. The sample is mounted between the poles of a powerful electromagnet and simultaneously is exposed to a superimposed radio-frequency signal.

In principle one can either sweep the field H at constant ν or sweep the frequency at constant H. For experimental convenience, the field is varied and the applied frequency is maintained constant. Most of the earlier generation of nmr spectrometers used an applied frequency of 60 megacycles per second and a field strength in the region of 14,000 gauss. Newer, more sophisticated spectrometers have used 100 megacycles per second (100 MHz) and 220 MHz applied frequencies, which allow greater resolution. Instruments utilizing even higher applied frequencies are currently being built.

The Chemical Shift

If all protons were to undergo magnetic resonance at identical field strengths and applied frequencies, no conformational information could be derived. Fortunately, however, the resonance frequency of each proton is sensitive to its electrical and magnetic environment. It can be seen from the diagram in Fig. 8.3 that the actual effective field experienced by atoms is the difference between the applied external field and that resulting from the molecular environment. Since the molecular field in the vicinity of different types of protons is different, a spectrum of absorption maxima is to be expected.

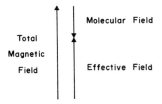

Fig. 8.3. Schematic diagram of the effective magnetic field experienced by a proton as a result of magnetic shielding by neighboring atoms.

Ideally these maxima should perhaps be indexed in terms of the difference between applied and effective field experienced by any given proton. In practice it is much simpler to use a reference, usually tetramethylsilane (in which all the protons experience an equivalent magnetic environment), and define the shift in absorption with respect to this reference. This shift in required field strength (the chemical shift) is then defined as

$$\delta = \frac{(\mathbf{H}_{sample} - \mathbf{H}_{reference}) \times 10^6}{\mathbf{H}_{reference}} \qquad (8.4)$$

and is thus expressed in parts per million.

It might be expected, therefore, that a liquid such as methanol, CH_3OH, would yield two absorption peaks caused by the two types of protons (CH_3 and OH), the former predominating in integrated intensity in the ratio 3:1. Similarly for ethanol, CH_3CH_2OH, three bands of ratio 3:2:1 might be expected. Indeed at low resolution this is exactly what is observed. At higher resolution, these peaks reveal substructure as shown in Fig. 8.4. For ethanol, the CH_3 peak resolves into three components, but the CH_2 peak resolves into four components!

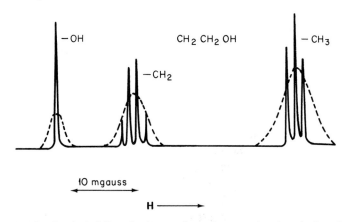

Fig. 8.4. The chemical shift and spin coupling of protons in ethanol; dotted line, low resolution; full line, high resolution. (From the same source as in Fig. 8.2. used with permission of McGraw-Hill Book Co.)

Spin Coupling

The reason for the splitting of nmr bands is that the spin of the nucleus in question can couple with the spin of another similar nucleus to cause a further (slight) magnetic field effect. In the ethanol molecule, for example, the CH_3 protons, which are all identical, are subjected to the coupling effect of the adjacent CH_2 protons. It can be seen in Fig. 8.5 that the two CH_2 protons may couple with spins paired with, or against, the external field, or they may have opposite spins in two different configurations. A proton in the neighborhood of the CH_2 protons will thus "feel" an additional magnetic effect which has three components (in the ratio $1:2:1$). Now in ethanol the CH_3 protons are in the "neighborhood" of these CH_2 protons, and thus the effective field to which they are exposed contains three components (see Fig. 8.4) in the ratio $1:2:1$.

Similarly the spins of the nuclei of the three CH_3 protons may couple as shown in Fig. 8.5, and protons in their neighborhood will be exposed to four

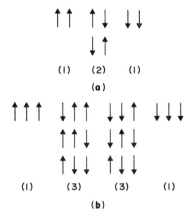

Fig. 8.5. Schematic of the spin combinations of (a) two protons giving rise to three discrete effective fields in their neighborhood, field strengths being in the ratio $1:2:1$ (see the splitting of CH_3 in Fig. 8.4), and (b) three protons giving rise to four coupled combinations, $1:3:3:1$ (see the splitting of CH_2 resulting from the coupling with CH_3 in Fig. 8.4).

spin components of the local field in the ratio $1:3:3:1$. The CH_2 band has four components in the ratio $1:3:3:1$, as can be seen in Fig. 8.4. Thus the number of components reflects the spin properties of adjacent protons.

It is relevant at this point to ask why the OH proton band is not split (into three components). The basic difference between OH and CH protons is that the former undergo rapid exchange with other OH protons, whereas the CH protons are nonexchangeable. This exchange rate shows up in the broadening of the resonance band, which obscures any spin coupling effects. The same is true for NH_2 protons. The $>NH$ protons, particularly when hydrogen bonded, exchange relatively slowly and there is, therefore, some hope of resolving components with high-resolution techniques.

Linewidth

If the mean probability of an upward and downward transition in nuclear spin transitions is W, then the kinetics of the conversion are characterized by a first-order decay process of rate constant $2W$. A relaxation time T_1, may be defined as

$$T_1 = 1/2W \qquad (8.5)$$

which is the so-called spin–lattice relaxation time.

The half-width of the nmr line resulting from the spin transition cannot, by the Heisenberg uncertainty principle, be less than $1/T_1$. There are, however, processes which can increase linewidths substantially over the value expected from spin–lattice relaxation; one of the most important arises from interaction with nearby nuclear moments.

When the linewidth is greater than that predicted from T_1 it is usual to define another time T_2 (shorter than T_1), known as the spin–spin relaxation time. For Lorentzian curves,

$$T_2 = (\pi \, \Delta\nu_{1/2})^{-1} \qquad (8.6)$$

where $\Delta\nu_{1/2}$ is the (frequency) width of the line at half-maximum intensity.

As mentioned previously, a third factor which can cause both line broadening and a shift in the resonance line is produced by chemically exchangeable protons. Let us, for example, imagine the situation represented in Fig. 8.6, involving the resonance peaks of two potentially exchangeable protons. If the exchange is very slow compared with the resonance frequency of the external field then narrow lines are produced, the half-width of which is given by T_2 and/or T_1. However, as the rate of exchange becomes comparable with the external field frequency, the spectrometer is unable to distinguish the separate protons and "sees" an average, as in Fig. 8.6b. Eventually if the

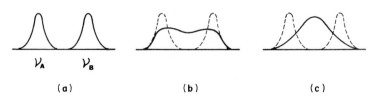

(a)   (b)   (c)

Fig. 8.6. Demonstration of the effect of temperature on the conversion of protons between two states. In (a), the protons are essentially nonexchanging between the two states (e.g., helix and coil). As the temperature is raised, a smearing is observed as appreciable exchange occurs (b) and eventually resonance frequencies ν_A and ν_B move together. At the coalescence point (c), the relative lifetime of the protons in each state τ may be assessed from Eq. (8.7).

exchange is faster than the field frequency an average such as that shown schematically in Fig. 8.6c is produced. We note that in the simple case of the liquid ethanol, the OH protons show a single broad maximum instead of the triplet expected from spin coupling, because of the exchange effect brought about by the labile −OH proton. The half-life of the conversion of the proton from one environment to another can be assessed from the curve in Fig. 8.6c at the "coalescence point" by using the relation

$$\tau = [2\pi(\nu_A - \nu_B)]^{-1} \tag{8.7}$$

We shall see later that this type of measurement can be useful in assessing the lifetime of portions of biopolymers in a particular conformation.

The Charge Distribution and Chemical Shifts of Amino Acids

It would be particularly useful if information required for the potential functions used in calculating polypeptide conformation could be extracted from nmr data. This problem resolves itself into two parts; the first relates to the chemical shift and charge environment, and the second concerns the relation between spin coupling constants and conformation. In the first of these areas, the approach has been to derive parameters from semiempirical quantum mechanical methods, the testing of these parameters in terms of predicting chemical shifts, and the subsequent use of the parameters in conformational calculations. It is an approach which seeks, therefore, to be self-consistent.

Although several theoretical approaches have been made, one of the most useful, for present purposes, has been the Del Re (3) method. This method assumes that the quantum mechanical behavior of electrons in a σ-bonded system may be treated formally, in a manner similar to the Hückel approach for π-bonded systems. The important aspects of the theory are that atoms in a σ-bonded molecule assume an effective charge as the result of mutually compatible electroinductive shifts. The inductive coefficient δ (not to be confused with the chemical shift) is a measure of the charge distribution and is given by

$$\delta_A = \delta_A^\circ + \Sigma \, v_{A\lambda} \, \delta_\lambda \tag{8.8}$$

where λ refers to adjacent atoms, δ_A° is the inductive coefficient of atom A, and the coefficient $v_{A\lambda}$ represents the influence on A of adjacent atoms. The summation is carried out only for adjacent atoms and is therefore a nearest-neighbor approximation.

Thus, given tables of coefficients, one can obtain and solve a series of equations in δ_A, δ_B ... δ_λ. Once these values are known, they can either be used in a series of secular equations or, more simply, used in the Mulliken formula, to find charge distribution coefficients. The Mulliken equation is

$$Q_{AB} = (\delta_A - \delta)_B / 2k_{AB} \qquad (8.9)$$

where Q_{AB} is the total charge on atoms A and B, and k_{AB} is another characteristic constant. The constants have been derived by Del Re (3) on the basis of the electronegativity of the atoms and observed dipole moments. The values are quoted in Table 8.1. With the use of this table, it is readily possible to calculate an effective charge distribution on the appropriate atoms in any σ- (single-) bonded system, including the appropriate portions of poly-saccharide and polynucleotide molecules. It should, however, be made clear that many approximations and assumptions have gone into this compilation, and the representation of quantitative partial charges residing on specific atoms cannot be regarded as being factually correct. Nonetheless, it probably does give us a feel for the effective situation and also provides a rational working basis for further development.

To show how the calculation is carried through, the following simple example is provided. Suppose we wish to calculate the effective charge distribution on the glycine molecule in its neutral form, with the atoms numbered as in Fig. 8.7; then we obtain the equations

$$\delta_{H_1} = \delta_H^\circ + v_{HC}\,\delta_C = \delta_{H_2}$$
$$\delta_{C_1} = \delta_C^\circ + 2v_{CH}\,\delta_{H_1} + v_{CC}\,\delta_{C_2} + v_{CN}\,\delta_N$$
$$\delta_{C_2} = \delta_C^\circ + v_{CC}\,\delta_{C_1} + v_{CO}\,\delta_{O_1} + v_{CO}\,\delta_{O_2}$$
$$\delta_N = \delta_N^\circ + 2v_{NH}\,\delta_{H_1} + v_{NC}\,\delta_{C_1}$$
$$\delta_{O_1} = \delta_O^\circ + v_{OC}\,\delta_{C_2}$$
$$\delta_{O_2} = \delta_O^\circ + v_{OC}\,\delta_{C_2} + v_{OH}\,\delta_{H_5}$$
$$\delta_{H_3} = \delta_H^\circ + v_{HN}\,\delta_N = \delta_{H_4}$$
$$\delta_{H_5} = \delta_H^\circ + v_{HO}\,\delta_{O_2}$$

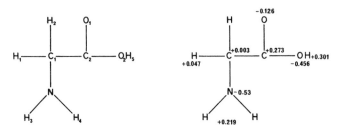

Fig. 8.7. Charge distribution on the atoms of glycine (neutral state) as calculated by the Del Re method.

TABLE 8.1

Del Re Parameters for Calculating Charge Distribution

Bond	C–H	C–C	C–N	C–O	C–O⁻	N–H	N⁺–H	O–H	C–N⁺	C–F	C–Cl	C–S	S–S	S–H
k_{AB}	1.00	1.00	1.00	0.95	0.80	0.45	0.60	0.45	1.33	0.85	0.65	0.75	0.60	0.70
$\nu_{A(B)}$	0.3	0.1	0.1	0.1	0.10	0.3	0.3	0.3	0.10	0.1	0.2	0.20	0.10	0.30
$\nu_{B(A)}$	0.4	0.1	0.1	0.1	0.10	0.4	0.4	0.4	0.10	0.1	0.4	0.40	0.10	0.40
δ_A°	0.07	0.07	0.07	0.07	0.07	0.24	0.31	0.4	0.07	0.07	0.07	0.07	0.07	0.07
δ_B°	0.00	0.07	0.24	0.40	0.40	0.33	0.00	0.00	0.31	0.57	0.35	0.07	0.07	0.00

Simultaneous solution of these equations, with the use of Table 8.1, gives

$$\delta_{H_1} = \delta_{H_2} = 0.064 \qquad \delta_{C_2} = 0.175$$
$$\delta_{C_1} = 0.159 \qquad \delta_{O_1} = 0.418$$
$$\delta_N = 0.337 \qquad \delta_{O_2} = 0.474$$
$$\delta_{H_3} = \delta_{H_4} = 0.135 \qquad \delta_{H_5} = 0.190$$

The charge distribution is then obtained from the Mulliken formula, Eq. (8.9), which gives

$$Q_{H_3N} = Q_{H_4N} = -0.22 \qquad\qquad Q_{C_2O_1} = -0.152$$
$$Q_{NC} = +0.89 \qquad\qquad Q_{C_2O_2} = -0.157$$
$$Q_{H_1C_1} = Q_{H_2C_1} = -0.048 \qquad\qquad Q_{O_2H_5} = +0.316$$
$$Q_{C_1C_2} = -0.008 \qquad\qquad Q_{H_3C} = +0.012$$

Then by simultaneous solution

$$Q_{H_3} = Q_{H_4} = (Q_{HN}) = +0.22 \qquad Q_{C_2} = +0.273$$
$$Q_N = 2Q_{HN} + Q_{NC} = -0.53 \qquad Q_{O_2} = -0.456$$
$$Q_{C_1} = +0.003 \qquad Q_{H_5} = +0.301$$
$$Q_{O_1} = 0.126 \qquad Q_{H_3} = +0.047$$

This charge distribution is also depicted in Fig. 8.7 and shows that the molecule is really quite polar. The charge distribution on oligomers and segments of macromolecules may be obtained in an exactly analogous manner, the complexity of the solution increasing with asymmetry.

In attempting to relate the preceding approach to chemical shifts in the pmr spectra of amino acids, peptides, etc., it is convenient to think of the internal magnetic field of a molecule at a specific proton as arising from (a) electron(s) around the proton $(\sigma)_{Loc}$ and (b) shielding effects arising from magnetic field anistropy of neighboring atoms $(\Sigma\sigma^1)$.

There are, of course, additional higher-order effects but for a first approximation we shall use the preceding assumptions. The magnetic shielding constant σ may then be written as the sum of these two effects.

$$\sigma_H = \sigma_{Loc} + \Sigma\sigma^1 \tag{8.10}$$

The relation between diamagnetic shielding constants and atomic charge has been described and may be represented by

$$\sigma_{Loc} = \beta^1(1 - Q_H) \tag{8.11}$$

where β^1 is a positive coefficient. Broadly speaking, this equation indicates that the larger the positive charge on the proton, the smaller is the shielding coefficient.

If we assume that magnetic anisotropy effects arise predominantly from the adjacent atoms, then, according to Pople (4), for one adjacent carbon atom

$$\Sigma\sigma^1 \sim \sigma_{Cadj} = \frac{1}{3R^3H} [\mu_x(1 - 3\cos^2\theta_x) + \mu_y(1 - 3\cos^2\theta_y)$$
$$+ \mu_z(1 - 3\cos^2\theta_z)] \qquad (8.12)$$

Here, μ_x, μ_y, and μ_z are magnetic dipole moments induced by (paramagnetic) electronic current around the carbon atom, H is the external field, and θ is the direction angle of vector R connecting C and H atoms.

The *ab initio* calculation of all the parameters in this equation would be intractable. However, following Del Re, the equation may be very roughly approximated by

$$\sigma_{Cadj} = -C^1 + A^1 Q_C \qquad (8.13)$$

where A^1 and C^1 are positive constants and Q_C is the charge on the appropriate carbon atom. Now, the chemical shift is a combination of shielding factors plus an arbitrary reference point, i.e.,

$$\delta = -AQ_C - BQ_H + C \qquad (8.14)$$

TABLE 8.2

Calculated and Experimental Chemical Shifts[a] for Some Amino Acids in Water

Compound	Group	δ_{obs}	δ_{calc} (RNH$_2$COOH)	δ_{calc} (RN$^+$H$_3$COO$^-$)
Gly	CH$_2$	+1.20	+1.35	+0.39
Ala	CH	+0.94	+1.04	+0.21
	CH$_3$	+3.36	+3.41	+3.22
Thr	CH	0	+0.08	−0.11
		+0.94	+0.81	−0.04
	CH$_3$	+3.35	+3.24	+3.23
Val	CH	+1.14	+1.07	+0.35
		+2.23	+2.79	+2.62
	CH$_3$	+3.73	+3.63	+3.62
Leu	CH	+0.96	+1.06	+0.34
		+3.03	+3.11	+2.74
	CH$_2$	+3.03	+3.13	+2.95
	CH$_3$	+3.63	+3.64	+3.63
Pro	CH(α)	+0.49	1.28	0.71
	CH$_2$ (β, γ)	+2.85	3.12	2.93
	γ			2.94
	δ	1.39	1.85	0.99

[a] In parts per million.

where *A*, *B*, and *C* are all positive and solvent effects are assumed to be included in *C*. Based on collected data, the quoted values of *A*, *B*, and *C* are $A = 9.92$, $B = 133.93$, and $C = 9.67$. Table 8.2 (5) contains a comparison of some experimental and calculated data for amino acids in aqueous solution. It can be seen that in view of the numerous assumptions, reasonably good correlations are achieved. It would be most interesting to attempt a correlation between the charge distribution of oligomers and their chemical shifts, although no work of this type has yet been reported.

Nuclear Magnetic Resonance Spectroscopy of Peptide Oligomers

Evidently the chemical shifts observed experimentally are a result of nuclear shielding, which in turn reflects the charge distribution, bonding, and conformation of molecules and macromolecules. The effects of the primary structure on the CH_2 resonance have been analyzed in some detail by Nakamura and Jardetzky (6, 7). They found that there were first- and second-order effects, of which the four first-order effects were caused by N- and C-terminal titration, N peptide and helix formation. Recently nmr spectroscopy has been used in studying the development of the α helix in γ-ethyl-L-glutamate oligomers (8). Nonhelical oligomers possess peptide-NH environments decidedly different from those occurring in the hydrogen-bonded $-NH \cdots O=C$ helix. Figure 8.8 (8) shows the chemical shift dependence of the highest-

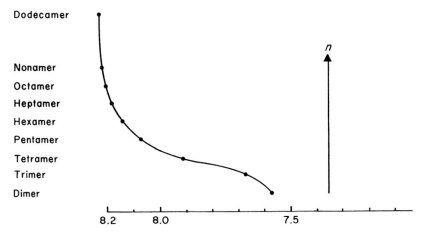

Fig. 8.8. Chemical shift of the high-field N—H resonance (220 MHz spectrum) of γ-ethyl-L-glutamate oligomers in trifluoroethanol, as a function of degree of polymerization *n*. The formation of a uniform conformation (α helix) appears to be complete by about the heptamer or octamer. (Taken, in part, from Ref. 8).

field N—H resonances (220 MHz spectrum) of the glutamate oligomers in trifluoroethanol. The data are taken as indicating that α-helix formation commences at about the heptamer for this given solvent–solute combination. In dimethyl sulfoxide (DMSO), on the other hand, no conformational development is observable as a function of molecular weight. The glutamate oligomers are therefore taken to be of random conformation in DMSO.

Nuclear Magnetic Resonance Spectroscopy of Polypeptides

Most work with polypeptides has concentrated on the observation of conformational changes. For example the α-helix → coil, polyproline I → II and β-sheet → random transitions, have each been studied.

α-Helix Coil Transition

The nmr spectrum of a polymer that can exist as an α helix or random coil generally contains peaks for two conformationally sensitive protons, the peptide N—H and the C_α—H. For several of the polyamino acids and esters studied by nmr, the N—H and C—H resonances have been resolved into two components, which various authors (cited below) have ascribed to helix and coil forms. There is, however, some considerable dispute over whether this is actually the case. Roberts and Jardetzky, in their recent review (2), advance the view that a heterogeneous molecular weight distribution leads to a heterogeneous distribution of conformation, some molecules being predominantly helical, and others predominantly random coil, in the transition range. Thus, it is argued, two peaks representing two separate populations of molecules, both rapidly exchanging between helix and coil, should be observed. This argument is said to be the only one consistent with the relatively narrow bandwidth of the two peaks. Table 8.3 lists the observations (up to early 1969) of single and double peaks in α-helical polypeptides. There is, however, also strong evidence that the doublets are in fact related to helix and coil content. Figure 8.9 (9) shows the 220 MHz spectra of the NH and C_αH region of the spectrum for poly-L-methionine in deuterochloroform/trifluoroacetic acid mixtures. In the former solvent, poly-L-methionine is α helical and only a single proton band is observed for both the —NH and C_αH protons. As the TFA, which is a helix-breaking solvent, is added, a second component to each peak appears, presumably caused by increasing randomness of the chains, until at 60–70% TFA only the random components remain. Since, as we have seen, the areas under the curves are directly proportional to the number of protons experiencing that particular environment, a direct estimation

TABLE 8.3

The nmr Observation of Single and Double Peptide Proton Bands

Polypeptide	DP	Double peak	Single peak	Peak frequency (MHz)	Reference[a]
Poly-L-alanine	40	+		100	1
	40		+		2
	280		+		2
	≥300	+		100/220	1
	700		+		3
Poly-D-alanine	?		+		4
Poly-β-benzyl-L-aspartate	760		+		2
	?	+			5
Poly-γ-benzyl-L-glutamate	13	+			6
	21	+			6
	92	(+)			6
	340		+		2
	640		+		6
Poly-L-leucine	?		+		3
	?	+			5
	?		+		4
Poly-L-methionine	280		+		2
Poly-D-norleucine	?	+			4
Poly-L-phenylalanine	?	+			7

[a] Key to references:

1. Text Ref. 10.
2. J. L. Markley, D. H. Meadows, and O. Jardetzky, *J. Mol. Biol.* **27**, 25 (1967).
3. W. E. Stewart, L. Mandelkern, and R. E. Glick, *Biochemistry* **6**, 143 (1967).
4. Text Ref. 17.
5. J. A. Ferretti, *Chem. Commun.*, 1030 (1967).
6. E. M. Bradbury, C. Crane-Robinson, H. Goldman, H. W. E. Ralthe, *Nature* **217**, 812 (1968).
7. F. Conti and A. M. Liquori, *J. Mol. Biol.* **33**, 953 (1968).

of α-helix content is possible. Furthermore, the total area under the random and helix bands should be constant. Oddly enough the helical component of the —NH band is less than the helical component of the C_α—H band. Although this effect is not fully understood, Haylock and Rydon (9) have suggested that the helix form of poly-L-methionine is partially protonated by the solvent prior to disruption, and the rapidly exchanging N—H protons are thus "lost" from the nmr signal. The areas of the proton bands are shown in Table 8.4.

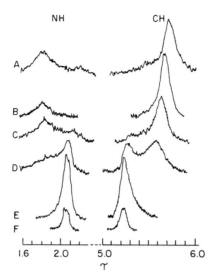

Fig. 8.9. The NH and $C_\alpha H$ region of the 220 MHz pmr spectrum of poly-L-methionine in deuterochloroform/trifluoroacetic acid mixtures: A–F, 10, 20, 30, 40, 50, 60% TFA, respectively. (Reprinted from Ref. 9 by permission of North-Holland Publishing Co. Amsterdam.)

TABLE 8.4

Poly-L-methionine in $CDCl_3$/TFA Mixtures: Areas of Proton Bands[a]

TFA (%)	NH		$C_\alpha H$	
	Helix	Coil	Helix	Coil
5	0.60	0.11	0.89	0.11
10	0.60	0.19	0.81	0.19
20			0.84	0.16
30	0.58	0.31	0.69	0.31
40	0.49	0.27	0.73	0.27
50	0.24	0.67	0.33	0.67
60	0.00	1.00	0.00	1.00

[a] At 220 MHz and 10°C.

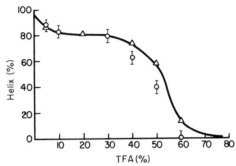

Fig. 8.10. By assuming that the area under the CH resonance peak reflects the helix content (see text), the results for nmr data may be compared with ORD data. In this figure ORD data ($\triangle$) and nmr data ($\bigcirc$) are compared for poly-L-methionine. (Based on data in ref. 9 and reprinted by permission of North-Holland Publishing Co.)

It is convenient and interesting to compare the α-helix content as derived from the C_α—H protons with values derived from ORD b_0 measurements. A comparison is shown in Fig. 8.10. The nmr data for this figure are derived from 60 MHz measurements at 33.5°C, and surprisingly good agreement between the two techniques is evident.

Studies with other polyamino acids (esters) have indicated that the behavior of poly-L-methionine is fairly typical. The helix/coil ratios for —NH and C_α—H protons in three such cases are shown in Table 8.5. Once again

TABLE 8.5

Helix–Coil Ratios from nmr Data (CDCl$_3$/TFA Solutions)

Polymer	NH		C_αH		Temp. (°C)	TFA (%)
	Helix	Coil	Helix	Coil		
Poly-γ-methyl-L-glutamate	1.9	2.08	5.59	5.29	10	40
Poly-β-benzyl-L-aspartate	1.17	2.11	5.68	5.27	33.5	3
Poly-γ-methyl-L-glutamate	1.89	2.17	5.90	5.13	50	10

it is apparent that there is some "missing" helical component in the NH band, again due, presumably, to protonation. Poly-L-alanine as studied by Ferretti and Paolillo (10) has revealed additional information. Because of the distinctly different environment of the terminal NH proton,

$$[\overset{+}{H_3N}-\underset{\underset{CH_3}{|}}{CH}-\overset{\overset{O}{\parallel}}{C}-(\underset{\underset{H}{|}}{N}-\underset{\underset{CH_3}{|}}{CH}-\overset{\overset{O}{\parallel}}{C})_n-O^-]_1$$

they were able to compare the concentrations of the N-terminal and "internal" protons and thus estimate molecular weights up to DP = 300, i.e., $M_w \simeq 30,000$, certainly a useful addition to methods for molecular weight determination. Attempts to induce coalescence of the NH or CH protons for the helix and coil by thermal methods have not been successful, and thus it has not proven possible to estimate the rate of conversion directly. However, for the peaks to be resolvable into spin components,

$$\tau \geq (2\pi J_{AB})^{-1}$$

where J_{AB} is the spin–spin splitting. From Eq. (8.7), for the chemical shifts of the helix and coil protons to be resolvable,

$$\tau \geq (2\pi \, \Delta \, \delta)^{-1}$$

From this latter relation it has been deduced (10) that the lifetimes of the predominant portions of the helix and random coil (in poly-L-alanine) are greater than 10^{-1} sec. This interpretation is not, however, in agreement with other methods (2), which indicate a relaxation time of $0.2–1.0 \times 10^{-6}$ sec for the helix–coil transition. [The situation thus complex and is the basis for the Roberts–Jardetzky argument (2).]

The Polyproline Transition

As noted in the previous section, changes in the nmr spectrum of α helices undergoing transition to the random form can readily be detected in the proton bands due to peptide-NH and $C_\alpha H$ protons. Poly-L-proline, however, does not possess —NH protons, but the pyrrolidine ring contains one C_α proton and two sets of protons on C_β, C_γ, and C_δ.

It might be expected that all of the protons are subjected to a very different environment in the polyproline I and II structures, since the former is more compact and the peptide backbone is less exposed. The primary effect, however, can be expected to arise from the *cis/trans* isomerization. The transition of the as-polymerized material from I to form II in acidic solvents has been studied by the nmr technique (11). The $C_\beta H_2$ and $C_\gamma H_2$ protons are found to be indistinguishable at ~ 2.05 ppm and are conformation independent; the band for the $C_\delta H_2$ protons occurs at ~ 3.70 ppm and also appears to be conformation independent. For form I the $C_\alpha H$ band lies at

4.6 ppm but changes to 4.8 ppm as the mutarotation to form II occurs. The 100 MHz spectra of polyproline in acetic acid solutions at various stages of the transition are shown in Fig. 8.11. A quantitative analysis of the conversion of form I to II has not apparently been carried out. It is to be expected that the transition involves protonation of the peptide carboxyl with a possible assist from protonation of the nitrogen, which causes an electronic shift, thus promoting the *cis/trans* isomerization. Infrared evidence indicates that the *trans* form II is indeed more strongly solvated (hydrogen bonded) than the *cis* form. The inductive shift of electrons may be partially transmitted to the $C_\alpha H$, causing at least part of the change in chemical shift.

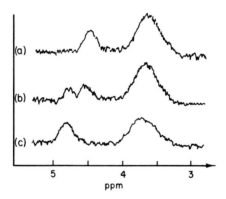

Fig. 8.11. The nmr spectra of poly-L-proline at 100 MHz. (a) Form I produced by dissolving the polymerization product in acetic acid, the spectrum being obtained immediately after dissolution. (b) A mixture of forms I and II obtained after 40 min in acetic acid. (c) Form II produced after mutarotation is essentially complete (after 1 hr in acetic acid). [Based on data from Conti *et al.* (11).]

β-Sheet → Random Coil Transition

It is thought that poly-*O*-acetyl-L-serine (POALS) and poly-*O*-benzyl-L-serine (POBLS) exist in a β form in weak solvents, such as chloroform, but undergo an order/disorder transition on addition of strong acidic solvents. This transition has been studied by nmr in $CDCl_3$/dichloroacetic acid (12) solvent mixtures. Figure 8.12 (12) shows plots of chemical shifts for the various protons of the two biopolymers. For POBLS the $C_\alpha H$ and peptide-NH protons both undergo a change in chemical shift with solvent composition, which seems to have an inflection point at approximately 10–12% DCA. Optical measurements made on the same polymer indicated a transition at about 8% DCA. In contrast with POBLS, POALS shows a slight shift of all the protons with solvent composition, although the major effect again

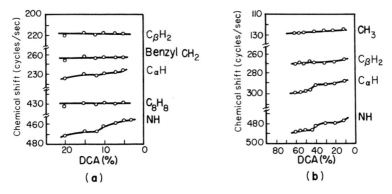

Fig. 8.12. (a) Plot of chemical shift data for various protons of poly-*O*-benzyl-L-serine in deuterochloroform/DCA mixtures. (b) Chemical shift data for poly-*O*-acetylserine in CDCl$_3$/DCA. (Reproduced from Ref. 12, with permission of the Association for Science Documents Information).

shows up with the C$_\alpha$H and NH protons. In the latter case, the transition appears to take place at about 40% DCA compared with about 55% by optical measurements. Although there is some discrepancy between the data from the two techniques, the nmr method detects local environmental changes which need not coincide with the actual transition. In any case, the acetyl ester seems to be the more stable β former in solution.

Coupling Constants and Conformational Analysis

As previously demonstrated, the spin–spin coupling of nuclear moments causes a splitting of the resonance bands. The separation of the split components in cycles per second (hertz) is represented by the coupling constant J. Further, the influence of group B on the A band leads to the splitting J_{AB}. Now in many nmr spectra, particularly those of large molecules, there is overlap of the bands and it can be extremely difficult to delineate the coupling constants. Usually, accurate values of J_{AB} may be obtained in a first-order approach only if $(v_A - v_B) \gg J_{AB}$.

The coupling constant J depends on electron density at the nucleus and is consequently also related to the s character of the bond (e.g., sp, sp^2, sp^3 hybridized X—H bonds). Of particular interest with regard to conformational calculations is the fact that J is related to the relative configuration of the interacting protons. For example, Karplus (13) has shown that for H$_a$—C$_a$—C$_b$—H$_b$ the coupling constant J_{HH} may be represented in terms of the dihedral angle ϕ (see Chapter 2).

$$J = A + B \cos \phi + C \cos 2\phi \qquad (8.15)$$

The constants A, B, and C have been evaluated (14) as 7, -1, and 5 Hz, respectively. Large values of J are predicted for *cis* (0°) and *trans* (180°) configurations; small values result from the *gauche* (60° and 120°) configurations. In an interesting extension of this type of approach to peptide systems Ramachandran and co-workers (15) have found that the $NH-C_aH$ coupling constant for small peptides may be represented approximately by the equation

$$J = A\cos^2\phi + B\cos\phi + C\sin^2\phi \qquad (8.16)$$

where $A = 7.9$, $B = 1.5$, and $C = 1.4$ Hz. The calculated and experimental values of the coupling constants for diglycine, di-L-alanine and di-L-leucine are, for example, identical with values of 5.8, 7.8, and 8.2 Hz, respectively. Thus, measurement of the coupling constant gives an immediate indication of the favored ϕ angle of rotation about the $N-C_\alpha$ bond. Unfortunately, as one proceeds to higher oligomers and polymers the nmr spectra become more diffuse due to spin–lattice relaxation, and the coupling constants cannot be delineated. However, for small peptides the above approach seems to have much to offer in determining favored conformations.

Application of nmr Spectroscopy to Native Proteins

There are two main categories of nmr study that are applied to native macromolecules, particularly proteins in solution. These are concerned with (a) the structure of the protein itself and (b) the nature of bound molecules and binding sites.

In the first category, we may recognize that the primary structure and its ionic state are reflected through the nmr spectrum. It has proven possible to use nmr purely as an amino acid analyzer for such proteins as gelatin (16). The sequence of peptides in a protein is somewhat outside the scope of nmr measurements, but in principle the conformation and possibly the tertiary structure could be defined if complete resolution were obtained via the conformational rules already expounded (and extensions thereto). In practice, proteins yield extremely complicated, overlapping spectra from which detailed coupling constants are virtually impossible to extract. Although spectra are improved with 100 and 220 MHz instruments, there is little hope of achieving sufficiently detailed spectra for complete structure determination, even with a new generation of instruments.

Other information that can be extracted from large macromolecules is related to conformational changes such as those occurring on denaturation.

In this case, changes in specific portions of the spectrum caused by change of local environment (induced chemically or thermally) may be followed. Evidently it is necessary to identify the species giving rise to the local spectrum, but this is often quite feasible.

A number of denaturation studies have been carried out with globular proteins (17–24). All of the spectral lines become narrower on denaturation as a result of the increased motional freedom of polypeptide chain segments. In addition, the local environment of the same type of protons becomes more nearly equivalent. By following the chemical shifts of protons identified with specific peptides it is possible to infer the nature of the unfolding (denaturation) process. For example, Markley *et al.* (25) have followed the denaturation of staphylococcal nuclease as a function of pH, as shown in Fig. 8.13 (25). The response of tyrosine and tryptophan aromatic protons is shown, and evidently some of the tyrosine protons experience environmental changes at pH values much below the denaturation pH and those at which tryptophan is affected. The process occurring during denaturation appears to be a combination of ionization (deprotonation) of the tyrosine hydroxyl group and the

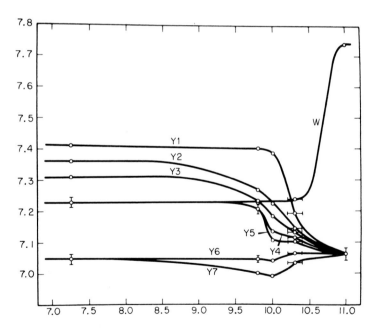

Fig. 8.13. Chemical shifts of the tyrosine (Y1–7) and tryptophan (W) aromatic protons of a selectively deuterated analog of staphylococcal nuclease in the pH range 7.0–11.0. [Reproduced from "Advances in Protein Chemistry," **24**, 527(1970) by G. C. K. Roberts and O. Jardetzky with permission of Academic Press, Inc.]

environmental changes evidenced by the original nonequivalence of the Y1–7 protons.

Although it is relatively straightforward to identify the proton bands of residues in the random denatured conformation, it is extremely difficult to identify resonance positions in native proteins. Apart from shifts caused by conformational rearrangements, the presence of neighboring phenyl groups, possibly from adjacent chains, can induce large and specific effects. Progress in aspects of identification and interpretation of resonances in native proteins has recently been reviewed by McDonald and Phillips (26). The expected resonance positions of protons in random-coil proteins in neutral D_2O at 40°C is presented in Table 8.6.

In the second category, a considerable amount of work has been performed concerning bound molecules, particularly water and binding sites—notably, when ions are involved. Since this topic is somewhat outside the scope of this book, the reader is referred to recent reviews of the subject (2, 26).

TABLE 8.6

Expected Resonance Positions of Proton Resonances of Random-Coil Proteins in Neutral D_2O at 40°C

Proton type	Resonance position (Hz from internal DSS at 220 MHz)	Estimated half-width (Hz)	Relative peak height per residue
Leucine CH_3	195	15	40.0
Isoleucine CH_3	183	20	30.0
Valine CH_3	205	17	35.2
Alanine CH_3	310	18	16.7
Threonine CH_3	270	16	18.7
Lysine $C_\delta H_2 + C_\beta H_2$	370	30	13.3
Methionine CH_3	454	10	30.0
Cystine $C_\beta H_2$	665	12	16.6
Histidine C-4	1555	10	10.0
C-2	1740	10	10.0
Tyrosine *ortho* to OH	1500	17	11.8
meta to OH	1560	17	11.8
Phenylalanine aromatic	1598	30	16.7
Tryptophan indole C-2	1584	10	10.0
indole C-5, C-6	1549	15	6.7
	1566	15	6.7
indole C-4, C-7	1638	18	5.6
	1658	18	5.6

Nuclear Magnetic Resonance Spectroscopy of Polynucleotides

It is to be expected that nmr studies of polynucleotides might throw further light on the nature of conformational transitions. The environment of the protons in the bases should reflect base stacking and pairing. In practice, these two effects are self-compensating. Base stacking causes an upfield shift, predicted theoretically to lie in the range 0.5–2.0 ppm, whereas base pairing through hydrogen bond formation should cause a downfield shift of (−) 0.15–0.3 ppm. The effects of stacking are readily observable in concentrated solutions of the bases. Figure 8.14 (27) shows the concentration dependence of chemical shifts in aqueous solutions of purine at pH 8.0.

The nmr spectra of single-chain polynucleotides are of high resolution, and it has been concluded that base stacking along the chain is short-lived. On the other hand, the spectra of double-stranded helices are broad and unresolved with linewidths of the order of several hundred cycles per second. Such a result is not unexpected if the double-stranded helices form rigid structures.

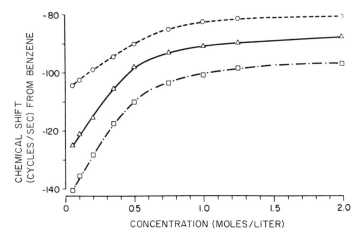

Fig. 8.14. Concentration dependence of chemical shifts in aqueous solutions of purine, pH 8.0 ± 0.1, at 27°C: □, H-6; △, H-2; ○, H-8. Shifts are in cycles per second with respect to benzene as an external standard. (Reproduced from Ref. 27, with permission.)

Dielectric Dispersion by Biopolymers

We have seen that the charge distribution and "electronic state" of a biopolymer are very fundamental to its physical properties. Not only is its conformation dependent to some extent on this distribution, but its behavior in electromagnetic fields is certainly strongly influenced by the internal force field. In the previous sections, the tentative links between the dipolar properties and chemical shifts of nmr spectroscopy have been explored. As mentioned in Chapter 6, biopolymers can be represented for some purposes as a series of connected dipoles, and certainly we would expect these dipoles to respond to external electrical fields.

The dipole moment, which is a vector quantity, is a measure of the magnitude of charge displacement and is defined by

$$\boldsymbol{\mu} = e\mathbf{d} \qquad (8.17)$$

where $\mathbf{d}$ is the vector distance separating positive and negative charges of magnitude e. In practice e is measured in electrostatic units and $\mathbf{d}$ in angstroms, the conversion factor D (1 debye) $= 10^{18}$ esu Å being used to render the data onto a convenient scale. The dipole moment of the peptide group, based on quantum mechanical calculation, is 3.71 D and is in the direction indicated in Fig. 8.15. The orientation has been verified experimentally by microwave spectroscopy experiments on gaseous formamide (28).

Some general, fundamental considerations of the properties of polymers lead to the belief that the cumulative effect of the dipoles will be conformationally sensitive. In Fig. 8.16 (29) a general polymer chain is depicted in which the angle between backbone linkages is constant (θ) but free rotation about the bond can occur. Similar permanent dipole moments are imagined to be associated with alternate atoms or groups in the backbone, and the absolute magnitude of these dipole moments is $\boldsymbol{\mu}_0$. For a chain of high degree of polymerization and free rotation, the mean square of the (total) moment is given by

$$\langle \mathbf{M}^2 \rangle = n\mu_0^2[1 + 2\cos^3 \theta/(1 - \cos^2 \theta)] \qquad (8.18)$$

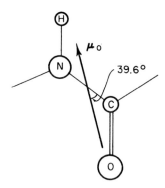

Fig. 8.15. Direction of the permanent dipole moment in a peptide unit (ground electronic state).

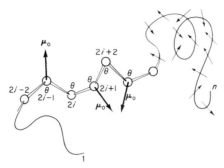

Fig. 8.16. Representation of a random polymer chain with free rotation in backbone bonds. (Following A. Wada, Ref. 29.)

If θ is the tetrahedral angle, $\cos \theta = -1/3$ and

$$\langle \mathbf{M}^2 \rangle = 0.92 \times n\mu_0^2 \tag{8.19}$$

which would be the value for a polymer with a carbon backbone in random-coil conformation. Note that the moment of the molecule is a function of its degree of polymerization n.

If rotation (in the backbone) is hindered, then the relation between $\langle \mathbf{M}^2 \rangle$ and μ_0 depends on the detailed nature of the potential barrier to rotation; enhancement, partial cancellation, or total cancellation of the total moment may result.

Of course, the dipole moment of a biopolymer molecule cannot be obtained in free space and, consequently, in order to relate the conformation to the dipolar properties, a consideration of the dielectric properties of biopolymer solutions is required.

Dielectric Properties of Biopolymer Solutions

If a polarizable solution is placed between the plates of an electrical condenser and a field is applied, the charge distribution may be treated in terms of the schematic model in Fig. 8.17. If the charge on the plates is σ and the polarization (charge) of the solution (dielectric) per unit area is p, then the effective field in solution is, from electrostatic theory,

$$E = 4\pi(\sigma - p) \tag{8.20}$$

or

$$E_0 - E = 4\pi p \tag{8.21}$$

where E_0 is the external field.

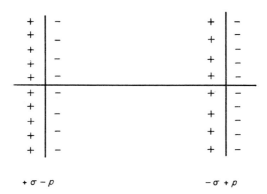

Fig. 8.17. The electrical behavior of the dielectric material of a condenser.

The dielectric constant of the solution D_{12} is defined by the ratio of the applied and effective fields, i.e.,

$$D_{12} = E_0/E \qquad (8.22)$$

Thus

$$D_{12} - 1 = 4\pi P_{12}/E \qquad (8.23)$$

where P_{12} is the molar polarization of the solution per unit volume. Now if it is assumed that the polarization of the solution is the simple sum of the contributions from solvent and (polymer) solute, designated P_1 and P_2 respectively,

$$D_{12} - 1 = 4\pi(P_1 + P_2)/E \qquad (8.24)$$

where P_1 and P_2 are related to molecular polarizabilities α_1 and α_2 and effective field strengths F_1 and F_2,

$$P_1 = n_1\alpha_1 F_1 \qquad (8.25)$$

$$P_2 = n_2 \alpha_2 F_2 \qquad (8.26)$$

where n is the number of molecules per unit volume, i.e., $n = \rho N W/M$, where ρ is the density, N is Avogadro's number, W is the weight fraction, and M is the molecular weight.

Furthermore, the effective field strength, from the Clausius–Mosotti equation is given by

$$F_1 = F_2[(D_{12} + 2)/3]E \qquad (8.27)$$

The application of the classical approach to biopolymer systems has been studied extensively by Wada (29) who obtained the equation

$$P_2 = 4\pi N \frac{\alpha_2}{M_2} \frac{F_2}{E} \tag{8.28}$$

The molecular polarizability of a biological macromolecule will depend on both the electronic polarization α and polarization resulting from the orientation of the molecule; thus

$$\alpha_2 = \alpha_0 + (\mathbf{M}^2/3kT) \tag{8.29}$$

and

$$P_2 = 4\pi \left(\frac{N}{M_2}\right)\left(\alpha_0 + \frac{\mathbf{M}^2}{3kT}\right) \tag{8.30}$$

This equation is analogous to the Debye equation for low molecular weight solutes,

$$P = \left(\frac{D-1}{D+2}\right)\left(\frac{M}{\rho}\right) = \frac{4\pi N}{3}\left(\alpha + \frac{\mathbf{\mu}^2}{3kT}\right)$$

For a flexible molecule of high degree of polymerization n, one has

$$\langle \mathbf{M}^2 \rangle = n\mathbf{\mu}_0^2 \tag{8.31}$$

$$\alpha_0 = \alpha_{00} n \tag{8.32}$$

$$M_2 = nM_0 \tag{8.33}$$

where $\mathbf{\mu}_0$, α_{00}, and M_0 are the dipole moment, polarizability, and molecular weight of individual residues, respectively. Thus for a random coil

$$P_2^{\text{coil}} = \frac{4\pi N}{M_0}\left[\alpha_{00} + \frac{\mathbf{\mu}_0^2}{3kT}\right] \tag{8.34}$$

Since the polarization can be obtained experimentally, it is of some interest to note that Eq. (8.34) does not include the degree of polymerization n and we may conclude, therefore, that the polarizability of the random coil is not molecular weight dependent. On the other hand, for a rigid helical molecule, the components of the group moments along the helical axis are additive and the polarizability P_2 becomes

$$P_2^{\text{helix}} = \frac{4\pi N}{M_0}\left[\alpha_{00} + \frac{n\mathbf{\mu}_{0\|}^2}{3kT}\right] \tag{8.35}$$

where $\mathbf{\mu}_{0\|}$ is the component of the dipole moment parallel to the helix axis. When n is large, the first term α_{00} becomes negligible compared with the second.

Dielectric Dispersion and Rotatory Diffusion Coefficient

It is usual, and convenient, to measure the dielectric constant of a medium as a function of frequency. At low frequencies, the molecules are able to orient themselves with the alternating field, and it is the value of D, extrapolated to zero frequency (D_0), that is the commonly quoted value of the dielectric constant. However, as the frequency is increased the molecular motion is no longer able to follow the field, and dielectric dispersion occurs. The frequency at which the molecules can no longer follow the field must be related to the rotatory diffusion rate, which may be specified in terms of a "relaxation time" τ.

$$D = D_\infty + \frac{D_0 - D}{1 + (2\pi v \tau)^2} \tag{8.36}$$

where D_∞ is the dielectric constant at very high frequency and v is the applied frequency.

The relaxation time τ is actually defined as the time in which polarization decreases to e^{-1} of the original value. Now τ is simply related to the rotatory diffusion coefficient θ.

$$\tau = \frac{\theta}{2} \tag{8.37}$$

If the biopolymer molecule is treated in terms of its hydrodynamically equivalent ellipsoid of revolution (see Chapter 7) with an axial ratio $\rho = a/b$, and the orientation of the net moment is parallel to the long axis, then θ is given by

$$\theta = \frac{3kT}{32\pi a^3 \eta} [2 \ln(2/\rho) - 1] \qquad a/b > 5 \tag{8.38}$$

$$\theta = \frac{3kT}{32\pi b^3 \eta} \qquad a/b < 5 \tag{8.39}$$

where η is the viscosity.

For a rigid α helix, $a = n \times 1.5$ Å and, in principle, the degree of polymerization n can be obtained from measurement of θ. In practice, however, frequencies in the range 1000–1 cycle or less are required to yield expected θ values in the 10^4–10 sec region. An alternative method of measuring rotatory diffusion coefficients, which is gaining in popularity, is that of laser heterodyne spectroscopy. In this technique, the frequency distribution of scattered light is analyzed and may be related to both the translational and rotatory diffusion of macromolecules.

Experimental Information

From the preceding equations it can be seen that the "shape" of a bio-polymer molecule can be obtained by measuring the dielectric constant D_0 (hence P_2) as a function of molecular weight or independently by measuring dielectric dispersion (hence θ) as a function of molecular weight. Alternatively, the molecular weight of a molecule of known conformation may be obtained, and by either approach the component of its individual group moment $\mu_{0\parallel}$ may be assessed.

The most extensive studies conducted so far have been for poly-γ-benzyl-L-glutamate in a variety of solvents (29, 30). Data for PBLG in 1,2-dichloroethane are shown in Fig. 8.18 which indicates in agreement with

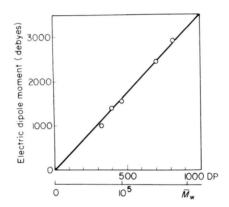

Fig. 8.18. Dependence of the electric dipole moment of the α helix of poly-γ-benzyl-L-glutamate polymers in 1,2-dichloroethane as a function of molecular weight. [From Wada (29) by courtesy of Marcel Dekker, Inc.]

other methods, that the molecule is extended (helical) in this solvent. From the slope of the line in Fig. 8.18, $|\mathbf{M}| = 3.4n$ (debyes).

It is concluded, therefore, that the contribution of the residue dipoles in the α helix is 3.4 D along the helical axis, a value in reasonable accord with expectation based on the previous considerations of the peptide dipole as represented in Fig. 8.19.

However, it must be remembered that the experimentally measured moment includes effects from the side chain and solvent. Since solvent effects and hydrogen bonding can be expected to enhance the experimental value, it has been suggested that the value of $\mu_{0\parallel}$ can be rationalized only if there is an opposing dipole in the side chain (Fig. 8.20]. The technique offers, then, the

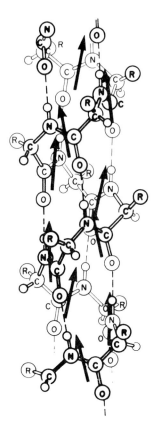

Fig. 8.19. Orientation of residue dipoles in the α helix. (From Ref. 29. by permission of Marcel Dekker, Inc.)

potential of determining side-chain orientation in solution, a very useful addition to other conformational methods.

The dielectric dispersion measurements are usually presented in terms of a critical frequency v_c, defined in terms of the relaxation time τ by

$$v_c = 2\tau/\pi = \theta/\pi \qquad (8.40)$$

Figure 8.21 shows a double logarithmic plot of v_c versus molecular weight for PBLG in ethylene dichloride. As expected from Eq. (8.38), an essentially linear plot is produced, indicating a rodlike behavior of the polymer in this solvent.

Although the polarization and dispersion methods have been fairly successful when applied to nonionic systems, additional complications arise

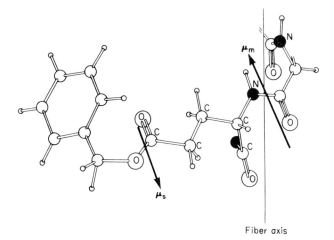

Fig. 8.20. A model for the side-chain conformation of PBLG and orientation of its dipole moment. (From ref. 29 with permission.) μ_m is the dipole moment of the backbone peptide and μ_s the component vested in the side chain.

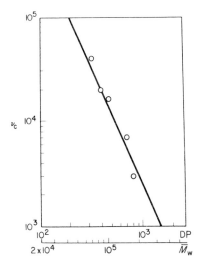

Fig. 8.21. Double logarithmic plot of the critical frequency of PBLG in EDC as a function of molecular weight. [From Wada 29 with permission of Marcel Dekker, Inc.]

when aqueous or acidic solvents are used. Under these circumstances electrical double layers that are highly polarizable are obtained, giving often rise to anomalously high apparent dielectric constants. It has, for example, been known for many years that inorganic colloidal dispersions in water

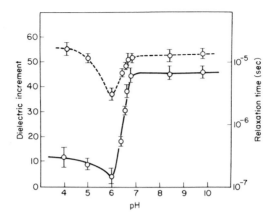

Fig. 8.22. Dielectric increment and relaxation time of poly-L-glutamic acid at various pH values. Solid line, dielectric increment; dashed line, relaxation time. It is noteworthy that the effect appears to reflect changes in solution viscosity. [From A. Wada, *J. Chem. Phys.* **30**, 328 (1958) with permission]

yield apparent dielectric constants in the region of $D_0 \sim 10^4$. Nevertheless qualitative measurements on aqueous solutions show transitions in the region generally associated with the helix/coil transition. Figure 8.22 shows the results of polarization and dispersion measurements on poly-L-glutamic acid. It is unfortunate that interpretation of these data is hindered by the previous considerations, since the method could, in principle, be used to resolve the question of whether the charged form is truly extended, as suggested by the ORD/CD data of Chapter 6. A fairly large number of poly-α-amino acids, their esters and copolymers have now been studied using dielectric methods (these are reviewed in reference 28). It can be seen that despite many interpretational difficulties, the technique does provide conformational information on biopolymers in solution, including information which is often very difficult or impossible to obtain by other methods.

REFERENCES

1. J. A. Pople, W. G. Schneider, and H. J. Bernstein, "High Resolution Nuclear Magnetic Resonance." McGraw-Hill, New York, 1959.
2. G. C. K. Roberts and O. Jardetzky, *Advan. Protein Chem.* **24**, 447 (1970).
3. G. Del Re, *J. Chem. Soc.*, 4031 (1958).
4. J. A. Pople, *Proc. Roy. Soc. London* **A239**, 541, 550 (1957).
5. G. Del Re, *Biochim. Biophys. Acta* **75**, 153 (1963).
6. A. Nakamura and O. Jardetzky, *Proc. Nat. Acad. Sci. U.S.* **58**, 2212 (1967).
7. A. Nakamura and O. Jardetzky, *Biochemistry* **7**, 1226 (1968).

8. M. Goodman, A. S. Verdini, C. Toniolo, W. D. Phillips, and F. A. Bovey, *Proc. Nat. Acad. Sci. U.S.* **64**, 444 (1969).
9. J. C. Haylock and H. N. Rydon, *Peptides*, 19 (1968).
10. J. A. Ferretti and L. Paolillo, *Biopolymers* **7**, 155 (1969).
11. F. Conti, M. Piatelli, and P. Viglino, *Biopolymers* **7**, 411 (1969).
12. M. Ishikawa and S. Sugai, *Rep. Progr. Polym. Phys. Jap.* **12**, 517 (1969).
13. M. Karplus, *J. Chem. Phys.* **30**, 11 (1959); *J. Amer. Chem. Soc.* **85**, 2870 (1963).
14. A. Bothner-By, *Advan. Magn. Resonance* **1**, 195 (1965).
15. G. N. Ramachandran, personal communication.
16. P. I. Rose, *Science* **171**, 573 (1971).
17. E. M. Bradbury, C. Crane-Robinson, H. Goldman, H. W. E. Ralthe, and R. M. Stevens, *J. Mol. Biol.* **29**, 507 (1967).
18. M. Saunders and A. Wishnia, *Ann. N.Y. Acad. Sci.* **70**, 870 (1958).
19. A. Kowalsky, *J. Biol. Chem.* **237**, 1807 (1962).
20. M. Mandel, *Proc. Nat. Acad. Sci. U.S.* **52**, 746 (1946).
21. C. C. MacDonald and W. D. Phillips, *J. Amer. Chem. Soc.* **89**, 6333 (1967).
22. J. S. Cohen and O. Jardetzky, *Proc. Nat. Acad. Sci. U.S.* **60**, 92 (1968).
23. J. L. Markley, I. Putter, and O. Jardetzky, *Z. Anal. Chem.* **243**, 367 (1968).
24. D. P. Hollis, G. McDonaldson, and R. L. Biltonen, *Proc. Nat. Acad. Sci. U.S.* **58**, 758 (1967).
25. J. L. Markley, M. N. Williams, and O. Jardetzky, *Proc. Nat. Acad. Sci. U.S.* **65**, 645 (1970).
26. C. C. McDonald and W. D. Phillips, *in* "Fine Structure of Proteins and Nucleic Acids" (G. D. Fasman and S. N. Timasheff, eds.), Chapter 1. Dekker, New York, 1970.
27. O. Jardetzky, *in* "Quantum Aspects of Polypeptides and Polynucleotides" (M. Weissbluth, ed.), p. 501. Wiley, (Interscience), New York, 1964.
28. R. J. Kurland and E. B. Wilson, *J. Chem. Phys.* **27**, 585 (1957).
29. A. Wada, *in* "Poly-α-Amino Acids" (G. D. Fasman, ed.), Chapter 9. Dekker, New York, 1967.
30. A. Wada, *J. Chem. Phys.* **30**, 328 (1958).

9

CONFORMATION OF POLYPEPTIDES

The principles underlying the conformation of polypeptide chains and the experimental means available for studying them have been presented in previous chapters. This chapter correlates the experimental information relating to conformation and conformational stability.

Homobiopolymers in the Solid State

In the solid state, polyamino acids and derivatives may be expected to obey the equivalence condition mentioned in Chapter 2, i.e., all pairs of backbone rotation angles are equal ($\phi_1 = \phi_2 = \ldots \phi_n; \psi_1 = \psi_2 = \ldots \psi_n$) unless the solid is precipitated in such a manner that a random or amorphous phase is produced. The equivalence condition limits the chain conformation to a symmetrical repeating structure of which the best known examples are, of course, the α and β structures. It has been shown, in previous chapters, on the basis of both theoretical and experimental evidence, that branching on the β carbon tends to favor the β structure whereas nonbranched (at C_β) polymers tend to be α helical. We shall see later, with sequential polypeptides, that other specific factors may also influence conformation.

Of the methods applied to determining conformation in the solid state, infrared spectroscopy is perhaps the most common; thereafter, x-ray techniques have been used, which are time consuming but often provide more definitive information. Other methods, mainly of more recent origin, include Raman and film circular dichroism spectroscopy, and electron diffraction. Table 9.1 contains a summary of the available information concerning poly-α-amino acids and their derivatives in the solid state. Clearly the α-helix and β-sheet forms are the most common. The antiparallel, and cross-β forms have been grouped together and labeled simply as β since in some cases infrared dichroism information is not available and in others the method of sample preparation seems to influence which form is produced. There appears to be

TABLE 9.1

Solid-State Conformation of Poly-α-Amino Acids and Derivatives

Biopolymer	Side chain, R	Conformation	Method of determination[a]
Polyglycine	—H	Polyglycine I (PGI), β sheet	a (1), b (2), e (3)
		Polyglycine II (PGII)	a (1), b (4), d (5)
Poly-L-alanine	—CH$_3$	α$_R$	a (6), b (7), c^b (8), d (5), e (9), f (10)
		β on stretching; also low mol wt.	
Poly-L-valine	—CH(CH$_3$)$_2$	β	a (6, 11), b (11), e (12)
Poly-L-leucine	—CH$_2$CH(CH$_3$)$_2$	α$_R$	a (6), c^b (13), e (14)
Poly-L-isoleucine	—CH(CH$_3$)—(C$_2$H$_5$)	β	a (15), c (16), e (15)
Poly-L-serine	—CH$_2$OH	β	a (17), b (18), c (19), e (15)
O-Acetylserine	—CH$_2$O—COCH$_3$	β	a (20), b (21), c^b (22), f (23)
O-Benzylserine	—CH$_2$O—COCH$_2$—⬡	β	a, b (18, 24), c^b (24), f (23)
Poly-L-threonine	—CH(CH$_3$)OH	β	a^c (6)
Poly-L-aspartic acid	—CH$_2$COOH	α$_R$	c^b (25)
β-Benzyl (ester)	—CH$_2$COOCH$_2$—⬡	ω	a, b (26)
		α$_L$	a, b (26), c^b (27)
		β	a (28)
Poly-L-glutamic acid[d]	—CH$_2$CH$_2$COOH	α$_R$	a (29), c^b (30), e (31), f (32)
Esters			
γ-Methyl	—CH$_2$CH$_2$COOCH$_3$	α$_R$	a (33), b (34), c^b (35), d (36, 37)
		β	b (38), d (36, 37)

(Continued)

TABLE 9.1 (*continued*)

Biopolymer	Side chain, R	Conformation	Method of determination[a]
γ-Ethyl	$-CH_2CH_2COOC_2H_5$	α_R	a, b, c, (39, 40)
γ-Benzyl	$-CH_2CH_2COOCH_2-$	α_R	a (41), b (42), c[b] (43), d (44, 45)
		β	a (46), b (47)
Salts (Ca, Sr, Ba)		β	b, d (48)
Poly-L-lysine[d]	$-(CH_2)_4NH_2$	α_R	a (6), c[b] (49, 30), d (50)
ε-Carbobenzoxy-L-lysine	$-(CH_2)_4NHOCOCH_2-$	α_R	a (50), a[b] (51), c[b] (52), d (50)
Salts	HPO_4	β	b, e (53)
Poly-L-histidine	$-CH_2-C-NH$ (histidine ring)	α_R	c[b] (54)
Poly-L-phenylalanine	$-CH_2-$	α_R	a (6), c[b] (55)
Poly-L-tyrosine	$-CH_2-$ $-OH$	α_R	a (6), b (5), c (56), 3 (5)
Poly-L-tryptophan	$--CH_2-$ (indole)	α_R	c[b] (57)
Poly-L-cysteine	$-CH_2-SH$		
Poly-S-methyl-L-cysteine	$-CH_2-SCH_3$	β	a (6), c[b] (58)
Poly-S-benzyl-L-cysteine	$-CH_2-S-CH_2-$	β	a (39, 58), b (39), c (39)
Poly-L-methionine	$-(CH_2)_2SCH_3$	α_R	a (6), c[b] (59)
Poly-L-proline	$-(CH_2)_3$	Polyproline I (PPI), 10_3 *cis* RH	a (60), b (61), c[b] (62), d (63), e (60), f (64)
	$-(CH_2)_3$	Polyproline II (PPII), 3_1 *trans* LH	a (60), b (65), c[b] (62), d (63), e (60), f (64)
Poly-L-hydroxyproline	$-CH_2CH(OH)CH_2-$	PPII 3_1	b (66), c[b] (62), e (67)
O-Acetyl derivative	$-CH_2CH(OCOCH_3)CH_2-$	PPII 3_1	b (66), c[b] (62), d (68)
		PPI 10_3	b (66), c[b] (62)

[a] Methods of determination: a, infrared; b, x ray; c, ORD/CD; d, electron diffraction; e, Raman; f, nmr. Key to representative references (numbers in parentheses):

1. T. Miyazawa, *in* "Poly-α-Amino Acids" (G. D. Fasman, ed.), Chapter 2. Dekker, New York, 1967.
2. C. H. Bamford, L. Brown, A. Elliott, W. E. Hanby, and I. F. Trotter, Nature **171**, 1149 (1953).
3. M. Smith, A. G. Walton, and J. L. Koenig, *Biopolymers* **8**, 29 (1969).

4. F. H. C. Crick and A. Rich, *Nature* **176**, 780 (1955).
5. F. J. Padden, and H. D. Keith, *J. Appl. Phys.* **36**, 2987 (1965).
6. E. R. Blout, *in* "Polyamino Acids, Polypeptides and Proteins" M. A. Stahmann, ed.), p. 275. Univ. of Wisconsin Press, Madison, 1962.
7. A. Elliott, *in* "Poly-α-Amino Acids" (G. D. Fasman, ed.), Chapter 1. Dekker, New York, 1967.
8. G. D. Fasman, *in* "Polyamino Acids, Polypeptides and Proteins" (M. A. Stahmann, ed.), p. 221. Univ. of Wisconsin Press, Madison, 1962.
9. J. L. Koenig and P. L. Sutton, *Biopolymers* **8**, 167 (1969).
10. E. M. Bradbury, *Aust. J. Chem.* **22**, 357 (1969).
11. R. D. B. Fraser, B. S. Harrap, T. P. MacRae, F. H. C. Stewart, and E. Suzuki, *J. Mol. Biol.* **12**, 482 (1965).
12. J. L. Koenig and P. Sutton, *Biopolymers* **10**, 89 (1971).
13. A. R. Downie, A. Elliott, W. E. Hanby, and B. R. Malcolm, *Proc. Roy. Soc. London* **A242**, 325 (1957).
14. J. T. Yang and P. Doty, *J. Amer. Chem. Soc.* **79**, 761 (1957).
15. P. L. Sutton, M.S. thesis, Case Western Reserve University, Cleveland, Ohio (1969).
16. E. R. Blout and E. Shechter, *Biopolymers* **1**, 565 (1963).
17. K. Imahori and I. Yahara, *Biopolym. Symp.* **1**, 424 (1964).
18. Z. Bohak and E. Katchalski, *Biochemistry* **2**, 228 (1963).
19. B. Davidson, N. Tooney, and G. D. Fasman, *Biochem. Biophys. Res. Commun.* **23**, 156 (1966).
20. G. D. Fasman and E. R. Blout, *J. Amer. Chem. Soc.* **82**, 2262 (1960).
21. I. Yahara, K. Imahori, Y. Iitaka, and M. Tsuboi, *J. Polymer. Sci.* **B1**, 47 (1963).
22. I. Yahara and K. Imahori, *J. Amer. Chem. Soc.* **85**, 230 (1963).
23. M. Ishikawa and S. Sugai, *Rep. Progr. Polym. Phys. Jap.* **12**, 517 (1969).
24. E. M. Bradbury, A. Elliott, and W. E. Hanby, *J. Mol. Biol.* **5**, 487 (1962).
25. J. Brahms and G. Spach, *Nature* **200**, 72 (1963).
26. E. M. Bradbury, L. Brown, A. R. Downie, A. Elliott, R. D. B. Fraser and W. E. Hanby, *J. Mol. Biol.* **5**, 230 (1962).
27. E. M. Bradbury, A. R. Downie, A. Elliott, and W. E. Hanby, *Proc. Roy. Soc. London* **A259**, 110 (1960).
28. E. M. Bradbury, L. Brown, A. R. Downie, A. Elliott, R. D. B. Fraser, W. E. Hanby and T. R. R. McDonald, *J. Mol. Biol.* **2**, 276 (1960).
29. E. R. Blout and M. Idelson, *J. Amer. Chem. Soc.* **78**, 497 (1956).
30. G. Holzwarth and P. Doty, *J. Amer. Chem. Soc.* **87**, 218 (1965).
31. J. L. Koenig and B. G. Frushour (unpublished data).
32. D. I. Marlborough, K. G. Orrell, and H. N. Rydon, *Chem. Commun.*, 518 (1965).
33. T. Miyazawa, Y. Masuda, and K. Fukushima, *J. Polymer. Sci.* **62**, 562 (1962).
34. C. H. Bamford, L. Brown, A. Elliott, W. E. Hanby, and I. F. Trotter, *Nature* **169**, 157 (1952).
35. S. Sakurai, *in* "Polyamino Acids, Polypeptides and Proteins" M. A. Stahmann, ed.), p. 225. Univ. of Wisconsin Press, Madison, 1962.
36. S. Ishikawa, T. Kurita, and E. Suzuki, *J. Polym. Sci.* **A2**, 2349 (1964).
37. S. Ishikawa, T. Kurita, and E. Suzuki, *J. Polym. Sci.* **B1**, 127 (1963).
38. S. Ishikawa, *J. Polym. Sci.* **3B**, 959 (1965).
39. R. D. B. Fraser, B. S. Harrap, T. P. MacRae, F. H. C. Stewart, and E. Suzuki, *J. Mol. Biol.* **14**, 423 (1965).
40. R. D. B. Fraser, B. S. Harrap, T. P. MacRae, F. H. C. Stewart, and E. Suzuki, *Biopolymers* **5**, 251 (1967).
41. M. Tsuboi, *J. Polym. Sci.* **59**, 139 (1962).
42. M. F. Perutz, *Nature* **167**, 1053 (1951).
43. For a summary, see G. D. Fasman, ed., "Poly-α-Amino Acids," p. 514. Dekker, New York, 1967.
44. S. H. Carr, A. G. Walton, and E. Baer, *Biopolymers* **6**, 469 (1968).

45. F. Rybnikar and P. H. Geil, *Bull. Amer. Phys. Soc.* **16**, 320 (1971).
46. E. R. Blout and A. Asadourian, *J. Amer. Chem. Soc.* **78**, 955 (1956).
47. C. H. Bamford, L. Brown, W. E. Hanby, and I. F. Trotter, *Nature* **169**, 357 (1953).
48. H. D. Keith, G. Giannoni, and F. J. Padden, *Biopolymers* **7**, 775 (1969).
49. E. Breslow, S. Beychok, K. Hardman, and F. R. W. Gurd, *J. Biol. Chem.* **240**, 304 (1965).
50. J. L. Koenig and P. L. Sutton, *Biopolymers* **9**, 1229 (1970).
51. G. R. Bird and E. R. Blout, *J. Chem. Phys.* **25**, 798 (1956).
52. G. D. Fasman, M. Idelson, and E. R. Blout, *J. Amer. Chem. Soc.* **83**, 709 (1961).
53. F. J. Padden, H. D. Keith, and G. Giannoni, *Biopolymers* **7**, 793 (1969).
54. S. Beychok, M. N. Pflumm, and J. E. Lehmann, *J. Amer. Chem. Soc.* **87**, 3990 (1965).
55. H. E. Auer and P. Doty, *Biochemistry* **5**, 1708 (1966).
56. S. Beychok and G. D. Fasman, *Biochemistry* **3**, 1675 (1964).
57. G. D. Fasman, M. Landsberg, and M. Buchwald, *Can. J. Chem.* **43**, 1588 (1965).
58. E. R. Blout, C. de Lozé, S. M. Bloom, and G. D. Fasman, *J. Amer. Chem. Soc.* **82**, 3787 (1960).
59. G. E. Perlmann and E. Katchalski, *J. Amer. Chem. Soc.* **84**, 452 (1962).
60. W. B. Rippon, J. L. Koenig, and A. G. Walton, *J. Amer. Chem. Soc.* **92**, 7455, (1970).
61. P. M. Cowan and S. McGavin, *Nature* **176**, 501 (1955).
62. I. Z. Steinberg, W. F. Harrington, A. Berger, M. Sela, and E. Katchalski, *J. Amer. Chem. Soc.* **82**, 5263 (1960).
63. J. C. Andries and A. G. Walton, *Biopolymers* **8**, 523 (1969).
64. F. Conti, M. Piatelli and P. Viglino, *Biopolymers* **7**, 411 (1969).
65. W. Traub and U. Shmueli, *Nature* **198**, 1165 (1963).
66. V. Sasisekharan, *J. Polym. Sci.* **47**, 391 (1960).
67. M. J. Deveney, A. G. Walton, and J. K. Koenig, *Biopolymers*, **10**, 615 (1971).
68. See text Ref. 3.

[b] Solution determination.

[c] Postulated.

[d] Charged coil forms also known and studied.

no good example of a parallel β structure for synthetic homopolypeptides and in general it is to be expected that the cross-β form can be produced for any of the antiparallel β formers under the right conditions of crystallization and molecular weight. Also most of the α-helical structures can probably be rendered into a β-form by extreme treatment (steam-stretching, rolling) and/or use of low molecular weight material.

The most unusual conformations are those of the *cis* 10_3 poly-L-proline I helix, already discussed at some length, and the so-called ω helix of poly-β-benzyl-L-aspartate. The ω helix of the aspartate is prepared by heating a film of material, which is cast from CCl_4 and is initially in the left-hand α-helix conformation, to about 160°C *in vacuo*. The x-ray studies (1) of this material show it to be a 4_1 helix, or, according to the old symmetry nomenclature (a 4.0_{13} helix) i.e., consisting of 13 member hydrogen-bonded rings as shown in Fig. 9.1 (1). Since this conformation arises from a left-hand α helix by re-arrangement, the inference is that this ω helix is also left-handed. Another interesting aspect of this unusual conformation is the infrared absorption spectrum (1). The amide A band is at 3296 cm^{-1}, which is quite common for

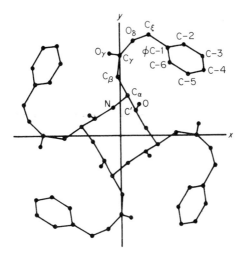

Fig. 9.1. Poly-β-benzyl-L-aspartate ω helix; projection down the c (molecular) axis showing the fourfold nature of the helix. [After A. Elliott in "Poly-α-Amino Acids" Ed. G. D. Fasman (1967) p. 39 and reprinted by courtesy of the publisher, Marcel Dekker, Inc., New York.]

α helices; however the Amide I band occurs at 1675 cm^{-1}, which is probably higher than any other known conformation (for the most intense band), and the Amide II is at 1536 cm^{-1}, which is comparable with the β conformers. The helix pitch of the aspartate ω form is about 5.3 Å, i.e., close to the 5.4 Å of the α helix. However, the peptide repeat is 1.3 Å, considerably smaller than the 1.5 Å of the α form.

The most useful method for determining whether the solid-state conformer is left- or right-handed appears to be film circular dichroism. This method has some disadvantages in that crystalline or oriented samples tend to scatter light and give spurious results. Also, if the polymer has optically active side groups, corrections of the type outlined in Chapter 6 must be made. Nevertheless the method does yield rapid and useful results and is, by now, in conventional use (2).

Homobiopolymers in Solution

Homobiopolymers in solution have been studied predominantly by CD/ORD and light-scattering or viscometric techniques, however nmr, Raman spectroscopy, dielectric relaxation, x-ray diffraction, and flow-birefringence methods are also now in common use. Most of these methods do not detect the conformation *per se* but the data may be related to known solid-state conformations. There is, for example, no *a priori* reason for believing the α, β, and polyglycine II structures should be maintained in solution, particularly since in the last two, intermolecular hydrogen bonding is involved.

It is evident that in strong solvents such as trifluoroacetic acid (TFA), dichloroacetic acid (DCA), and formic acid (FA) the solvent tends to hydrogen bond to and/or protonate the peptide backbone, thus disrupting intramolecular bonds. In weak solvents, where intramolecular hydrogen bonding can be maintained, concomitant molecular aggregation is not unusual.

The interaction of polypeptides with solvent molecules not only determines the solubility of the polymer but is to a large degree the directing force which controls conformation in solution. In general, homopolymers with charged side groups ($-COO^-$, $-NH_3^+$), or secondary amino groups, are water soluble. However, conformation and the accessibility of the peptide group to the surrounding solvent are also important. For example, poly-L-proline II is water soluble with the peptide $\diagdown C{=}O$ group accessible to, and hydrogen bonded with, hydrating molecules, but the more compact hydrophobic form I is insoluble. Similarly, both homopolymers and sequential polymers containing carboxyl or free amino groups are much more soluble in water in the relatively open charged-coil form than in the neutral, intramolecularly bonded closed α conformations.

The approximate pH values at which helix/coil transitions are observed for several poly-α-amino acids are listed in Table 9.2. Although the transition

TABLE 9.2

Conformational Transitions of
Poly-α-Amino Acids in Water

Biopolymer	pH (50%)[a]
Poly-L-glutamic acid	5.7[b]
Poly-L-lysine	10.1[c]
Poly-L-histidine	6.0
Poly-L-aspartic acid	(~7)
Poly-L-tyrosine	11.5
Poly-L-ornithine	9.5[d]

[a] At which 50% of the polymer is transformed.
[b] In 0.0050 M NaCl.
[c] In 2 M NaCl.
[d] 40% methanol 60% H_2O.

is associated with the ionization of the side chains, it does not correspond exactly to 50% ionization, different degrees ionization are required in the side chain to overcome the α helix-stabilizing forces for different polymers.

The neutral poly-α-amino acids are generally insoluble in water but in some cases the presence of concentrated salts, such as lithium bromide, solubilize the material; thus polyglycine and its sequential polymers poly (Ala-Gly) and poly (Ala-Gly-Gly) can be solubilized in this manner. The reason for this dissolution is not entirely clear since the solution conformation of polyglycine is probably the 3_1 polyglycine II helix, which in the solid state is intermolecularly hydrogen bonded.

Materials that have a β sheet conformation in the solid state are almost invariably insoluble in aqueous solution. If the side chains are large and contain esters, then solvents such as chloroform, toluene, dioxane, etc., are the most suitable dispersing media; for polypeptides with small hydrophobic side chains, formic acid, dichloroacetic acid, difluoroethanol, and trifluoroacetic acid are generally the best solvents.

It is clearly of some interest to determine whether the conformations of molecules in solution are the same as or similar to those found in the solid state. Two of the spectroscopic methods, Raman spectroscopy and circular dichroism, are suitable for examining relative conformations in solid and solution, although the former is mainly limited to aqueous solutions. In general, the conformations appear to be similar in both states but are not necessarily identical, as was thought a few years ago. Several biopolymers which appear to be of random conformation in strong solvents tend to crystallize from these solvents in the α or β forms, as described in Chapter 4. Certainly if the polymer has an ordered conformation in solution, it generally precipitates in that form. Even this rule does not hold entirely, an exception being poly-O-acetyl-L-hydroxyproline, which in solution appears to have the poly-L-proline I conformation yet precipitates in the poly-L-proline II form (3).

Nonspectroscopic techniques applied to solution conformation include x-ray methods, flow birefringence, and the hydrodynamic and light-scattering measurements described in Chapter 7. For polypeptides that are α helical in the solid state, the solution techniques yield data which indicate that the peptide repeat of the "α helix" in solution lies between 1.4 and 2.2 Å in weak solvents (4,5). Unfortunately, the result obtained is to some extent dependent upon the model chosen (rigid, flexible, wormlike, etc.), and although the preceding values are not inconsistent with the typical 1.5 Å repeat of the α helix they also are in the range of the 3_{10} helix, which has a 2.0 Å repeat. The 3_{10} nomenclature actually implies an intramolecularly bonded helix with three peptides per turn or, in modern nomenclature a 3_1 helix. The old nomenclature is maintained here to avoid confusion with the 3_1 polyglycine or polyproline helices, which are not intramolecularly bonded and have a peptide repeat of approximately 3.1 Å. Early studies of the x-ray diffraction properties of poly-γ-benzyl-L-glutamate in weak solvents also suggested that a 3_{10} helix was likely (4), but recently more refined x-ray measurements (6) have shown that several polypeptides are truly α helical in solution. Table 9.3 contains the data for three such materials.

TABLE 9.3

Some x-Ray Diffraction Data from Polypeptide Solutions[a, b]

Polypeptide	M_w	$R_c(\text{Å})$	$h(\text{Å})$
Poly-γ-benzyl-L-glutamate	100,000	4.7	1.54
	40,000	4.8	1.53
Poly-ε-CBZ-L-lysine	370,000	6.7	1.55
Poly-L-glutamic acid	82,500	4.5	1.53

[a] The data for solutions in dimethylformamide extrapolated to zero concentration; R_c is the radius of gyration (rod axis) and h the peptide repeat.

[b] Taken from Ref 6., reprinted by permission of Marcel Dekker, Inc.

Whereas the α helix now seems well established as a stable solution conformation, the evidence for β structures is less convincing. There appears to be no x-ray evidence for solution β forms and the evidence rests on spectroscopic data—mainly circular dichroism and infrared. A similar situation exists with other conformations, i.e., similarity among solid conformations is demonstrated by spectroscopic methods, but x-ray data are not available.

Conformation in Mixed Solvents

Because of the mutual solubility of a large number of poly-α-amino acids and their derivatives in chloroform–dichloroacetic acid–trifluoroacetic acid mixtures, and the suitability of such systems for study by ORD or nmr (using deuterated solvents) techniques, it has been possible to analyze the relative stability of the α-helical forms. Basically, if the α-helical peptide backbone is stabilized by the same forces in all polypeptides, then differences among various polymers must reflect side-chain interactions.

It is usual to study the change of the Moffitt–Yang parameter b_0 (see Chapter 6) as a function of increasing solvent polarity: $CHCl_3 \rightarrow DCA \rightarrow TFA$. For α-helical molecules the b_0 value generally commences in the region of

+600 in near 100% CHCl$_3$ and drops fairly sharply toward zero as the polarity of the mixture is increased by increasing the content of DCA or TFA. The material is thus said to undergo the helix–random coil* conformational change as a function of solvent environment. Typical data for polypeptides are presented in Fig. 9.2. The relative stabilities of some familiar polypeptides in this type of solvent system are poly-L-alanine > poly-L-leucine > poly-L-methionine > poly-γ-benzyl-L-glutamate > poly-γ-ethyl-L-glutamate > poly-γ-methyl-L-glutamate > poly-γ-carbobenzoxy-L-lysine > poly-γ-benzyl-L-aspartate. The solvent composition corresponding to the collapse of the α helix

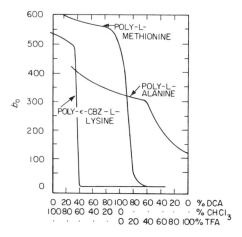

Fig. 9.2. Values of b_0 versus solvent composition: DCA, dichloroacetic acid; TFA, trifluoroacetic acid. [From G. D. Fasman, ed., *in* "Poly-α-Amino Acids," Dekker, New York, 1967 with permission.]

(i.e., 50% of original b_0 values) for several polyamino acid derivatives is displayed in Table 9.4.

There are essentially two ways of regarding the stabilizing or destabilizing influence of the side chains. Loss of helix stability for a series of polypeptides could arise either from unfavorable side-chain interactions or from unfavorable solvent interaction. The latter would, for example, show up in variable solubility, which is to some extent the case. On this basis it might be expected that the larger and more lyophobic the side chains, the less would be the helix stability; such an effect might explain the alanine→leucine→methionine

---

* Unlike the conformation of the charged (ionized) form of poly-L-glutamic acid (PLG) and poly-L-lysine (PLL) there is no current evidence for any regular conformation. The two cases are distinguished, therefore, as *charged* or *extended* coil (PGA, PLL) and *random* coil (present case in mixed solvents).

→lysine ester-equence. On the other hand, the glutamate series seems to indicate side chain stabilizing effects, particularly with the "stacking" of the aromatic rings in the benzyl derivative. The aspartate ester is expected to be only weakly stable since it has a left-hand helix within which the backbone is less energetically favorable than its right-handed counterpart.

TABLE 9.4

α-Helix Stability in Mixed Solvents (from ORD Data)

Polymer	Solvent compositions (%) of increasing helix-breaking ability				
	CHCl₃	DMF	DCA	TFA	Reference[a]
β-Benzyl-L-aspartate	94	6	6		1
ε-Carbobenzoxy-L-lysine	63		37		2
L-Methionine	50			50	3
L-Leucine	40			60	4
L-Alanine	20			80	5
L-Tyrosine		25	75		6
γ-Methyl-L-glutamate	35		65		7
γ-Ethyl-L-glutamate	60		40		8
γ-Benzyl-L-glutamate	20		80		9

[a] Key to references:

1. R. H. Karlson, K. S. Norland, G. D. Fasman, and E. R. Blout, *J. Amer. Chem. Soc.* **82**, 2268 (1960).
2. G. D. Fasman, M. Idelson, and E. R. Blout, *J. Amer. Chem. Soc.* **83**, 709 (1961).
3. G. E. Perlmann and E. Katchalski, *J. Amer. Chem. Soc.* **84**, 452 (1962).
4. S. Hanlon and I. M. Klotz, *Biochemistry* **4**, 37 (1965).
5. G. D. Fasman, in "Polyamino Acids, Polypeptides and Proteins" (M. A. Stahmann, ed.), p. 221. Univ. of Wisconsin Press, Madison, 1962.
6. E. Katchalski, *Proc. Int. Congr. Biochem. 4th* **8**, 21 (1960).
7. T. Yoshida, S. Sakurai, T. O. Kuda, and Y. Takagi, *J. Amer. Chem. Soc.* **84**, 3590 (1962).
8. M. Goodman and Y. Masuda, *Biopolymers* **2**, 107 (1964).
9. E. M. Bradbury, A. R. Downie, A. Elliott, and W. E. Hanby, *Proc. Roy. Soc. London* **A259**, 110 (1960).

Temperature-Induced Transitions

Further insight into the influence of solvating molecules on helix stability can be obtained from studies of temperature-induced transitions. In the random form we may suppose that the solvent molecules have overcome the intramolecular bonding forces and either are bound to, or have protonated, the peptide backbone. At solvent compositions close to those at which the helix/random conformation is induced, there is in fact competition between intramolecular and extramolecular bonding. At elevated temperatures it is not unusual for the solvating molecules to be "loosened" from the backbone and a random to helix transition induced. This phenomenon, which can be treated in terms of the entropic terminology of thermodynamics, is often referred to as an inverse temperature transition. An extensive review of the theories of helix/coil or helix/random transitions has been published recently (7). Table 9.5 contains a brief summary of the conditions pertaining to some temperature-induced conformational transitions in glutamate and aspartate esters. The transition temperature is found to be molecular weight dependent and may, to a certain extent, be regarded as the counterpart to thermal denaturation in proteins.

Apart from the transitions between helix and coil, certain heat-induced $\alpha \to \beta$ transitions are known, of which that of poly-L-lysine is best characterized. The β structure may be obtained at high pH (>11) by heating to about 50°C and has been characterized in D_2O by infrared measurements (8) and in aqueous solution by ORD/CD measurement (9). Both methods indicate a β structure, the IR amide I bands at 1612 and 1680 cm^{-1} providing evidence for the antiparallel conformation. It is not known at this stage whether the antiparallel arrangement arises from the folding of a single molecule, as in the cross-β-structures, or through side-by-side antiparallel aggregation. Nevertheless the driving force stabilizing the β form seems to arise from polymer–polymer interactions.

TABLE 9.5

Temperature-Induced Conformational Transitions

Polymer	Solvent (%)[a]	Temp. (°C)	Transition	Reference
Poly-γ-benzyl-L-glutamate	EDC, 24; DCA, 76	25–40	Coil → helix	1
	m-Cresol, 24; DCA, 76	20–60	Coil → helix	2
	TBM, 57; DCA 43	20–60	Coil → helix	3
Poly-β-benzyl-L-aspartate	*m*-Cresol	20–60	Helix → coil	4
	m-Cresol, 13; DCA, 87	20–50	Helix → coil	2
	EDC, 93; DCA, 7	20–50	Coil → helix	2

[a] Solvents: EDC, ethylene dichloride; DCA, dichloroacetic acid; TBM, tetrabromomethane.
[b] Key to references:

1. P. Doty and J. T. Yang, *J. Amer. Chem. Soc.* **78**, 498 (1956).
2. G. D. Fasman, ed., *in* "Poly-α-Amino Acids," p. 522. Dekker, New York, 1967.
3. J. Y. Cassim and E. W. Taylor, *Biophys. J.* **5**, 533 (1965).
4. E. M. Bradbury, A. R. Downie, A. Elliott and W. E. Hanby, *Proc. Roy. Soc. (Lond.)* **A259**, 110 (1960).

Copolymers

Polypeptide copolymers may be considered under three classifications: (a) *random* copolymers, in which various amounts of two or more monomers are copolymerized; (b) *block* copolymers, in which either small blocks are copolymerized or one block is grafted onto another; and (c) *multichain* copolymers, in which a second peptide is grafted onto the side chains of an appropriate homopolymer.

Random Copolymers

Several objectives lie behind the study of random copolymers. These include analysis of helix stabilization or destabilization by incorporating a second L residue into an L sequence; study of the effect inserting D residues (which would normally form helices of the opposite sense), into an L sequence; or rendering an insoluble polar polymer soluble, by incorporating ionic residues.

In the last-named area, copolymers of L-glutamic acid with L-leucine, L-tyrosine, L-serine, L-glutamine, and L-lysine have been studied. Leucine stabilizes the α helix of poly-L-glutamic acid at low pH, both with respect to the ionization of the carboxyl side groups and the temperature-induced conformational transition (10). Thus it is necessary to go to higher pH and higher temperature before the normally observed helix → coil transition occurs. The pH (ionization) effect would seem to arise from the reduced electrostatic repulsion in the carboxyl side chains at any given pH, compared with poly-L-glutamic acid itself. However, there appears to be some evidence that, although the carboxyls are further separated, the electrostatic repulsion is actually increased as a result of the decreased dielectric constant of the intervening medium (11). The temperature-induced transition appears to conform to theoretical calculations (12,13) but can also be accounted for by hairpin superfolding of the molecule (14).

Similar results were found for the glutamic acid/phenylalanine copolymer, in which the apolar residue was found to stabilize the glutamic acid helix (15). Other charged copolymers which show increased helix stability include those of tyrosine and glutamic acid, tyrosine and aspartic acid, and tyrosine and lysine (16).

Since L-leucine, L-phenylalanine, and L-tyrosine are α-directing residues, they may be expected to cause helix stabilization, which they do; other residues such as L-serine, a β former, should cause helix destabilization. Studies of random copolymers of L-serine/L-glutamic acid indicate that indeed helix destabilization does occur (17). Table 9.6 shows the drastic effect

TABLE 9.6

Stability of L-Glutamic Acid–L-Serine Random Copolymers in Dimethyl Sulfoxide, 16% 1 M HCl

L-Serine (%)	L-Glutamic acid (%)	α Helix (%)
0	100	90
8	92	67
16	84	41
33	67	23
42	58	14

of L-serine on the α-helix stability of poly-L-glutamic acid. The fact that incorporation of small amounts of serine has a disproportionately large effect on helix stability suggests that the destabilization arises predominantly from unfavorable interactions between the serine side chain and the glutamic acid side chain and/or solvent, rather than serine-serine interactions. However, it is never entirely certain that random copolymers have a truly random sequence, and small blocks may induce specific destabilizing effects.

Another copolymer that destabilizes the poly-L-glutamic acid helix is that with L-glutamine, although the influence is less marked than with serine. Copolymers of high glutamine content undergo extensive aggregation in aqueous solution (18).

A particularly interesting example of a glutamic acid copolymer is that with L-lysine. In the pH range 5–10, the glutamic acid residue is normally

TABLE 9.7

Stability of α Helix of Poly-L-glutamic Acid at low pH 3 in Terms of L-Lysine Copolymer Content[a]

L-Lysine (%)	α Helix (%)
0	100
30	94
40	70
50	50
60	20
100	0

[a] From E. R. Blout and M. Idelson, *J. Amer. Chem. Soc.* **80**, 4909 (1958).

negatively charged, and lysine is positively charged. Some of the more polar proteins often have a rather large concentration of these two residues and it is therefore of some interest to study their interaction. In the acidic region below pH 4, the incorporation of the charged L-lysine residue into the neutral poly-L-glutamic acid destabilizes the α helix as expected (19). The helix content is shown in Table 9.7 (17). At higher pH, where both polylysine and polyglutamic acid are normally in the charged coil form and the α-helix content is zero, the 1:1 copolymer has an α-helix content of approximately 15% (at pH 7). Presumably the attractive interaction between residues in the $i + 3$ and $i + 4$ positions tends to stabilize the short peptide repeat of the α helix as opposed to the pseudoextended from of the charged coil. At high pH, (~ 12) there is no α-helix content, thus indicating that the polylysine helix is weaker than that of polyglutamic acid.

The effect of helix destabilization by insertion of disrupting residues is also readily assessed by the use of solvent mixtures. For example the β-directing properties of L-valine and S-methyl-L-cysteine are evident when copolymerized with the normally α-directing L-methionine. Table 9.8 shows, as expected, a decrease in α-helix stability with increase in the β-directing component. Evidently S-methyl-L-cysteine has a greater destabilizing effect than the L-valine.

Copolymers of D and L residues have also been studied in terms of helix stability. Perhaps rather surprisingly, the helix content of DL-glutamate esters [methyl (20) and benzyl (21)] is quite high. The inference is that the formation of right-hand helices by the L residues, tends to overcome the formation of left-hand helices by the D residues.

TABLE 9.8

α-Helix Stability of Copolymers in Mixed Solvents[a, b]

Component (%)		CHCl$_3$ (%)	DCA (%)	TFA (%)
L-Methionine	L-Valine			
100	0		80	20
90	10		80	20
80	20		90	10
70	30	10	90	
L-Methionine	S-Me-cys			
90	10			100
80	20	10	90	
70	30	20	80	

[a] From S. M. Bloom, G. D. Fasman, C. de Lozé and E. R. Blout, *J. Amer. Chem. Soc.* **84**, 458 (1962).

[b] Composition at which approx. 50% helix exists.

Block Copolymers

The main objective of synthesizing block copolymers has been to solubilize (in water) the normally apolar and insoluble polyamino acids. This can be achieved by forming half-sandwich polymers of the type $(A)_x (B)_y$, where A is the soluble ionic portion and B is the apolar portion, or full sandwiches of the type $(A)_x(B)_y(A)_z$. Normally, the peptide chosen for A is the DL-glutamyl or DL-lysyl residue. The choice of a racemic optical mixture surrounding an optically active B section enables the conformation of the B grouping to be delineated. Table 9.9 contains a summary of some of the block copolymers which have been prepared, as well as their molecular weights. All of the examples studied seem to contain α formers in the B block and all, as expected, are α helical in aqueous solution. However, the most intriguing aspect of block copolymers of this type lies in their potential as very primitive models for globular proteins. It seems possible that in some cases extensive molecular aggregation occurs in aqueous solution to form micellar globules in which the interior is hydrophobic and α helical, the outside surface being charged (coil) and hydrophilic. For large block copolymers, it seems probable that the apolar section folds back on itself to maintain a superfolded spherical structure with the DL charged residues again residing on the outer surface.

TABLE 9.9

Block Copolymers of Type $(A)_x (B)_y (A)_z$

A	B	Mol. wt.	Comments	Reference[a]
DL-Glutamic acid	L-Alanine	$x = 175$, $y = 325$, $z = 175$ ($M_n = 62{,}000$)	α Helical	1
DL-Lysine	L-Alanine	$y = 10$ $y = 30$ $y = 40$	55% α Helix 68% α Helix 73% α Helix	2
DL-Glutamate	L-Phenylalanine	$y = 60$	α Helical	3
	L-Leucine		α Helical	4
Ethyl-DL-glutamate	L-Tryptophan	$x + z = 160$, $y = 30$	α Helical	5

[a] Key to references:

1. W. B. Gratzer and P. Doty, *J. Amer. Chem. Soc.* **85**, 1193 (1963).
2. N. Lupu-Lotan, A. Berger, E. Katchalski, R. I. Ingwall, and H. A. Scheraga, *Biopolymers* **4**, 239 (1966).
3. See text Ref. 15.
4. H. E. Auer and P. Doty, *Biochemistry* **5**, 1708 (1966).
5. A. Cosini, E. Peggion, M. Terbojevich, and A. Portolan, *Chem. Commun.* 930 (1967).

Multichain Copolymers

Multichain polyamino acids consist of a homopolymer backbone, such as polylysine, polyornithine, polyglutamic acid, or polyaspartic acid, and co-polymerized side chains, often neutral amino acids, sometimes in the DL form. Although such polymers have no analog in the native biological state they are interesting from several points of view. First, the branched-chain arrangement usually prevents crystallization and provides a biopolymer analog of the commercial glassy polymers; second the influence of large side chains on backbone conformation can be studied; and third, some of the hydrodynamic properties of globular proteins are duplicated (charged end groups protecting an apolar backbone), and some immunological and antigenic activity (22) has been observed. Relatively few of the physical properties of these branched-chain polypeptides have been examined. Solution studies have shown (23), however, that multichain polyamino acids containing glutamic and aspartic acid side chains show low viscosity and high sedimentation and diffusion coefficients (indiciting compact shape) compared with the corresponding linear polyamino acids of similar molecular weight.

Probably the most extensively studied multichain polypeptide is poly-DL-alanyl–poly-L-lysine. Samples are available commercially with an average side-chain length (DL-alanine) in the region of 5–20 residues. Circular dichroism studies should reflect the conformation of the backbone. The single negative band at 196 mμ is indicative of the true random form, which is observed for this polymer both in solution and when film cast from a number of different solvents (24). Infrared spectra of the solid are typical of a random form, with amide A and B bands indicating that the side-chain peptide N—H's are hydrogen bonded—a fact that almost suggests the presence of at least some ordered structure. The amide V band at 660 cm^{-1} seems to support the concept of the random form. Both x-ray and electron diffraction patterns have been obtained (24) and show typical amorphous rings, the former technique yielding two rings at 7.7 and 4.4 Å, the latter yielding only the 4.4 Å ring. Although the morphology might be expected to be granular or globular, the definition of any tertiary structure in the solid state has proven elusive (24).

Sequential Polypeptides

In Chapter 10 it is shown that the primary sequence of fibrous proteins appears to involve repeating units that lead to specific structural features. There is thus some immediate motivation for studying sequential polypeptides in an attempt to discover which features of the sequence promote the specific conformation. The past few years have seen a dramatic increase in the effort to prepare and characterize a large number of different sequential polypeptides. In fact, the majority of materials reported as being synthesized have not yet been subjected to any extensive structural characterization. Emphasis in this section is placed upon those peptides which have been subjected to conformational or structural analysis and the principles which have emerged.

Polydipeptides

The majority of polydipeptides that have been prepared and characterized contain glycine. There are perhaps three good reasons why this should be so. Glycine, being optically inactive, does not racemize during the synthetic process and consequently usually yields a polymer of considerable configurational and conformational integrity. Second, glycyl polydipeptides are believed to be suitable models for several of the silk fibroin proteins as reported in Chapter 10. The third, and perhaps least important, point is that as the simplest amino acid, glycine represents a logical, but not necessarily the only logical, starting point for conformational studies.

POLY (GLY-ALA)*

This polymer, the simplest in the series, has received considerable attention. Since both glycine and alanine homopolymers have one of their two known conformations, the β form, in common, it is not surprising that poly (Gly-Ala), of $M_w \sim 10,000$, is found in the antiparallel cross-β form (25), as precipitated from strong (acidic) solvents. Characterization has been carried out using most of the conventional methods including x-ray (powder) diffraction, infrared dichroism, electron diffraction, and film CD/ORD (26). It may be put in the Fasman class II CD of β forms. (See Chapter 6).

In addition to this common form, poly (Gly-Ala) also has a unique conformation in concentrated aqueous salt solutions, a situation in which the polyglycine II form might have been expected. Both the CD/ORD curves of the solution and the x-ray analysis of the precipitated form show, however, that the material is not, apparently, of PGII conformation. The circular dichroism curves (27) of suspensions show positive bands at 223 and 200 mμ. The x-ray diffraction pattern (28) is interpreted in terms of a unit all with dimensions $a = 4.72$ Å, $b = 14.40$ Å, and c (chain axis) = 9.6 Å. The actual conformation is said to be a peculiar "crankshaft" arrangement.

---

* The polymers are synthesized with glycine in the C-terminal position to avoid racemization.

Poly (Gly-Ser)

Both of the residues here favor the β form, and once again the antiparallel cross-β form has been observed by a combination of techniques (26). No additional conformation has been detected.

Poly [Gly-Glu(OEt)]

Although the glutamyl ester is an α former, only the β form of this polymer has been noted to date (both in solution and in the solid state). Both β- and cross β forms have been reported (26,29) but this material is poorly crystalline and difficult to orient.

Poly [Gly-Glu(OBz)]

The benzyl ester of glutamic acid is known to be a rather stronger α former than the ethyl ester, and one might expect to observe two different conformations. Although relatively little characterization work has been carried through with this polymer it has been reported as α helical in the solid state (29).

Poly (Gly-Glu)

The dipeptide containing glutamic acid itself is a rather interesting polymer since it is water soluble and may be rendered into the protonated or charged forms as a function of pH. In addition, the charged-coil form should be similar to polyglycine II in its optical properties. Although there is indeed a conformational change in the pH 2 region, the solid precipitated below pH 2 appears to be an α form (30). The solution form, identified as random in the pH 2–9 range, probably does possess the pGII-like coil form described in Chapter 2, although further work is needed to clarify this point.

Poly (Gly-Pro) and Poly (Gly-Hypro)

Both of these polymers are water soluble and it might be expected that both would possess the polyproline II (PPII) form since glycine and proline are readily compatible with this conformation. Although only circular dichroism evidence is available, poly (Gly-Hypro) does indeed appear to have an ordered (PPII) form in water. However, somewhat surprisingly a similar form has yet to be identified for the nonhydroxylated form (31).

Other Glycyl Polydipeptides

Several other glycyl polydipeptides have been synthesized including poly (Gly-Leu), poly (Gly-DL-Phe), poly (Gly-Tyr), poly (Gly-DL-Ala), poly (Gly-DL-Leu), and poly (Gly-Phe), but little or no characterization has been reported. The little evidence that exists seems to support the contention that most, including DL forms, may be crystallized in the β form.

NONGLYCYL POLYDIPEPTIDES

Probably the most extensively studied polypeptide in this category is poly (Ala-Glu) (24). Above pH 7, this water-soluble polymer may be crystallized in its salt form (e.g., the ammonium salt) in the β conformation. Presumably, salts of poly (Gly-Glu) would also be of β form.) Between pH 3 and 7 poly (Ala-Glu) stays in solution as the charged-coil form, the CD curves being very similar to those of polyglutamic acid. Below pH 3 the polymer is insoluble in water and although definitive evidence is still lacking it appears to be random rather than in the α form which might have been expected.

The CD spectrum of poly (Ala-Pro) indicates that this material is ordered in aqueous solution and presumably has a PPII conformation (31). Although polyalanine is not found in the PPII conformation, it is energetically favorable and alanyl residues are known to be compatible with this structure (e.g., in collagen).

Several polydipeptides containing glutamyl esters have been characterized (29), including poly [Val-Glu(OMe)], poly [Cys(SBz)Glu(OEt)], and poly [Ser(OAc)Glu(OMe)]. Each of these is found by x-ray diffraction to be in the β form in the solid state, although the cysteine-containing peptide shows some evidence of α structure by infrared and ORD/CD of films, and in solution there can be 50% α-helical content (ORD). This evidence suggests that the cysteine derivative is a much weaker β former than valine, serine, or glycine.

Most of the other synthetic polydipeptides have not been characterized, although the interesting poly (D-Ala-L-Ala) seems to be β (23).

Polytripeptides

As mentioned in Chapter 10, the fibrous protein collagen may be well represented as a glycine-containing polytripeptide, and many synthetic poly tripeptides have been synthesized and characterized. Most of the collagen model polypeptides contain proline, and considerable advancement in understanding the collagen structure has resulted from a detailed study of these materials. The simplest polytripeptides are probably those containing two glycine residues and some other residue.

POLY (GLY-GLY-ALA)

This polymer has two known conformations. The first, obtained by precipitation from strong acidic solvents, is the β form; the second, obtained in water (very sparingly soluble), in aqueous salt solutions, or in weak solvents, is the polyglycine II conformation (33,34). The latter structure is particularly interesting because it is probably the simplest polypeptide in the PGII conformation containg an optically active residue. This permits the ORD/CD curves of the PGII helix to be investigated.

In the solid state the x-ray powder patterns indicate a sheetlike structure with the helices hydrogen bonded to four (35) or perhaps five (36) nearest-neighbor helices. The alanyl residues lie on one side of the 3_1 helix, thus preventing hydrogen bonding of at least one adjacent helix into the hexagonal structure.

In aqueous solution, the CD curves are found to be similar to the left-hand polyproline II spectrum but displaced to shorter wavelength (see Chapter 6). It appears, therefore, that poly (Gly-Gly-Ala) forms a left-handed PGII conformation in aqueous solution (37). In concentrated calcium chloride solution or in water at elevated temperatures, the polyglycine helix melts into the true random form. The melting curve for poly (Gly-Gly-Ala) is shown in Fig. 9.3.

The infrared spectra of the tripeptide in PGII and random forms also show interesting differences which depend upon conformationally sensitive vibrational modes (see Chapter 5).

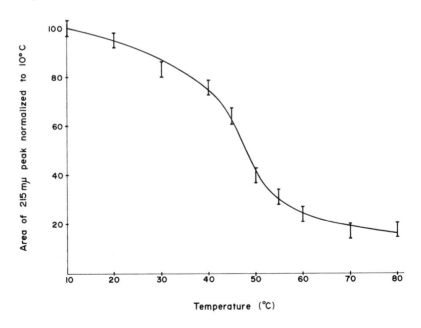

Fig. 9.3. Melting curve for the polyglycine II → random conformational change of poly (Gly-Gly-Ala) as determined from the 215 mμ CD band (aqueous solutions).

POLY (GLY-GLY-PRO)

Originally synthesized as a collagen analog, this polymer is found to have a sheet structure in the solid state somewhat similar to poly (Gly-Gly-Ala) with the chains in the PPII conformation (38). Although only one conformation has been determined experimentally, an extensive theoretical analysis (39) of this polypeptide has pointed out the possibility of a unique conformation and more than one efficient packing mode, as noted in Chapters 2 and 10. In aqueous solution, the material aggregates and it is not possible to assign its conformation unambiguously (40).

OTHER DIGLYCYL POLYTRIPEPTIDES

In this category poly (Gly-Gly-Phe), poly (Gly-Gly-Ser), poly [Gly-Gly-Asp(OMe)] poly (Gly-Gly-His), and poly[Gly-Gly-Glu(OEt)], as well as some DL isomers, have been synthesized. Very little is known of the conformation of these biopolymers, although there is some indication of a PGII conformation of poly [Gly-Gly-Glu(OEt)] in aqueous solution.

POLY (GLY-ALA-ALA)

Of the polytripeptides containing only one glycine residue, poly (Gly-Ala-Ala) is one of the more interesting. It has three known conformations (41). The first, and best characterized, is the β structure, which is produced by precipitation from strong acidic solvents. Thus, although the L-alanyl residues are nominally α directing, the most general structure seems to be dominated by the β influence of the glycyl residue.

Polymer precipitated from weak solvents including water shows x-ray evidence of an α-helix form; in aqueous solution, however, (Gly-Ala-Ala)$_n$ is apparently true random. Efforts to find a conformational transition at very low temperatures give indication that in glycol/water mixtures there may be a fourth form (PGII) developing at around $-112°$.

POLY (GLY-PRO-PRO)

Poly (Gly-Pro-Pro) in the solid state gives an x-ray pattern very similar to collagen and is believed to be an excellent model for sections of this protein. It is found to be in a triple-strand coiled-coil conformation, the individual chains taking on a left-handed, distorted PPII conformation intertwined into a right-hand superhelix. The x-ray diffraction (42), electron diffraction, and morphology (43), and the spectroscopic (44) and solution properties have been studied quite intensively because of this polymer's fundamental importance in collagen-structure studies. As with collagen, it "melts" in aqueous solution, the triple-strand structure converting to true random, the analog in fact of the collagen/gelatin transition. The melting curve is, however, displaced to a much higher temperature than collagen as a result of the high proline content.

OTHER COLLAGENLIKE POLYTRIPEPTIDES

The collagenlike polytripeptides and hexapeptides are discussed in some detail in Chapter 10. They undoubtedly represent the most comprehensively studied set of sequential polypeptides yet prepared. The predominant methods used to study these materials have been x-ray diffraction and ORD/CD methods. The triple-helix peptide repeat, which in collagen is 2.85 Å, varies somewhat among these biopolymers even in the triple-helical arrangement. Recently, there has been an interesting study of the relation between the peptide repeat, which presumably reflects the supercoil period and the interchain bonding, and the infrared spectrum, particularly the $-N-H$ stretch in the 3300 cm^{-1} region of the spectrum (44). Table 9.10 shows this relation and indicates that the strength of interchain hydrogen bonding controls, to a high degree, the extent of supercoiling and the formation of the triple helix.

TABLE 9.10

Relationship between the Amide A Band and Residue Repeat for Some Collagen Model Polytripeptides and Polyhexapeptides[a]

Compound	Amide A (cm^{-1})	Peptide repeat (Å)
(Gly-Hypro-Hypro)$_n$	3360	2.75
(Gly-Pro-Hypro)$_n$	3350	2.82
(Gly-Pro-Pro)$_n$	3350	2.87
(Gly-Pro-Ala-Gly-Pro-Pro)$_n$	3350	2.87
(Gly-Pro-Ala)$_n$	3345	2.88
(Gly-Ala-Pro-Gly-Pro-Pro)$_n$	3335	2.95
(Gly-Ala-Pro-Gly-Pro-Ala)$_n$	3325	2.95
(Gly-Ala-Ala-Gly-Pro-Pro)$_n$	3320	2.95
(Gly-Pro-Gly)$_n$	3308	3.10

[a] From Ref. 44.

NON-PROLINE-CONTAINING POLYTRIPEPTIDES

Some of the glycine-led non-proline-containing polytripeptides are also mentioned in Chapter 10. Most thoroughly studied of these has been poly [Gly-Ala-Glu(OEt)], which from morphology, x-ray and electron diffraction, infrared dichroism, and CD/ORD has been found to be antiparallel cross-β

in conformation (45,46). DeTar and co-workers have prepared an interesting series of similar compounds: poly [Gly-Ser(OAc)Asp], poly [Gly-Ser-Asp (OMe)], and poly (Gly-Ser-Asp) (47). With serine present in the tripeptide it is to be expected that β structures are even further stabilized, and indeed the esters do appear to have a β structure. However, the free acids in aqueous solution may be in the charged coil form. Poly (Gly-Ser-Glu) is water soluble in the pH range 2–9 in the charged form and has been classified as random (30). It would be of some interest to know whether this is the true random form or the Krimm–Mark charged coil. If it is the latter, the stabilizing forces must be, at least partially, London van der Waals since the juxtaposition of a charged residue in every third position of an approximately threefold helix is likely to be electrostatically unfavorable. The resolution of this point has implications in collagen-structure work since charged glycine-led tripeptide sequences are common.

Other non-glycine-containing polytripeptides which have been prepared and characterized include poly [Val(Glu(OMe))$_2$], poly [Cys(S Bz)(Glu(OEt))$_2$], and poly [Ser(OAc)(Glu(OMe))$_2$] (29). These compounds are considered in terms of their conformational integrity later, but are predominantly β in conformation.

Higher Orders of Sequential Polypeptides

Polytetra; penta- and hexapeptides have been prepared, but with increasing length and complexity of synthesis there are fewer detailed reports of characterization (and synthesis). Exceptions are the polyhexapeptides of the form [(Gly-A-B)(Gly-C-D)]$_n$ which, when containing proline, are again collagen models (see Chapter 10) and have been studied fairly extensively.

Of the polytetrapeptides, poly [Gly$_3$Ala] has been prepared and characterized (35) and has the PG II-like conformation and structure. The x-ray diffraction pattern may be indexed on the basis of a hexagonal unit cell, dimensions $a = b = 4.89$ Å and $c = 9.15$ Å (fiber axis). These data may be compared with the hexagonal unit cell of polyglycine II which is $a = b = 4.8$ Å and $c = 9.33$ Å (fiber axis).

Other polytetrapeptides containing triglycyl sequences which have been reported are poly [Gly$_3$Pro] and poly [Gly$_3$Hypro]. The former is said to be "disordered" and the latter "ordered" in aqueous solution; presumably the ordered phase has the polyproline conformation (31). A number of glycine-containing polytetrapeptides such as poly (Glu-His-Lys-Tyr) and poly [Glu(OEt)Glu(OMe)Val-Glu(OMe)] have been reported but without characterization.

Poly [Ser-HisLeu$_3$] and poly (Leu-His-Leu-Leu-Ser-Leu) of molecular

weight $\sim 10,000$ have been prepared as models for the active site in chymotrypsin (48), where a juxtaposition of histidine and serine (and aspartic acid)
is believed to form the active moiety. The polymers were formed with the
α-directing L-leucine in the expectation that the resulting polymer would be
α helical and that the positioning of the serine and histidine in the $i + 3$ and
$i + 4$ positions, which lie adjacent in the α helix, would resemble the native
protein. Although the polymers are insoluble or only sparingly soluble in
water, some enzymatic activity (substrate hydrolysis) was observed in aqueous
mixtures and the molecules were shown to have a maximum α helicity of about
40 and 50%, respectively.

Series of sequential polypeptides of the type poly (A_xB_y), with x being
0–5, $y = 1$ or 2, and the residues variously distributed, have been prepared
and characterized (29). The residues used for A were various esters of glutamic acid and B was L-valine, S-benzyl-L-cysteine, glycine, or O-acetyl-L-
serine. These series provide a basis for studying various aspects of conformational stability, which are reported in the following section.

Conformational Stability of Sequential Polypeptides

In the sequential poly- (A_xB_y) peptides mentioned above, the A residues
are α-helix forming, and the study of a series in which B, a β former, is placed
at various positions along the chain should give some insight into the factors
that might induce helix instability. The first such series that we shall consider is
poly- $[Glu(OMe)_xVal_y]$.* Structural data for this series are shown in Table
9.11. Residue positions are shown for valine such that if a residue occurs in the
ith position in the chain, it also occurs in the position $i + z$, the values of z
being quoted.

In studying the effect of valine residues on the formation of α helices of
poly-γ-methyl-L-glutamate, we note that the homopolymer is only weakly α
helical ($\sim 60\%$), yet the introduction of 20% L-valine does not perturb this
helical content. When valine is introduced in position $i + 4$, there is a small
decrease in solution α content but a strong decrease in the solid; for the
tripeptide with valine in the i, $i + 3$ positions, there is a marked decrease in
solution and solid-state helix content compared with the homopolymer.
However, if the polyhexapeptide with the same valine content (33%) and
adjacent valyl residues is examined, it can be seen that there is an increase in
helicity if valine does not occur in the $i + 3$ or $i + 4$ position both in the solid
and solution. Increase of valine content above one-third renders the polymer
into the β form.

---

* For the purpose of simplifying nomenclature the subscripts $_x$ and $_y$ are written
referring to the total residue rather than the ester portion alone.

TABLE 9.11^a

Copolymers of L-Valine (V) and γ-Methyl-L-glutamic Acid (G)

Polymer	G Content (mole %)	Positions of V residues	Max helix content (%)		Conformation in solid	
			Solution^b	Solid^c	x-Ray^d	Infrared^e
$(G_3)_n$	100		62	58	α, β	$\alpha, X\beta$
$(G_3VG)_n$	80	$i, i+5, ...$	62	54	α, β	α, β
$(G_2VG)_n$	75	$i, i+4, ...$	52	33	β	β, α
$(GVG)_n$	67	$i, i+3, ...$	44	32	β	β, α
$(G_3V_2G)_n$	67	$i, i+1; i+6,$ $i+7, ...$	54	38	β	β, α
$(VG)_n$	50	$i, i+2, ...$	25	16	β	β
$(V)_n$	0	$i, i+1, ...$	0	0	β	$X\beta$

^{a} From Ref. 29, p. 58, with permission.
^{b} Calculated assuming $b_0 = -630$ corresponds to 100% helix.
^{c} Includes random material.
^{d} Dominant crystalline phase given first.
^{e} β denotes chain axes parallel to direction of stroking; $X\beta$ denotes chain axes perpendicular to this direction.

Since the α helix has approximately 3.6 residues per turn we might suppose that bulky residues placed in the $i+3$ or $i+4$ positions would interact sterically, destabilizing the helix. Conformational studies indicate that for the valylglutamate polymer the $i+4$ interaction is indeed unfavorable (short contact), as shown in Fig. 9.4. The situation with the $i+3$ position is not as clearly defined.

Comparison of the $i+4$, poly [Glu(OMe)$_3$Val$_1$] solution and solid data suggests that the solution conformation may be relaxed from the regular 18_5 helix. Although there are no definitive data available for this situation, various suggestions of wormlike or 10_3 helices have been made, on the basis of light-scattering data, for nominally α-helical materials in various solutions. Certainly a decrease in the number of peptide residues per helical turn from 3.6 would help to explain the diminished $i+4$ effect in solution. In any case, it may be deduced that the β-directing properties of the L-valyl residue arise predominantly from non-nearest-neighbor steric effects.

The results obtained for the poly [Glu(OEt)$_x$Cys(SBz)$_y$] series are noticeably different, as is shown in Table 9.12. It can be seen that the homopolymer is a substantially more helical polymer than its methyl ester counterpart under the prescribed conditions, yet, in contrast to the effect of valine, small amounts (20%) of S-benzyl-L-cysteine cause a considerable decrease in α content in both solution and solid states. Further increase in Cys(SBz)

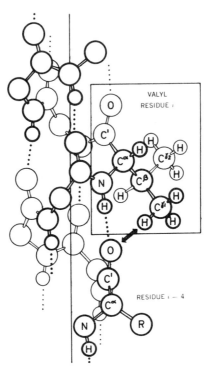

Fig. 9.4. Conformation of an L-valyl residue in an α helix calculated with $\chi_1 = 180°$ and $300°$ for $C_i^{\gamma 2}$ and $C_i^{\gamma 1}$, respectively. A short contact between a hydrogen atom on $C_i^{\gamma 1}$ and O_{i-4} is indicated by an arrow. (From R. D. B. Fraser *et al.*, *in* " Symposium on Fibrous Proteins" (W. G. Crewther, ed.), Plenum, New York, 1968, with permission of Plenum Press Publishing Corp.)

causes a steady decrease in α content, independent of the exact location of the residues. It seems likely, from these data, that the effect must arise predominantly from the unfavorable side chain/solvent interaction for the cysteine derivative.

O-Acetyl-L-serine, another well-known β-directing residue, also has a large effect upon the stability of the α helix when applied " in small doses." The data for poly[Glu(OMe)$_x$Ser(OAc)$_y$] are presented in Table 9.13. The introduction of 33% of the seryl derivative renders the α content zero. It is noteworthy that a gel phase is formed, at this and higher serine content, which eventually undergoes transformation into the β form. The molecular structure of this transition phase is believed to involve a conformation (listed as random) in which the seryl oxygen is hydrogen bonded to the peptide N—H, as shown in Fig. 9.5. Such a tendency to hydrogen bond would certainly disrupt the intramolecular hydrogen bonding of the α helix and can adequately explain the distruptive effect of the Ser(OAc) peptide residues.

TABLE 9.12[a]

Copolymers of S-Benzyl-L-cysteine (C) and γ-Ethyl-L-glutamic Acid (G)

Polymer	G Content (mole %)	Positions of C residues	Max helix content (%)		Conformation in solid	
			Solution[b]	Solid[c]	x-Ray[d]	Infrared[e]
$(G_2)_n$	100		108	84	α	α, Xβ
$(G_3CG)_n$	80	$i, i+5, \ldots$	66	57	β, α	α, Xβ
$(GCG_2)_n$	75	$i, i+4, \ldots$	64	52	β, α	α, β
$(GCG)_n$	67	$i, i+3, \ldots$	64	48	α, β	Xβ, α
$(GCGCG)_n$	60	$i, i+2;$ $i+5, i+7, \ldots$	59	34	β	Xβ, α
$(GC_2G_2)_n$	60	$i, i+1;$ $i+5, i+6, \ldots$	51	28	β	Xβ, α
$(CG)_n$	50	$i, i+2$	50	21	β	β, α
$(C)_n$	0	$i, i+1$	0	0	β	Xβ

[a] From Ref. 29, p. 58, with permission.
[b] Calculated assuming $b_0 = -630$ corresponds to 100% helix.
[c] Includes random material.
[d] Dominant crystalline phase given first.
[e] β denotes chain axes parallel to direction of stroking; Xβ denotes chain axes perpendicular to this direction.

TABLE 9.13[a]

Copolymers of O-Acetyl-L-serine (S) and γ-Methyl-L-glutamic Acid (G)

Polymer	G Content (mole %)	Positions of S residues	Max helix content (%)		Conformation in solid	
			Solution[b]	Solid[c]	x-Ray[d]	Infrared[e]
$(G_3)_n$	100		62	58	α, β	β, Xβ
$(G_3SG)_n$	80	$i, i+5, \ldots$	50[f]	35–40	β	Xβ, α
$(G_2SG)_n$	75	$i, i+4, \ldots$	41[f]	30	β	Xβ, α
$(GSG)_n$	67	$i, i+3, \ldots$	0[g]	0	β	β
$(SG)_n$	50	$i, i+2, \ldots$	0[g]	0	β	β

[a] From Ref. 29, p. 59. with permission.
[b] Calculated assuming $b_0 = -630$ corresponds to 100% helix.
[c] Includes random material.
[d] Dominant crystalline phase given first.
[e] β denotes chain axes parallel to direction of stroking; Xβ denotes chain axes perpendicular to this direction.
[f] Decreases to <10% helix; solution gels on standing.
[g] Solution gels on standing.

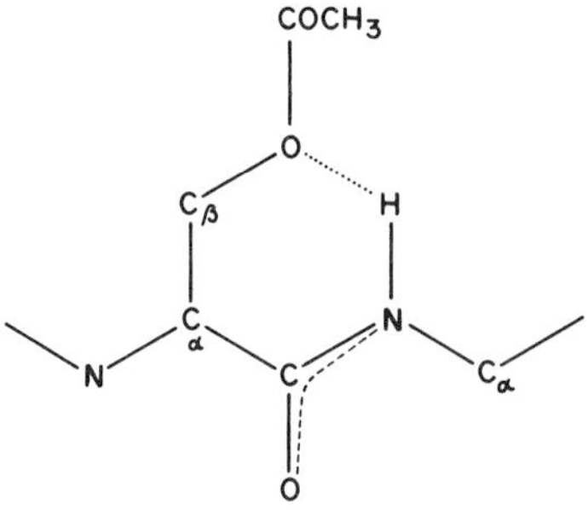

Fig. 9.5. Possible arrangement of a hydrogen-bonded side chain for poly-*O*-acetyl-L-serine in an intermediate gel phase.

One other series of sequential polypeptides, is poly[Glu(OEt)$_x$Gly$_y$] for which the data are more difficult to explain, as shown in Table 9.14. The glycyl residue can be seen to have a strong destabilizing effect on the glutamate α helix which is not position dependent. The effect is, in fact, as strong as that for the *S*-benzyl-L-cysteine substitution, yet one can hardly invoke steric or unfavorable side-chain interactions. Two explanations of this effect have been suggested. The first could involve the solvation of the amide, which is sterically more available for glycine than for peptides with side chains; it is not immediately obvious, however, how the solvent interaction would favor a β structure. The second possibility is more subtle; it is supposed

TABLE 9.14[a]

Copolymers of Glycine (g) and γ-Ethyl-L-glutamic Acid (G)

Polymer	G Content (mole %)	Positions of g residues	Max helix content (%)		Conformation in solid	
			Solution[b]	Solid[c]	x-Ray[d]	Infrared[e]
$(G_2)_n$	100		108	84	α	α, Xβ
$(G_2gG_3)_n$	83	$i, i+6, \ldots$	65	63	β^f	α, β
$(G_4g)_n$	80	$i, i+5, \ldots$	62	30	β	β, α
$(G_3g)_n$	75	$i, i+4, \ldots$	56	28	β	β, α
$(G_2g)_n$	67	$i, i+3, \ldots$	10	0	β	β
$(Gg)_n$	50	$i, i+2, \ldots$	5	0	β	β

[a] From Ref. 29, p. 59, with permission.
[b] Calculated assuming $b_0 = -630$ corresponds to 100% helix.
[c] Includes random material.
[d] Dominant crystalline phase given first.
[e] β denotes chain axes parallel to direction of stroking; Xβ denotes chain axes perpendicular to this direction.
[f] The failure to detect the α phase in the x-ray pattern indicates that it is not crystalline.

that the glycyl residues introduce more degrees of conformational freedom to the structure, thus affecting the entropy and free energy of stabilization. Neither argument appears entirely satisfactory at this stage but it is clear that in a number of polypeptides, glycine does have strong β-directing properties.

REFERENCES

1. E. M. Bradbury, L. Brown, A. R. Downie, A. Elliott, R. D. B. Fraser, and W. E. Hanby, *J. Mol. Biol.* **5**, 230 (1962).
2. L. Stevens, R. Townend, S. N. Timasheff, G. D. Fasman, and J. Potter, *Biochemistry* **7** 3717 (1968).
3. J. C. Andries and A. G. Walton, *Biopolymers* **8**, 465 (1969).
4. V. Luzzati, M. Cesari, G. Spach, F. Masson, and J. M. Vincent, *J. Mol. Biol.* **3**, 566 (1961).
5. G. Spach, L. Freund, M. Daune, and H. Benoit, *J. Mol. Biol.* **7**, 468 (1963).
6. P. Saludjian and V. Luzzati, *in* "Poly-α-Amino Acids" (G. D. Fasman, ed.), Chapter 4. Dekker, New York, 1967.
7. H. A. Scheraga and D. Poland, "Helix–Coil Transitions." Academic Press, New York, 1970.
8. P. Doty, K. Imahori, and E. Klemperer, *Proc. Nat. Acad. Sci. U.S.* **44**, 424 (1958).
9. P. K. Sarkar and P. Doty, *Proc. Nat. Acad. Sci. U.S.* **55**, 981 (1966).
10. G. D. Fasman, C. Lindblow, and E. Bodenheimer, *J. Amer. Chem. Soc.* **84**, 4977. (1962).
11. E. Iizuka and J. T. Yang, *Biochemistry* **4**, 1249 (1965).
12. H. A. Scheraga, G. Némethy, and I. Z. Steinberg, *J. Biol. Chem.* **237**, 2506 (1962).
13. M. E. Baur and L. H. Nosanow, *J. Chem. Phys.* **38**, 578 (1963).
14. D. C. Poland and H. A. Scheraga, *Biopolymers* **3**, 335 (1965).
15. H. J. Sage and G. D. Fasman, *Biochemistry* **5**, 286 (1966).
16. M. Sela and E. Katchalski, *J. Amer. Chem. Soc.* **78**, 3986 (1956).
17. G. D. Fasman, ed., "Poly-α-Amino Acids," p. 562. Dekker, New York, 1967.
18. L. H. Krull and J. S. Wall, *Biochemistry* **5**, 1521 (1966).
19. E. R. Blout and M. Idelson, *J. Amer. Chem. Soc.* **80**, 4909 (1958).
20. T. Miyazawa, *in* "Aspects of Protein Structure" (G. N. Ramachandran, ed.), p. 257. Academic Press, New York, 1963.
21. T. Tsuboi, Y. Mitsui, A. Wada, T. Miyazawa, and N. Nagashima, *Biopolymers* **1**, 297 (1963).
22. M. Sela, S. Fuchs, and R. Arnon, *Biochem. J.* **85**, 223 (1962).
23. A. Yaron and A. Berger, *Biochim. Biophys. Acta* **107**, 307 (1965).
24. H. H. Chen, M.S. thesis. Case Western Reserve University, Cleveland, Ohio, 1971.
25. R. D. B. Fraser, T. P. MacRae, F. H. C. Stewart, and E. Suzuki, *J. Mol. Biol.* **11**, 706 (1965).
26. J. M. Anderson, H. H. Chen, and A. G. Walton, *Bull. Amer. Phys. Soc.* **16**, 320 (1971).
27. A. Brach and G. Spach, *Comptes Rend.* **271C**, 916, (1970).
28. B. Lotz and H. D. Keith, *J. Mol. Biol.* **61**, 195 (1971).
29. R. D. B. Fraser, B. S. Harrap, R. Ledger, T. P. MacRae, F. H. C. Stewart, and E. Suzuki, *in* "Fibrous Proteins" (W. G. Crewther, ed.), p. 57. Plenum, New York, 1968.
30. D. F. DeTar and T. Vajda, *J. Amer. Chem. Soc.* **89**, 998 (1967).
31. W. L. Mattice and L. Mandelkern, *J. Amer. Chem. Soc.* **92**, 5285 (1970).
32. F. H. C. Stewart, *Aust. J. Chem.* **18**, 887 (1965).

33. A. Brach and G. Spach, *in* "Peptides" (E. Borias, ed.), p. 45. North-Holland Publ., Amsterdam, 1968.
34. J. C. Andries, J. M. Anderson, and A. G. Walton, *Biopolymers*, **10**, 1049 (1971).
35. B. Lotz and H. D. Keith, *J. Mol. Biol.* **61**, 195 (1971).
36. J. C. Andries, Ph.D. thesis. Case Western Reserve University, Cleveland, Ohio, 1970.
37. W. B. Rippon and A. G. Walton, *Biopolymers*, **10**, 1207 (1971).
38. W. Traub and A. Yonath *in* "Conformation of Biopolymers" (G. N. Ramachandran, ed.), Vol. 2, p. 449. Academic Press, New York, 1967.
39. A. J. Hopfinger and A. G. Walton, *J. Macromol. Sci. Phys.* **B4**, 185 (1970).
40. P. J. Oriel and E. R. Blout, *J. Amer. Chem. Soc.* **88**, 2041 (1966).
41. B. B. Doyle, W. Traub, G. P. Lorenzi, F. R. Brown, and E. R. Blout, *J. Mol. Biol.* **51**, 47 (1970).
42. A. Yonath and W. Traub, *J. Mol. Biol.* **43**, 461 (1969).
43. J. C. Andries and A. G. Walton, *J. Mol. Biol.* **54**, 579 (1970).
44. B. B. Doyle, Ph.D. thesis. Harvard University Medical School, Boston, Massachusetts, 1970.
45. W. B. Rippon, J. M. Anderson and A. G. Walton, *J. Mol. Biol.* **56**, 507 (1971).
46. J. C. Andries and A. G. Walton, *J. Mol. Biol.* **56**, 515 (1971).
47. D. F. DeTar, F. F. Rogers, and H. Bach, *J. Amer. Chem. Soc.* **89**, 3039 (1967).
48. D. Hudson, Ph.D. thesis. Exeter University, Exeter, England, 1970.

10

FIBROUS PROTEINS AND BIOPOLYMER MODELS

Introduction

One of the avowed objectives of biopolymer research is to study the synthetic entity as a model for the more complex native structure. There has, therefore, been some emphasis on models for proteins, particularly those of the fibrous variety. The reason for this emphasis is that, in general, the fibrous proteins contain chains in conformations which are more or less analogous to the conformations discussed in Chapter 2, and furthermore, the treatment of fibrous materials by x-ray techniques is much the same as that for synthetic polymers. Globular proteins, however, which often contain recognizable conformations, must be treated by more sophisticated x-ray methods, and so far little progress has been made in synthesizing useful models. The fibrous proteins generally form part of the skeletal and structural materials in higher animals, i.e., they have a predominantly mechanical function, whereas the globular proteins have a predominantly chemical function.

It is evident that the more closely a synthetic polypeptide resembles the protein primary sequence, the more likely it is that the model will take on the detailed characteristics of the protein. Yet to a certain extent wide variations in amino acid content can be tolerated without discernible changes of conformation and gross structure.

The first piece of information that is generally required before a model biopolymer can be formulated is an outline of the primary chemical sequence of the protein. Unfortunately, except for collagen and certain silk fibroins, details of the sequence and structure of fibrous proteins are sadly lacking. Many of them give very poor x-ray fiber diagrams, and consequently there would be real value in preparing simpler biopolymer models—if the outline of the primary sequence were known.

Most reports in the literature concerning fibrous proteins have concentrated on the relation between conformation, morphology, and function. Chemical purification and dissection for analytical purposes have met with considerable difficulties. Furthermore, many of the fibrous proteins are insoluble and chemically intractable. In such cases effort has often been directed toward discovering the precursor phase in the biosynthetic pathway, which is generally soluble to some extent.

TABLE 10.1[a]

Amino Acid Composition[b] of Sericin and Fibroin from Bombyx mori

Amino acid	Sericin	Fibroin
Aspartic acid	148.0	13.3
Threonine	86.0	9.2
Serine	373.0	121.3
Glutamic acid	34.1	10.2
Proline	7.6	3.1
Glycine	147.0	445.3
Alanine	43.0	293.5
Valine	35.4	22.4
Cystine/2	5.1	
Isoleucine	7.6	7.1
Leucine	13.9	5.1
Tyrosine	25.3	52.0
Phenylalanine	3.8	6.1
Lysine	24.0	3.1
Histidine	11.4	1.6
Arginine	35.4	4.6
Tryptophan		1.5
(NH$_3$)	86.0	

[a] Reprinted from S. Seifter and P. M. Gallop, in "The Proteins," Vol. IV, ed. H. Neurath (1966), Academic Press, with permission.
[b] In this table and those that follow, amino acid compositions are given in residues/1000 of hydrolysis product.

A detailed discussion of biological function is not pertinent at this point, but the structures of six important groups of proteins will be discussed. These are the silk proteins (fibroin), collagen, elastin, resilin, myosin, and the keratins. A more extensive review of this field has been made by Seifter and Gallop (1).

Silk Proteins

Silk is a fibrous material secreted by certain species of insects. It may be predominantly proteinaceous or chitinous (polysaccharide), although the latter is rare and only the former is considered here. The protein component of silks may be subdivided into the silk fibroin, generally composed of crystalline fibers (or fibers which can be rendered crystalline to some extent), and the embedding medium, a water-soluble component known as sericin. A comparison of the amino acid composition of sericin and fibroin from the species *Bombyx mori* (from which commercial silk is obtained) is shown in Table 10.1(1). It can be seen that the sericin abounds in polar amino acids, particularly serine and aspartic acid, whereas the fibroin from *B. mori* is high in nonpolar (glycine and alanine) amino acids. Sericin solutions form a gel on cooling, and no crystalline component has been observed. It may be concluded therefore that sericin is an amorphous, random-chain protein, probably cross-linked (note 0.5% cystine/2), which acts predominantly as a supporting matrix.

The Fibroins

The silk fibroins have been investigated in some detail. Over one hundred distinct, chemically different fibroins have been identified and analyzed. Although most possess β structures, some have α-helical, polyglycine II, and collagenlike conformations. However, it should be pointed out that fibroins are never 100% crystalline and local regions of disorder are present.

ANTIPARALLEL β-FIBROINS

The antiparallel β structure is found in most of the silk from different sources so far examined. These structures were divided into five groups by Warwicker (2) on the basis of their x-ray diffraction patterns. A sixth type of structure has been recognized by Lucas and Rudall (3), and examples of all six structures and their unit cell dimensions are given in Table 10.2 (2). All of these structures are pleated β sheets and have in common a period of

TABLE 10.2

Values for Dimensions c of the Parallel-β-Pleated Fibroin Unit Cell[a]

Group	Example	$c(\text{Å})$
1	*Bombyx mori*	9.3
2a	*Anaphe moloneyi*	10.0
2b	*Clania* sp.	10.0
3a	*Antheraea mylitta*	10.6
3b	*Dictyoploca japonica*	10.6
4[b]	*Digelansinus diversipes*	13.8
5	*Thaumetopoea pityocampa*	15.0
6	*Nephila senegalensis*	15.7

[a] In each example, the values for dimensions a and b are 9.44 and 6.95 ± 0.05, respectively.

[b] Data of Lucas and Rudall (3).

approximately 6.95 Å (b) along the fiber axis and a repeat of approximately 9.44 Å (a) corresponding to the hydrogen-bonded direction. However, in the c direction, corresponding to the separation of the pleated sheets, there is considerable difference among species such that the c value varies between 9.3 and 15.7 Å. Although there originally seemed to be a rather poor correlation between bulky-peptide content and chain spacing it is now becoming apparent that the structural classes do in fact reflect subtle differences in primary structure.

GROUP 1. The primary sequence of *Bombyx mori* fibroin is probably composed of three types of segments:

Type I. Segments of type I are of sequence

Gly-Ala-Gly-Ala-Gly-[Ser-Gly(Ala-Gly)$_n$]$_3$-Ser-Gly-Ala-Ala-Gly-Tyr

In this sequence n is usually 2 and has a mean value of 2.

Type II. Segments of type II are largely tetra- and octapeptides in which glycine alternates with other peptides, particularly alanine, tyrosine, and valine.

Type III. Segments of type III contain about 230 residues and include all the bulky and ionic side residues.

The characteristic x-ray pattern probably originates in crystalline regions of type I, there being a distinct axial period of 21 Å corresponding to the repeating hexapeptide Ser-Gly-Ala-Gly-Ala-Gly. A schematic diagram of the chain arrangement of *Bombyx mori* fibroin is shown in Fig. 10.1 (3).

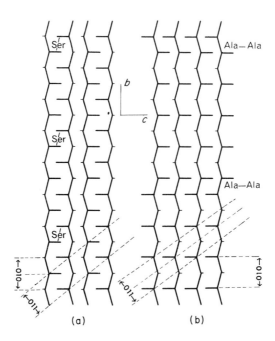

Fig. 10.1 (a) Chain arrangement of *Bombyx mori* fibroin with the sequence of hexa-peptides (Gly-Ala)$_2$ Gly-Ser. Reflection 010 is absent. Axial period at 21 Å. (b) Arrangement of chains composed of repeating sequences (Gly-Ala)$_n$Ala, such as may exist in fibroin from *Anaphe moloneyi*, where n is 4, 5, or 6. Reflection 010 is present. (Reprinted from Ref. 3, with permission of Plenum Press Publishing Corp.)

GROUPS 2A and B. These fibroins are composed to a large extent of alanine and glycine and probably have repeating sequences of the type -[(Gly-Ala)$_x$-Ala]$_n$ where x is possibly 4, 5, or 6. The appearance of an 010 reflection in the x-ray pattern is in accord with the proposed arrangement of chains shown in Fig. 10.1. Group 2b differs in equatorial intensities but this effect is not understood in terms of primary structure.

GROUP 3. This group has been found to contain long sequences very rich in alanyl residues and has an x-ray pattern similar to that of synthetic poly-L-alanine.

GROUP 4. This group of fibroins, obtained from certain species of South American sawflies, is characterized by a very high content (57%) of alanine and glutamic acid, the latter of which almost certainly arises from glutamine residues in the intact protein. Hydrolysis results in the conversion of gluta-mine to glutamic acid. There is only a small amount of glycine present, and the x-ray data are in accord with a structural unit based on a polydipeptide of form (Ala-Gln)$_n$.

GROUPS 5 AND 6. Sequential information for these categories is quite limited. Examples are not particularly common but there is a continuing trend toward an increased content of peptides with long side chains (group 1, 30%; group 5, 38%; group 6, 44%).

CROSS-β-FIBROINS

A good example of the cross-β structure is found in the egg stalk of the lacewing fly *Chrysopa flava* (4). The x-ray pattern of this material and a diagram of the proposed structure are shown in Figs. 10.2 and 3. The cross-β conformation is indicated by the presence of the 4.6 Å meridional reflection corresponding to the repeat along the axis in the hydrogen bond direction. An important characteristic of the primary structure is the presence of a high proportion of seryl residues (43%) along with glycine (25%) and alanine

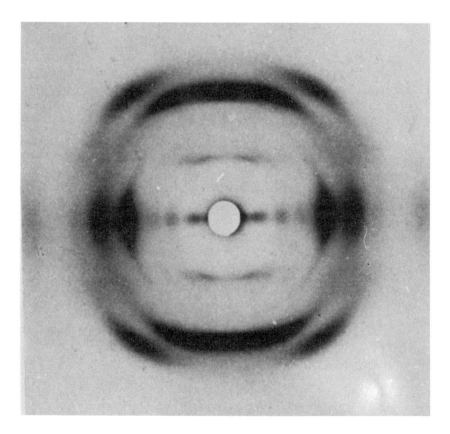

Fig. 10.2 The x-ray fiber pattern of *Chrysopa* egg stalk (Taken from Ref. 4, with permission.)

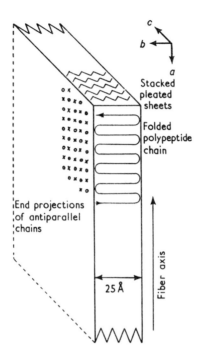

Fig. 10.3. Proposed cross-β structure of *Chrysopa* egg stalk silk fibroin. (Taken from Ref. 4, with permission.)

(21 %). The high serine content appears to be related to the folding process, and the polar residues are believed to occur predominately in the "bends" of the polypeptide chains. If this material is stretched to five or six times its original length then a "parallel" β structure of type 3 above is produced (4).

α-FIBROINS

The α structure is relatively unusual but is found in silks produced by bees, wasps, and ants. A β form may be produced on stretching. Detailed chemical analyses have yet to be made but high proportions of alanine, serine, and the dicarboxylic acids are present.

POLYGLYCINE II STRUCTURE IN SILK

The fibroin obtained from the Solomon seal sawfly shows an unusually high glycine content (66 %) with no other residue in excess of 8 %. Although this fibroin is normally amorphous, after steam treatment an x-ray powder pattern typical of the PGII structure may be obtained. With such a high proportion of glycine it seems possible that long runs are present which promote the PGII structure. However, recent studies with synthetic polytripeptides of the form (Gly-Gly-X)$_n$ also show PGII conformation, suggesting that polyglycine blocks do not necessarily have to be present.

COLLAGEN STRUCTURE IN SILK

Another interesting form of silk has been obtained from larvae of the gooseberry sawfly. This material has 6–9% proline and 20–27% glycine and gives a collagenlike x-ray fiber pattern. Since the the collagen pattern can arise only from polytripeptide chains (Gly-X-Y)$_n$ where X or Y is frequently proline, such sequences must be present in this fibroin (cross-β structure may also be present).

SYNTHETIC SILKLIKE BIOPOLYMERS

In general, the β-fibroins contain major amounts of glycine, alanine, and serine, the most usual percentages being 20–45%, 20–50%, and 10–40%, respectively. Thus, alternating copolymers of these amino acids might provide useful models for studying silk structure. One of the early models suggested for *Bombyx mori* fibroin was, for example, the polydipeptide of L-alanylglycine. This polymer has been synthesized (5) with a molecular weight of $\sim$ 7000 and, although it has a β conformation, the synthetic material was found to have a cross-β structure with $a = 9.42$ Å, $b = 6.95$ Å, and $c = 8.78$ Å. The low molecular weight may give rise to the difference in chain orientation, but on the other hand the presence of specific fold-blocking regions in the native protein could also be involved, as could the mode of biosynthesis.

More recently, different laboratories (6, 7) have reported physical studies of poly (Ala-Gly) of M_w 10,000–20,000, confirming that the cross-β structure is formed when the polymer is precipitated from strong solvents. However, a very unusual form is produced by precipitation from concentrated (aqueous) solutions of certain salts (e.g., lithium chloride). The unit cell dimensions of this second form have been reported as $a = 4.72$ Å, $b = 14.40$ Å, and $c = 9.6$ Å (chain axis). Circular dichroism and ORD studies (5) seem to reveal a totally new conformation for this biopolymer in solution, one which is not resolved at this time.

Other silklike biopolymers that have been prepared and at least partially characterized are poly (Ser-Gly), poly [Glu(OEt)Gly] (7), and poly (Glu-Ala) (8). All three may be obtained in the β form (cross-β for the first two). The acidic polydipeptide forms a β structure when precipitated as a salt from alkaline solution, but at neutral pH in aqueous solution it resembles the polyglutamic acid charged form.

The morphology and structure of the glycyl dipeptides have also been examined. The fibrous morphology of poly (Ala-Gly) and poly (Ser-Gly) in their cross-β forms is evident from the electron micrographs shown in Figs. 10.4A and B (9). These fibers appear to consist of primary structures $\sim$ 60 Å in thickness. Electron diffraction from these fibers reveals that the molecular axis lies predominantly flat on the substrate and perpendicular to the fiber axis. The molecules are undoubtedly superfolded into an antiparallel cross-β form, the fold period being 60 Å.

Infrared dichroism and CD of films of poly [Glu(OEt)Gly] indicate that the cross-β form is again the most prominent when the polymer is crystallized from strong solvents.(9)

Collagen

Collagen is the most common protein of connective tissue and forms a major part of the organic matrix of bones. Probably more is known about the sequence, conformation, and structure of collagen than of any other fibrous protein. Many biopolymer models have been synthesized which emulate some of its structural characteristics. Although collagens from different animals, both invertebrates and vertebrates, differ slightly in amino acid composition, the general structural features are the same.

Tropocollagen

The collagen molecule (tropocollagen) may be extracted in dilute acid, salt, or citrate solutions, and although molecular aggregates are generally obtained, the basic unit has a molecular weight of $\sim$ 300,000. Light scattering shows the molecule to be rodlike with a length of approximately 2800 Å. At temperatures in excess of about 40°C, the tropocollagen molecule dissociates to form three separate units of equal molecular weight. One unit, known as the α_2 chain, has an amino acid composition slightly different from that of the other two, the so-called α_1 chains. The melting of the ordered tropocollagen molecule to the randomized chains is known as the collagen/gelatin transition.

Two notable aspects of the amino acid composition of collagen (10, 11) (Table 10.3) are (a) the high ($\sim 33\%$) content of glycine and (b) the high imino acid (proline and hydroxyproline) content ($\sim 20\%$). Sequential analysis of the collagen chains has been greatly facilitated by the discovery that cyanogen bromide cleaves the chains into eight fragments (CB1–8) of molecular weights in the range 250–25,000. Thus a substantial portion of the α_1 chain has been sequenced (12) and is shown in Fig. 10.5. It can be seen that apart from the first few peptides occurring in the tail or "telopeptide" region,

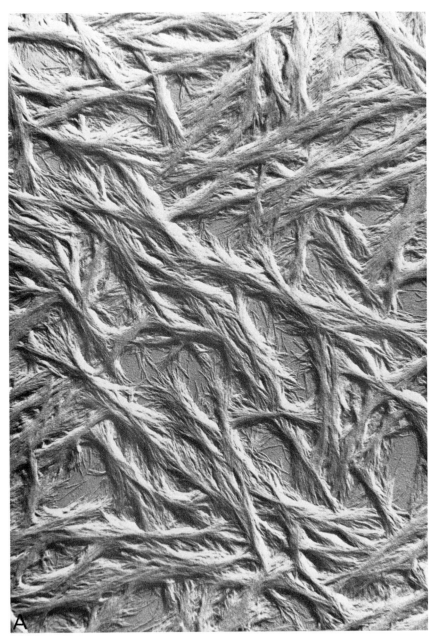

Fig. 10.4A. Electron micrograph of poly (Ala-Gly) film cast from formic acid (×19,600).

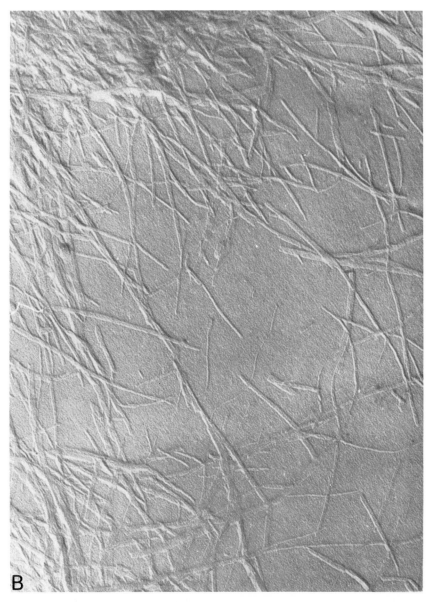

Fig. 10.4B. Fibers of poly (Ser-Gly) precipitated from formic acid with ethanol (×57,750).

TABLE 10.3

Comparison of the Amino Acid Content of Individual and Cross-linked Chains of Rat Tail and Human Skin Collagen

Amino acid	Rat tail tendon (10)					Human skin (11)					
	Orig.	α_1	β_2	β_1	α_2	Orig.	α_1	β_{11}	β_{12}	α_2	β_{22}
3-Hydroxyproline	4.2	4.1	3.9	3.8	3.8	1.1	0.8	1.0	1.0	0.9	1.0
4-Hydroxyproline	90	93	91	87	82	93	91	91	82	82	83
Aspartic acid	45	47	47	46	44	45	43	43	46	47	48
Threonine	19.9	19.7	20.1	19.8	19.8	17.5	16.5	16.3	17.9	19.2	19.3
Serine	43	41	42	43	43	35.6	36.8	36.7	35.2	35.1	34.1
Glutamic acid	72	74	74	70	68	73	77	76	72	68	68
Proline	122	129	126	119	115	128	135	136	123	120	118
Glycine	331	329	334	336	335	330	333	332	338	337	339
Alanine	107	110	111	106	103	110	115	116	112	105	104
Valine	22.9	19.2	19.8	25.7	30.4	24.4	20.5	20.6	28.8	33.3	31.1
Methionine	8.4	9.0	9.1	7.5	6.7	6.1	4.9	5.0	5.0	5.2	5.4
Isoleucine	9.6	6.3	6.3	11.6	15.1	9.5	6.6	6.5	11.6	14.8	13.7
Leucine	23.6	18.1	18.8	25.7	30.9	24.3	19.5	19.1	26.1	30.1	30.7
Tyrosine	3.9	3.7	3.3	3.3	3.9	2.8	2.1	2.0	3.8	4.6	4.6
Phenylalanine	11.9	11.9	11.3	11.2	11.3	12.0	12.3	12.5	12.3	11.7	12.2
Hydroxylysine	6.6	4.8	4.1	7.4	9.8	5.8	4.4	4.4	6.1	7.6	8.0
Lysine	26.9	30.1	28.7	24.3	21.0	26.9	30.0	30.1	25.2	21.6	21.3
Histidine	4.1	1.5	1.4	4.9	7.4	4.8	3.0	2.0	6.3	9.7	10.7
Arginine	50	49	48	49	51	51	50	51	49	51	50
Amide N	(40)	(39)	(40)	(41)	(41)	(36.9)	(37.9)	(37.2)	(44)	(45)	(45)

CB1   H-Gly -Tyr-Asp-Glu -Lys-Ser -Ala -Gly -Val -Ser -Val -Pro-Gly -Pro-Met

               1   5   10   15

CB2
   -Gly-Pro-Ser-Gly-Pro-Arg-Gly-Leu-Hypro-Gly-Pro-Hypro-Gly-Ala-Hypro-
               20   25   30
   -Gly-Pro-Gln-Gly-Phe-Gln -Gly-Pro-Hypro-Gly-Glu -Hypro-Gly-Glu-Hypro-
               35   40   45
   -Gly-Ala -Ser-Gly-Pro-Met-
               50

CB4
   -Gly -Pro-Arg-Gly -Pro-Hypro-Gly -Pro-Hypro-Gly -Lys-Asn-Gly -Asp-
               55   60   65
   -Asp-Gly -Glu -Ala -Gly -Lys-Pro-Gly -Arg-Hypro-Gly -Gln -Arg-Gly -Pro-
                  70   75   80
   -Hypro-Gly-Pro-Gln-Gly-Ala-Arg-Gly-Leu-Hypro-Gly-Thr-Ala-Gly-Leu-
                   85   90   95
   -Hypro-Gly-Met-

CB5
   -Hylyl - Gly - His -Arg-Gly -Phe-Ser -Gly -Leu-Asp-Gly -Ala -
   Glucosyl   100   105   110
   Galactosyl
   -Lys-Gly -Asn-Thr-Gly -Pro-Ala -Gly -Pro-Lys-Gly -Glu -Hypro-Gly -Ser-
                            120   125
   -Hypro-Gly -Glu -Asn-Gly -Ala -Hypro-Gly -Gln -Met-
               130   135

Fig. 10.5. The primary sequence of the N-terminal end of α_1 rat skin collagen. Note the linkage of a saccharide moiety at residue 99. The telopeptide region (residues 1–12) does not contain the Gly-X-Y tripeptide sequence. Cyanogen bromide cleavage between each Met-X peptide pair produces fragments labeled CB1, 2, 4, 5, which is the order of their emergence from a methyl cellulose column. Regions of the α_1 chain deficient in L-proline and L-hydroxyproline are underlined.

glycine occurs every third residue and the chain alternates in regions rich or poor in imino acid residues.

Electron micrographs of the tropocollagen molecules affirm that they are flexible rods about 2800 Å in length and, as a consequence of the three chain components, the molecule is expected to be of triple-strand structure. In view of the high proline and hydroxyproline content, the structure is unlikely to be α or β in conformation. x-Ray fiber diagrams are readily obtained, and a typical pattern is shown in Fig. 10.6A (13). The pattern is indeed incompatible with either the α or the β form.

When a large proportion of the peptide residues favors a certain conformation, it is not unreasonable to look for such a conformation in the protein.

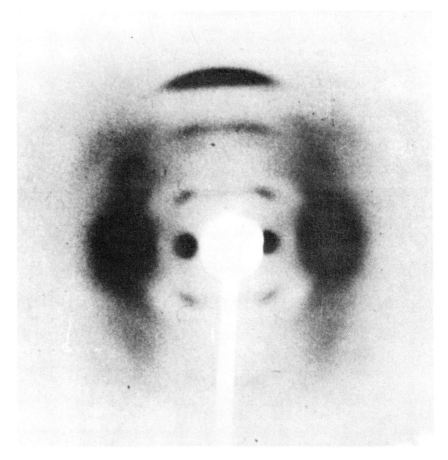

Fig. 10.6A. Inclined x-ray diffraction pattern of collagen. Note the flat bottom and the appreciable thickness of the spot on the 3 Å meridional layer. The measurement of the Bragg angle corresponding to the bottom edge of this gives the spacing of this layer line. [Taken from "Collagen," p. 113, ed. G. N. Ramachandran (1967), Academic Press, with permission.]

Since > 50% of the residues (glycine + proline + hydroxyproline) favor a left-hand polyglycine II or polyproline II conformation, it seems reasonable that this conformation occurs in the collagen structure.

A meridional reflection at 2.9 Å in the collagen pattern shows a peptide repeat distinctly different from that of other proteins; the decrease from the 3.1 Å repeat of polyproline II has been interpreted as resulting from supercoiling. The x-ray pattern arising from the "crystalline" regions of collagen is consistent with a supercoiled triple helix, each of the chains having a conformation resembling the PPII helix. The observed meridional reflection occurs on the tenth layer line, with strong intensity on the third and seventh layers, indicating a 10_3 helix. The repeat along a single chain is ten tripeptide units, i.e., $10 \times 3 \times 2.86$ Å $= 85.6$ Å, with a repeat for the triple helix of 28.6 Å [see Fig. 10.6B (14)]. It is now generally accepted that this structure is correct, although the relative arrangement of the three strands has been in some contention. Rich and Crick presented a model (15) in which the NH of residue 1 in chain I could hydrogen bond with the first or second carbonyl of chain II (see Fig. 10.7). These structures were referred to as the collagen I and collagen II structures, each of which has one hydrogen bond per tripeptide. Ramachandran and co-workers postulated a different one-bond model (16) and also a model containing two hydrogen bonds per tripeptide (13, 17). A comparison of the chain arrangements in these models is shown in Fig. 10.7A (15) and B (13).

Fig. 10.6B. A schematic drawing of the three-stranded helical structure of collagen. Detail is shown in one strand only. Note that each strand is itself a helical structure (small circle, glycine; large circle, proline or hydroxyproline). [Taken from *"Engström-Finean Biological Ultrastructure,"* by J. B. Finean (1967), Academic Press, with permission.]

More recently, Traub and co-workers have made an intensive study of a collagen-model polytripeptide $(Gly-Pro-Pro)_x$ and have deduced a somewhat different one-hydrogen-bond model, similar to collagen II but with the non-bonded carbonyls rotated slightly outward (18, 19). A model of the Traub helix is shown in Fig. 10.8. On the basis of these studies, it is now thought that the collagen molecule has one hydrogen bond per tripeptide directed between strands and a second via water bridging.

Some interesting spectroscopic features of collagen include an amide A band (N—H stretch) at 3330 cm^{-1}, indicating a rather longer than normal hydrogen bond (3.0–3.1 Å), and an ORD b_0 value of zero.

The x-ray diffraction from collagen fibers indicates a "crystallinity" of only 20–40%, and some interest has centered around the question of whether the tropocollagen molecule is uniform in structure. Until recently it was felt that misorientation or wrinkling of the tropocollagen molecule within the collagen fiber was the basis of the poor ordering. More recent electron microscope observation of the deformation of collagen fibers suggests (20), however, that there may be regions of different or amorphous structure in the tropocollagen molecule, possibly associated with the polar, non-proline-containing

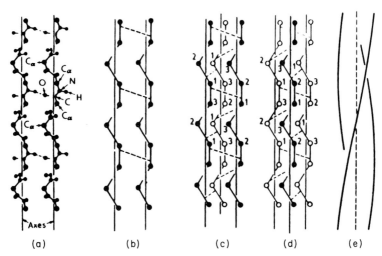

(a)   (b)   (c)   (d)   (e)

Fig. 10.7A. Chain backbone arrangements in polyglycine II and related collagen structures based on the single-hydrogen-bond model. (a) Two strands of polyglycine with black dots representing various atoms and dashed lines as hydrogen bonds. (b) Two chains of the polyglycine II lattice in which only the C_α atoms and hydrogen bonds are shown. (c) The third chain, shown with open circles, lies behind the two shown in (b) to make a collagen I arrangement. The numbers indicate the phasing of the residues on the poly-peptide chains. (d) The third chain is added in front of the two in (b). The chain in front is shown by solid lines. This gives rise to the collagen II model. (e) Solid lines represent the axes around which the polyglycine chains are coiled. These axes are gently coiled around each other in the collagen molecule.

Two-bonded structure					One-bonded structure				
C	A	B	C	A	C	A	B	C	A
...	...	...	...	...	...	...	...	...	...
N-6	N-8	N-7	N-6	N-8	N-6	N-8	N-7	N-6	N-8
O-6	O-8	O-7	O-6	O-8	O-6	O-8	O-7	O-6	O-8
N-5	N-7	N-6	N-5	N-7	N-5	N-7	N-6	N-5	N-7
O-5	O-7	O-6	O-5	O-7	O-5	O-7	O-6	O-5	O-7
N-4	N-6	N-5	N-4	N-6	N-4	N-6	N-5	N-4	N-6
O-4	O-6	O-5	O-4	O-6	O-4	O-6	O-5	O-4	O-6
N-3	N-5	N-4	N-3	N-5	N-3	N-5	N-4	N-3	N-5
O-3	O-5	O-4	O-3	O-5	O-3	O-5	O-4	O-3	O-5
N-2	N-4	N-3	N-2	N-4	N-2	N-4	N-3	N-2	N-4
O-2	O-4	O-3	O-2	O-4	O-2	O-4	O-3	O-2	O-4
N-1	N-3	N-2	N-1	N-3	N-1	N-3	N-2	N-1	N-3
O-1	O-3	O-2	O-1	O-3	O-1	O-3	O-2	O-1	O-3
...	...	...	...	...	...	...	...	...	...

Fig. 10.7B. The hydrogen bond sequence in the Ramachandran "two-bond" model compared with one-bond sequence. [Taken from A. Rich and F. H. C. Crick, *J. Mol. Biol.*, **3**, 492(1961), with permission.]

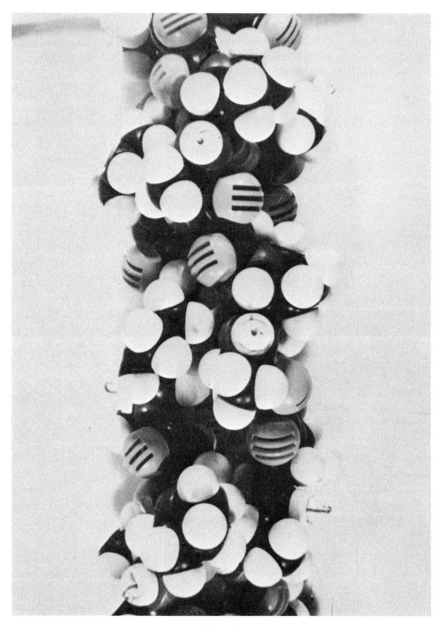

Fig. 10.8. Model of collagen triple helix (based on Gly-Pro-Ala sequence) in the conformation proposed by Traub and co-workers. (Taken from Ref. 19, with permission.)

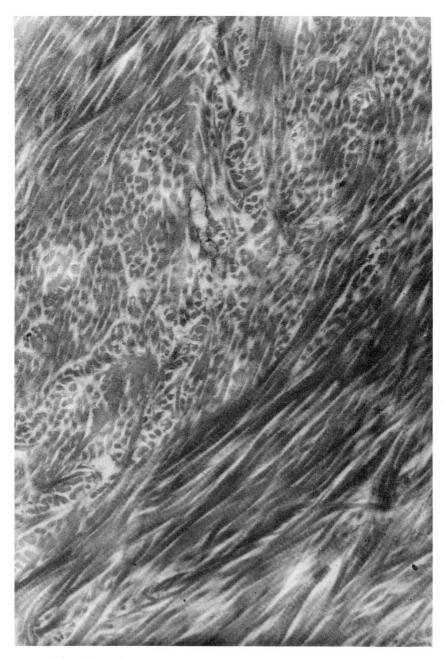

Fig. 10.9. A low-resolution micrograph of collagen fibers ($\times 20{,}000$).

regions. Although it has yet to be proven that the proline-rich and proline-poor regions "match" in the phasing of all three chains, a likely model for the tropocollagen molecule would be that containing regions of relatively uniform conformation, interspersed with "defect" regions containing many of the bulky, charged, polar residues.

Tertiary Structure

Tropocollagen molecules may be reconstituted into various forms of collagen fibers. Simple precipitation at pH 7 and staining reveals fibers resembling those collected from connective tissue. Figure 10.9 shows typical (native) collagen fibers stained with phosphotungstic acid (PTA). Typical is the 640 Å banding period, which is also observed in low-angle x-ray diffraction studies. High-resolution electron microscope studies reveal a number of different internal bands, which are given the symbols shown in Fig. 10.10 (21).

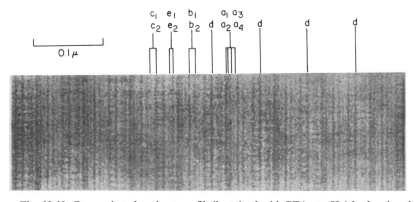

Fig. 10.10. Reconstituted native-type fibril, stained with PTA at pH 4.2, showing the axial period of ~ 700 Å, the characteristically "polarized" intraperiod band pattern, and the generally used band nomenclature for the native-type pattern ($\times 113,750$). (Taken from Ref. 13, by permission of A. J. Hodge and Academic Press, Inc.)

The relation between the position of the tropocollagen molecules and the structure of the fibers is clarified to some extent if the tropocollagen molecules are reconstituted in the presence of adenosine triphosphate (ATP). Under these conditions "blocks" of "segment long-spaced" (SLS) collagen approximately 2800 Å long [Fig. 10.11 (22)] are produced. They are believed to result from a parallel aggregation of the tropocollagen molecules, which themselves contain parallel peptide chains. On the other hand, reconstitution in the presence of glycoprotein produces "fibrous long-spaced" (FLS) collagen

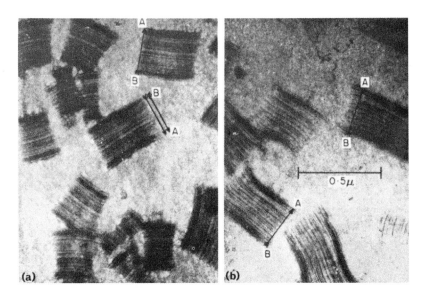

Fig. 10.11. Segments (i.e., monomeric SLS-type aggregates) obtained by addition of ATP (2 mg/ml) to an aged solution of calf skin collagen in 0.008 N acetic acid: (a) stained with PTA; (b) stained with cationic uranium. The orientation of the tropocollagen macromolecules is indicated by the arrows labeled A and B. Note the differences in the two band patterns. (Reprinted from Ref. 21, with permission of the National Academy of Sciences, U.S.)

[Fig. 10.12 (22)] in which the major staining period is again 2800 Å but the inner bands suggest an antiparallel array of tropocollagen molecules.

Since the staining patterns observed in SLS and FLS collagens are transverse to the molecular axis, it is reasonable to assume that the same is true in native and normal reconstituted collagens with their 640 Å period, and

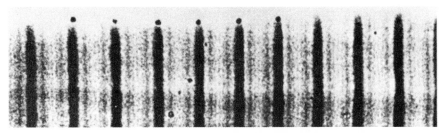

Fig. 10.12. An FLS-type fibril (produced by dialysis against distilled water) from an acid solution of calf skin collagen with added serum glycoprotein and stained with PTA. The symmetrical axial period of $\sim$2800 Å (indicated by dots) is somewhat less than the length of the collagen macromolecule. (Taken from Ref. 13, by permission of Prof. A. J. Hodge and Academic Press, Inc.)

consequently a "quarter-stagger" theory has been introduced. Negative staining of reconstituted collagen seems to indicate that there is a hole zone ~400 Å long adjacent to the end of the tropocollagen molecule. On this basis the molecules are presumed to be approximately 2800 Å long and to pack in the manner shown in Fig. 10.13 (23).

Most native collagen is not readily extractable into aqueous solution because of intra- and intermolecular cross-linking, which appears to develop as a function of age. Two α_1 chains may, for example, cross-link to form a β_1, α_1 and α_2 may form β_{12}, and higher orders of cross-linking may result in γ-collagens, etc. (Thus, the α-, β-, and γ-collagen species would have molecular weights in the ratio 1 : 2 : 3.) Bone collagen, for example, is a particularly insoluble form which is believed to be highly cross-linked. Artificial cross-linking can be induced by chemical treatments, e.g., with formaldehyde and glutaraldehyde, such processes being of considerable commercial significance in the leather industry.

The rate and energetics of the recombination of tropocollagen to form collagen fibers have been studied (24) and there seems to be little doubt that the intermolecular arrangement is controlled by ionic forces emanating from local areas.

The collagen molecule and fiber are strong, fairly inextensible, and rigid structures that have a predominantly mechanical function in mammalian tissue. The rigidity arises from the compact triple helix, but the suspicion remains that some chemical function may arise from the polar, less compact regions. Forces controlling the formation of triple helical structures have proven a fruitful field for synthetic biopolymer research. The synthetic analogs are explored in the following section

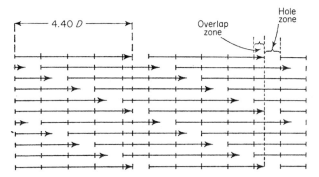

Fig. 10.13. A two-dimensional representation of the packing arrangement of tropocollagen macromolecules in the native-type fibril. Longitudinal displacement of nearest neighbors by a distance D results in formation of a fibril of period D, with each period comprising an overlap zone of $\sim 0.4\ D$ and a hole zone of $\sim 0.6\ D$. (Reprinted from Ref. 13, with permission of Prof A. J. Hodge and Academic Press, Inc.)

Biopolymer Models

PROLINE-CONTAINING BIOPOLYMERS

We have seen that the basic chain conformation of the tropocollagen molecule is believed to be closely related to polyproline II and, hence, much attention has been paid to polyproline itself. It has been shown in Chapter 2 that polyproline II is a "floppy" conformation which is readily distortable, but it nevertheless crystallizes in regular 3_1 helices.

In view of the known sequencing of the collagen molecule, attention was soon focused on synthetic polytripeptides of the form $(Gly-Pro-X)_n$ and $(Gly-X-Pro)_n$. A list of the reported structures of such materials is given in Table 10.4. In general, a fairly simple test is applied to determine whether a

TABLE 10.4

*Structure of Some Glycine- and Proline-Containing
Polytripeptides in the Solid State*

Polymer	Structure
$(Gly-Pro-Pro)_n$	Triple helix
$(Ala-Pro-Pro)_n$	Single helix, PPII conformation
$(Gly-Pro-Hypro)_n$	Triple helix
$(Gly-Pro-Ala)_n$	Triple helix
$(Gly-Pro-Ser)_n$	Triple helix
$(Gly-Pro-Gly)_n$	Single helix, PPII conformation
	Sheet structure
$(Gly-Hypro-Hypro)_n$	Triple helix
$(Gly-Pro-Lys)_n$	Triple helix
$(Gly-Ala-Pro)_n$	Triple helix
	PPII Sheet structure I
	PPII Sheet structure II

triple helical conformation is achieved. A 3.1 Å peptide repeat is taken as an indication of single-chain PPII helices, whereas 2.75–2.95 Å repeats are taken as the triple-stranded supercoil. It can be seen that all $(Gly-Pro-X)_n$ polymers, with the exception of $(Gly-Pro-Gly)_n$ and $(Gly-Pro-Leu)_n$*, have been found to form triple helices in the solid state. On the other hand, no compound in which glycine is absent [e.g., $(Ala-Pro-Pro)_n$] has ever been found to form a collagenlike triple helix. This information is totally in accord with the collagen

---

* It seems probable that racemization of the L-leucine prevented this biopolymer from achieving a collagenlike triple helix.

models presented in the previous section, since only glycyl tripeptides can be accommodated in the triple-strand structure.

It is interesting to note that whereas most residues X in the third position can exist in the PPII conformation, certain residues destabilize it (see, for example, results of theoretical calculations in Table 10.5). The melting characteristics of these synthetic biopolymers in aqueous solution are in accord with the view that (Gly-Pro-Pro)$_n$ forms the most stable triple helix (mp $> 80°$C, high molecular weight) and the introduction of other residues in position 3 reduces the melting temperature to an extent that is dependent upon the residue involved.

The biopolymers (Gly-X-Pro)$_n$ have been found to possess a number of different structures. Some are collagenlike, others have single strands in the PPII conformation, and yet others have a different chain conformation. Three such forms of poly (Gly-Ala-Pro) are reported in Table 10.4, in terms of their x-ray diffraction parameters.

The case of (Gly-Pro-Gly)$_n$ requires some special mention. Although it apparently fulfills the requirements for forming a triple-stranded collagenlike conformation, such a structure has not been observed. For this reason the

TABLE 10.5

Comparison of the Experimental and Theoretical[a] Melting Temperatures for Collagenlike Polytri- and Polyhexapeptides in Solution[b]

			T_m (°C)	
Polymer	Solvent	DP	Experimental	Theoretical
(Gly-Pro-Pro)$_n$	H_2O	22	67	57
	H_2O	100		78
	PD[c]	22	89	90
(Gly-Pro-Ala)$_n$	PD[c]	55	69	79
	H_2O	∞		25
(Gly-Pro-Ser)$_n$	PD[c]	42	51	56
	H_2O	∞		-20
(Gly-Ala-Pro)$_n$	H_2O	40		-60
(Gly-Ser-Pro)$_n$	H_2O	40		-160
(Gly-Pro-Gly)$_n$	H_2O	40		-142
(Gly-Ala-Pro-Gly-Pro-Pro)$_n$	H_2O	24	32	20
(Gly-Pro-Ala-Gly-Pro-Pro)$_n$	H_2O	26	49	40
(Gly-Ala-Ala-Gly-Pro-Pro)$_n$	H_2O	25	35	25
(Gly-Ala-Ala)$_n$	H_2O	40		-102

[a] Method of Hopfinger.

[b] Taken from F. R. Brown, A. J. Hopfinger, and E. R. Blout, *J. Mol, Biol.* **63**, 114(1972)

[c] 1:3 Propanediol.

biopolymer has been studied extensively both theoretically and structurally. It appears that the presence of glycine in the first and third positions of the polytripeptide allows a particularly favorable hydrogen-bonded sheet structure. In the experimentally determined structure the chains have the polyproline II conformation, although the theoretical study suggests the possibility of an unusual zig-zag chain conformation. The chain arrangement in the sheet structure in shown in Fig. 10.14.

An interesting series of rational structures can thus be made up for polyglycine, poly (Gly-Pro-Gly), poly (Gly-Pro-X), and polyproline, as shown in Fig. 10.15. Although it is readily understandable that poly (Gly-Pro-X) (X is proline, hydroxyproline, or a fairly bulky nonpolar group) should form clusters of three, the underlying energetic driving force to supercoiling has not yet been fully explored.

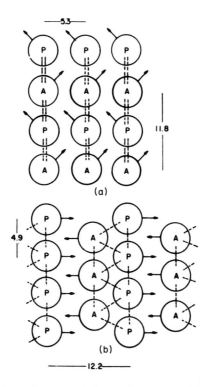

Fig. 10.14. A schematic representation of the two possible crystal structures of (Gly-Pro-Gly)$_n$. (a) Unit cell: $a = 5.3$, $b = 11.8$, $c = 9.3$ Å; (b) unit cell: $a = 4.9$, $b = 12.2$, $c = 9.3$ Å. Dashed lines represent the hydrogen bonds and the arrows show the orientation of the proline rings. [From A. J. Hopfinger and A. G. Walton, *J. Macromol. Sci. Phys.* **B4**, 185(1970), by courtesy of Marcel Dekker, Inc.]

Fig. 10.15. The packing of various proline- and glycine-containing polymers. Long heavy protrusions represent proline rings; light lines represent hydrogen bonds. (After J. Engel, private communication.)

The optical rotatory and circular dichroic properties of the synthetic biopolymers and collagen also form a rather interesting series. As can be seen from Table 10.6 (25) a decreasing proline content for poly proline helices in the order poly (Gly-Pro-Pro), poly (Gly-Pro-Ala), poly (Gly-Ala-Gly), causes an increasing shift in the CD curves. Collagen, with a proline content of about 20%, falls between the last two in curve position.

The question of supercoiling and antiparallel chains also bears further

TABLE 10.6

Circular Dichroism Data for Polypeptides with $\sim 3_1$ *Helical Conformation*[a]

λ	(Pro)$_n$ and (Hypro)$_n$	(Pro-Pro-Gly)$_n$	(Pro-Ala-Gly)$_n$	Collagen	(Ala-Gly-Gly)$_n$
Maximum	226	222	220	220	213
Crossover	219	213	203	203	202
Minimum	206	201	197	196	192

[a] Values indicate wavelength in millimicrons.

examination. Collagens (with the exception of *Ascaris* collagen) are believed to possess parallel peptide chain structures, based on enzyme cleavage studies. It is known, however, that polyglycine and poly-L-proline II can have folded antiparallel structures (from the morphological evidence in Chapter 4). The crystal structure of poly (Gly-Pro-Gly) also requires antiparallel chains. It would perhaps be surprising if the (Gly-Pro-X)$_n$ polymers could not be crystallized in the antiparallel chain array. Indeed evidence was presented in Chapter 4 that (Gly-Pro-Pro)$_n$ can form antiparallel triple-stranded helices. It does not however follow that (Gly-Pro-Pro)$_n$ always crystallizes with an antiparallel arrangement of chains. Arguments have been made (on the basis of x-ray diffraction and solution properties) that it often forms a parallel array (18). Evidently such a question is highly relevant to the solution of the collagen structure, since the arrangement and conformation of chains in (Gly-Pro-Pro)$_n$ are thought to be identical with those in the crystalline regions of collagen. An antiparallel chain model for collagen can in fact be built and has been described in the literature (26). It may also be the basis of the *Ascaris* collagen structure.

It is fairly clear that polymers in which 66% of the residues are proline, e.g., (Gly-Pro-Pro)$_n$, or 33% are proline, e.g., (Gly-Pro-X)$_n$, readily, form triple helices in the solid state, which are presumably uniform in conformation (relatively poor x-ray patterns are generally obtained). However, (Gly-X-Pro)$_n$ and (Gly-X-Hypro)$_n$ are somewhat neutral in their triple-helix-forming powers. It is of some interest then to inquire whether the collagen chains, containing $\sim 20\%$ imino acids adopt a uniform conformation. In an effort to explore the effect of nonuniform polytripeptides, a number of polyhexa-peptides have been synthesized and characterized. Table 9.10 lists these polyhexapeptides and some of their physical characteristics. It can be seen that all of the examples show triple-helical character, and it is evident that the Gly-Pro-Pro sequence, which "wants" to be in the polyproline II conformation and triple helical, prevails over Gly-Ala-Ala, which would "prefer" to be β sheet or α helical. Furthermore, there are no signs of polymorphism for the [(Gly-X-Pro) (Gly-Pro-X)]$_n$ polymers. These properties can be quite well understood when it is realized that the PPII conformation is not unfavorable for most α and β formers whereas the α helix and β sheet are untenable for proline residues. However, the effect of bulky, charged amino acids in runs in a proline-containing polytripeptide, has not yet been tested.

Non-Proline-Containing Collagen Models

The peculiarities of the proline residue are clearly responsible for many of the properties of the collagen triple helix. The low "crystallinity," the unusual, localized deformation properties, and the presence of runs of non-proline-containing residues raise the possibility of regions of different conformation. Almost all attention in the synthesis of collagen models has focused on the proline-containing polytri- or polyhexapeptides.

Recently, several biopolymers of the type $(Gly-X-Y)_n$ have been synthesized, where X and Y are glycine, alanine, or esters or derivatives of glutamic acid, and where X and Y may be the same. Of these materials, relatively few have been subjected to physical characterization. Three exceptions are $(Gly-Gly-Ala)_n$, $(Gly-Ala-Ala)_n$, and $[Gly-Ala-Glu(OEt)]_n$. The first of these may take on the PGII, cross-β, or true random forms and has been mentioned previously as a silk fibroin model. Poly $(Gly-Ala-Ala)$ may also take on two conformations, the α helix and a β-sheet form. The third polytripeptide, poly $[Gly-Ala-Glu(OEt)]$ is probably the most relevant because of its bulky glutamic ester group. This material was also mention in Chapter 4. It crystallizes rather easily into a superfolded β sheet, no other conformations being observed. Two other glycine-led polytripeptides have been characterized in a preliminary manner. Poly $[Gly(Glu(OEt))_2]$ has been shown by x-ray and infrared measurements to be in the β form and, similarly, poly $(Gly-Ala-Phe)$ has been found to be in β form by infrared measurements. No information is available concerning the detailed nature of the conformations, which are probably antiparallel superfolded β sheets.

Although the "neutral" glycine-containing polytripeptides are predominantly of β form it is unlikely that such a conformation can contribute to the structure of collagen; nonetheless, collagenlike and β structures coexist in one of the silk fibroins, as has been mentioned. Charged species such as polyglutamic acid appear to favor left-hand helices with a distorted PGII or PPII conformation, as shown in Chapter 2. It may be, therefore, that charged polytripeptides containing glycine, cause local perturbations and strain in the collagen molecule. The role of such regions in the molecule is open to question. Apart from helping direct the tertiary structure, it may be that the response of the charged regions and the varying strain induced as a response to electrolytes in solution may be related to the calcification process.

Elastin

As its name indicates, elastin is an elastic protein, generally present in tissues where elastic function is necessary, e.g., aorta, blood vessels, lung tissue, and ligaments. Some confusion in the field of elastic studies has arisen because of the use of heterogeneous material. Most work seems to have been performed on material that probably contains two components, the minor component (5–10%) of which is fibrous and the major component, amorphous. The amino acid composition of these two components is shown in

Table 10.7. Very little work has been performed on these individual components, although the nonfibrous form is probably that which should truly be termed elastin. For the purposes of this chapter we shall refer to elastic tissue, recognizing that it probably contains the two components.

Perhaps the most remarkable feature of the chemical composition is that the small nonpolar residues glycine, alanine, and valine make up 50–70% of the protein and very few polar residues are present. Approximately one-third of the residues are glycine, which is strongly reminiscent of collagen, but the elastic properties and resistance to attack by collagenase are decidedly uncollagenlike.

The x-ray patterns from elastic tissue are generally poor (27), with four diffuse rings at 1.1, 2.2, 4.4, and 8.9 Å. Although oriented patterns are very

TABLE 10.7

Amino Acid Analysis of the Amorphous and Fibrillar Portions of Elastic Tissue and the Precursor (Tropoelastin) Phase[a]

Amino acid	Elastin	Microfibrils	Tropoelastin
Gly	324	110	334
Hypro	26	154	39
Cationic residues (Asp, Glu)	21	228	21
Anionic residues (His, Lys, Arg)	13	105	55
Nonpolar residues (Pro, Ala, Val, Met, Leu, Ile, Phe, Try)	595	356	541
Cys/2	4	48	0

[a] Taken from R. Roth and P. Bornstein, *Sci. Amer.* **224**, No. 6, 44 (1971) and *J. Cell Biol.* **40**, 366 (1969).

difficult to obtain, the 4.4 and 8.9 Å reflections appear to be on the equator. Thus recent measurements seem to preclude the presence of a meridional 2.8–3.0 Å reflection, which might be indicative of a collagenlike structure. Elastin may readily be stretched 100% or more; however, the x-ray pattern does not sharpen with such a procedure, with the possible exception of the 4.4 Å reflection. Electron micrographs of elastic tissue show a complicated network of fibers, the smallest of which seem to be about 100 Å in width [see Figs. 10.16 (28)]. Some intertwining can be observed, but on the whole the fibers seem remarkably lamellar in structure.

The network arrangement is highly suggestive of cross-linking but evidently, from the amino acid analysis data in Table 10.7, this does not occur through cystine, which is essentially absent. Instead, the cross-linking

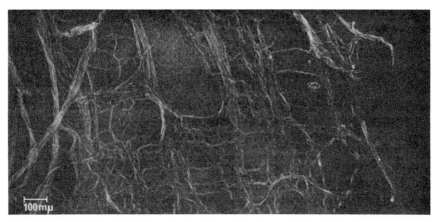

Fig. 10.16. Electron micrograph showing the fibrillar structure of an elastic matrix from bovine ligamentum nuchae. The elastic fibers were treated with dilute sodium hydroxide at 98°C for 1 hr, washed with hot water, and disintegrated ultrasonically at 40 kc/sec. Experimental details are given in the text. The specimen has been shadowed with chromium at 20°C. (Taken from Ref. 28, by permission of the American Chemical Society.)

is believed to occur through a sequence which converts lysine side chains to an aldehyde, four side chains condensing to form desmosine or isodesmosine (29) [see Fig. 10.17 (30)]. The conversion of lysine to the desmosines has been followed during maturation and aging of chickens, the results being presented in Table 10.8 (31).

Presumably, these cross-links give elastic tissue its rubberlike qualities. Assuming only 1 % cross-linking, then an average of one hundred peptide units would lie between each cross-link, giving a molecular weight of ~ 10,000. Diffusion measurements suggest that cross-linking of one form or another is rather extensive, giving only 10–20 peptide units between cross-links. Such a figure is, however, difficult to visualize in terms of the large fraction of inert amino acids.

Attempts at high-resolution electron microscopy have to some extent been hindered by the fact that elastic tissue has such a low polar amino acid content that it does not stain well. However, efforts in this direction, combined with a number of other techniques, have led Partridge and co-workers (30) to a picture of elastin that is rather different from the fibrous network described above. They feel that elastin behaves essentially as a three-dimensional gel into which 50 Å subunits are built, each containing folded chains (1) (see Figs. 10.18A and B). This model is consistent with current views of the nature of the major component of elastic tissue, i.e., true elastin.

Elastic tissue in its native form is cross-linked, insoluble, and chemically intractable. Precursor elastin (preelastin) may be extracted from young lathyritic animals, and some effort has been made to obtain sequential infor-

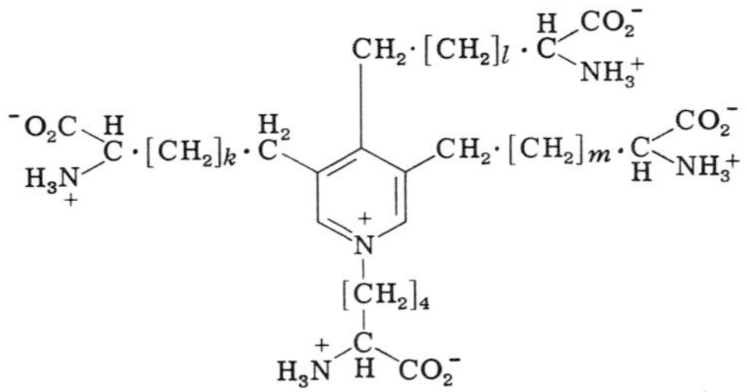

($k = m = 1,\ l = 2$)
Desmosine
(I)

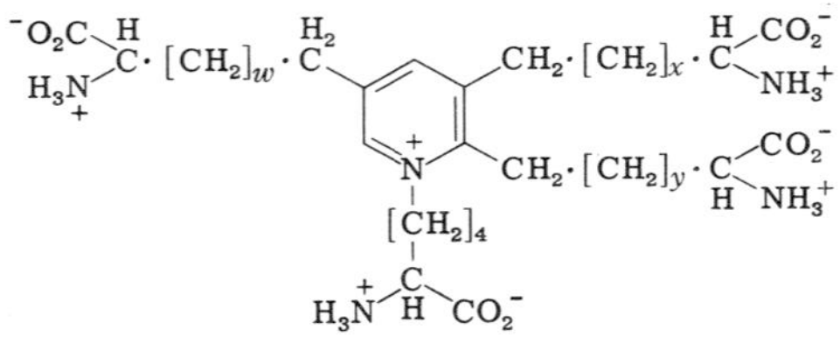

($w = x = 1,\ y = 2$)
Isodesmosine
(II)

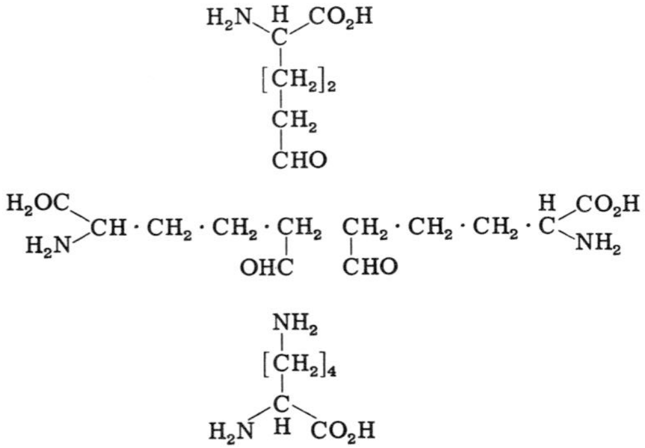

(III)

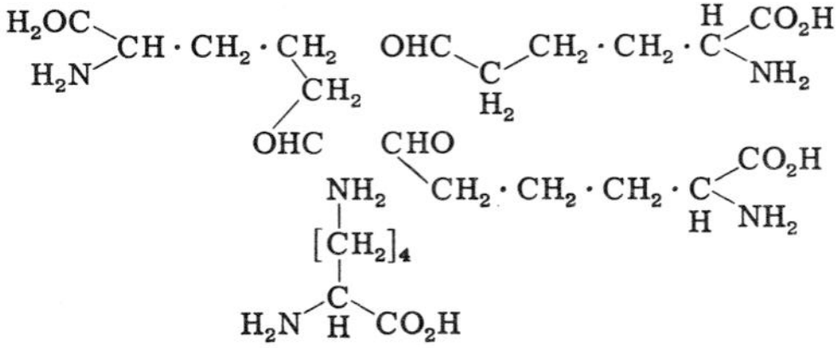

(IV)

Fig. 10.17. Hypothetical scheme of biosynthesis of desmosines; III and IV represent the probable arrangement of intermediates.

TABLE 10.8[a]

Amino Acid Composition of Elastin from Chicken Aortas

Amino acid	Embryo (12-day)	Embryo (16-day)	Embryo (20-day)	Chick (3-week)	Chick (4-week)	Chicken (1-year)
Hydroxyproline	24	25	22	23	22	23
Aspartic acid	1.9	1.9	1.8	1.9	1.9	1.8
Threonine	4.2	3.4	3.6	3.1	3.5	4.6
Serine	5.4	4.1	3.2	5.1	5.7	4.1
Glutamic acid	12	12	12	12	11	12
Proline	122	123	124	128	126	124
Glycine	352	352	351	352	352	352
Alanine	172	176	180	175	177	177
Cystine/2	0.5	0.5	0.6	0.4	0.3	0.6
Valine	177	177	177	176	176	174
Isoleucine	19	19	19	19	20	20
Leucine	62	59	56	58	57	58
Tyrosine	11	11	12	12	12	12
Phenylalanine	22	22	22	22	22	22
Desmosine/4[b]	4.3	5.7	6.7	6.8	7.1	10.9
Lysine	5.7	4.6	3.9	3.6	3.5	1.6
Histidine	<0.2	<0.2	<0.2	<0.2	<0.2	<0.2
Arginine	4.9	4.2	5.6	4.5	4.1	4.5
Amide N	(37.7)	(21.4)	(25.1)	(25.4)	(21.3)	(19.6)

[a] Taken from Ref. 1, with permission.
[b] Desmosine plus isodesmosine. The term desmosine/4 indicates that each molecule has four amino groups that can react with ninhydrin.

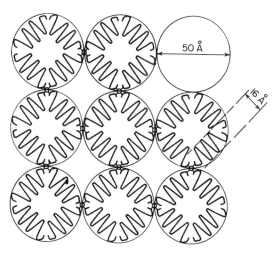

Fig. 10.18A. Two-dimensional representation of a model for the structure of elastin incorporating measurements from gel filtration experiments. (From Ref. 1, with permission.)

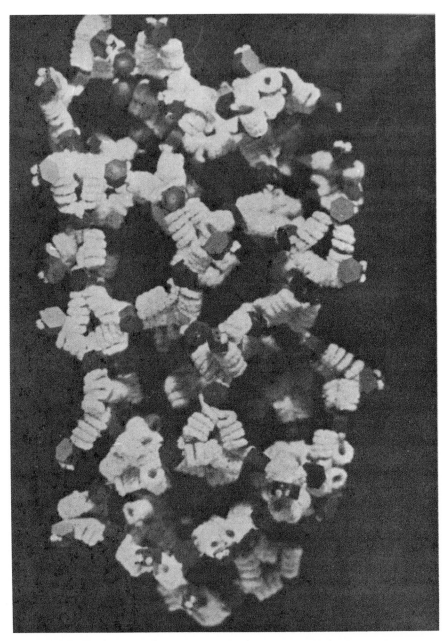

Fig. 10.18B. Photograph of a three-dimensional model constructed on the basis of Fig. 10.18A. (From Ref. 1.) [Fig. 10.18A and B taken from S. M. Partridge in "Symposium on Fibrous Proteins," ed. W. G. Crewther, p. 228 (1968), by permission of Plenum Press.]

mation for this soluble form of elastin. It has been reported (32) that two types of sequence occur regularly, those containing Ala_2-Lys or Ala_3-Lys peptides and tetrapeptides (33) of the form Gly-Gly-Val-Pro. The lysine-containing peptides may occur in local α-helical regions, which would explain how they enter into juxtaposition prior to forming half of the desmosine or isodesmosine cross-links. It is notable, however, that intact elastin contains, at best, a very small ($<10\%$) content of α helix. The tetrapeptide sequences could occur in regions of PGII conformation but a detailed analysis of the circular dichroism spectrum is difficult because of the intractable nature of elastin and the presence of the chromophoric desmosine and isodesmosine groups.

Resilin

Resilin is contained in the rubberlike exoskeleton of many insects and in many ways has properties similar to those of elastin. Like elastin, about $\frac{1}{3}$ of its residues are glycine and about 55% have small side chains, i.e., glycine, alanine, and serine [Table 10.9 (34)]. Unlike elastin, however, it has a fairly

TABLE 10.9[a]

Amino Acid Composition of Resilin

Amino acid	Content
Aspartic acid	102.0
Threonine	29.6
Serine	78.6
Glutamic acid	50.4
Proline	79.4
Glycine	376.0
Alanine	111.0
Valine	25.6
Methionine	Nil
Isoleucine	20.4
Leucine	25.6
Tyrosine	29.2
Phenylalanine	27.4
Lysine	4.9
Histidine	6.5
Arginine	33.6
(NH_3)	69.2

[a] Taken from Ref. 1, with permission.

large proportion of polar amino acids ($\sim 25\%$). The material is highly cross-linked with about 100 residues or so between cross-links. The chains themselves appear to be random in conformation, and no x-ray information has yet been obtained.

As we have seen in the previous section, elastic tissue contains virtually no cystine, and the cross-links are formed by a unique process involving desmosine or isodesmosine. Resilin also has no detectable cystine nor, for that matter, any desmosine. Evidence is now emerging (35) that dityrosine and trityrosine bridges are involved in the cross-links. This interesting piece of information suggests that nature reserves cystine for forming inter- and intrachain cross-links in predominantly globular proteins and uses specialized mechanisms to produce elastic fibers.

The Keratins

The keratins are fibrous proteins derived from hair, wool, feathers, etc., and may, on the basis of x-ray diffraction data, be divided into two general categories, the α- and β-keratins. Chemical analyses indicate that no particular amino acid predominates [Table 10.10 (1)] but that there is a high content of polar residues, cystine, and proline. In view of the presence of the last two, it is perhaps surprising that the keratins give such "crystalline" x-ray fiber patterns. However, the heavy cross-linking, presumably derived from S—S bridges, renders keratins in the native state difficult to characterize. There are essentially four methods of preparation for physical characterization:

Keratohyalin Granules

Extracted from the skin of newborn rats, this material contains no filaments, sulfhydryl groups, or disulfide bridges. The keratohyalin may contain a soluble precursor phase. The intact epidermis of mammals, reptiles, birds, and some fish gives wide-angle x-ray diffraction patterns similar to that from α-keratin.

Prekeratin

During formation of fibrous horny material, the fibrous protein often becomes insolubilized by cross-linking to an amorphous matrix. Some progress has been made in extracting the protein prior to cross-linking by

TABLE 10.10

Amino Acid Composition of Keratins and Modified Keratins[a]

Amino acid	Whole wool	α-Keratose 1	SCMKA-1	SCMKA-2	γ-Keratose 1	SCMKB-1	SCMKB-2
Cystine/2	106	74.4	50.5	62.9	197	237	215
Aspartic acid	70.5	80.6	98.7	92.9	28.5	7.6	10.8
Threonine	61.7	48.0	42.1	47.6	102	114	108
Serine	112	90.3	70.2	85.7	134.8	148	154
Glutamic acid	109	131	201	157	75.1	98.3	114
Proline	85.2	42.7	24.8	33.9	133	124	109
Glycine	84.8	107	58.4	65.9	64.1	63.3	88.4
Alanine	61.4	56.3	68.6	71.9	30.9	30.3	34.1
Valine	39.0	58.0	57.6	62.8	58.2	42.2	40.0
Methionine	4.7	4.6	5.4	4.9	Trace	nd[b]	nd[b]
Isoleucine	24.6	31.5	35.8	36.8	30.2	27.2	41.4
Leucine	71.2	93.3	121	104	38.9	18.3	20.0
Tyrosine	38.6	43.3	31.1	32.9	16.1	20.9	19.0
Phenylalanine	26.7	31.5	23.2	25.9	20.6	7.5	13.7
Lysine	27.3	28.6	39.8	31.6	6.0	4.8	0
Histidine	7.1	6.1	6.0	6.0	7.0	5.7	0
Arginine	69.6	72.2	66.2	77.0	58.2	50.7	32.6
(NH$_3$)		(121)	(122)	(131)	(112)	(113)	(120)

[a] From Ref. 1, with permission.
[b] Not detected.

solubilization in acid citrate. This prekeratin can be precipitated as fibers in $NaHCO_3$ solution or by extrusion. The molecular weight in solution is 640,000.

Keratoses

Products formed from intact keratins by oxidation, which breaks the disulfide linkages, are partly soluble. The α-keratoses, give an α-helical wide-angle x-ray diffraction pattern and have molecular weights in the 40,000–80,000 range. The γ-keratoses give poorly defined x-ray patterns and have lower molecular weights (10,000–30,000).

Kerateines

S-Carboxymethylcysteine kerateines (SCMK) can be formed by reduction of the disulfide bonds in mature keratins. Two fractions precipitated from solution, known as SCMKA and SCMKB, respectively, are found to have low and high sulfur contents, respectively. The SCMKA has a molecular weight of 60,000–70,000, is low in proline residues, and has about 50% α-helix content (ORD). The polypeptide chains can be split into two fragments, one part α helical, the other nonhelical. The SCMKB has a molecular weight of 22,000–28,000, has a high proline content and essentially no α-helix content.

STRUCTURE OF α-KERATINS

The principal wide-angle reflections of α-keratin are given in Table 10.11 (36). The α form has a fiber axis repeat of 5.1 Å and a chain diameter of 9.8 Å. Appearance of the 5.1 Å reflection on the meridian suggests the coiled-coil form. The low-angle x-ray pattern of highly oriented wool keratin is shown in Fig. 10.19 (37). The equatorial reflections at 77, 41, and 26 Å are interpreted as arising from the packing of rods with a diameter of about 70 Å in an amorphous matrix.

Although little is known of the sequential nature of the peptides in α-keratins, certain structural features of α-helix coiled coils must impose severe restrictions on the underlying sequence. First, the geometry is such that every residue has the same environment as the seventh one preceding it and the seventh one following it. Furthermore, the residues lying adjacent to each other on neighboring helices, i.e., on the "inside" of the coiled coil, are likely to be apolar, whereas those on the "outside" are likely to be polar such that favorable interactions with surrounding solvent can be attained. On the basis of the previous assumptions, the sequence of the helical portions of keratins might be expected to show a periodic distribution of residues with

TABLE 10.11

Principal Wide-Angle Reflections of α-Keratin

Reflection	Spacing (Å)	Intensity[a]
Equator	27	s
	9.8	vs
	3.5	vw
Meridian (first layer)	10.3	vw
(second layer)	5.1	vvs
	3.9	s
	3.4	w–m
	3.0	m–s
	2.5	w–m
	1.48	m–s
First layer (nonmeridian)	4.5	w
Second layer (nonmeridian)	4.1	m
Third layer (nonmeridian)	3.0	w

[a] Abbreviations: v, very; w, weak; m, medium; s, strong.

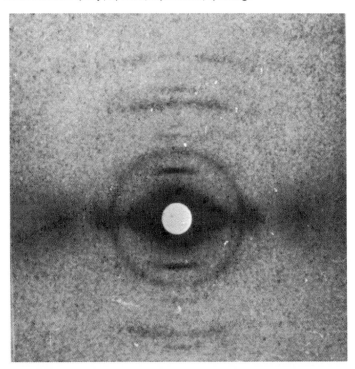

Fig. 10.19. An x-ray fibre diagram of wool keratin. (Taken from Ref. 35, with permission.)

polar residues in positions 0, 3, and 4 and nonpolar residues in positions 2 and 5. The limited data currently available (38, 39) can indeed be fitted to such a pattern, as shown in Fig. 10.20. It would also be expected that adjacent polar residues would not contain side chains of the same charge because disruptive electrostatic repulsion would occur. Again, the general features of this hypothesis are borne out in the incomplete sequential analyses.

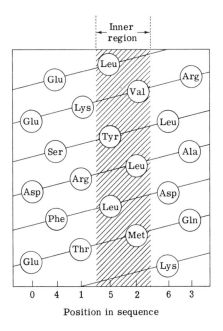

Fig. 10.20. Environment of residues in one of the component α-helices of a two-strand rope structure. Residues 2 and 5 are situated in the interior of the rope and are likely to be hydrophobic. Charged residues would be expected to preponderate in position 0. Hydrophobic and charged residues would therefore be expected to be periodically distributed in the sequence. The sequence shown was obtained from wool. (Kindly supplied as a preprint of Ref. 42, by Dr. R. D. B. Fraser, and reprinted by permission of John Wiley and Sons, Publishers.)

It is now possible, on the basis of the preceding assumptions, to understand how high contents of polar residues can be accommodated in α helices, provided that approximately the same concentrations of positive and negative side groups are achieved. Indeed, we shall see that almost all α-helix coiled-coil structures do contain high polar amino acid content. Evidently, even relatively short sections of each chain must be essentially neutral in ionic character; otherwise the extended coil forms of poly-L-glutamic acid and poly-L-lysine would be produced.

Electron micrographs of cross sections of wool keratin (40) indicate the presence of microfibrils of about 70 Å diameter spaced about 100 Å apart, as shown in Fig. 10.21. The substructure appears to consist of a 9 + 2 arrangement of the triple-strand coiled coils, as shown in Fig. 10.22 (41). There is, however, continuing discussion on the validity of this model. It is interesting to note that the microfibrils are somewhat less than 1 μ in length, which would give a molecular weight of $\sim$ 600,000 to each polypeptide chain based on the α-helix model. This value is in fairly good agreement with that noted previously for prekeratin.

The most recent analysis of electron micrographs (42) of porcupine quill α-keratin indicates that the two central coiled coils may result from imaging artifacts and that the combination of x-ray and electron microscope evidence may best be represented by a core structure, as shown in Fig. 10.23 (42). The x-ray pattern shows a near-equatorial layer line spacing of 70–85 Å, implying a pitch of 140–170 Å for a two-strand rope, or 210–255 Å for a three-strand rope. The lateral distribution of intensity in this region is very similar to that for dried molluscan catch muscle. Since the latter is believed to contain two-strand supercoils, it is suggested by analogy that this is also the structure in α-keratin.

The α-keratins may be converted to a β form by stretching in steam, aqueous alkali, hot water, etc. The tabulated data for the stretched α forms are shown in Table 10.12 (36).

TABLE 10.12

x-Ray Data for the β Form Produced by Stretching α-Keratin

Reflection	Spacing (Å)	Intensity[a]
Equator	9.7	s
	4.65	vs
	2.4	w
Meridian (second layer)	3.33	s
(third layer)	2.2	w
First layer (nonmeridian)	4.7	m
	3.75	s
	2.2	vw
Second layer (nonmeridian)	2.7	w
Third layer (nonmeridian)	2.0	w

[a] Abbreviations: v, very; w, weak; m, medium; s, strong.

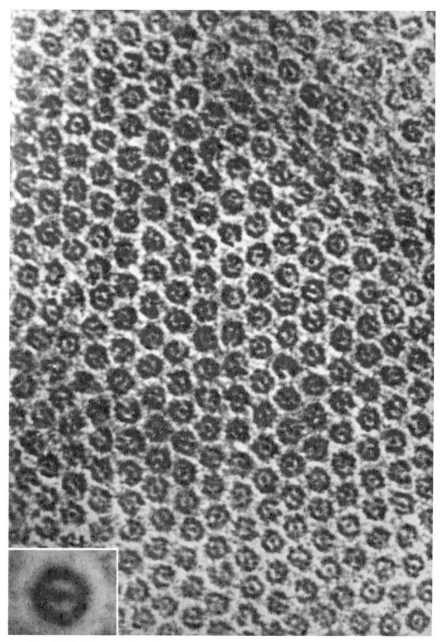

Fig. 10.21. Cross section of Australian fine merino wool. The insert is an enlargement of a microfibril which suggests a group of nine outer and two inner protofibrils (× ~ 500,000). (G. E. Rogers' original micrograph appearing in Ref. 1, p. 337.)

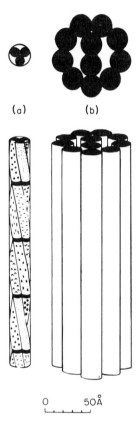

(a)   (b)

0   50Å

Fig. 10.22. Early model of the molecular structure of α-keratin in wool. (a) Protofibril consisting of three α helices coiled into a rope. Within the axial repeat of 200 Å, three similar but not identical subunits are shown. Between each subunit there is a major interruption in electron density. (b) Microfibril containing eleven protofibrils, each consisting of a three-strand rope. [Taken from R. D. B. Fraser, T. P. Macrae, and G. E. Rogers, *Nature*, **193**, 1054(1962), with permission of MacMillan and Co., Ltd.]

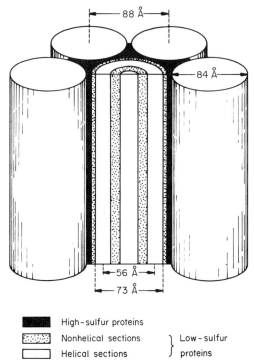

Fig. 10.23. Newer model of α-keratin based on a combination of x-ray diffraction and electron microscopy of dry porcupine quill. The spacing of the innermost equatorial x-ray reflection (75 Å) leads to an estimate of 86½ Å for the center-to-center distance between microfibrils. The microfibril "diameter" estimated from x-ray data is 72 Å, but this is less than that required to accomodate all the low-sulfur protein (82 Å). (Taken from Ref. 42, with permission; see caption to Fig. 10.20.)

THE β-KERATINS

Feather keratins have an amino acid composition apparently similar to that of the α-keratins, but they show a typical well-ordered β pattern by x-ray diffraction. The pattern is much more detailed than that produced by stretching α-keratins.

Feather keratin can be dispersed in strong detergent solutions to give particles with a molecular weight of 40,000. More recently, particles with a molecular weight of approximately 10,000 have been obtained; these should probably be regarded as the basic units. On this basis there are only about 100 peptides per chain which must be ~300 Å long.

Electron micrographs of cross sections of feather keratin reveal 30 Å diameter microfibrils spaced approximately 35 Å apart. Attempts to resolve the detailed structure have centered around efforts to reconcile the x-ray and ultrastructural observations. The x-ray pattern reveals a unit cell with $a = 34$ Å, b (fiber axis) $= 95$ Å, and $c = 4.7$ Å.

Several types of models have been presented, including a superfolded β structure (43) and a "β helix" in which the sheet twisting is caused by the presence of equally spaced proline residues (44). The most recently proposed structure is that shown in Fig. 10.24 (42). In this model it can be seen that each feather keratin microfibril consists of a pair of pleated sheets twisted into a degenerate (twofold) left-hand superhelix. Although this structure is fully

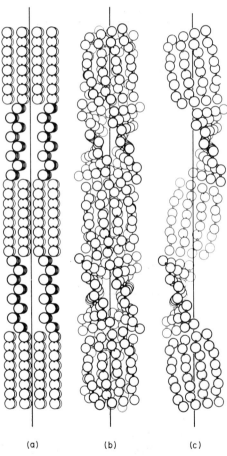

(a)   (b)   (c)

Fig. 10.24. Stages in the development of a model for the antiparallel-chain pleated-sheet core of the microfibril of feather keratin. (a) At each level of a right-handed molecular helix, the pleated-sheet portions of two molecules, related by a horizontal diad, form a two-sheet crystallite. (b) The sheets are twisted so as to follow a left-handed ruled surface giving a structure consisting of two strands of pleated sheets. (c) A single strand of pleated sheets. (Taken from Ref. 42, with permission; see caption to Fig. 10.20.)

consistent with the wide-angle x-ray diffraction pattern, the low-angle pattern of feather (rachis) keratin is not explained. Since the low-angle pattern can be removed by treatment of the protein with denaturing agents without disturbing the wide-angle pattern, it appears that the former originates with the rather large ($\sim 70\%$) non-β-sheet portion of the native material.

The Muscle Proteins

If this chapter were concerned primarily with ultrastructure and biological function rather than the biopolymeric nature of fibrous proteins, the extensive and painstaking research performed on muscle structure would merit major attention. However, only a brief review of the protein components is offered here; reviews of the ultrastructure and proposed functional mechanism for muscle are given in Refs. 45 and 46. The three major proteins of vertebrate striated muscle are myosin, actin, and tropomyosin, and in living rabbit skeletal muscle they occur as about 53, 20, and 15%, respectively, of the total weight.

Myosin

The individual myosin unit appears in the electron microscope as a thin rod of about 1500 Å length and 20–30 Å diameter with a globular head region. The molecular weight is approximately 500,000. On enzymatic treatment with trypsin (47) the molecule cleaves into two fragments, heavy meromyosin (HMM), molecular weight 320,000–360,000, and light meromyosin (LMM), molecular weight 120,000–160,000. The LMM gives a well-defined α-helical x-ray pattern, and ORD measurements show 70–100% α-helix content depending upon preparation. The wide-angle x-ray pattern has the 5.1 Å meridional d-spacing consistent with the supercoiled α helix. Light meromyosin may be reconstituted into fibers which give low-angle meridional spacings that are orders of 428 Å. This same periodicity may be observed in unstained specimens examined in the electron microscope [see Fig. 10.25 (48)] with a subperiod of 70 Å. The available information is then in accord with an α coiled coil for the LMM section of myosin with a superhelix repeat of 70 Å. Studies based on the cleavage of myosin by the enzyme papain have led to the conclusion that LMM is a two-strand coiled coil and that HMM contains a portion of two-strand coiled coil and a globular (ATPase) region. A schematic model is shown in Fig. 10.26 (49).

Heavy meromyosin gives a poor wide-angle x-ray diffraction pattern of the α type, and ORD measurements suggest it contains about 45% α-helical structure. It would seem then that the myosin molecule has both a structural function expressed in its fibrous form, and a chemical function in the globular head.

The secondary structure is to some extent understandable on the basis of

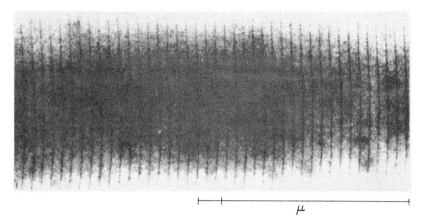

Fig. 10.25. Electron micrograph of an unstained and unshadowed preparation of LMM. (Taken from Ref. 48, with permission.)

the amino acid composition [Table 10.13 (1)]. Particularly noticeable is the large (38%) contribution of the polar amino acids (glutamic, aspartic acids, lysine) in the total myosin molecule. It is possible that of this amount some 10% arises during hydrolysis of glutamine. Proline (2.5% of the amino acid content) is found, as expected, entirely in the HMM fraction and is absent in the α-helical LMM. However, the polar residue fraction is an incredible 50% in LMM (less perhaps 13% from glutamine). It seems likely that the polar residues in the α-helix coiled-coil section of the molecule are present in a sequence such as that suggested for α-keratin in the previous section, i.e., in the 0, 3, 4 positions of a polyheptapeptide sequence. The low

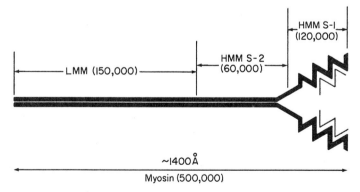

Fig. 10.26. Schematic representation of the myosin molecule based on papain digestion. The LMM and the major portion of HMM subfraction 2 consist of the double-stranded α-helix coiled coil, whereas the HMM subfraction 1 lies in the "head" of the myosin molecule and is of nonuniform conformation. (Taken from Ref. 50, with permission.)

TABLE 10.13

Amino Acid Composition of Rabbit Muscle Proteins and Meromyosin Fractions[a]

Amino acid	Myosin	LMM Fraction 1	HMM	Tropomyosin	Actin
Aspartic acid	98.5	98.4	96.9	107	86.9
Threonine	51.0	39.1	52.0	31.1	79.9
Serine	45.2	40.3	46.0	47.9	59.4
Glutamic acid	182	249	162	254	107
Proline	25.5	0	37.8	2.0	46.6
Glycine	46.4	21.3	59.1	14.4	71.0
Alanine	90.4	96.1	86.3	129	75.2
Valine	49.8	45.1	56.7	32.3	44.5
Cystine/2	10.2	4.7	8.7	7.8	11.9
S-Carboxymethyl-cysteine	—	—	—	—	—
Methionine	26.6	22.5	30.7	19.1	31.8
Isoleucine	48.7	46.3	52.0	35.9	60.4
Leucine	93.9	114	86.3	114	66.8
Tyrosine	23.2	10.7	24.8	18.0	34.9
Phenylalanine	33.6	4.7	42.5	4.0	30.7
Lysine	107	112	102	128	55.1
Histidine	18.5	24.9	16.5	6.6	20.1
Arginine	49.8	71.2	40.2	49.1	40.3
Tryptophan	—	—	—	—	10.6
(NH$_3$)	(107)	(128)	(106)	(77)	(70)

[a] From Ref. 1, with permission.

α-helical content in the HMM fraction may arise more from the absence of residue repeat sequences than from the presence of proline and cysteine (cf. α-keratin fragments). In all probability, therefore, the major differences in conformational properties of the head and tail sections of myosin will not be found to be in the total amino acid content but rather in the sequential array in the tail and lack thereof in the head.

Tropomyosin

Two different tropomyosins have been separated and characterized. The first, tropomyosin A, appears only in certain invertebrate muscles. Tropomyosin B appears in all muscles, vertebrate and invertebrate.

Tropomyosin A (Paramyosin)

Paramyosin is a water-insoluble protein for which a molecular weight of 220,000 has been reported. In the electron microscope the molecules appear to be rod shaped with a length of 1400 Å. x-Ray diffraction studies show the material to be a α helical, ORD data showing the molecules to be ~80% helical.

The above data are consistent with a double-stranded α coiled coil. As with the meromyosins a large proportion of the peptide content is polar. The α helix is more stable for paramyosin at low pH (<2.5) showing that polar-side-chain interactions are important in stabilizing the helix. There is very little proline and no cystine to disturb the α-helical character. Amino acid analyses are given in Table 10.14 (1).

TABLE 10.14

Amino Acid Composition of Various Paramyosins

Amino acid	Annelid		Mollusk	
	Lumbricus	*Arenicola*	*Busycon*	*Venus*
Aspartic acid	116	149	140	140
Threonine	48.0	41.0	50.0	44.1
Serine	61.2	54.6	63.6	47.7
Glutamic acid	203	194	205	207
Proline	2.0	5.0	4.6	1.8
Glycine	30.0	31.0	23.7	19.6
Alanine	103	108	113	132
Valine	34.8	37.2	36.2	34.3
Cystine/2	—	—	—	—
Methionine	1.8	16.1	13.7	13.5
Isoleucine	42.0	31.0	24.9	26.9
Leucine	149	132	123	130
Tyrosine	12.0	12.4	15.0	22.0
Phenylalanine	14.4	7.4	10.0	7.3
Lysine	67.2	69.5	68.6	72.2
Histidine	12.0	7.4	7.5	4.9
Arginine	103	105	101	99.1
(NH₃)	(118)	(140)	(140)	(135)

[a] From Ref. 1, with permission.

Tropomyosin B

Tropomyosin B has an amino acid composition similar to that of light meromyosin and paramyosin, but in contrast to paramyosin it is soluble in water at neutral pH. The tabulated chemical analysis is given in Table 10.13. It can be seen that about 24% of the peptides are acidic and 16% are basic, giving the protein the largest "zwitterion" density of any known protein. Tropomyosin B has a molecular weight of ~ 54,000; it is almost entirely α helical in content and is probably a double-stranded coiled coil about 340 Å long and 14 Å in diameter.

The difference in solubility between paramyosin and tropomyosin B may arise from the fact that the former contains approximately the same number of negatively and positively charged residues (after correcting for glutamine hydrolysis), which in the coiled-coil sequence might result in an effectively neutral molecule. Tropomyosin B, on the other hand, has a strong excess of negatively charged carboxyl groups. The sequencing of tropomyosin B seems feasible and could represent an important step forward in understanding sequence/conformation relationships for the α coiled coil. Without such information, it seems likely, once again, that a repeating polar heptapeptide could be involved, such as that previously suggested for the α-helix coiled-coil portions of α-keratin and myosin.

Actin

Since the basic units of actin are extracted in globular form (G-actin), it may seem strange that it merits discussion as a fibrous protein. However, in aqueous solution (0.1 M KCl plus trace Mg^{2+}) a fibrous form (F-actin) is produced. Native G-actin contains bound ATP, which is apparently the active ingredient causing the polymerization. Removal of the ATP renders G-actin incapable of undergoing polymerization and allows the properties to be more readily studied.

G-Actin

The "monomer" component, G-actin, has a molecular weight of 57,200. The amino acid composition (Table 10.13) indicates the presence of about 1% half-cystine and a composition otherwise somewhat similar to heavy meromyosin. As might be expected from the globular nature of G-actin, no definable internal structure has been identified. As built into F-actin, the globules appear in the electron microscope to be spheres of about 55 Å diameter.

F-Actin

Although the wide-angle x-ray pattern of actin, which represents the internal structure of G-actin, is devoid of detail, showing only a diffuse ring at 4.6 Å, the small-angle pattern indicates two possible axial periods of 406 or 351 Å.

Electron micrographs of F-actin stained with uranyl acetate [see Fig. 10.27 (50)] indicate strands of two chains composed of globular elements with a period of approximately 350 Å. A model of the 350 Å period strands with a right-hand twist (not yet established) is shown in Fig. 10.28 (51). In such a model there are 13 subunits per turn of the two strands.

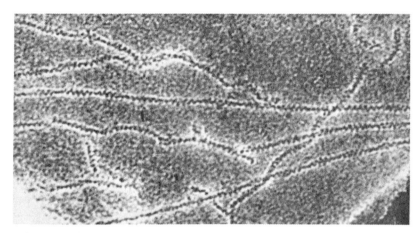

Fig. 10.27. Electron micrograph of a preparation of F-actin negatively stained with uranyl acetate. (Taken from Ref. 50, with permission.)

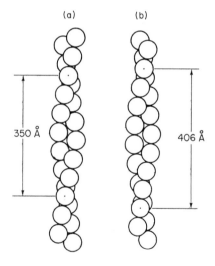

Fig. 10.28. Two possible arrangements of the subunits in F-actin filaments. Either model is consistent with the diffraction pattern from actin in intact muscle. Model (a) agrees with the results obtained by examining in the electron microscope filaments isolated from muscle or formed *in vitro* from the extracted protein. In model (a) the two "strands" of subunits cross over one another at intervals of 350 Å (this interval corresponds to the length of the nonprimitive unit cell deduced from diffraction data); there are 13 subunits per turn of the helix described by each strand. In model (b) the spacing of the crossover positions (the length of the nonprimitive unit cell) is 406 Å and there are 15 subunits per turn. (Taken from Ref. 51, with permission.)

Fibrinogen and Fibrin

FIBRINOGEN (52)

In the process of blood clotting, fibrinogen, a soluble plasma protein, is converted to fibrin, an insoluble fibrous protein. In solution, fibrinogen is found to be of molecular weight $\sim 340,000$, but it may be separated by chemical treatment into three pairs of duplicate chains, each of molecular weight in the 50,000 range. The three types of chains in bovine fibrinogen are identified by the residues at the N-terminal end, which are glutamine (G chain), tyrosine (T chain), and one masked with acetyl (O chain).

Human fibrinogen yields three pairs of chains of 47,000, 56,000, and 63,000 molecular weights each, respectively. In the native molecule these are

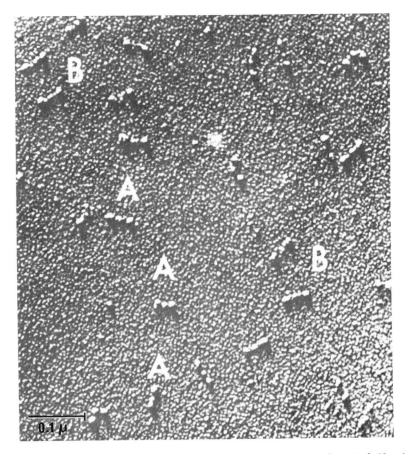

Fig. 10.29. Electron micrograph of fibrinogen molecules. (Taken from Ref. 52 with permission of Marcel Dekker, Inc.)

believed to be held together by sulfur bridges from cystine. The chains are folded into globules which appear in the electron microscope in sets of three (Fig 10.29). Although it is not possible to observe connecting linkages between the globules by direct microscopy, it is believed that rodlike segments join the globules. Light-scattering measurements yield values for the radius of gyration which are somewhat higher than the length of the molecule observed in the microscope. A rational model appears, therefore, to be that given in Fig. 10.30. It is not known whether the six chains run through all three segments; some suggestion that the pairs are mainly confined to each globule has been made.

The total amino acid composition of fibrinogen is given in Table 10.15 (1). It can be seen that a fairly large proportion of the peptides involved are acidic, there being an unusually large amount of aspartic acid (12 %).

TABLE 10.15

Amino Acid Composition of Bovine and Human Fibrinogen[a]

Amino acid	Bovine fibrinogen		Human fibrinogen	
	Residues in 10^5gm protein[b]	Amino acid residues (% dry wt)	Residues in 10^5 gm protein	Amino acid residues (% dry wt)
Lysine	58	63.5	8.14	66.0	9.2
Histidine	15	16.6	2.28	18.6	2.6
Ammonia	(76)	99.2	(1.69)		
Arginine	45	47.3	7.38	46.2	7.8
Aspartic acid	103	103.2	11.8	111.0	13.1
Threonine	59	56.9	5.76	51.0	6.1
Serine	67	67.1	5.85	64.2	7.0
Glutamic acid	95	93.0	12.01	99.0	14.5
Proline	39	46.1	4.47	42.7	5.7
Glycine	87	83.1	4.74	85.1	5.6
Alanine	40	36.6	2.61	45.6	3.7
Cystine/2	18	21.4	2.20	20.4	2.7
Valine	40	44.2	4.38	41.4	4.1
Methionine	15	14.8	1.94	20.4	2.6
Isoleucine	36	36.5	4.13	35.6	4.8
Leucine	50	49.7	5.62	51.0	7.1
Tyrosine	33	29.2	4.77	30.0	5.5
Phenylalanine	26	26.1	3.85	28.2	4.6
Tryptophan	16	(23.9)	(4.44)	16.2	3.3
	842	835.3	92.01	870.6	110.0

[a] From Ref. 1.
[b] The two columns represent data of different authors.

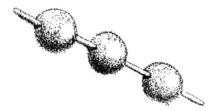

Fig. 10.30. Schematic representation of the fibrinogen molecule. (Taken from Ref. 52, with permission of Marcel Dekker, Inc.)

The x-ray diffraction of fibrinogen films yields an α pattern similar to α-keratin, which on deformation gives some indication of a β transition. On the other hand, ORD/CD measurements of fibrinogen solutions indicate a maximum of 34% α helix.

POLYMERIZATION

The conversion of fibrinogen to fibrin is a complicated process in which one important step is initiated by the enzyme thrombin. The reaction may be represented by

$$\text{Fibrinogen} \xrightarrow{\text{thrombin}} (\text{fibrin})_a + \text{peptide A} + \text{peptide B}$$

Thus the polymerization step involves the release of two peptide fragments, both of which are acidic. The size of the peptide fragments varies somewhat in different mammals. The A and B units from human fibrinogen involve 16 and 14 peptides, respectively, and the bovine A and B fragments contain 19 and 21 units, respectively. The sequence of these peptide fragments (and those of many other fibrinogens) is known, and that of the A fragment is shown in Table 10.16 (1).

For bovine fibrinogen, then, the thrombin step apparently cleaves four of the six peptide chains.

G Chains
$\begin{cases} \text{Glu-peptide A (19 residues) -Arg-} \mid \text{-Gly} \underline{\hspace{2cm}} \text{COOH} \\ \text{Glu-peptide A (19 residues) -Arg-} \mid \text{-Gly} \underline{\hspace{0.3cm}\mid\hspace{0.8cm}} \text{COOH} \end{cases}$

with $(S_2)_x$ bridges

O Chains
$\begin{cases} \text{Ac-peptide B (21 residues) -Arg-} \mid \text{-Gly} \underline{\hspace{0.3cm}\mid\hspace{0.8cm}} \text{COOH} \\ \text{Ac-peptide B (21 residues) -Arg-} \mid \text{-Gly} \underline{\hspace{0.3cm}\mid\hspace{0.8cm}} \text{COOH} \end{cases}$

with $(S_2)_x$ bridges

T Chains
$\begin{cases} \text{Tyr} \underline{\hspace{4cm}\mid\hspace{0.5cm}} \text{COOH} \\ \text{Tyr} \underline{\hspace{5cm}} \text{COOH} \end{cases}$

with $(S_2)_x$ bridge

TABLE 10.16

Sequence of Peptide A Fragment[a]

	19	18	17	16	15	14	13	12	11	10	9	8	7	6	5	4	3	2	1
Ox	Gly	Asp	Gly	Ser	Asp	Pro	Pro	Ser	Gly	Asp	Phe	Leu	Thr	Glu	Gly	Gly	Gly	Val	Arg
Sheep	Ala	Asp	Asp	Ser	Asp	Pro	Val	Gly	Gly	Glu	Phe	Leu	Ala	Glu	Gly	Gly	(Gly	Val)	Arg
Goat	Ala	Asp	Asp	Ser	Asp	Pro	Val	Gly	Gly	Glu	Phe	Leu	Ala	Glu	Gly	Gly	Gly	Val	Arg
Reindeer	Ala	Asp	Gly	Ser	Asp	Pro	Ala	Gly	Gly	Glu	Phe	(Leu	Ala	Glu	Gly	Gly	Gly	Val)	Arg
Pig			Ala	Glu	Val	Gln	Asp	Lys	Gly	Glu	Phe	Leu	Ala	Glu	(Gly	Gly	Gly	Val)	Arg
Rabbit				Val	Asp	Pro	Gly	Glu	Thr	Ser	Phe	Leu	(Thr	Glu	Gly	Gly)	Asp	Ala	Arg
Dog				Thr	Asp	Ser	Gln	Gly	Gly	Glu	Phe	Ile	Ala	Glu	Gly	Gly	Gly	Val	Arg
Man				Ala	Asp	Ser	Gly	Glu	Gly	Asp	Phe	Leu	Ala	Glu	Gly	Gly	Gly	Val	Arg
Cat				Gly	Asp	Val	Gln	Glu	Gly	Glu	Phe	Ile	Ala	Glu	Gly	Gly	Gly	Val	Arg

[a] Data taken from Ref. 1.

It is noticeable that thrombin always acts on the Arg-Gly linkage in all fibrinogens examined. How polymerization occurs following thrombin action is still unclear.

FIBRIN

The polymerization step causes both end-to-end and side-to-side aggregation of fibrinogen to form fibrin fibers. These fibers show various detail in the electron microscope depending upon the method of treatment. Micrographs of shadowed material show beaded character presumably caused by the fibrinogen globules. On the other hand, preparations of fibrin stained with phosphotungstic acid show prominent cross striations, mostly in the 210–250 Å region. There is also a less dense intraperiod line [Fig. 10.31 (53)]. At high resolution, 50 Å particles can be seen in the regions where the stain is not prominent (between bands).

In contrast to polymerization by thrombin, in which both A and B peptides are released, polymerization by snake venom, in which only A peptides are liberated, is limited to end-to-end aggregation. On this basis it has been assumed that peptide A is present in the end globules of the fibrinogen molecule. One of the most unfortunate aspects of the electron microscopy of stained fibrous proteins is that whereas an apparently straightforward repeat pattern is often revealed in the microscope, it is not necessarily an easy process to deduce the molecular packing from this pattern. The tertiary structure of fibrin is no exception; however, it would appear that the polar residues on

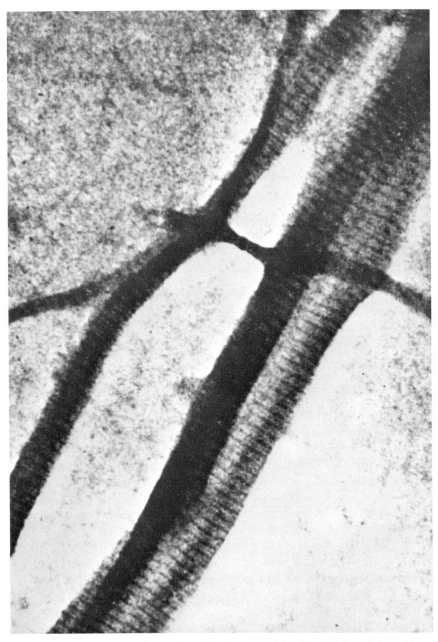

Fig. 10.31. Electron micrograph of fibrin prepared by the addition of thrombin to fibrinogen. (Taken from Ref. 53, with permission of the American Society of Biological Chemists, Inc.)

the exterior of the connecting chains, which may be two-stranded α coiled coils, attract the phosphotungstic acid stains. The degree of shrinkage, caused by polymerization and by artifact in the microscope, is still not entirely clear.

Summary of Fibrous Proteins

Even though the fibrous proteins are usually regarded as structural in function, we have seen that biochemical control generally requires that at least part of the molecule have a chemical function. In the blood clotting process, which presumably requires a rapid chemical response, a large proportion of the fibrous phase (fibrin) and its precursor (fibrinogen) are globular and have a chemical function. Muscular action also involves chemical processes initiated in an active portion of the basic molecular units. Large portions of the muscle proteins are devoted to chemical activity and thus have a globular structure. At the other extreme, the silk fibroins, which once synthesized are probably biochemically inert, are devoid of globular structure, although they do contain structurally amorphous regions. The function of collagen is mainly structural, but the matrix properties of this protein indicate that calcification may involve some chemical activity.

Although the relationship between overall amino acid content and conformation is not always obvious, there are often some important clues. In the mainly structural fibers, the overall amino acid analysis may readily indicate a preferred conformation; for example, the one-third glycine and high proline content in collagen, and the high glycine, alanine, and serine content in fibroins, suggest a regular conformation. The thesis put forward here is that the ordered, nonglobular portions of fibrous proteins probably possess repeat sequences which can potentially be modeled by synthetic polyamino acids and sequential polypeptides. Synthetic modeling of the globular proteins is much more difficult, particularly if chemical function is also required.

Synthetic biopolymers are used, first, as models that allow structural deductions to be made concerning the more complicated native materials and, second, as materials interesting in their own right but usually designed for chemotherapeutic or biomaterials purposes. The emphasis in this book lies in the former area, and it is of some interest to review very briefly the type of information that has evolved from polypeptide modeling. Virtually all sequential polymers produced so far have had seven or less members in the repeating sequence and usually contain one or more β formers (counting glycine as β forming). In consequence the two major conformations produced have been β (or cross-β) and polyproline II (or collagen triple helix). Thus the factors influencing the formation of β or collagenlike structures are fairly well understood, particularly the latter. Some aspects of superfolding of β forms still require much more detailed study. On the other hand, sequences containing charged (acidic or basic) polypeptides of α-forming sequences have not been studied to any great extent. Since many of the interesting fibrous proteins apparently contain α coiled coils and certainly contain polar residues, we may regard the synthesis of such polymers as a particularly interesting challenge. Polytri-, tetra-, and heptapeptides of α formers comprise a series in which the entity placed in the third, fourth, or seventh position traces a superhelix around the basic α helix. The efficient packing of such helices in the solid state may induce supercoiling.

Potentially, some aspects of the tertiary structure of the globular portions of fibrous proteins e.g., superfolding of the peptide chains, can also be studied in synthetic polypeptides by methods similar to those described in Chapter 4. Such a model would require apolar/polar block copolymers in which the polar portion would be expected to lie outside and the folded apolar portion to lie inside the globular structure.

Structural studies of the amorphous elastic proteins such as elastin and resilin and their models present a novel and difficult challenge since the available techniques for ascertaining detailed molecular sequence and conformation are not adequate. Analogies between the biological world and that of commercial polymer properties abound and, as advances occur in the study of cross-linked rubber and elastically deformable plastics, similar advances are to be expected with the native cross-linked proteins.

REFERENCES

1. S. Seifter and P. M. Gallop, *in* "The Proteins" (H. Neurath, ed.), Vol. 4, Chapter 20. Academic Press, New York, 1966.
2. J. O. Warwicker, *J. Mol. Biol.* **2**, 350 (1960).
3. F. Lucas and K. M. Rudall, *in* "Symposium on Fibrous Proteins" (W. G. Crewther, Ed.). Plenum, New York, 1968.
4. A. J. Geddes, K. D. Parker, E. D. T. Atkins, and E. Beighton, *J. Mol. Biol.* **32**, 343 (1965).
5. R. D. B. Fraser, T. P. MacRae, F. H. C. Stewart and E. Suzuki, *J. Mol. Biol.* **11**, 706 (1965).
6. B. Lotz and H. D. Keith, *Bull. Phys. Soc.* **15**, 305 (1970).
7. J. M. Anderson, H. H. Chen, and A. G. Walton, *Bull. Amer. Phys. Soc.* **16**, 320 (1971).
8. A. Brach and G. Spach, *C. R. Soc. Biol.* **271**, 916 (1970).
9. H. H. Chen, M. S. thesis. Case Western Reserve University, Cleveland, Ohio, 1971.
10. K. A. Piez, E. Eigner, and M. S. Lewis, *Biochemistry* **2**, 58 (1963).
11. P. Bornstein and K. A. Piez, *J. Clin. Invest.* **43**, 1813 (1964).
12. W. T. Butler, *in* "Chemistry and Molecular Biology of the Intracellular Matrix" (E. A. Balaz, ed.), Vol. 1, p. 149. Academic Press, New York, 1970.
13. G. N. Ramachandran, ed., "Collagen." Academic Press, New York, 1967.
14. F. H. C. Crick and J. C. Kendrew, *Advan. Protein Chem.* **12**, 133 (1957).

15. A. Rich and F. H. C. Crick, *J. Mol. Biol.* **3**, 483 (1961).
16. G. N. Ramachandran and R. Chandrasekharan, *Biopolymers* **6**, 1649 (1968).
17. G. N. Ramachandran, *in* "Collagen" (Ramanathan, ed.), p.3. Wiley, New York, 1962.
18. W. Traub and A. Yonath, *J. Mol. Biol.* **16**, 404 (1966); **43**, 461 (1969).
19. W. Traub, A. Yonath, and D. M. Segal, *Nature* **221**, 914 (1969).
20. A. Schwartz, P. H. Geil, and A. G. Walton, *Biochim. Biophys. Acta* **194**, 130 (1969).
21. A. J. Hodge and F. O. Schmidt, *Proc. Nat. Acad. Sci. U.S.* **46**, 186 (1960).
22. J. A. Petruska and A. J. Hodge, *Proc. Nat. Acad. Sci. U.S.* **51**, 871 (1964).
23. A. J. Hodge, J. A. Petruska, and A. J. Bailey, *in* "Structure and Function of Connective and Skeletal Tissue" (S. Fitton-Jackson, R. D. Harkness, S. M. Partridge, and G. Tristram, eds.). Butterworth, London, 1965.
24. H. B. Bensusan, *J. Amer. Chem. Soc.* **22**, 4995 (1960).
25. W. B. Rippon and A. G. Walton, *Biopolymers* **10**, 1207 (1971).
26. G. N. Ramachandran, B. B. Doyle, and E. R. Blout, *Biopolymers* **6**, 1771 (1968).
27. L. Gotte, M. Mammi, and G. Pezzin, *in* "Symposium on Fibrous Proteins" (W. G. Crewther, ed.), p. 236. Plenum, New York, 1968.
28. L. Gotte, P. Stern, D. F. Elsden, and S. M. Partridge, *Biochem. J.* **87**, 344 (1963).
29. S. M. Partridge, J. Thomas, and D. F. Elsden, *in* "Structure and Function of Skeletal and Connective Tissue" S. Fitton-Jackson, R. D. Harkness, S. M. Partridge, and G. Tristam, eds.), p. 88. Butterworth, London, 1965.
30. S. M. Partridge, *Advan. Protein Chem.* **17**, 227 (1962).
31. E. J. Miller, G. R. Martin, and K. A. Piez, *Biochem. Biophys. Res. Commun.* **17**, 248 (1964).
32. L. B. Sandberg, N. Weissman, and W. R. Gray, *Biochemistry* **10**, 52 (1971).
33. L. B. Sandberg, personal communication, 1971.
34. K. Bailey and T. Weis-Fogh, *Biochim. Biophys. Acta* **48**, 452 (1967).
35. S. O. Anderson, *Biochim. Biophys. Acta* **93**, 223 (1964).
36. J. C. Kendrew, *in* "The Proteins" (H. Neurath and K. Bailey, eds.), Vol. 2, Part B. Academic Press, New York, 1954.
37. R. D. B. Fraser and T. P. MacRae, *J. Mol. Biol.* **3**, 640 (1961).
38. I. J. O'Donnell, *Aust. J. Biol. Sci.* **22**, 471 (1969).
39. M. C. Corfield and J. C. Fletcher, *Biochem. J.* **115**, 323 (1969).
40. G. E. Rogers, *in* "The Proteins" (H. Neurath, ed.), Vol. 4. Academic Press, New York, 1966.
41. T. P. MacRae and G. E. Rogers, *Nature* **193**, 1052 (1962).
42. R. D. B. Fraser, T. P. MacRae, G. R. Millard, D. A. D. Parry, E. Suzuki, and P. A. Tulloch, *J. Appl. Polym. Sci.*, in press.
43. R. D. B. Fraser and T. P. MacRae, *J. Mol. Biol.* **1**, 387 (1959).
44. R. Schor and S. Krimm, *Biophys. J.* **1**, 467 (1961); **1**, 489 (1961).
45. G. H. Bourne, ed., "The Structure and Function of Muscle." Academic Press, New York, 1960.
46. J. Gergeley, ed., "Biochemistry of Muscle Contraction." Little, Brown, Boston, Massachusetts, 1964.
47. E. F. Woods, S. Himmelfarb, and W. F. Harrington, *J. Biol. Chem.* **238**, 2374 (1963).
48. D. E. Philpott and A. G. Szent-Györgyi, *Biochim. Biophys. Acta* **15**, 165 (1954).
49. S. Lowey, H. S. Slater, A. G. Weeds, and H. Baker, *J. Mol. Biol.* **42**, 1 (1969).
50. J. Hanson and J. Lowy, *J. Mol. Biol.* **6**, 46 (1963).
51. J. Hanson and J. Lowy, *Proc. Roy. Soc. London* **B160**, 449 (1964).
52. K. Laki, "Fibrinogen." Dekker, New York, 1968.
53. C. E. Hall, *J. Biol. Chem.* **179**, 857 (1949).

11

THE POLYSACCHARIDES

Introduction

The polysaccharides form part of the group of molecules known as carbohydrates. The latter term was applied originally to compounds with the general formula $C_x(H_2O)_y$ such as glucose, $C_6H_{12}O_6$, and cellulose, $(C_5H_{10}O_5)_n$, but now it is also used to describe a variety of derivatives including nitrogen- and sulfur-containing compounds. The polysaccharides discussed in this chapter fall mainly into three groups, based on their function. First, there are the structural polysaccharides, such as cellulose and xylan in plant cell walls, and chitin in insect cuticles, where extended chains form long fibrils or sheets which support the organism. The second group is comprised of the storage polysaccharides such as amylose, which occurs in starch granules, and glycogen, an animal storage polysaccharide. In these materials the chains are highly branched and are thought to be folded back on themselves to give compact structure. Third, there are gel-forming polysaccharides, such as agar and the connective tissue mucopolysaccharides, where the straight-chain molecules are believed to be coiled or otherwise associated together, giving large cagelike spaces which hold water.

The structure of the component monomers of polysaccharides are discussed in Chapter 1; some commonly occurring examples are shown in Fig. 1.8. The possible conformations for the pyranose ring have been discussed in

Chapter 2, and we shall now consider how these can be linked together in polymer chains. In aqueous solution, the monomers form an equilibrium mixture of different ring configurations as well as the open chain forms. This would be a most unsatisfactory property for a structural polymer, which is required to have a regular conformation. However, when a *glycoside* derivative of a monomer is formed, such as β-methyl-D-glucoside (Fig. 11.1),

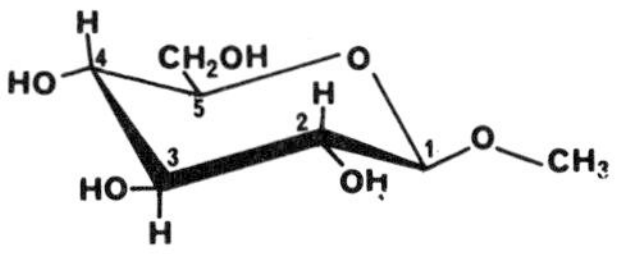

(a)

$$HO-X-OH + HO-X-OH + HO-X-OH \rightarrow -O-X-O-X-O-X-$$

(b)

Fig. 11.1. (a) β-Methyl-D-glucoside. (b) Condensation to form glycosidic polymer chain.

where the OH group on the anomeric carbon C-1 is replaced by $-OCH_3$, then the ring compound does not mutarotate in solution. This property is utilized in polysaccharide structures so that the pyranose rings are stable in solution. The monomers are linked together by condensation such that the sugar rings (X) are joined by glycosidic oxygen bridges. In a straight-chain polymer of D-glucose, two of the five OH groups are required for the linkage and one of these is at position 1 on the ring. A particular glycosidic linkage is described as (1,2) or (1,3), etc., which defines the adjacent carbon atoms for the two rings. In addition, the configuration at the anomeric carbon needs to be defined and the full description of a linkage would be, for example, β-(1,3), α-(1,4). Examples of polysaccharide chains are shown in Fig. 11.2. Figure 11.2a shows a cellulose chain, where position 1 on one ring is linked to position 4 on the next; the configuration at C-1 is β and the structure is described as poly-β-(1,4)-D-glucose [or β-(1,4)-D-glucan]. Similarly, amylose (Fig. 11.2b) is poly-α-(1,4)-D-glucose. Considering the variety of the naturally occurring sugar monomers and the eight possible linkages, the possible permutations for polysaccharide structures are enormous. However, most of the natural polysaccharides are homopolymers or alternating copolymers of the type $(A-B)_n$. Furthermore, these are mainly polymers of glucose and a few other sugar monomers such as galactose, mannose, xylose, and arabinose and their derivatives (amino-, amido-, sulfates, uronic acids, etc.).

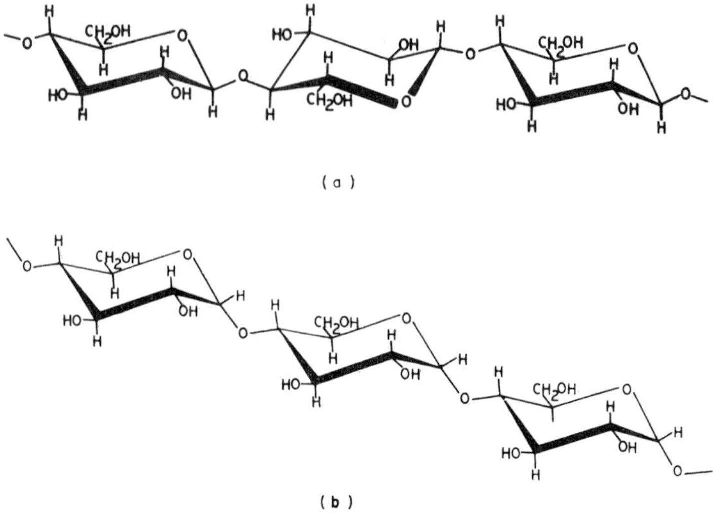

(a)

(b)

Fig. 11.2. (a) Cellulose: poly-β-(1,4)-D-glucopyranose. (b) Amylose: poly-α-(1,4)-D-glucopyranose.

Polysaccharide Structures

The work on the physical structures of the carbohydrates has been comprised mainly of studies of the molecular conformations in the crystalline state by x-ray methods and examination of the crystalite morphology using the electron microscope. Whenever possible, other techniques are used to obtain additional qualifying information. These structural studies are the subject of this chapter. The reader is referred to the review by Marchessault and Sarko (1) for further discussion of the present material.

Naturally, there are many polysaccharides, such as dextran and heparin, which are not discussed here since they have yielded very little information concerning their chain conformation. Furthermore, the large volume of work on the sequences and chemistry of the polysaccharides is outside the scope of this book, and the reader is referred to the many excellent reviews, e.g., in *Advances in Carbohydrate Chemistry.*

Cellulose

Cellulose is probably the world's most abundant natural polymer, and it has been estimated that between 10^{10} and 10^{11} tons are synthesized (and also destroyed) each year. It forms the skeletal material of plant cell walls and its structure and morphology have been the subject of a large body of work. The polymorphic forms of cellulose and its derivatives are the starting materials for many of the textile and paper industries. The cellulose molecule is a polymer of β-(1,4)-linked D-glucose residues. In nature it generally occurs in the presence of other polysaccharides such as xylan and mannan, which are often known as "hemicelluloses." Cellulose is purified by removal of these other materials, as well as proteins and lipids, etc., by boiling in dilute alkali. Accurate figures for the DP of cellulose are difficult to obtain, since dissolving the material probably involves some degradation. Examples of figures quoted are 3900 for ramie cellulose (2) and 16,000–18,000 for *Valonia* cellulose (3).

At least five polymorphic forms have been recognized so far on the basis of their x-ray diffraction patterns (4). Native plant cellulose is usually in the form known as cellulose I. Treatment with sufficiently concentrated aqueous sodium hydroxide converts this to cellulose II. This is the form found in rayon fibers, which are obtained by extruding a solution of cellulose in sodium hydroxide into water; in general any cellulose that is reprecipitated in this manner is known as *regenerated* cellulose. The change from I to II can be effected without dissolving simply by swelling the native fibers with concentrated NaOH. This process is known as *mercerization*, after Mercer, who first used it in 1845 to increase the strength of cotton fibers; such a conversion can also increase the luster and dye affinity of the fibers. The change from I to II is believed to be irreversible, and no one has yet found a way of reprecipitating the native form. For this reason cellulose II is thought to be energetically more stable, which leaves open the question of why nature prefers a less stable form for an inert skeletal material. Cellulose III is obtained when another form (I or II) is treated with liquid ammonia, monomethylamine, or monoethylamine. The structure known as cellulose IV is produced by heating form I or form II in glycerol at 280°C or by boiling the cellulose–ethylenediamine complex in dimethylformamide. Treatment of cotton or wood pulp with concentrated phosphoric or hydrochloric acids yields the structure known as cellulose X. The unit cell proposed for this form is similar to that for cellulose IV, although distinct intensity differences between the two x-ray patterns have been reported. The x-ray patterns for oriented fibers of cellulose I–IV are shown in Fig. 11.3 A–D (1).

The first detailed structure work was performed by Meyer and Misch (5), who determined the unit cell dimensions of the native form. The unit cell dimensions of the polymorphs are given in Table 11.1. These cells are monoclinic and each contains two cellulose chains. In all cases the space group is taken to be $P2_1$, with the twofold screw axis parallel to the fiber axis $(b)^*$.

---

* Traditionally, b is taken as the fiber axis for cellulose and chitin; in most other polymer work, c is so used.

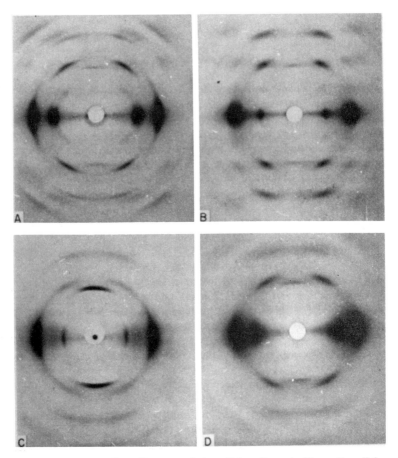

Fig. 11.3. The x-ray fiber diagrams of A, cellulose I ramie fibers; B, cellulose II regenerated Fortisan fibers; C, cellulose III ammonia-treated ramie fibers; D, cellulose IV ramie fibers after treament in glycerol at 280°C. [Taken from Ref. 1, with permission of Academic Press.]

This conclusion stems from the absence of odd-order $0k0$ reflections, although for some specimens a weak 010 reflection is present. The most likely structure for each polymorphic form has chains with twofold screw axes running through the center and corners of the ac projection of the unit cell, with their axes parallel to b. Such structures satisfy the requirements of the $P2_1$ space group, irrespective of whether the center and corner chains have the same or opposite sense.

Highly crystalline cellulose is obtained from the cell walls of the marine algae *Valonia ventricosa*. Electron diffraction patterns from oriented sections indicate that the a and c of unit cell dimensions are twice those for other native

TABLE 11.1

Unit Cell Parameters for the Polymorphic Forms of Cellulose

Form	a (Å)	b (Å)	c (Å)	β	Reference[a]
Cellulose I (native)	8.17	10.34	7.85	96.4°	1
Cellulose II (mercerized, regenerated)	7.92	10.34	9.08	117.3°	1
Cellulose III (ammonia)	7.74	10.3	9.9	122°	1
Cellulose IV (high temperature)	8.11	10.3	7.9	90°	1
Cellulose X	8.1	10.3	8.0	90°	2
Algal cellulose	16.43	10.33	15.70	96°58′	3

[a] Key to references:
1. H. J. Wellard, *J. Polym. Sci.* **13**, 471 (1954).
2. P. H. Hermans and A. Weidinger, *J. Appl. Phys.* **19**, 491 (1949).
3. Text Ref. 7.

celluloses (6), which means that the unit cell contains eight chains. The same structure has been found in the cell walls of some other marine algae, and the accurate unit cell dimensions determined by Nieduszynski and Atkins (7) for *Chaetamorpha melagonium* cellulose are given in Table 11.1. It is not possible to say whether this larger cell is peculiar to this material or whether it is general to all native celluloses but can be resolved only for highly crystalline specimens.

The cellulose chain in all four polymorphic forms has the same repeat distance of 10.3 Å and contains at least an approximate twofold screw axis. Thus it is likely that the main features of the chain conformation are the same in each case. However, the different unit cell dimensions indicate that the chain packing is different for the individual forms, while the bands due to O—H stretching in the infrared spectra show that each structure has a different hydrogen-bonding network (see Chapter 5).

The early models of Meyer and Misch were proposed before ideas of stereochemistry limited the possible conformations. However, space-filling models were used for a cellulose chain as early as 1943 by Hermans *et al.* (8,9) with the result that the "straight" chain of Meyer and Misch was replaced by the "bent" or "Hermans" conformation, as shown in Fig. 11.4. The bent conformation eliminates the bad contacts between the hydrogens on C-4′ and C-1 and between O-2′ and the C-6 side chain, and has the added feature of a hydrogen bond between O-3′—H and O-5. x-Ray crystallography has shown that the two glucose residues of the cellobiose molecule adopt this conformation.

Figure 11.5 shows one of the possible structures of native cellulose proposed by Liang and Marchessault. (10) It is essentially the same as the

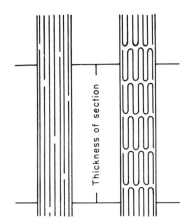

Molecular weight distribution
(GPC chromatogram)

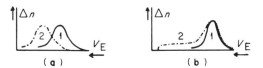

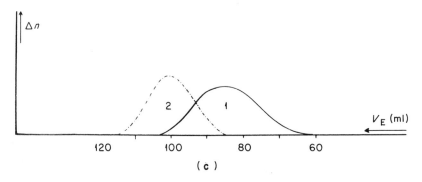

Fig. 11.6. Schematic diagram showing effect of sectioning a fiber made up of (a) extended chains, (b) folded chains, and the subsequent predicted molecular weight distributions in each case. Curve 1, before sectioning; curve 2, after sectioning. For the extended chain conformation a reduction in chain length occurs, while for the folded conformation only a small fraction of the molecules are cut. The experimental results obtained from gel permeation chromatography (GPC) of the tricarbanilate of ramie cellulose before (1) and after (2) sectioning are shown in (c). [Ref. 13; diagrams taken from Ref. 2, with permission of John Wiley and Sons, Inc.]

with the average DP reduced to approximately half the original figure. However, if the chains were folded as shown in Fig. 11.6b, then most chains would be unaffected by the cutting. The molecular weight distribution would be much the same except for an increase in the short-chain fraction, due to multiple breaks in the chains at the cutting points. The ramie fibers were sectioned on a microtome, and the resulting molecular weight distribution, as determined by gel permeation chromatography, is shown in Fig. 11.6c. The average DP was reduced to 1600, and the results clearly suggest extended rather than folded chains in native cellulose. Other physical properties of cellulose fibers, such as the high tensile strength, cannot be explained on the basis of highly folded chains in the microfibrils.

When the morphology of native cellulose is examined in the electron microscope, long *microfibrils* are observed with different cross-sectional dimensions, depending on the source of the specimen. However, in recent years, high-resolution work using negative-staining techniques has shown the presence of regular fibrils with a width of 35 Å in all celluloses. These have been termed *elementary fibrils* by Frey-Wyssling and Mühlethaler (14), and it has been proposed that these are the fundamental structural units produced in cellulose biosynthesis. Figure 11.7a,b (15) shows electron micrographs of sections cut parallel and perpendicular to the fiber axis of ramie cellulose. The negative staining reveals the presence of 35 Å fibrils.

While electron microscopy reveals the 35 Å fibril as the structural unit in native cellulose, x-ray measurements usually indicate larger crystallite

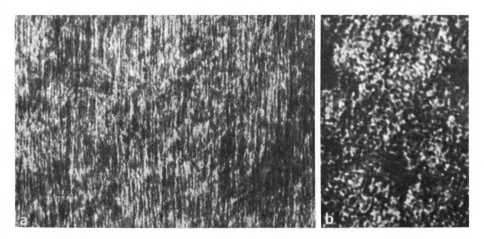

Fig. 11.7. Electron micrographs of elementary fibrils of cellulose. (a) Longitudinal section of ramie fiber; (b) cross section of ramie fiber. The elementary fibrils with a regular cross section of 35 Å diameter have a tendency to form clusters. [Taken from Ref. 15, with permission of Academic Press.]

widths, often in the range 50–100 Å. This suggests coagulation of the elementary fibrils into larger microfibrils. The micrograph of ramie cellulose (Fig. 11.7b) shows the tendency of the elementary fibrils to form clusters. Electron micrographs of the highly crystalline celluloses of *Valonia* and *Chaetamorpha* reveal the presence of microfibrils with a width of ~ 210 Å and thickness ~ 100 Å. Crystallite widths determined for *Valonia* (16) and *Chaetamorpha* (17) indicate that these microfibrils are single crystals. Nevertheless, it has been reported (18) that these microfibrils can be broken into smaller 35 Å subunits, and the regularity of these subunits has been demonstrated by deformation on Mylar films (19). Figure 11.8A (19), an electron micrograph of the end of a *Valonia* microfibril that has been broken by sonication, reveals the presence of regular substructure. Figure 11.8B (19) shows a microfibril that has been broken by deformation on Mylar. Of interest is how these elementary units are arranged in a crystalline array to produce the microfibril with dimensions 210 Å × 100 Å. In addition, the arrangement of the elementary fibrils may account for the larger unit cell detected in *Valonia* and *Chaetamorpha* celluloses.

An electron micrograph of a section of the *Chaetamorpha* cell wall is shown in Fig. 11.9 (20). The microfibrils are arranged parallel to each other in layers; successive layers have different fiber directions. The x-ray work of Frey-Wyssling (21) shows that the microfibrils are oriented with their $10\bar{1}$ planes [with respect to the unit cells in Table 11.1] parallel to the surface of the cell wall. In this, algal cellulose differs from that of higher plants, where the 101 planes are parallel to the surface. It has been proposed that synthesis of the cellulose fibrils occurs at spherical enzyme complexes at the cell surface which facilitate the orientation of layers of microfibrils in different directions (21).

When cellulose is regenerated from solution, crystalline fibrils with a width of approximately 35 Å are obtained. Measurements by Manley and Ionue (22) have shown that the average fibrillar length is between 15 and 20 times *less* than that predicted for a straight chain of average molecular weight. Chain folding seems to be the most likely explanation for this observation, although this problem remains in dispute at the present time.

Chitin

Chitin forms the skeletal material of many of the lower animals and, as such, is analogous to cellulose in plants and to collagen in higher animals. It is abundantly present in the arthropods, e.g., in their exoskeletons and tendons, in mollusks, and in annelids, but it is not found in the vertebrates.

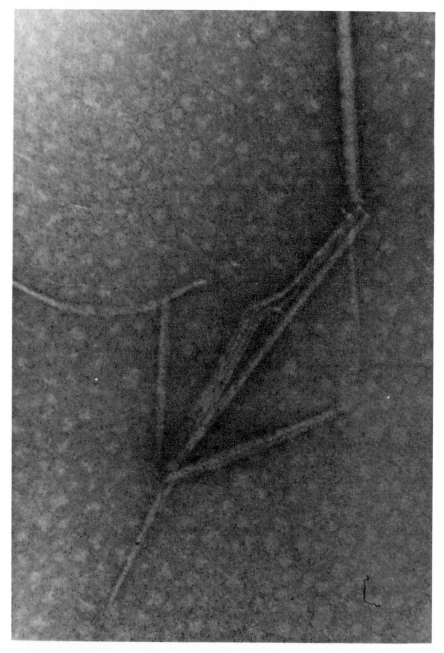

Fig. 11.8A. Electron micrograph of *Valonia* cellulose suspended in water by use of ultrasonic vibrations. Specimen was negatively stained with uranyl acetate. Microfibril has been disrupted by the ultrasonic treatment. Note the subunits, particularly the four side by side, which have equal widths of 35 ± 5 Å. [Taken from Ref. 19, with permission of Academic Press.]

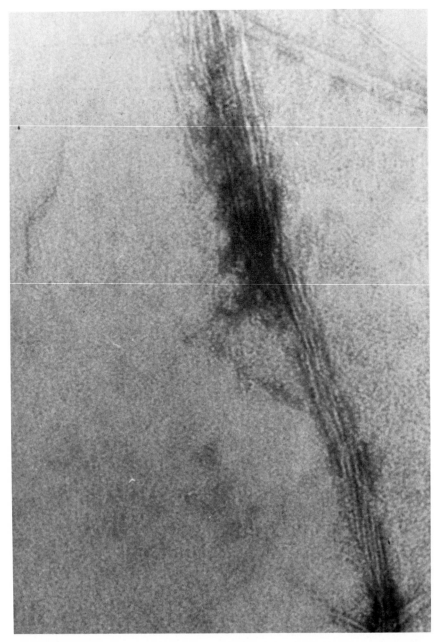

Fig. 11.8B. Electron micrograph of a microfibril of *Valonia* cellulose which has been deformed by stretching on Mylar, and then negatively stained with uranyl acetate. The microfibril appears to have been broken into at least 12 subunits, which are ~35 Å wide or multiples thereof. [Taken from Ref. 19, with permission of Academic Press.]

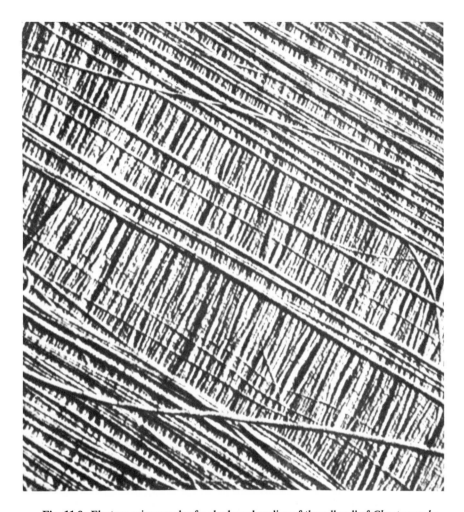

Fig. 11.9. Electron micrograph of a shadowed replica of the cell wall of *Chaetamorpha*. The microfibrils are arranged parallel to each other in layers such that the fibrillar axes in successive layers are approximately perpendicular. [Taken from Ref. 20, with permission of the British Association for the Advancement of Science.]

In plants, chitin is relatively rare but occurs in the cell walls of certain fungi and the extracellular fibrils of some diatoms. The occurrence of chitin in nature has been reviewed by Richards (23) and Rudall (24).

Chitin is considered to be a homopolymer of β-(1,4)-N-acetyl-D-gluco-samine (see Fig. 11.10) and is thus very similar to cellulose, except that an acetamido group replaces the hydroxyl group on C-2. It occurs generally as a chitin–protein complex, the protein content of which is in the range 50–95 % (25). The complex itself may also be calcified, and the $CaCO_3$ content is as high

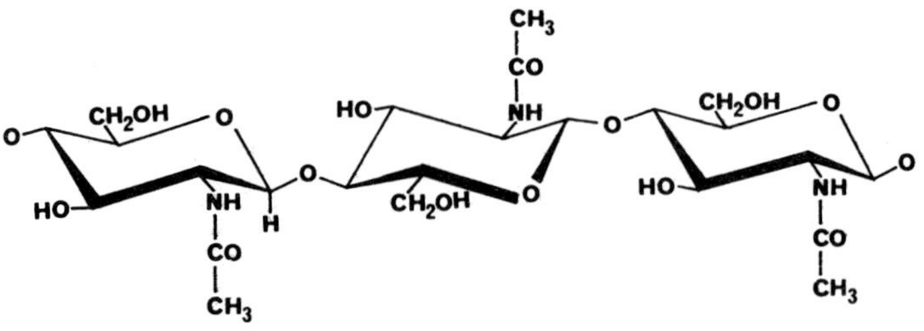

Fig. 11.10. Chitin: poly-β-(1,4)-N-acetyl-D-glucosamine.

as 75% in the shells of crustaceans. Purification of chitin requires treatment with dilute acid to remove the carbonate and boiling in dilute alkali to dissolve the protein. The acetyl content of the resulting chitin is often less than one per sugar residue; analyses by Giles *et al.* (26) of crab shell chitin indicated that one residue in six or seven was deacetylated. Hackman and Goldberg (27) consider that some of the protein is linked by covalent bonds, which may require the absence of certain side groups. The chitin from the diatom spines is apparently protein free in nature, and has been shown to consist of almost 100% N-acetyl-D-glucosamine residues (28).

Three polymorphic forms have been identified by x-ray methods (25). By far the most abundant is α-chitin, which is found in the arthropods and also, for example, in certain fungi. The so-called β-chitin has been discovered in *Aphrodite* aculeatas and subsequently in the pen of the squid *Loligo*, in pogonophore tubes (29), and in diatom spines (30). A third form, γ-chitin, has been identified in the stomach lining of *Loligo* and the brachiopod *Lingula* (25). α-Chitin has also been found in the beak of *Loligo* and thus all three forms exist in the same animal, indicating that the differences relate to function rather than animal grouping (25). The x-ray patterns for the three forms of *Loligo* chitin are shown in Fig. 11.11 (25).

The most satisfactory structure suggested for α-chitin is that proposed by Carlström [Fig. 11.12 (31)]. The x-ray data were obtained from the x-ray pattern of lobster mandibular tendon. The unit cell is orthorhombic with dimensions $a = 4.76$ Å, $b = 10.28$ Å, and $c = 18.85$ Å and space group $P2_12_12_1$, i.e., three mutually perpendicular twofold screw axes. This space group requires neighboring chains to be antiparallel. The chains adopt the bent "Hermans" conformation similar to those of cellulose. Successive chains along the a axis are linked by hydrogen bonds between the amide groups, and thus the structure consists of a series of sheets of chains. The three polymorphic forms of chitin are all made up of this type of sheet, but they differ in the sense of the chains in adjacent sheets.

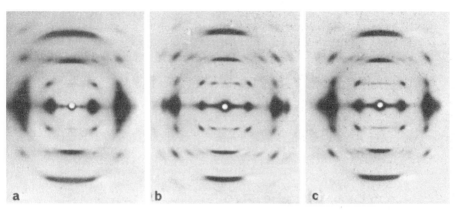

Fig. 11.11. The x-ray fiber diagrams of the three forms of chitin from the squid *Loligo*. (a) β-Chitin (pen); (b) α-chitin (beak); (c) γ-chitin (stomach lining). [Taken from Ref. 25, with permission of Academic Press.]

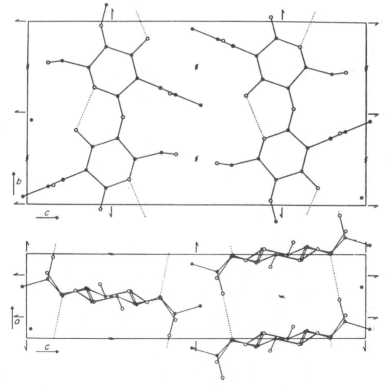

Fig. 11.12. Structure proposed for α-chitin by Carlström. [Taken from Ref. 31, with permission of The Rockefeller Institute for Medical Research.]

The specimens of β-chitin obtained from the spines of the diatoms *Thalassiosira fluviatilis* and *Cyclotella cryptica* (28,30), and also from pogonophore tubes (29), are very crystalline and are analogous to *Valonia* cellulose. Oriented specimens can be prepared by suspending these fibrils in a poorly crystalline fibrin fiber (32). The x-ray diffraction pattern obtained for *Thalassiosira* spines in shown is Fig. 11.13 (32). The chitin reflections can be indexed by a monoclinic unit cell with dimensions $a = 4.85$ Å, $b = 10.38$ Å, $c = 9.26$ Å, and $\beta = 97.5°$ and with space group $P2_1$. The proposed structure for this material is shown in Fig. 11.14 (32). The unit cell can accommodate

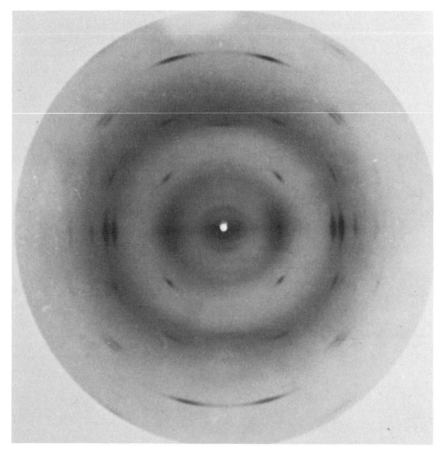

Fig. 11.13. The x-ray fiber diagram of the chitin spines of the diatom *Thalassiosira fluviatilis* oriented in a stretched fibrin fiber. The sharp reflections are due to the chitin. The diffuse rings at ~9 Å and ~4 Å and the small-angle meridionals are due to the fibrin. [Taken from Ref. 32, with permission of John Wiley and Sons, Inc.]

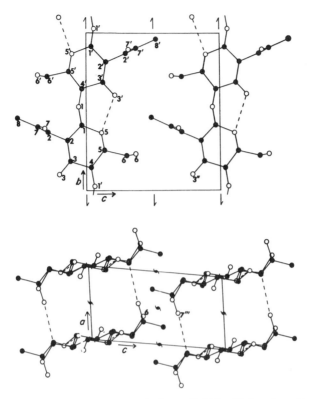

Fig. 11.14. Proposed structure for anhydrous β-chitin. [Taken from Ref. 32, with permission of John Wiley and Sons, Inc.]

only one chitin chain, and, like α-chitin, the structure consists of sheets of chains, except that all the chains are parallel. The increase in the *a* dimension of the cell over that in α-chitin indicates a longer hydrogen bond between the amide bonds for the β form, which has been confirmed by infrared spectroscopy (33). α-Chitin appears to be the most stable form and may be obtained by treating β-chitin with 6 *N* hydrochloric acid. No method is known to reverse this change. Oriented specimens of β-chitin from *Loligo* pen contract to half their original length during the process, and since this also involves conversion from a parallel to an antiparallel chain system, a chain-folding mechanism has been proposed (25).

β-Chitin is able to form a series of hydrates, the only change in the unit cell being an increase in the *c* dimension corresponding to separation of the chitin sheets to accommodate water molecules (32). The chitin in pogonophore tubes is believed to consist of a mixture of two similar structures, both of which probably consist of parallel chains. The first of these structures is

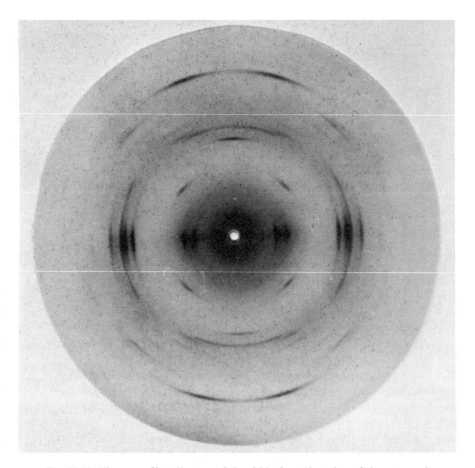

Fig. 11.15. The x-ray fiber diagram of the chitin from the tubes of the pogonophore *Zenkravichiana longissima.* The chitin was suspended in water using ultrasonic vibrations and oriented in fibrin. [Taken from Ref. 32, with permission of John Wiley and Sons, Inc.]

identical with the chitin of diatom spines and has been named β-chitin A. The second form, β-chitin B, gives rise to larger unit cells of two or four chains and exhibits a different series of hydrates. The diffraction pattern of oven-dried pogonophore chitin is shown in Fig. 11.15 (32). It shows all the reflections of β-chitin A in Fig. 11.13, plus additional reflections due to β-chitin B. A similar interpretation can be made for the less crystalline β-chitin from *Loligo* pen.

Figure 11.11c shows the x-ray pattern for γ-chitin from the stomach lining of *Loligo.* The reflections are indexed by a three-chain monoclinic unit cell with dimensions $a = 4.7$ Å, $b = 10.3$ Å, $c = 28.4$ Å, and $\beta = 90°$ and with space group $P2_1$. Rudall (25) has suggested that two parallel chains are followed by one antiparallel chain in this structure, as shown in Fig. 11.16. This structure could arise by the folding mechanism shown in Fig. 11.17a

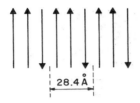

Fig. 11.16. Relative chain directions in γ-chitin as proposed by Rudall (25). The unit cell contains three chains in a repeat of 28.4 Å.

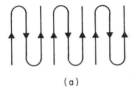 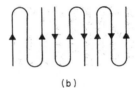

(a)   (b)

Fig. 11.17. (a) Possible folded-chain structure in γ-chitin to produce a "two-up–one-down" structure. (b) Similar folded-chain structure for α-chitin giving successive up and down chains. Layers, three chains wide, have been observed in blowfly laval cuticle (see Fig. 11.19).

Sheets, three chains thick, are observed in some chitin protein complexes (see below) which yield α-chitin on purification. The arrangement of three-chain layers in α-chitin could be as shown in Fig. 11.17b, i.e., antiparallel "threes" as opposed to parallel "threes" for γ-chitin.

A variety of morphology has been reported for chitin–protein complexes. In some cases, fibrous structures are observed in materials where cylindrical chitin fibers are embedded in a protein matrix. Figure 11.18A (34) is an electron micrograph of a cross section of the ovipositor of *Megarhyssa nortoni*, where an approximately hexagonal array of rods ~50 Å in diameter can be seen. The x-ray diffraction pattern of this material [Fig. 11.18B (34)] shows three equatorial reflections corresponding to hexagonal packing with a center to center distance of 69 Å. Thus the ideal structure has rodlike chitin fibrils with a diameter of ~50 Å packed hexagonally in a protein matrix such that the center-to-center distance is 69 Å. The layer line repeat for this complex is 31Å and suggests an association of the protein with every sixth sugar residue. In other similar fibrous complexes, such as from the leg apodemes of locust,

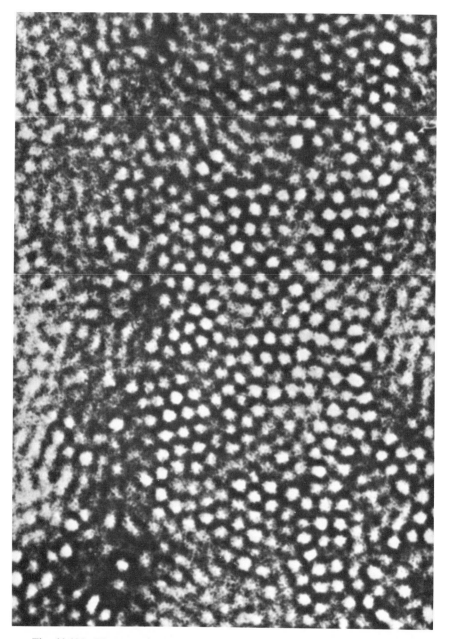

Fig. 11.18A. Electron micrograph of negatively stained cross section of the chitin–protein complex of the ovipositor of *Megarhyssa nortoni.* An approximately hexagonal array of unstained chitin rods with diameter ~50 Å can be seen in a stained protein matrix. [Taken from Ref. 34, with permission of Academic Press.]

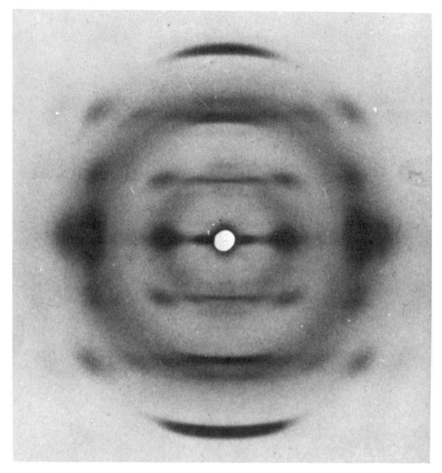

Fig. 11.18B. The x-ray diffraction pattern of the *Megarhyssa* ovipositor. The small-angle region shows three equatorial reflections corresponding to hexagonal packing with a center-to-center distance of 69 Å. [Taken from Ref. 34, with permission of Academic Press.]

the fiber repeat is 41 Å, indicating a possible association every eighth sugar residue.

Another type of chitin–protein system is the "layered" complex. Figure 11.19A (34) is an electron micrograph of a section from blowfly laval cuticle. The section has been cut parallel to the chain axis and shows layers of chitin approximately 25 Å wide interspaced by stained protein. The x-ray diffraction pattern of this material [Fig. 11.19B (34)] shows an equatorial reflection of 33 Å. Thus a model for the layers would have three chitin chain sheets arranged in an antiparallel manner with a thickness of 25–27 Å, followed by a

Fig. 11.19A. Electron micrograph of negatively stained blowfly laval cuticle. Unstained chitin layers ~25 Å wide can be seen in a stained protein matrix. [Taken from Ref. 34, with permission of Academic Press.]

layer of protein 6–8 Å wide, which is presumably in the extended β form. The 33 Å layer separation can be increased to as much as 45 Å by treatment with uranyl and zirconium salts, etc.; the 45 Å staining pattern is observed in the electron microscope.

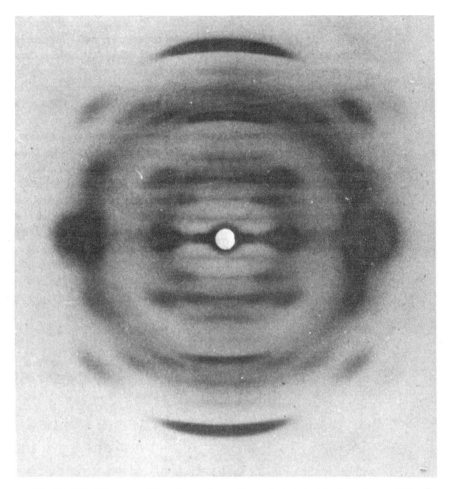

Fig. 11.19B. The x-ray diffraction pattern of an oriented stretched specimen of blowfly laval cuticle. An equatorial reflection at ~33 Å corresponds to the separation of the layers, and the fiber repeat is 31 Å, i.e., every six sugar residues. [Taken from Ref. 34, with permission of Academic Press.]

Electron micrographs (35) of the highly crystalline diatom spines show them to be approximately 1000 Å wide and to be tapered to a point at one end. On treatment with ultrasonic vibrations, these spines break into lamellae with widths in the range 200–300 Å. Twisted sections of these lamellae show their thickness to be only ~20 Å. Such a sheetlike morphology is compatible with the known molecular structure and analogous to the chitin layers in blowfly laval cuticle (34). Figure 11.20 (35) is a micrograph of the chitin of

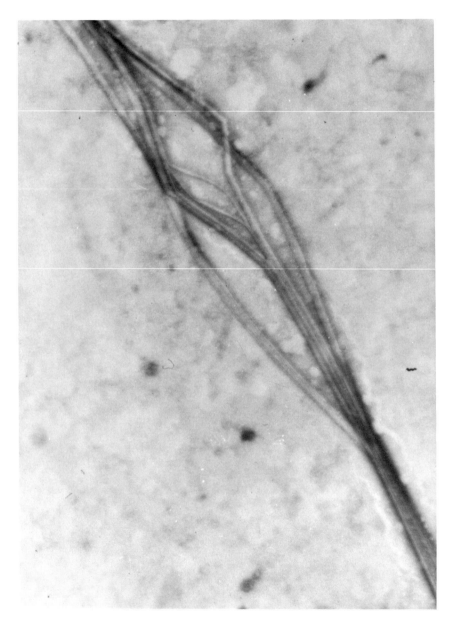

Fig. 11.20. Electron micrograph of negatively stained *Thalassiosira* spine chitin after ultrasonic treatment. This figure shows a lamella fragment of a spine which can be seen to be breaking into smaller subunits with widths of ~35 Å. [Taken from Ref. 35, with permission of John Wiley and Sons, Inc.]

Thalassiosira spines after treatment with ultrasonic vibrations. Part of one lamella can be seen to have broken into smaller subunits with widths of ~35 Å. Striations can be seen on the intact lamellae analogous to those of the *Valonia* cellulose microfibrils. A substructure of this type may well account for the variety of larger unit cells observed in pogonophore β-chitin (32).

Xylans

β-(*1,4*)-*Xylan*

β-(1,4)-Xylan and β-(1,4)-mannan form the molecular backbone of the group of plant polysaccharides known as hemicelluloses. These substances are often partially acetylated and have side chains of single sugar rings; they are also branched to varying degrees. The side chains in β-(1,4)-xylan are usually L-arabinose and 4-*O*-methyl-D-glucuronic acid. A section of such an arabinoglucuronoxylan chain in shown in Fig. 11.21 (36). The actual physical

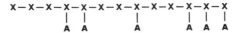

Fig. 11.21. Section of an arabinoglucuronoxylan chain (36) showing the random positions of the side chains (A) attached to the main xylan chain (X).

state of the xylans and mannans in wood is unknown. However, they do not crystallize without chemical modification, i.e., removal of the side chains, and they probably serve as an amorphous matrix between the cellulose fibrils. Nevertheless, the polarized infrared spectra of wood sections (37) show that the xylan and mannan chains are oriented parallel to the cellulose fibrils. The ratios of cellulose, xylan, mannan, etc., in woods of different species have been reviewed by Timell (36). Higher proportions of xylan are found in hard-woods, from which xylan is extracted with dilute alkali.

Oriented crystalline fibers have been obtained from β-(1,4)-xylan whose sugar side chains and *O*-acetyl groups were removed with dilute acid (38). Nieduszynski and Marchessault (39) have found that β-(1,4)-xylan from hardwood occurs in at least three forms, depending on the degree of hydration. The fiber repeat is approximately the same in all three structures. Meridional reflections are observed on layer lines with $l = 3n$, and the chain conformation is considered to be a left-hand helix with three D-xylose residues per turn. The form known as xylan hydrate exists at low humidities and is thought to contain one water molecule per sugar residue. The unit cell has dimensions $a = b = 9.16$ Å, $c = 14.85$ Å (fiber axis), and $\gamma = 120°$, with space group $P3_121$. This unit cell contains two xylan chains: one at a corner and

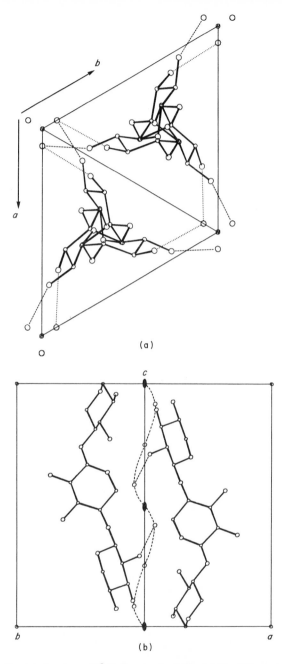

(a)

(b)

Fig. 11.22. Crystal structure of β-(1,4)-xylan hydrate. (a) The 001 projection (perpendicular to the fiber axis); (b) 110 projection [looking along the short diagonal of the unit cell in (a) with the 001 planes horizontal]. The water molecules are represented by the circles linked only by the broken lines (hydrogen bonds) to each other and to the oxygens of the xylan chain. [Kindly supplied by Dr. R. H. Marchessault prior to publication; taken from Ref. 39 with permission of John Wiley and Sons, Inc.]

490

the other on one of the threefold screw axes of the space group, as shown in Fig. 11.22 (39). Intensity calculations indicate that the six water molecules in the unit cell are arranged in a hydrogen-bonded chain. Consideration of the potential energy due to intramolecular interactions of the β-(1,4)-xylobiose residue (see Chapter 2) predicts that a 2_1 helix, like that of cellulose, should be the most favorable conformation. It is argued (39), however, that the formation of hydrogen bonds between the water molecules and O-2 and O-3 of those xylose residues, as well as the weak O-3'—H $\cdots$ O-5 bonding, leads to a stabilizing of the 3_1 helix.

At 100% relative humidity, a different structure is obtained with unit cell dimensions $a = b = 9.64$ Å, $c = 14.95$ Å (fiber axis), and $\gamma = 120°$. It is thought probable that this form has two water molecules per xylose residue. The x-ray pattern of this structure, known as xylan dihydrate, is shown in Fig. 11.23 (39). The third form, "dry xylan," is obtained from the hydrated specimens by drying *in vacuo*. The diffraction pattern is of low quality, and a "pseudo" unit cell is suggested with dimensions $a = b = 8.4$ Å, $c = 14.85$ Å (fiber axis), and $\gamma = 120°$. Polymer single crystals of β-(1,4)-xylan, consisting

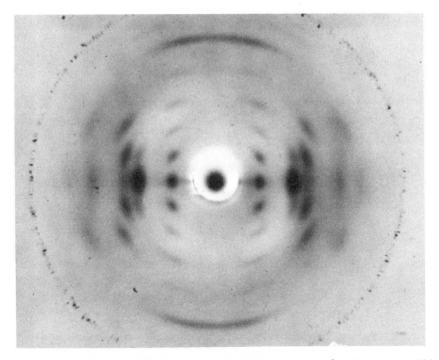

Fig. 11.23. The x-ray diffraction pattern of an oriented fiber of β-(1,4)-xylan at 100% relative humidity. The structure adopted under these conditions is known as β-(1,4)-xylan dihydrate. [Kindly supplied by Dr. R. H. Marchessault prior to publication; taken from Ref. 39, with permission of John Wiley and Sons, Inc.]

of layers approximately 50 Å thick, have been observed in the electron microscope (40). The molecular weight of the material studied corresponds to an
average chain length of 150 Å and thus the xylan chains, which run perpendicular to the surface of the crystals, must be folded in these preparations.

Complete acetylation of β-(1,4)-xylan yields the diacetate, which has also
been studied by x-ray diffraction and conformational analysis (41). The chain
conformation is a 2_1 helix, like that in cellulose, rather than 3_1, as in the
unacetylated xylan. Potential energy calculations indicate no preference on
an intramolecular basis for 2_1 over 3_1 for the diacetate, and it is argued that
the former is probably preferred because of intermolecular packing effects.
The unit cell for xylan diacetate contains two chains or four residues; it is
monoclinic, with dimensions $a = 7.64$ Å, $b = 12.44$ Å, $c = 10.31$ Å (fiber
axis), and $\gamma = 85°$; and the space group is $P2_1$.

β-(1,3)-Xylan

In certain algae and seaweeds, which form a significant part of the plant
kingdom, cellulose is replaced by β-(1,3)-xylan as the skeletal material. Unlike the β-(1,4)-xylan discussed above, which is an amorphous hemicellulose,
the β-(1,3)-linked polymer forms crystalline fibrils. In the electron microscope, sections of the cell walls of certain siphoneous green algae appear to be
made up of an array of xylan microfibrils, with a cross-sectional width of
approximately 200 Å (42), as can be seen in Fig. 11.24. These microfibrils,
and their arrangement in oriented layers, are analogous to the cellulose fibrils
in *Valonia* cell walls (see Fig. 11.9).

The x-ray diffraction pattern of an oriented specimen of β-(1,3)-xylan
from the algae *Penicillus dumetosis*, at 98% relative humidity, is shown in
Fig. 11.25a (42). Atkins *et al.* (43) have shown that this pattern results from
a triple-helical structure in which three xylan chains are coiled together with
$6_1 3$ symmetry. Each chain is a right-hand helix, with six D-xylose residues
per turn, repeating in 18.36 Å. The unit cell contains one triple-helix unit,
has dimensions $a = c = 15.4$ Å, $b = 6.12$ Å (fiber axis), and $\beta = 120°$, and
the space group is $P6_2$. In addition, the structure is believed to contain one
water molecule per sugar residue. Due to the symmetry of the triple helix,
only every third layer line is present for the single-chain repeat of 18.36 Å,
i.e., the repeat for the triple helix is one-third that for a single chain. The
proposed triple-helical conformation is shown in Fig. 11.25b (43). The
O-2—H groups on each chain point toward the helix axis, and form a hydrogen-bonded triad, as shown in Fig. 11.25c (43). Polarized infrared spectra
indicate the presence of hydrogen-bonded O—H groups directed perpendicular to the fiber axis, which is in accord with this proposed network.
Under vacuum conditions, the fiber axis decreases slightly to $b = 5.97$ Å.

Fig. 11.24. Electron micrograph of the xylan fibers in the cell walls of the siphoneous green algae *Penicillus dumetosis*. [Taken from Ref. 42, with permission.]

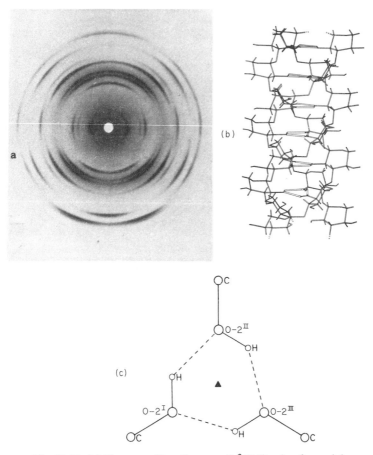

Fig. 11.25. (a) The x-ray fiber diagram of β-(1,3)-xylan from siphonous green algae at 98% RH. (b) Diagram of triple helical structure determined for β-(1,3)-xylan by Atkins *et al.* (43). (c) Hydrogen-bonding triplet linking the three chains of the triple helix. [Taken from Ref. 43, with permission.]

A similar triple-helical conformation has been proposed by Rees and Scott (44) for β-(1,3)-glucan from a number of sources, based on preliminary x-ray data and potential energy analysis.

Mannans

The mannan hemicellulose found in plant cell walls has a backbone of poly-β-(1,4)-D-mannose, probably with some (1,6) branching. A number of single sugar ring side chains are present, and some *O*-acetyl groups also occur (36). These mannans probably form an oriented but otherwise amorphous structure between the crystalline cellulose fibrils.

Mannans are frequently found as storage carbohydrates in bulbs and the endosperm of seeds such as ivory nuts. The first crystalline specimens were prepared from ivory nut mannan (45), and the x-ray patterns suggested that the mannan chain has a conformation similar to that of cellulose, with two mannose residues repeating in 10.3 Å. Treatment with alkali produced an apparent conversion to a second structural form, analogous to the transformation of cellulose I to II. This last result has been questioned (1) since it was thought that the transformation could be simply the conversion of a cellulose impurity. However, seaweeds have been reported in which pure poly-β-(1,4)-D-mannose forms the crystalline skeletal material in the cell walls, and

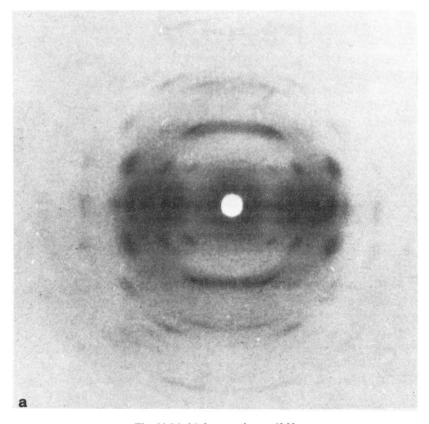

Fig. 11.26. (a) (see caption p. 496.)

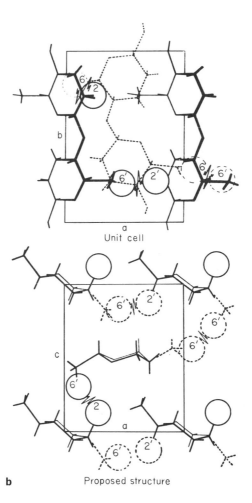

Unit cell

b   Proposed structure

Fig. 11.26. (a) The x-ray fiber diagram of an oriented specimen of the mannan from the cell wall of *Codium* seaweed (mannan I). (b) Unit cell and proposed structure for mannan I. [Taken from Ref. 46, with permission.]

cellulose is completely absent. The x-ray photograph of an oriented sample of the mannan from *Codium* seaweed is shown in Fig. 11.26a (46) and can be indexed in terms of an orthorhombic unit cell, with dimensions $a = 7.21$ Å, $b = 10.27$ Å, and $c = 8.82$ Å containing two mannan chains. The structure proposed for this material is shown in Fig. 11.26b (46). This natural form of mannan, mannan I, can be converted into a second polymorphic form, mannan II, as reported previously for ivory nut mannan. This new structure has a monoclinic unit cell with dimensions $a = 18.8$ Å, $b = 10.2$ Å, $c = 18.7$ Å, and $\beta = 57.5°$. The x-ray pattern for mannan II is shown in Fig. 11.27 (46),

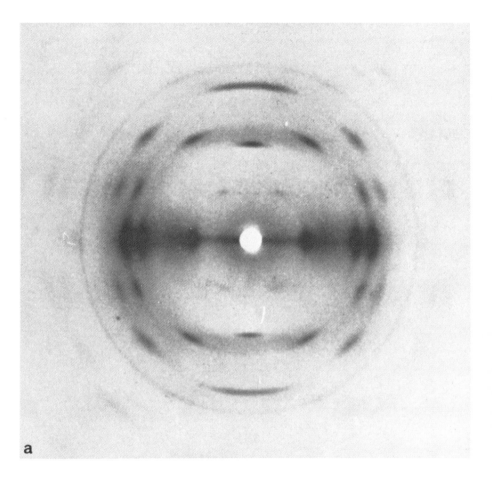

a

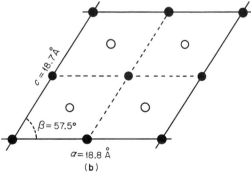

(b)

Fig. 11.27. (a) The x-ray fiber diagram of an oriented specimen of mannan II. (b) Unit cell projection perpendicular to the fiber axis for mannan II. [Taken from Ref. 46, with permission.]

along with a projection of the unit cell perpendicular to the fiber axis. Electron micrographs of the cell walls have not revealed definite evidence for a microfibrillar structure, but short rodlets with a width of approximately 100 Å have been observed (46). Microfibrils have been observed in samples of ivory nut mannan, but it is not clear whether or not cellulose is also present.

Ordered specimens of a galactomannan from guar seeds have been prepared. Chemical analyses indicate a 2 : 1 ratio for mannose: galactose, and the chain is believed to be poly-β-(1,4)-mannan with (1,6)-linked galactose side chains, as shown in Fig. 11.28a (47). The fiber diagram of guar galactomannan indicates that this material has a fiber repeat of 10.3 Å, like mannan I and II, and is indexed by an orthorhombic unit cell with dimensions $a = 15.5$ Å, $b = 10.3$ Å, and $c = 8.65$ Å. The chains are thought to be arranged as shown schematically in Fig. 11.28b (47). Along the c axis, the chains form sheets with all the galactose rings on the same side, while adjacent sheets have these side chains on opposite sides.

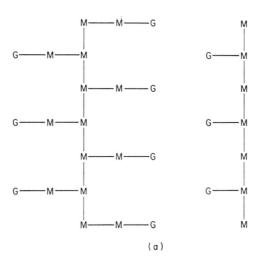

(a)

Fig. 11.28. (a) Schematic of the galactomannan chains from guar seeds. Galactose and mannose residues are indicated by G and M, respectively. (b) Schematic arrangement of the galactomannan chains in the proposed unit cell. [Taken from Ref. 47, with permission of the American Chemical Society.]

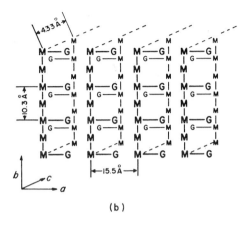

(b)

Fig. 11.28. (b)

Amylose

Starch can generally be separated into two polysaccharide components, amylose and amylopectin. Amylose, the minor component, consists of poly-β-(1,4)-D-glucose. Amylopectin is mainly α-(1,4) but has varying amounts of α-(1,6) branching, depending on the source. Amylose is separated from amylopectin as a solution in butanol and purified by successive recrystallization. It is unlikely that a sharp dividing line exists between the components, and amylose probably contains some (1,6) branches.

The DP's determined for amylose go up to 3800 for potato and 800 for corn amylose. Figures as high as 220,000 have been obtained for the DP of branched amylopectin molecules from potato starch (48). The animal storage polysaccharide glycogen is similar to amylopectin, but is more branched.

Unlike the polysaccharides considered so far, those in starch do not form fibrous structures. The first x-ray work was done with powder patterns from unoriented specimens of starch (49). It was soon found that starches from different sources gave different x-ray patterns which could be divided into two groups; the so-called "A" pattern was given by cereal starches, while the "B" pattern came from tuber starches. The *d*-spacings observed in typical A and B starch patterns are given in Table 11.2 (50). A third "C" pattern

TABLE 11.2

Observed d-Spacings and Intensities for the x-Ray Powder Patterns of Potato (B) and Corn (A) Starches[a]

Potato		Corn	
Spacing (Å)	Intensity[b]	Spacing (Å)	Intensity[b]
15.8	s		
8.9	w	8.8	vw
7.9	vw	7.9	vw
6.36	m		
5.92	ms	5.88	s
5.22	s	5.22	s
		4.85	s
4.56	w		
		4.45	w
4.40	vw		
4.04	ms		
		3.87	s
3.70	ms	3.70	m
3.42	m		
		3.37	m

[a] Based on table of Bear and French (50).
[b] Abbreviations: v, very; w, weak; m, medium; ms, medium strong; s, strong.

has been recorded, although it has been suggested that this is due to a mixture of A and B structures in granules of rice, tapioca, etc. After separation, amylose can be crystallized in both the A and B forms, depending on the preparation conditions. It is likely that the A and B structures are very similar and that the crystalline regions of starch are made up of amylose or straight-chain sections of amylopectin.

Precipitation of amylose from the organic solvents used in the purification process gives a crystalline structure different from the native A, B, and C forms. This form, called "V-amylose," can be reconverted into the naturally occurring forms under conditions of high humidity (51). The material crystallized from solution is usually a complex of amylose and solvent molecules, and each complex gives a somewhat different x-ray pattern. When the complexing agent is butanol or dimethyl sulfoxide (DMSO), this can be removed by treatment with a methanol–water (3 : 1) mixture. The resulting structure has been named V_a for "anhydrous," although it probably contains some water molecules. A higher hydrate, designated V_h, can also be formed. Structural work has focused on these hydrates, the complexes with DMSO and butanol, and the V-amylose–iodine complex, which is the basis of the starch–iodine test. Rundle and co-workers (50) in the early 1940's determined the unit cell dimensions of a number of V structures from powder diagrams and proposed that the chain conformation in all cases is a helix with six glucose residues repeating in approximately 8.0 Å. They realized that such a helix would have a hole in the center large enough to accommodate the complexing agent. In the V-amylose–iodine complex, the iodines are thought to form long polyiodide chains along the axis of the helix.

Oriented specimens of V-amylose have been prepared by stretching films cast from dimethyl sulfoxide solution. An x-ray pattern of the V_a structure is shown in Fig. 11.29 (52) and is indexed in terms of an orthorhombic unit cell with dimensions $a = 13.0$ Å, $b = 18.5$ Å, and $c = 8.0$ Å. The probable space group is $P2_12_12_1$, which requires antiparallel amylose chains. The chains are cylindrical with a diameter of 13 Å and a hole down the center which can accommodate at least one water molecule for every two glucose residues. A cylindrical projection of the proposed V chain conformation is shown in

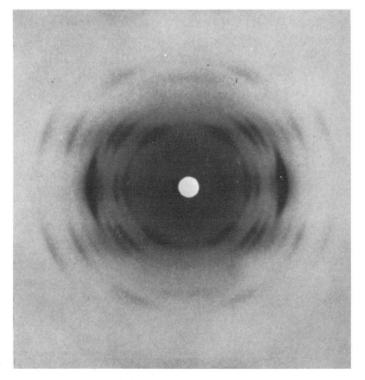

Fig. 11.29. The x-ray diffraction pattern of an oriented film of V_a-amylose. [Taken from Ref. 52, with permission of John Wiley and Sons, Inc.]

Fig. 11.30a (53). Successive residues along the chain form intramolecular hydrogen bonds between the O-2 and O-3′ hydroxyls; successive turns are linked by an O-6—H ··· O-2″ hydrogen bond between residue 1 and residue 7, which leads to a very stable 6_1 helix with a repeat of 8.0 Å. This chain conformation is believed to be present in all the V structures.

Single crystals of V-amylose have been obtained (54) which show layers ~50 Å in thickness. The chains had a DP of ~2000, and electron diffraction indicated that they were arranged perpendicular to the crystal surface. A fully extended V chain would be approximately 2500 Å in length and thus the chains must be folded in the crystals. On these grounds, it seems likely that the chains are antiparallel in the film specimens of V-amylose.

Studies of polysaccharides in solution have not been discussed in this chapter since most structural work has been related to the solid-state chain conformation. The native polysaccharides in general do not possess a suitable chromophore to facilitate ORD–CD type of measurements. However, evidence for the chain conformation of amylose has come from the UV spectra of amylose–iodine solutions. The iodine is believed to form long polyiodide sequences in the central cavity of the helix, and it is argued that the length of the iodine chain will correspond to the length of the helical-chain section in solutions. Wolf and Schutz (55) have shown that the UV spectra of amylose–iodine solutions correspond to iodine chains of an average length of 14 atoms or approximately 55 Å. Thus the amylose molecule in solution with iodine is believed to form short helical segments of approximately seven turns of the chain, presumably linked by sections of random coil.

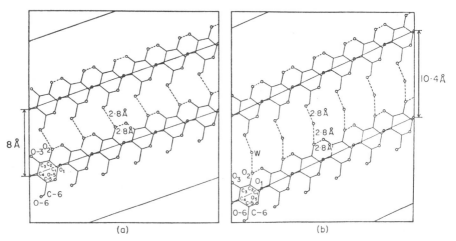

Fig. 11.30. Cylindrical projections of the proposed chain conformations for (a) V-amylose, (b) B-amylose. [Taken from Ref. 53, with permission of Academic Press.]

The amylose chain is able to adopt a number of different conformations. Acetylation of the three OH groups gives amylose triacetate (the x-ray pattern of an oriented fiber of this material is discussed in Chapter 3). The structure has been determined by Sarko and Marchessault (56), and the chain conformation is a left-hand helix with 14 sugar residues in three turns repeating in 51.4 Å. The rise per residue is 3.67 Å, and the chain is much more extended than that of V-amylose, where the rise is 1.33 Å per residue. Deacetylation of the oriented fiber with dilute alcoholic potassium hydroxide (57) yields the alkali–amylose complex. The chain conformation is undetermined, but it probably involves six residues per turn repeating in 22.6 Å. The most extended chain is that obtained when the KOH is replaced by KBr by soaking the specimen in an alcoholic solution of the salt. This salt complex has a fiber repeat of 16.0 Å, and the chain conformation is a 4_1 helix with a rise per residue of 4.0 Å (57,58). Conformational analysis (see Chapter 2) has indicated that the structure is energetically unfavorable, but the effect of the ionic forces has not been considered. Removal of the salt by treatment with 75% aqueous ethanol converts the structure to the V form. Humidification followed by boiling in water converts the V structure to the natural B form, and the A structure can be obtained as an intermediate (57).

The most highly oriented specimens of B-amylose have been prepared in the above manner rather than from oriented V-amylose films cast from solution, since amylose triacetate is more easily obtained in a highly oriented state. Figure 11.31a (53) shows an x-ray pattern of such a specimen of B-amylose. Tilting the specimen to the correct angle reveals the presence of a sixth layer line meridional reflection, which is the only one observed [Fig. 11.31b (53)]. The conformation is thus thought to be a 6_1 helix with a fiber repeat of 10.4 Å. This is similar to the V helix except that the fiber repeat has increased by approximately 2.4 Å. A cylindrical projection of the proposed chain conformation is shown in Fig. 11.30b; residues on successive turns are linked by hydrogen bonds through a water molecule between residues 1 and 7. The intramolecular O-2 $\cdots$ O-3' hydrogen bond proposed for V-amylose can also be formed by the B structure.

Conversion of form V to form B is effected by humidification, and the mechanism probably involves breaking the intramolecular hydrogen bond between successive turns of the V helix and remaking the bonds via water molecules, as the water is absorbed and the fiber repeat increases (53).

Concerning the structure of the actual starch granules, early work with the polarizing microscope showed that they are spherulitic, and x-ray diffraction indicated that at least some of the chains are ordered. Growth of the granule is thought to occur by apposition of spherical layers about a central nucleus. The work of Kreger (59) involving microbeam x-ray diffraction from individual large granules of *Phajus* starch established that the linear portions

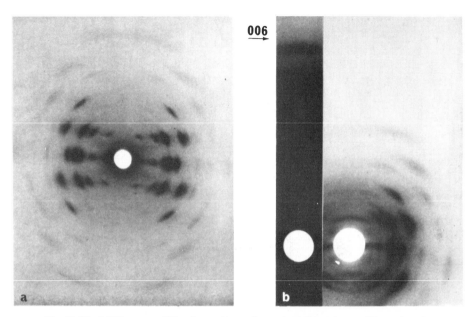

Fig. 11.31. (a) The x-ray diffraction pattern of an oriented specimen of B-amylose in an atmosphere of helium at 98% RH. (b) The x-ray diffraction pattern of the same specimen tilted for the sixth layer line. The strip on the left-hand side is the meridional section of a longer exposure showing the 006 reflection. [Taken from Ref. 53, with permission of Academic Press.]

of the amylose and amylopectin chains are likely to be oriented perpendicular to the growth layers. Electron microscopy (60) has recently revealed the presence of lamellae which can be subdivided into elements with thicknesses as low as 100–200 Å. The chain axes are believed to be oriented perpendicular to these lamellae, and in the proposed model shown in Fig. 11.32 (61), both amylose and amylopectin chains are folded back on themselves. It is suggested that amylopectin folds at the branch points, and thus the side chains

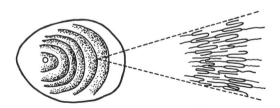

Fig. 11.32. Schematic diagram of the proposed organization of the growth layers in the starch granule. The detail shows the proposed chain folding within the layers. [Taken from Ref. 61, with permission of John Wiley and Sons. Inc.]

would not distort the ordered regions. Such a layered model is in accord with results from solid-state light scattering (62).

Alginic Acids

The polysaccharides known as alginic acids are obtained from the cell walls and intercellular regions of certain algae and seaweeds. They are believed to be copolymers of D-mannuronic acid (M) and D-guluronic acid (G) and to contain chain sequences of the type $(M)_n$, $(G)_n$, and $(M-G)_n$. Isolates from different sources contain different ratios of the two types of residue and have been shown to give different x-ray patterns (63). The fiber diagram given by a specimen containing 96% of mannuronic acid residues has been interpreted as being the x-ray pattern of polymannuronic acid (64). This polysaccharide thus has an orthorhombic unit cell with dimensions $a = 7.58$ Å, $b = 10.35$ Å (fiber axis), and $c = 8.58$ Å, with a probable space group $P2_12_12_1$. The fiber repeat of 10.35 Å corresponds to a helix with each turn containing two β-(1,4)-linked D-mannuronic acid residues in the stable C1 chain conformation. Therefore, it is very likely that this material has a chain conformation like that in cellulose and chitin, as has been confirmed by intensity calculations (65).

The structure of polyguluronic acid has been determined (64,65) from a specimen containing 73% of this material. The unit cell is orthorhombic, with dimensions $a = 8.6$ Å, $b = 8.72$ Å (fiber axis), and $c = 10.74$ Å, and probable space group $P2_12_12_1$. Again the fiber axis contains two D-guluronic acid residues per turn, which is the repeat obtained when two α-(1,4)-linked residues in the stable 1C conformation are arranged in a 2_1 helix.

A-B-polysaccharides

A number of polysaccharides with a general repeating sequence [A-(1,3)-B-(1,4)]$_n$, where A and B are residues of different sugar monomers, can be isolated from animal connective tissues and seaweeds. The common members of this group are listed in Table 11.3 (66). The first six of these are known as *mucopolysaccharides* and are isolated from animal connective tissues. These tissues consist mainly of cells, tough protein fibers, which are usually collagen or elastin, and an amorphous matrix material known as ground substance (67). The mucopolysaccharides occur in the ground substance and in some cases are also attached to the protein fibers by covalent bonds. The mucopolysaccharide fraction is thought to be involved in the formation of

TABLE 11.3

Basic Structures of Some Polysaccharides from Animal and Seaweed Tissues: Variations on a General Structure $[A\text{-}(1 \to 3)\text{-}B\text{-}(1 \to 4)\text{-}]_n$[a]

Polysaccharide	Residue A	Residue B
Chondroitin	β-D-Glucopyranuronic acid	2-Acetamido-2-deoxy-β-D-galactopyranose
6-Sulfate (chondroitin sulfate C)	β-D-Glucopyranuronic acid	2-Acetamido-2-deoxy-β-D-galactopyranose 6-sulfate
4-Sulfate (chondroitin sulfate A)	β-D-Glucopyranuronic acid	2-Acetamido-2-deoxy-β-D-galactopyranose 4-sulfate
Hyaluronic acid	β-D-Glucopyranuronic acid	2-Acetamido-2-deoxy-β-D-glugopyranose
Dermatan sulfate (chondroitin sulfate B)	α-L-Idopyranuronic acid and β-D-glucopyranuronic acid	2-Acetamido-2-deoxy-β-D-galactopyranose 4-sulfate and, sometimes, 6-sulfate
Keratan sulfate	2-Acetamido-2-deoxy-β-D-glucopyranose and its 6-sulfate	β-D-Galactopyranose and its 6-sulfate
Agarose	3,6-Anhydro-α-L-galactopyranose	β-D-Galactopyranose
Porphyran	3,6-Anhydro-α-L-galactopyranose and L-galactopyranose 6-sulfate	β-D-Galactopyranose and its 6-methyl ether
κ-Carrageenan	3,6-Anhydro-α-D-galactopyranose, its 2-sulfate, and α-D-galactopyranose 6-sulfate	β-D-galactopyranose 4-sulfate
ι-Carrageenan	3,6-Anhydro-α-D-galactopyranose 2-sulfate and α-D-galactopyranose 2,6-disulfate	β-D-Galactopyranose 4-sulfate
λ-Carrageenan	α-D-Galactopyranose 2,6-disulfate	β-D-Galactopyranose and its 2-sulfate
μ-Carrageenan	3,6-Anhydro-α-D-galactopyranose and α-D-galactopyranose 6-sulfate	β-D-Galactopyranose 4-sulfate
κ-Furcellaran	3,6-anhydro-α-D-galactopyranose and α-D-galactopyranose 6-sulfate	β-D-Galactopyranose, some of it possibly representing branch points, and its 4-sulfate

[a] Taken from Ref. 66, with permission of Academic Press. Sources for individual polysaccharides are given therein.

open water-containing networks, which provide flexibility and complement the strength of the protein fibers. As an example, chains of chondroitin sulfates A and C are attached as side chains to the collagen molecules in cartilage and are thought to contribute to the springiness of this material. Keratan sulfate occurs in the cornea of the eye and is believed to facilitate changes in the water content, thereby allowing the required variation in refractive index. The function of the seaweed A-B-polysaccharides, which comprise the rest of Table 10.4, appears to be in the formation of gels, particularly in the young cell wall. Agar and carrageenan are isolated commercially as gel formers. Agar, a family of polysaccharides of which agarose and porphyran are well-characterized members, forms gels when the concentration in aqueous solution is less than 1% and is the substrate for gel electrophoresis and gel chromatography. The carrageenans are now used in a large number of food products, where they provide a creamy texture.

Rees (68) has argued that many of these polysaccharides have relatively similar backbones, which differ mainly in the nature of the side groups, and that they should adopt similar chain conformations. Figure 11.33a (66) shows the generalized backbone for the six mucopolysaccharides, which is similar to that for the series from seaweed (Fig. 11.33b). Conformational analysis, using hard-sphere criteria, indicates that the possible conformations for these backbones are severely limited (68). Less than 1% of the possible backbone conformations lie in the allowed region, and this limitation is even more severe for the connective tissue mucopolysaccharides, since they have bulky equatorial groups. Any crystalline structures obtained for these materials are artificial since the structures in native tissues are all believed to be amorphous, although not necessarily random. Nevertheless, determination of the molecular structure in the crystalline state should assist in determination of the conformation in aqueous gels and in solution.

Carrageenans

Oriented crystalline specimens have been obtained for alkali metal salts of ι, κ, and λ carrageenans (69). The x-ray pattern for the ι-carrageenan is shown in Fig. 11.34 (68). For κ-carrageenans, the layer line spacing is 24.6 Å, and meridionals are observed only on the third, sixth, and ninth layers, suggesting a helical structure with three disaccharide units repeating in 24.6 Å. ι-Carrageenan, however, shows a simpler diffraction pattern with a fiber repeat of 13.0 Å and only one meridional on the third layer. In this case the reflections can be indexed by a hexagonal unit cell with side 22.6 Å, and adjacent helices would have a center-to-center distance of 13.0 Å. The pattern of κ-carrageenan is consistent with a similar cell with side ~ 20 Å. The unit cell dimensions and the observed density for κ-carrageenan require the chains to be arranged in double helices, and a double helix was thus also proposed for the ι structure. In the latter case, the two chains are coiled together with

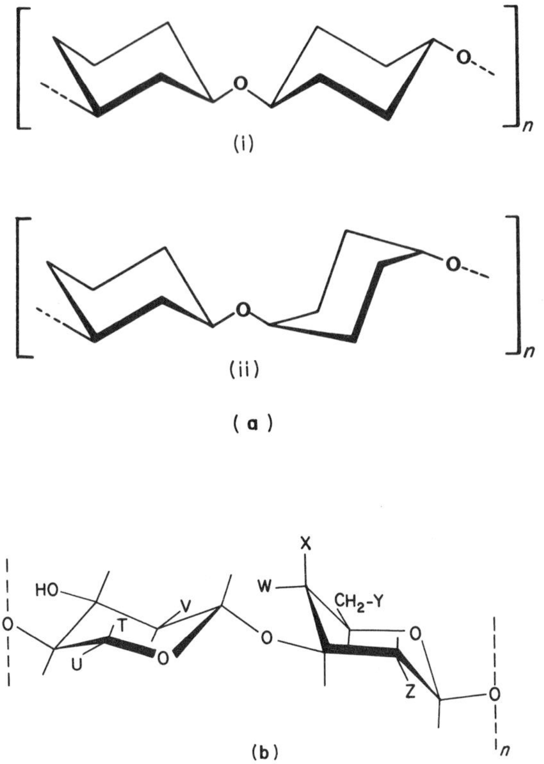

(i)

(ii)

(a)

(b)

Fig. 11.33. (a) Generalized backbones for (i) the connective tissue mucopolysaccharides; (ii) the seaweed A-B-polysaccharides. (Taken from Ref. 66, with permission.) (b) Skeletal formulas for the six connective tissue mucopolysaccharides. Key to the substituted groups:

Mucopoly-saccharide	T	U	V	W	X	Y	Z
Chondroitin	H	COOH	OH	H	OH	OH	$NHCOCH_3$
Chondroitin sulfate A	H	COOH	OH	H	OSO_3^-	OH	$NHCOCH_3$
Chondroitin sulfate C	H	COOH	OH	H	OH	OSO_3^-	$NHCOCH_3$
Keratan sulfate	H	$CH_2OSO_3^-$	$NHCOCH_3$	H	OH	OSO_3^-	OH
Hyaluronic acid	H	COOH	OH	OH	H	OH	$NHCOCH_3$
Chondroitin sulfate B	CH_2OH	H	OH	H	OSO_3^-	OH	$NHCOCH_3$

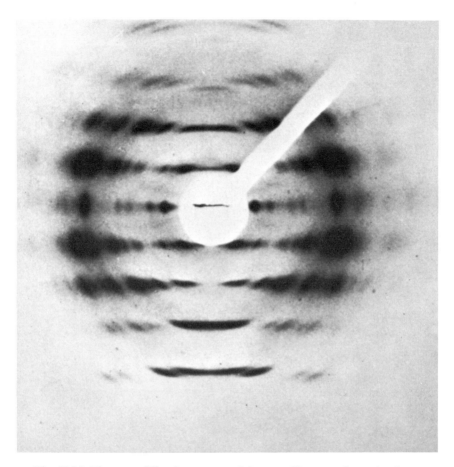

Fig. 11.34. The x-ray diffraction pattern of the crystalline potassium salt of ι-carrageenan. [Taken from Ref. 68, with permission of Academic Press.]

the same sense and a fiber repeat of 26.0 Å. If the chains are arranged so that the second chain is displaced half the repeat distance along the axis, odd-order layer lines for the repeat of 26.0 Å would have zero intensity; only the even layer lines are seen in Fig. 11.34. A photograph of a model of this structure is shown in Fig. 11.35 (66). In κ-carrageenan the chains are not arranged in the special position found for the ι-structure—they may well be antiparallel—and thus all the layer lines predicted for the single chain are observed. Apparantly this form has a shorter fiber repeat. λ-Carrageenan is believed to form a similar structure.

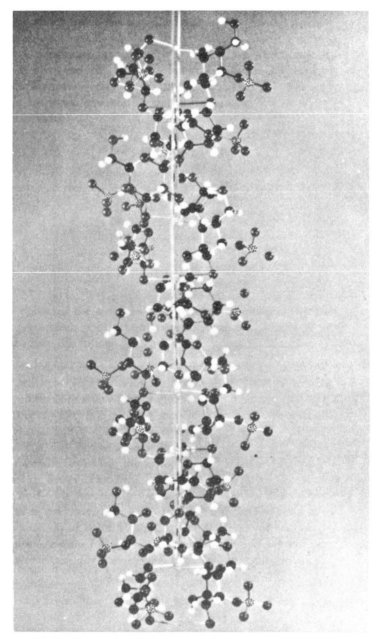

Fig. 11.35. Model of the proposed double-helical structure for ι-carrageenan. [Taken from Ref. 66, with permission of Academic Press.]

Hyaluronic Acid

Details of the crystalline structures of hyaluronic acid have been determined by Atkins and Sheehan (70). The high-quality x-ray patterns for sodium hyaluronate indicate a *single*-chain 3_1 helical conformation, with three disaccharides per turn repeating in 28.5 ± 0.1 Å. The repeat per disaccharide of 9.5 Å along the helix axis is only 0.7 Å less than the maximum extension, and thus this is a highly extended conformation. When the sodium salt is converted to the free acid, the layer line spacing changes to 19.6 ± 0.1 Å. It is argued that the chain conformation has changed from 3_1 to 2_1, and thus the rise per disaccharide repeat has increased to 9.8 Å.

REFERENCES

1. R. H. Marchessault and A. Sarko, *Advan. Carbohyd. Chem.* **22**, 421 (1967).
2. K. Mühlethaler, *J. Polym. Sci.* **C28**, 305 (1969).
3. M. Marx-Figini, *Biochim. Biophys. Acta* **177**, 27 (1969).
4. See, for example, D. W. Jones, *in* "Cellulose and Cellulose Derivatives" (N. M. Bikalis and L. Segal, eds.), Part IV, p. 117. Wiley (Interscience), New York, 1971.
5. K. H. Meyer and L. Misch, *Helv. Chim. Acta* **20**, 232 (1937).
6. G. Honjo and M. Watanabe, *Nature* **181**, 326 (1958); see also D. G. Fisher and J. Mann, *J. Polym. Sci.* **42**, 189 (1960).
7. I. A. Nieduszynski and E. D. T. Atkins, *Biochim. Biophys. Acta* **222**, 109 (1970).
8. P. H. Hermans, J. de Booys, and C. Maan, *Kolloid-Z.* **102**, 169 (1943).
9. P. H. Hermans, "The Physics and Chemistry of Cellulose Fibers." Amer. Elsevier, New York, 1949.
10. C. Y. Liang and R. H. Marchessault, *J. Polym. Sci.*, **37**, 385 (1959).
11. D. W. Jones, *J. Polym. Sci.* **32**, 371 (1958); **42**, 173 (1960).
12. R. St. J. Manley, *Nature* **204**, 1155 (1964); see also B. A. Tønnesen and Ø. Ellefsen, *Nors. Skogindu.* **14**, 266 (1960).
13. R. Muggli, *J. Cellul. Chem. Technol.* **2**, 549 (1969).
14. A. Frey-Wyssling and K. Mühlethaler, *Makromol. Chem.* **62**, 25 (1963).
15. A. J. Heyn, *J. Cell. Biol.* **29**, 181 (1966).
16. D. F. Caulfield, *Text. Res. J.* **41**, 267 (1971).
17. I. Nieduszynski and R. D. Preston, *Nature* **225**, 273 (1970).
18. A. Frey-Wyssling, K. Mühlethaler, and R. Muggli, *R. Holz Roh Werkst.*, **24**, 443 (1966).
19. K. H. Gardner and J. Blackwell, *J. Ultrastruct. Res.* **36**, 725 (1971).
20. R. D. Preston, *Advan. Sci.* **103**, 500 (1966).
21. A. Frey-Wyssling, *Biochim. Biophys. Acta* **18**, 166 (1955).
22. R. St. J. Manley and S. Inoue, *J. Polym. Sci.* **B3**, 691 (1965).
23. A. G. Richards, "The Integument of the Arthropods." Univ. of Minnesota Press, Minneapolis, 1951.
24. K. M. Rudall, *Symp. Soc. Exp. Biol.* **9**, 49 (1955).
25. K. M. Rudall, *Advan. Insect Physiol.* **1**, 257 (1963).
26. C. H. Giles, A. Hassan, M. Laidlaw, and R. Subramanian, *J. Soc. Dyers Colour.* **74**, 647 (1958).

27. R. H. Hackman and M. Goldberg, *J. Insect. Physiol.* **2**, 228 (1958).
28. M. Falk, D. B. Smith, J. McLachlan, and A. G. McInnes, *Can. J. Chem.* **44**, 2269 (1966).
29. J. Blackwell, K. D. Parker, and K. M. Rudall, *J. Mar. Biol. Ass. U.K.* **45**, 659 (1965).
30. J. Blackwell, K. D. Parker, and K. M. Rudall, *J. Mol. Biol.* **28**, 383 (1967).
31. D. Carlström, *J. Biophys. Biochem. Cytol.* **3**, 669 (1957).
32. J. Blackwell, *Biopolymers* **7**, 281 (1969).
33. K. D. Parker, cited by K. M. Rudall, Ref. 25.
34. K. M. Rudall, *in* "Conformation of Biopolymers," (G. N. Ramachandran, ed.), Vol. 2, p. 751, Academic Press, New York, 1967.
35. K. H. Gardner and J. Blackwell, *J. Polym. Sci.* **C32**, 327 (1971).
36. T. E. Timell, *Advan. Carbohyd. Chem.*, **19** 247 (1964); **20**. 410 (1965).
37. R. H. Marchessault and C. Y. Liang, *J. Polym. Sci.* **59**, 357 (1962).
38. R. H. Marchessault and W. Settineri, *J. Polym. Sci.* **B2**, 1047 (1964); see also W. Settineri and R. H. Marchessault, *J. Polym. Sci.* **C11**, 253 (1965).
39. I. A. Nieduszynski and R. H. Marchessault *Biopolymers*, in press.
40. R. H. Marchessault, F. F. Morehead, N. M. Walter, C. P. J. Glaudemans, and T. E. Timell, *J. Polym. Sci.* **51**, S66 (1961).
41. S. M. Gabby, M. S. thesis, Université de Montréal, Canada, 1971; S. M. Gabby P. R. Sundararajan, and R. H. Marchessault *Biopolymers*, **11**, 79 (1972).
42. E. Frei and R. D. Preston, *Proc. Roy. Soc. London* **B160**, 293 (1964).
43. E. D. T. Atkins, K. D. Parker, and R. D. Preston, *Proc. Roy. Soc. London*, **B173**, 209 (1969).
44. D. A. Rees and W. E. Scott, abstracts of papers presented at meeting of Amer. Chem. Soc., Los Angeles, 1971.
45. H. Meier, *Biochim. Biophys. Acta* **28**, 229 (1958).
46. E. Frei and R. D. Preston, *Proc. Roy. Soc. London* **B169**, 127 (1963).
47. K. J. Palmer and M. Ballantyne, *J. Amer. Chem. Soc.* **72**, 736 (1950).
48. C. T. Greenwood, *Advan. Carbohyd. Chem.* **11**, 335 (1956).
49. J. R. Katz and T. B. Van Itallie, *Z. Phys. Chem.* **A150**, 90 (1930).
50. R. E. Rundle, *J. Amer. Chem. Soc.* **69**, 1769 (1947); R. S. Bear and D. French, *J. Amer. Chem. Soc.* **63**, 2298 (1941); R. E. Rundle and D. French, *J. Amer. Chem. Soc.* **65**, 1707 (1943).
51. R. E. Rundle and F. C. Edwards, *J. Amer. Chem. Soc.* **65**, 2200 (1943).
52. H. F. Zobel, A. D. French, and M. E. Hinkle, *Biopolymers* **5**, 837 (1967).
53. J. Blackwell, A. Sarko, and R. H. Marchessault, *J. Mol. Biol.* **42**, 379 (1969).
54. R. St. J. Manley, *J. Polym. Sci.* **A2**, 4503 (1964); see also Y. Yamashita, *J. Polym. Sci.* **A3**, 3251 (1965).
55. R. Wolf and R. C. Schultz, *J. Macromol. Sci. Chem.* **A2**, 821 (1968).
56. A. Sarko and R. H. Marchessault, *J. Amer. Chem. Soc.* **89**, 6454 (1967).
57. F. R. Senti and L. P. Witnauer, *J. Amer. Chem. Soc.* **70**, 1438 (1948).
58. J. J. Jacobs, R. R. Bumb, and B. Zaslow, *Biopolymers* **6**, 1659 (1968).
59. D. R. Kreger, *Biochim. Biophys. Acta* **6**, 406 (1951).
60. K. Mühlethaler, *Stärke* **17**, 1965 (1965).
61. R. H. Marchessault and A. Sarko, *J. Polym. Sci.* **C28**, 317 (1969).
62. J. Borch, A. Sarko and R. H. Marchessault, *Stärke*, **21**, 279 (1969); R. S. Finkelstein and A. Sarko, *Biopolymers*, **11**, 881 (1972).
63. E. Frei and R. D. Preston, *Nature* **196**, 130 (1962).
64. E. D. T. Atkins, W. Mackie, and E. E. Smolko, *Nature* **225**, 626 (1970).
65. E. D. T. Atkins, W. Mackie, K. D. Parker, and E. E. Smolko, *Poly. Lett.*, **9** 311 (1971).

66. D. A. Rees, *Advan. Carbohyd. Chem.* **24**, 267 (1969).
67. J. S. Brimacombe and J. M. Webber, "Mucopolysaccharides," B.B.A. Library, Vol. 6. Amer. Elsevier, New York, 1964.
68. D. A. Rees, *J. Chem. Soc. B* 217 (1969).
69. N. S. Anderson, J. W. Campbell, M. M. Harding, D. A. Rees, and J. W. B. Samuel, *J. Mol. Biol.* **45**, 85 (1969).
70. E. D. T. Atkins and J. K. Sheehan, *Nature*, (*New Biology*) **235**, 253 (1972).

12

NUCLEIC ACIDS AND POLYNUCLEOTIDES

Introduction

The genetic information required for the function and replication of biological organisms is provided by nucleic acids. The chemical structures of the monomer constituents of these molecules are described in Chapter 1. Two types of nucleic acid are found to occur in nature, known as ribonucleic acid (RNA) and deoxyribonucleic acid (DNA). Ribonucleic acid has the general formula

$$\left[\begin{array}{c} \text{base} \\ | \\ -\text{ribose} - \text{phosphate} - \end{array} \right]_n$$

The base can be any one of four common unsaturated bases—adenine (A) guanine (G), cytosine (C), and uracil (U)—as well as a number of less common *minor* bases. In DNA, the situation is similar, except that the sugar residue is 2-deoxy-D-ribose, and thymine (T) replaces uracil among the common bases. The genetic information is carried as a code consisting of a specific sequence of bases, which dictates the sequence of amino acids, etc., during the synthesis of new biological macromolecules. The present knowledge of the nature of this code will be discussed later in this chapter. Extensive reviews of the structure and properties of nucleic acids can be found in References 1 and 2.

Deoxyribonucleic acid occurs in the nucleus of the cell as a nucleoprotein complex. Since DNA is highly negatively charged due to the presence of the phosphate groups, the associated proteins frequently contain a high proportion of basic residues, such as lysine in the case of histones. After removal of the protein, the molecular weight of the DNA is found to be in the range 10^6–10^9, depending on the source. It is thought that all the genetic information is carried by a single length of DNA in the cell nucleus, which consists of two associated DNA chains in a double-helical conformation. Deoxyribonucleic acid is also found in certain viruses, where it can be either single stranded or double helical. In some viruses, the ends of the DNA chains may be covalently linked together to give cyclic molecules. For example, ϕX174 phage (a bacterial virus) contains single-strand cyclic DNA with a molecular weight of 1.7×10^6; polyoma virus DNA is double stranded and cyclic with a molecular weight of 3×10^6. The acyclic double-stranded DNA from T2 phage has a molecular weight of 1.3×10^8.

Ribonucleic acid serves a variety of functions in the cell in the translation of the genetic message contained in the nuclear DNA. Certain RNA's with molecular weights in the range 0.6–2.0×10^6 are found along with protein in the cytoplasmic particles known as *ribosomes*. These structures serve as the sites for protein synthesis, and the RNA extracted therefrom is termed ribosomal RNA (rRNA). The so-called messenger RNA (mRNA), which has a base sequence equivalent to sections of the nuclear DNA, has molecular weights of the order of 5×10^5. Ribonucleic acids with small molecular weights are also extracted from cells. These include transfer RNA (tRNA) molecules, which consist of single chains of 75–85 nucleotides. Complete base sequences have been determined for some of these tRNA molecules, and what is known of their structure and function is discussed below. In addition, single- and double-stranded RNA's are found in virus particles.

The results of base fraction analyses for a number of DNA's and RNA's are listed in Tables 12.1 and 12.2; it can be seen that the common bases occur in approximately equal quantities, i.e., all four are generally in the range 20–30% of the total. It was first noted by Chargaff (3) that in many cases the percentages of adenine and thymine (or uracil) are more nearly equal, as are the percentages of guanine and cytosine. This observation proved to be very important in the determination of the double-helical structure for DNA by Watson and Crick (4). In the results for DNA (Table 12.1), it can be seen that the proportions of the bases follow Chargaff's rule fairly closely except for the DNA's wheat germ, T2 phage, and ϕX174 phage. In wheat germ DNA, the concentration of cytosine is low, but this is compensated by the 5.8% of 5-methylcytosine (5-MeC), so that $[C] + [5\text{-MeC}] \simeq [G]$. In the case of T2 phage, cytosine is completely replaced by 5-hydroxymethylcytosine. In the figures for ϕX174 phage, large deviations occur which are

TABLE 12.1

Molar Proportions of Bases in DNA's from Various Sources

Source	Adenine	Guanine	Cytosine	Thymine	5-Methyl-cytosine	Reference[a]
Man thymus	30.9	19.9	19.8	29.4		1
Pig spleen	29.6	20.4	20.8	29.2		1
Sheep liver	29.3	20.7	20.8	29.2		1
Calf thymus	28.2	21.5	21.2	27.8	1.3	2
Herring sperm	27.8	22.2	20.7	27.5	1.9	2
E. coli (K12)	26.0	24.9	25.2	23.9		3
M. tuberculosis (avian)	15.1	34.9	35.4	14.6		4
Wheat germ	26.5	23.5	17.2	27.0	5.8[b]	1
T2 phage	32.5	18.2		32.5	16.8[b]	5
φX174	24.6	24.1	18.5	32.8		6

[a] Key to references:
1. G. R. Wyatt, *Biochem. J.* **48**, 584 (1951).
2. E. Chargaff and R. Lipshitz, *J. Amer. Chem. Soc.* **75**, 3658 (1953).
3. B. Gandelman, S. Zamenhof, and E. Chargaff, *Biochim. Biophys. Acta* **9**, 399 (1952).
4. E. Vischer, S. Zamenhof, and E. Chargaff, *J. Biol. Chem.* **177**, 429 (1949).
5. A. D. Hershey and N. E. Melechen, *Virology* **3**, 207 (1957).
6. R. L. Sinsheimer, *J. Mol. Biol.* **1**, 43 (1959).
[b] Hydroxymethylcytosine.

TABLE 12.2

Molar Proportions of Bases in RNA's from Various Sources

Source	Adenine	Guanine	Cytosine	Uracil	Reference[a]
Calf liver	19.5	35.0	29.1	16.4	1
Rabbit liver	19.3	32.6	28.2	19.9	2
Chicken liver	19.5	33.3	26.5	20.7	2
Bakers yeast	25.1	30.2	20.1	24.6	3
E. coli	27.0	27.6	23.0	22.4	4
Serratia narcescens	27.0	27.6	23.0	22.4	4
Tobacco mosaic virus	29.9	25.4	18.5	26.3	5
Turnip yellow mosaic virus	22.6	17.2	38.0	22.2	6

[a] Key to references:
1. E. Volkin and C. E. Carter, *J. Amer. Chem. Soc.* **73**, 1516 (1951).
2. G. W. Crosbie, R. M. S. Smellie, and N. Davidson, *Biochem. J.* **54**, 287 (1953).
3. E. Chargaff, B. Magsanik, E. Vischer, C. Green, R. Doniger, and D. Elson, *J. Biol. Chem.* **186**, 51 (1953).
4. D. Elson and E. Chargaff, *Biochim. Biophys. Acta*, **17**, 367 (1955).
5. C. A. Knight, *J. Biol. Chem.* **171**, 297 (1948).
6. R. Markham and J. D. Smith, *Biochem. J.* **49**, 401 (1951).

not balanced by the presence of similar minor bases. However, this DNA is single stranded and thus does not need to have equal quantities of certain bases, which, as we shall see, are necessary for the double-helical species. Chargaff's rule appears to hold approximately for the various RNA's (Table 12.2), although the match is not quite as good as it is for the DNA's.

Double-Helical Conformation

x-Ray patterns of oriented fibers of DNA were first obtained in the 1930's. These were all of poorly crystalline fibers and were similar to, or of lower quality than, that shown in Fig. 3.22. Astbury and Bell (5) noted the presence of a meridional reflection at $d = 3.34$ Å. This spacing corresponds to the "thickness" of one of the planar conjugated bases; a similar figure is obtained for the separation of layers in structures such as graphite. Astbury and Bell proposed that the bases are stacked perpendicular to the fiber axis. It was thought that the presence of higher layer lines could indicate some sort of repeat in the nucleotide sequence.

Following the formulation of the α helix by Pauling and Corey in 1951 (6), Cochran *et al.* (7) developed the theory of the scattering of x rays by helical structures. It was soon recognized that the x-ray patterns from oriented fibers of DNA were indicative of a helical structure. Furthermore, the diameter of the helix required that it be a multiple-chain conformation. Pauling and Corey (8) first considered a triple-strand helix with the bases on the outside and the phosphate groups close to the fiber axis. This arrangement was discarded because it was inconsistent with the chemical properties of DNA and also with the x-ray data from more crystalline specimens. Watson and Crick (4) proposed a double-helical structure in which antiparallel chains are arranged together with 10 nucleotides per turn, repeating in 34 Å. Two deoxyribose–phosphate chains lie on the outside of the molecule, and the purine and pyrimidine bases are hydrogen bonded together in the center. The nature of the hydrogen bonding between the bases is crucial to the structure. Watson and Crick noted Chargaff's observation that the base concentrations follow the rule [A] $\simeq$ [T] and [C] $\simeq$ [G]. These bases must be stacked approximately perpendicular to the fiber axes because their thickness is ~ 3.4 Å. Arranging models of the bases side by side showed that the two most favorable hydrogen-bonding positions occur when A is bonded to T and C to G, as shown in Fig. 12.1. The distance between the linkage points for the sugar–phosphate chains is approximately 11 Å for each pair. Watson and Crick proposed that this type of base pairing occurs in the DNA molecule, leading to a symmetrical double helix. A typical imaginary sequence of bases

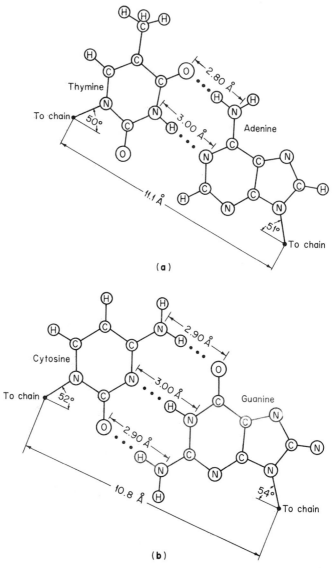

Fig. 12.1. Watson and Crick base-pairing scheme for hydrogen bonding between chains in the DNA double helix. (a) Adenine–thymine (A ··· T). (b) Guanine–cytosine (G ··· C). [Taken from A. Rich, *Rev. Mod. Phys.* **31**, 191 (1959), with permission.]

along one DNA chain is shown in the left-hand side of Fig. 12.2. It was proposed that the other chain must have the sequence shown on the right-hand side of Fig. 12.2. The bases on these chains hydrogen bond together and

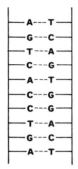

Fig. 12.2. Base pairing between complementary strands of DNA to form the double helix. Adenine is always opposite to thymine and guanine to cytosine.

the chains twist together into the double helix. This in no way limits the base sequence of one chain, but the second chain must have complementary bases. A diagram of the double helix is shown in Fig. 12.3 (1).

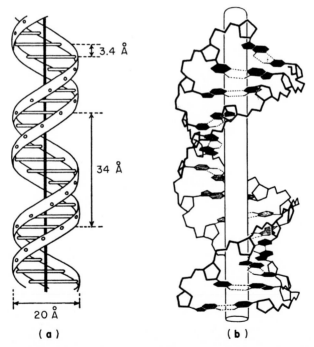

Fig. 12.3. Schematic representation of the DNA double helix. (a) Path of the main chain with 10 base pairs per turn repeating in 34 Å. (b) Section of (a) in greater detail. The purine and pyrimidine bases are shaded black, and the open pentagons are the pentose rings. [Taken from Ref. 4, with permission of MacMillan and Co. Ltd., and from Ref. 1, with permission of the publishers, Methuen and Co., Ltd.]

Evidence from a large number of techniques indicates that the double-helical structure is correct. In x-ray diffraction studies of highly crystalline salts of DNA, excellent agreement has been obtained between the observed data and that predicted for the double-helical model (9–11). Strands of DNA observed in the electron microscope have a thickness of ∼20 Å, which corresponds to the dimension of the model (12). Furthermore, the measured length of the molecules corresponds to the theoretical length of the double helix for that molecular weight. Many of the solution properties of DNA can be interpreted only in terms of a double-helical structure. In addition, pairs of bases, nucleosides, and nucleotides can be cocrystallized, and Watson and Crick base pairing has been found in many cases in these structures, as determined by x-ray diffraction (13).

Solid-State Structure

In the solid state, DNA has been found to exist with at least three different chain conformations, known as the A, B, and C structures. The form believed to occur in nature and in solution is DNA-B. This structure is obtained when the sodium salt is maintained at 90 % relative humidity, and the x-ray fiber diagram is that shown in Fig. 3.22. Under these conditions, DNA-B is not very crystalline, and this x-ray pattern is often used, as it is in Chapter 3, as an example of scattering by an isolated discontinuous helix. However, when DNA fibers prepared from 3 % lithium chloride solution are maintained at 66 % relative humidity, the B form takes up a very crystalline structure, and the resultant x-ray pattern is shown in Fig. 12.4 (9). This structure has been refined by Langridge *et al.* (9). A model of the DNA-B structure is shown in Fig. 12.5 (14). Each chain has 10 nucleotides per turn with a rise per nucleotide of 3.4 Å along the helix axis. The conjugated bases form planar Watson and Crick pairs and have relatively high electron density. Consequently, the theoretical x-ray pattern is very sensitive to changes in orientation of the planar base pairs. This orientation can thus be refined very accurately, and in DNA-B, the base planes are almost perpendicular to the fiber axis. The two sugar–phosphate chains wind around the stack of perpendicular base pairs. Since the two strands have opposite sense, these sugar–phosphate chains are not equally spaced along the molecule, with the result that the double helix possesses wide and narrow grooves. These grooves are thought to be important to the structure of nucleoprotein complexes, where the protein chains are believed to wrap around one of the grooves of the double helix.

In the absence of lithium chloride, but still at 66 % relative humidity, the chain adopts a new conformation known as DNA-C (10). The x-ray pattern

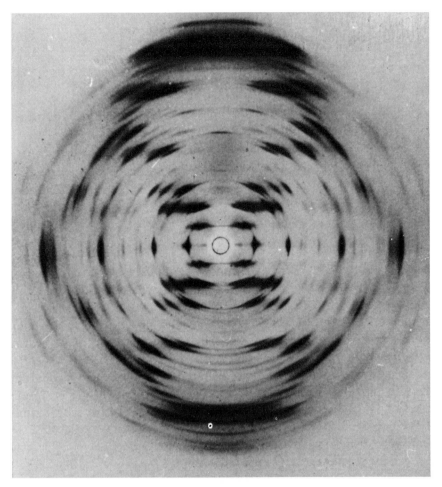

Fig. 12.4. The x-ray fiber diagram of an oriented crystalline specimen of DNA-B. [Taken from Ref. 9, with permission of Academic Press.]

obtained for such a specimen is shown in Fig. 12.6 (11). The chain conformation has been shown to be a nonintegral 28_3 double helix ($9\frac{1}{3}$ nucleotides per turn) with a rise per nucleotide of 3.3 Å along the helix axis. In this structure, base planes are tilted 5° away from the perpendicular position with a result that the widths of the two grooves are more pronounced than in DNA-B. The double helices are packed hexagonally in this DNA-C structure. When the specimen containing 3% lithium chloride (Fig. 12.4) is maintained at 44% relative humidity, conversion to DNA-C occurs, but under these conditions the chains are packed in an orthorhombic unit cell (10).

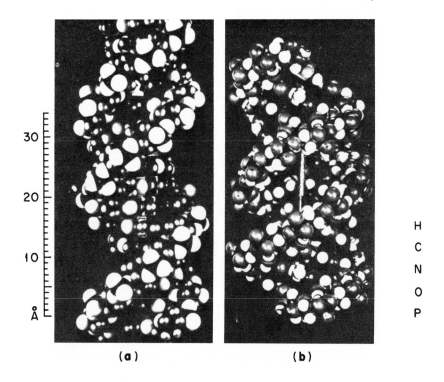

Fig. 12.5. Space-filling Courtauld models of (a) DNA-B, (b) RNA (11_1 helix; a conformation similar to DNA-A). [Taken from Ref. 14, with permission of John Wiley and Sons, Inc.]

A third pattern is obtained for the sodium, potassium, or rubidium salts at 75% relative humidity. This is the DNA-A structure; an x-ray fiber diagram of this structure is shown in Fig. 12.7. It has a double-helical conformation with 11 nucleotides per turn and a rise per nucleotide of 2.6 Å (11). The base planes in the A structure are inclined at 20° to the perpendicular, the tilt being in the opposite direction to that in DNA-C. One result of this tilt is that the grooves in the A helix are more nearly the same width. Model building and conformational analysis indicate that in the A form, the deoxyribose ring has the C-3'-*endo* conformation, while in the B and C forms the sugar is C-2'-*endo*.

Oriented specimens can also be prepared for ribosomal RNA. Two structures have been distinguished, known as α- and β-RNA, which apparently have the same chain conformation but different chain packing for the helical molecules (15). The x-ray patterns are not of the same quality as those of the crystalline DNA structures, and it is not possible to be so definite about the conformation. However, the patterns of RNA and DNA show the same characteristic features, and it is clear that RNA forms the same sort of Watson and Crick double helix. The fiber repeat is 30.4 Å and there is apparently a meridional on the tenth layer line. This suggests a double-helical structure with 10 nucleotides per turn and a rise per residue of 3.04 Å, which would require the bases to be tilted at 11° to the perpendicular. However, the systematic absences elsewhere on the photograph and chain-packing considerations indicate that, if the chains have a 10_1 helical conformation, the double helices must be arranged around a threefold screw axis and packed with space group $P3_2$. For this space group, a meridional

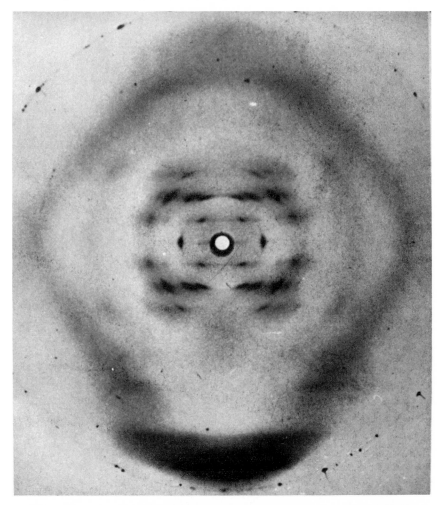

Fig. 12.6. The x-ray fiber diagram of an oriented crystalline specimen of DNA-C. [Taken from Ref. 11, with permission of Academic Press.]

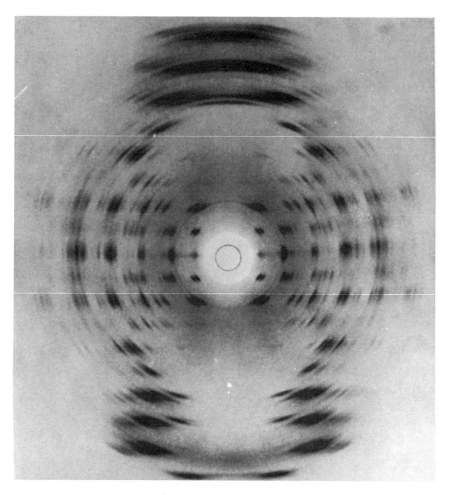

Fig. 12.7. The x-ray fiber diagram of an oriented crystalline specimen of DNA-A. [Taken from Ref. 11, with permission of Academic Press.]

on the tenth layer line should be absent. The fact that such a reflection is present indicates a distortion of the lattice, but it is not necessarily evidence for a 10_1 helix. Calculations for a second model with 11 base pairs per turn, and the base planes tilted by $14°$ from the perpendicular, give slightly better agreement, but the x-ray data are not of sufficient quality to distinguish between the models.

This problem for the RNA structure has been resolved to some extent by study of the analogous structures formed by synthetic polyribonucleotides (16). A considerable amount of research has been done on the synthetic *homo*polyribonucleotides: poly A, poly C, poly G, and poly U. Another base that is frequently used is hypoxanthine, which forms the ribonucleotide *inosinic acid*, I. Hypoxanthine is one of the minor bases found in tRNA, which can form base pairs I $\cdots$ C equivalent to C $\cdots$ G hydrogen bonding.

Additional experimental evidence in favor of the concept of base pairing came from the observation of complex formation by poly A and poly U. When equal quantities of these polynucleotides are mixed, the complex known as poly (A + U) is formed, in which a chain of poly A unites with one of poly U to form a double helix with A $\cdots$ U base pairs along its entire length. The x-ray diffraction pattern of fibers of this complex is shown in Fig. 12.8 (16). This pattern has the same general appearance as those for the crystalline DNA structures and is considerably more detailed than those for

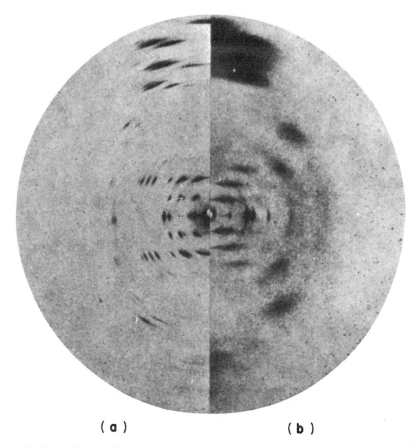

(a)   (b)

Fig. 12.8. The x-ray fiber diagrams of oriented crystalline specimens of (a) poly (A + U), a 1:1 complex of poly A and poly U; (b) poly (A + 2U), a 1:2 complex of poly A and poly U. [Taken from Ref. 16, with permission of MacMillan and Co., Ltd.]

ribosomal RNA. The structure appears to be isomorphous with β-RNA and has been considered in terms of 10_1 or 11_1 double helices. Meridional reflections are observed on the first, second, and tenth layer lines. However, other evidence for the chain packing indicates that these layer lines should have zero meridional intensity for either 10_1 or 11_1 helical structures, and the observed meridionals are thought to be spurious, probably due to imperfections in the structure. Comparison of the observed and calculated intensities shows a preference for the 11_1 helical conformation. By analogy it is argued that the x-ray patterns for native RNA are indicative of 11_1 helices. A model for the 11_1 helix of RNA is shown in Fig. 12.8 (14); this is essentially the same structure as that of DNA-A.

A similar 11_1 double-helical conformation has been proposed for the poly (I + C) complex (16). However, at 2% salt concentration, changes are observed in the x-ray pattern of this complex, which have been interpreted as resulting from conversion to a 12_1 double helix. The structure of the poly (G + C) complex is not well characterized, but a nonintegral double helix has been proposed with 11.3 nucleotides per turn (16).

It can be seen that all the structures determined for DNA, natural RNA, and base-paired complexes of synthetic RNA's are very similar. The chains form antiparallel double helices linked by Watson and Crick base pairing. However, the degree of twisting in the helix is variable.

Crystalline structures are also formed by some of the individual homopolyribonucleotides. In these structures base pairs of the type A $\cdots$ U and C $\cdots$ G cannot be formed. However, x-ray work has shown that poly C, poly A, and poly I form multiple-strand helices. In these structures the identical bases are hydrogen bonded together in planar complexes in the center of the helix. Figure 12.9 (17) shows the detailed x-ray pattern of a fibrous specimen of poly C prepared at acid pH. A meridional reflection is observed at $d = 3.11$ Å on the sixth layer line of this pattern. However, poorly crystalline poly C shows additional layer lines halfway between those for the crystalline specimens. Thus the meridional is really on the twelfth layer line and the structure is a 12_1 double helix with parallel chains arranged so that the odd layer lines have zero intensity. The x-ray pattern of poly A at acid pH (18) is of poorer quality than that of poly C. Nevertheless, the data indicate a multiple-strand helix. A meridional reflection is observed on the fourth layer line, and a double helix with eight nucleotides per turn is suggested for this structure. The rise per residue of 3.8 Å is the highest recorded for a polynucleotide structure. In the case of poly I, the pattern is interpreted (19) in terms of a triple helix of parallel chains, each of which is a 26_3 helix (i.e., $8\frac{2}{3}$ nucleotides per turn). The rise per nucleotide along the helix is 3.4 Å. The proposed hydrogen-bonding network between the bases of these three structures is shown in Fig. 12.10.

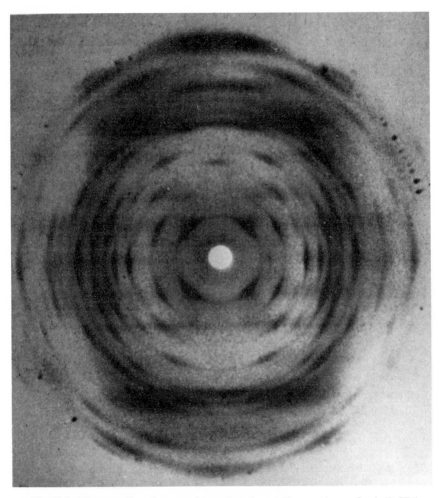

Fig. 12.9. The x-ray fiber diagram of an oriented crystalline specimen of poly C. [Taken from Ref. 17, with permission of MacMillan and Co., Ltd.]

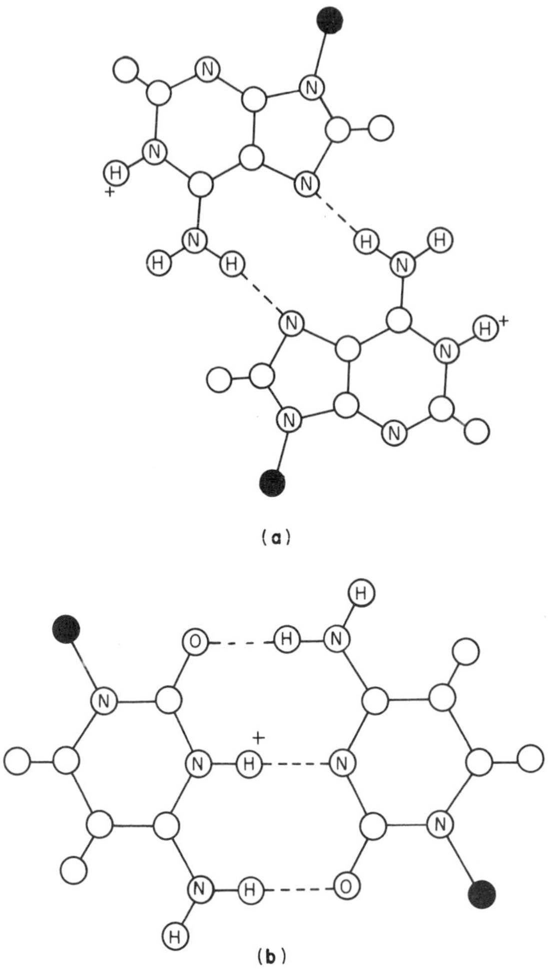

(a)

(b)

Solution Properties

All the evidence from solution properties indicates that DNA can retain its double-helical conformation in solution, unless definite measures are taken to break up the structure. One of the most interesting properties of DNA is the so-called helix–coil transition in solution; the breakup of the double helix into single random strands is achieved by heating in solution or by changes in pH or ionic strength. This phenomenon is also known as denaturation or *melting*, and is analogous to the breakup of the collagen triple helix.

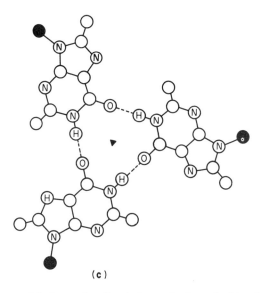

(c)

Fig. 12.10. Proposed hydrogen bonding between the bases in (a) poly A, (b) poly C, Poly I. [Taken from D. R. Davies, *Annu. Rev. Biochem.* **36**, 321 (1967), with permission of Annual Reviews, Inc.]

Simple light-scattering measurements for calf thymus gland DNA indicate a molecular weight of 3.5×10^6 and an R_G of 1166 Å (20). This corresponds to a rodlike species with a length of about one-third that predicted for the double helix with this molecular weight. It is postulated that this DNA in solution is almost all double helix, but there are one or two flexible regions, perhaps due to incorrect sequencing leading to base-pair imperfections.

One of the most useful ways of following a conformational change in DNA such as denaturation is by means of ultraviolet spectroscopy (see Chapter 6). The purine and pyrimidine bases are highly conjugated and absorb strongly in the UV region in the range 250–290 mμ. Deoxyribonucleic acid possesses a mixture of attached bases and absorbs ultraviolet light, giving a broad, strong peak at about 260 mμ. By combining the UV spectra of the constituent nucleotides in the ratio of the proportions of the bases in the DNA specimen, one can predict the spectrum of DNA, i.e., the position of the peak. However, the observed intensity for DNA is about 40 % less than the calculated figure. This is the same effect found for polypeptides and is known as hypochromism. It results from the regular side-by-side interactions between the stacked bases, i.e., interaction between the π orbitals above and below the planes of the rings. This regular structure breaks down on denaturation, and a rise in absorption occurs.

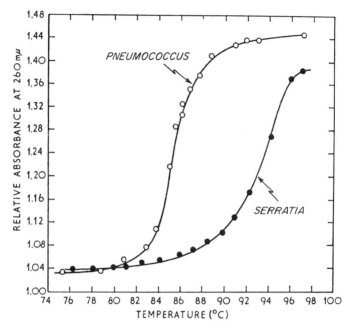

Fig. 12.11. Plot of the relative UV absorbance at 260 mμ against temperature for *Pneumococcus* and *Serratia* DNA's. The temperature at the point of inflection in curves of this type is known as the melting temperature, T_m. [Taken from Ref. 21, with permission of Academic Press.]

Figure 12.11 (21) shows a plot of the intensity of the absorption band at 260 mμ for bacterial DNA as a function of temperature. It can be seen that there is a sharp increase in absorption at about 85°C corresponding to a sudden breakdown of the helical structure at about this point. The temperature corresponding to the inflection point in the curve is known as the *melting temperature*, T_m. A similar plot of viscosity against temperature would show a sharp decrease in viscosity at this temperature. These curves are obtained by slow heating of a solution of DNA. If the solution is cooled slowly from temperatures above T_m the curve is reversed, although full hypochromicity is never restored, meaning that all the chains do not fit back together perfectly. With rapid cooling from above T_m there is no restoration of hypochromicity and the DNA stays as separate chains. This is also revealed by density gradient centrifugation (21). Native pneumococcal DNA has a density of 1.700. Denaturation followed by slow cooling gives a density of 1.704. Rapid cooling gives a density of 1.716 for the random coil.

That strand separation actually occurs was demonstrated (21) by the preparation of double strand DNA from *E. coli* with one strand labeled with ^{15}N. This was achieved by isolation of DNA from *E. coli* after allowing one cell division in an ^{15}N medium. After heating to allow denaturation, followed by rapid cooling, two fractions are detected, one with density 1.724 corresponding to ^{14}N *E. coli* DNA and one with density 1.740 corresponding to the ^{15}N species. Similarly, double strand ^{14}N DNA can be mixed with double strand ^{15}N DNA. After denaturation and slow cooling, a new peak for the hybrid ^{14}N^{15}N double helix is detected as well as those for ^{14}N^{14}N and ^{15}N^{15}N double strands, and it must be concluded that the chains are completely separated in solution and then recombine with any complementary chain.

The value of T_m varies for DNA's from different sources and is found to have a higher value the higher the proportion of [C + G] in the specimen. Figure 12.12 (21) shows that the variation of T_m is an approximately linear function of the percentage of [C + G] bases. Examination of the Watson and Crick base-pairing scheme in Fig. 12.1 shows that there are three hydrogen bonds for the C $\cdots$ G pair but only two for the A $\cdots$ T pair. Thus, the

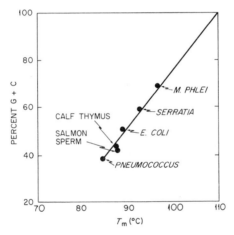

Fig. 12.12. Plot of the total guanine and cytosine content of various DNA's (expressed as a percentage of the total base content) against their respective melting temperatures. [Taken from Ref. 21, with permission of Academic Press.]

C $\cdots$ G pair is more strongly bonded, and the higher the [C + G] concentration the more difficult it becomes to separate the chains and the higher the value of T_m.

For RNA's in solution, most evidence is indicative of a globular molecule rather than a rod, except for some of the synthetic polynucleotides in complexes such as poly (A + U). Solutions of poly U are only slightly hypochromic (less than 10%) and this effect is not temperature dependent. The specific rotation is $-8°$, while double-helical DNA's have a large positive value. These results, and those from light-scattering and flow-birefringence measurements, are all in accord with a random chain in a highly coiled conformation (22). This ties in with the x-ray results, which indicate that the structure is not crystalline. Poly A, on the other hand, is known to form a double-helical structure in the solid state (18). However, this rodlike molecule has not been detected in solution. Light scattering indicates a random coil molecule (22). For a DP of about 9000, the radius of gyration is 690 Å, corresponding to an end-to-end separation of 1700 Å. For a double helix, the length would be 30,000 Å. Viscosity measurements are consistent with a random structure, and no fibrous structures are seen in the electron microscope. Nevertheless, some ordering must be present in this structure. At 20°C and 0.1 M ionic strength, the hypochromism of the solution is about 40%. This hypochromism is temperature dependent, but, unlike that of DNA, it decreases steadily with increasing temperature, as shown in Fig. 12.13 (23).

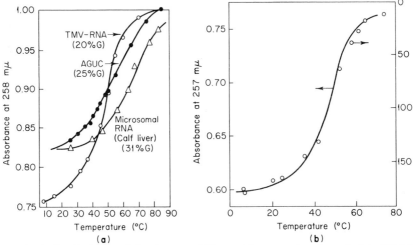

Fig. 12.13. Variation of absorbance at 257 mμ with temperature for poly A in 0.015 M citrate, 0.15 M NaCl at pH 7. [Taken from Ref. 23, with permission of Elsevier Publishing Company.]

A specific rotation of $+155°$ is observed for a solution of poly A with molecular weight 10^5, which strongly suggests some helix formation. The hypochromism and rotation are greatly reduced in solution in 8 M urea, which would break the hydrogen bonds and disrupt any helical regions. All this evidence (22) suggests that poly A at alkaline pH has regions of ordered helix separated by regions of random chain. It is argued that the molecule forms helix loops, as shown in Fig. 12.14, and that these are grouped together to form the globular molecule.

This type of conformation is also believed to occur for natural RNA (24). The RNA from tobacco mosaic virus (TMV) with a molecular weight of

2×10^6 has a radius of gyration of 300 Å in solution at pH 8.5 and ionic strength 0.06 M. This corresponds to about 1/25th of that for an extended double helix. However, the solution is highly hypochromic and this property declines slowly with temperature rise, as for poly A. The specific rotation is $+190°$ at 20°C and declines to zero at 80°C. A similar coiled structure is therefore proposed for RNA in solution, with short loops of double helix followed by random-chain conformation. These double-helical regions are probably of different lengths, which accounts for the gradual change in hypochromicity with temperature, unlike the sharp transition for DNA.

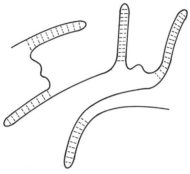

Fig. 12.14. Helix with loops model for RNA sections of regular double helix of varying length connected by regions with random conformation.

We have seen that a crystalline complex, poly (A + U), can be formed from a 50 : 50 mixture of poly A and poly U in solution (16). The absorption at 259 mμ for this solution is some 25 % lower than that predicted from values for separate solutions of poly A and poly U, which suggests that there is extensive interaction between the species in solution. A plot of the absorption at 259 mμ versus molar percentage of poly A and poly U is shown in Fig. 12.15a (25). The solid circles show the UV absorption measured immediately on mixing various fractions at low ionic strength (0.01 M at pH 6.5). The open circles indicate the absorption values after the solution has been allowed to stand for 24 hr. It can be seen that the absorption falls in each case and that the maximum hypochromism corresponds to a 1 : 1 mixture, i.e., poly (A + U). At higher ionic strength (0.1 M NaCl), this curve is obtained immediately on mixing. However, after the mixtures have been allowed to stand for 48 hr, a distinct bulge is detected on the side corresponding to higher poly U concentrations, and the absorption minimum is shifted toward higher [poly U]: this effect is dependent on ionic strength. At a concentration of 0.7 M sodium chloride, or in the presence of a much lower concentration of magnesium, the minimum of the curve (25) is at 67 % poly U [see Fig. 12.15b (25)]. This has been interpreted as a three-chain complex of one poly A and

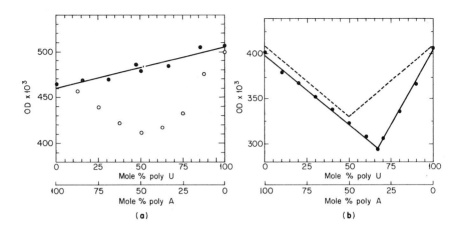

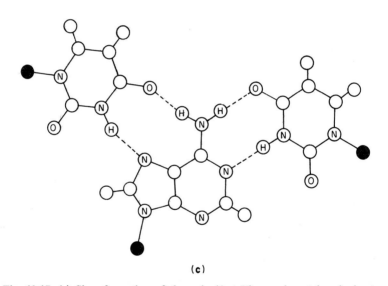

(c)

Fig. 12.15. (a) Slow formation of the poly (A + U) complex at low ionic strength (0.01 M sodium cacodylate, pH 6.5) is shown in this plot of optical density at 259 mμ against molar percentage. Solid circles: after zero time; open circles: after 24 hr. (b) The equilibrium absorbance (solid circles and lines) of mixtures of poly A and poly U in 0.1 M NaCl, 0.01 M glycylglycine, and 1.2 × 10^{-3} M MgCl$_2$, pH 7.2. The minimum at ∼67% poly U represents formation of the poly (A + 2U) complex. The dashed line corresponds to the mixing curve at very short reaction times and represents formation of the poly (A + U) complex only. The ordinate is the optical density at 259 mμ. [Parts (a) and (b) taken from Ref. 25, with permission of Elsevier Publishing Company.] (c) Proposed hydrogen-bonding scheme in the poly (A + 2U) complex. [Taken from D. R. Davies, *Ann. Rev. Biochem.*, **36**, 321 (1967); with permission of Annual Reviews, Inc.]

two poly U strands, known as poly (A + 2U). The x-ray diffraction pattern of fibers of this complex (Fig. 12.8) have been interpreted as arising from a triple-helical structure (16). The proposed hydrogen-bonding network between the bases on the three chains is shown in Fig. 12.15c.

Transfer RNA

The average cell contains a large number (at least 40) of different transfer RNA's. Each of these molecules consists of a single unbranched chain of 75–85 nucleotides and can be thought of as a "globular" nucleic acid, since it adopts a compact ellipsoidal structure. The role of tRNA in the cell is to link to a particular amino acid and carry it to the site of protein synthesis (see below). There is probably more than one different tRNA specific for a particular amino acid in each type of cell. The first known nucleotide sequence was established by Holley *et al.* (26) for a yeast transfer RNA specific for alanine, tRNA^{Ala}. This sequence, and that for a yeast tRNA^{Ser} (27), are shown in Fig. 12.16. The sequences of 16 different tRNA's are listed by Jukes and Gatlin (28). Present knowledge of the three-dimensional structure of tRNA has been reviewed by Cramer (29) and by Arnott (30).

Base fraction analysis of tRNA shows the presence of a significant proportion of the minor bases, although the relationships [A] ≃ [U] and [C] ≃ [G]

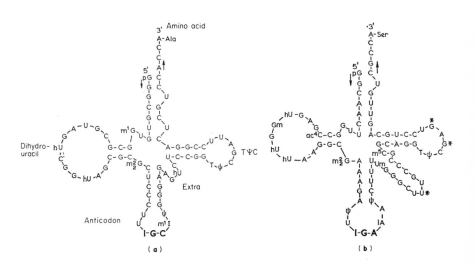

Fig. 12.16. Base sequence in (a) yeast tRNA^{Ala} (26) (b) yeast tRNA^{Ser} (27). [Taken from Ref. 1, with permission of the publisher, Methuen and Co., Ltd.]

still hold. Solutions of tRNA show high hypochromicity (31), which is indicative of substantial base pairing and regular stacking in the solution conformation. Optical rotatory dispersion studies show that tRNA molecules contain an average of approximately 26 base pairs (32). The melting behavior is similar to that of other RNA's, i.e., the UV absorption increases slowly with temperature. The melting curve for yeast tRNA[Phe] is shown in Fig. 12.17 (31). In addition, it is possible to draw oriented fibers of tRNA, which

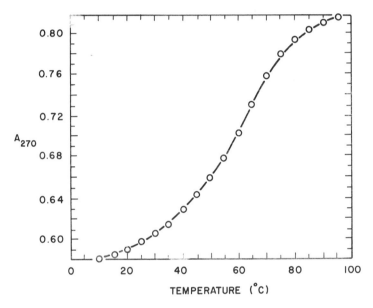

Fig. 12.17. Plot of UV absorption against temperature for a dilute solution of tRNA[Phe] [Taken from Ref. 31, with permission.]

show x-ray patterns similar to those of other RNA's (33), and indicate the presence of a regular double-helical conformation. The x-ray small-angle scattering from solution below 40°C gives a radius of gyration for tRNA[Phe] of 24 Å (34), which shows that the molecule has a very compact conformation.

The above data all suggest that the tRNA chain is folded back on itself to give a compact "helix with loops" conformation. When the first sequences became available, it could be seen that various sections of the chain could be arranged together in double-helical sections with Watson and Crick base pairing. The sequences were written in various ways to produce the maximum helical content, i.e., sequences made up of A $\cdots$ U and C $\cdots$ G units, assuming that the minor bases were not generally involved in base pairing. For a number of reasons, the most favorable arrangement was the so-called *cloverleaf* structure. The sequences in Fig. 12.16 are written in the schematic of the cloverleaf. Now that a number of sequences are known, it is clear that they can all be arranged in this manner (29,30), and it is very likely that the cloverleaf approximates the actual chain folding in tRNA.

In the cloverleaf structure, the chain is arranged in five folded sections, which are designated the *amino acid, TΨC, extra, anticodon,* and *dihydrouracil arms* (see Fig. 12.16a). The *amino acid arm* contains the two ends of the chain; in aminoacyl-tRNA, the amino acid is attached to the 3' end. The four unpaired nucleotides at the 3' end are known as the *tail*; in all the sequences so far determined, the last three nucleotides are -C-C-A. The *TΨC* and *anticodon arms* are of constant length. The loop at the end of the latter arm contains the triplet of bases known as the *anticodon*, which bonds to the codon of mRNA during protein synthesis (see below). This triplet can therefore be arranged in an exposed position, at the end of one of the helical rods. The *TΨC arm* is so named since its loop starts with that sequence of bases. The *dihydrouracil* arm is of variable length. The helical section may be three or four base pairs, and the loop, which always contains dihydrouridylic acid residues, is eight to ten nucleotides in length. However, most of the differences in chain length of the individual tRNA species are accommodated in the *extra arm*, which is comprised of from three to fourteen nucleotides.

As already mentioned, some indication of the three-dimensional conformation in solution is given by x-ray small-angle scattering. Results for *E. coli* tRNA^{Phe} (34) point to an elongated molecule with an axial ratio of almost 4 : 1. The fact that it is possible to draw oriented fibers of tRNA also suggests that the molecules have a rod-like conformation. When lyophilized specimens of tRNA are viewed in the electron microscope, threadlike stuctures are observed. Figure 12.18 is a micrograph obtained by Froholm and Olsen (35) of tRNA, positively stained with uranyl acetate. The threads are 80–100 Å in length and approximately 20 Å thick. The micrograph also reveals longer threads, several hundred angstroms in length, which are believed to be end-to-end aggregates of the shorter lengths. In positively stained specimens of yeast tRNA^{Ser}, which is known to form dimers in solution, short rods are seen with lengths of 85–125 Å and widths of 28–40 Å. These are thought to be dimers of tRNA^{Ser}, associated side by side.

Hydrated single crystals have now been obtained for a number of tRNA's, and x-ray diffraction data have been recorded to $d < 3$ Å in some cases. The unit cell dimensions and proposed space groups for some of these structures are listed in Table 12.3. These data strongly suggest that the different tRNA species have closely similar chain conformations. Some of the largest crystals reported so far have been obtained by Blake *et al.* (36) from an unfractionated

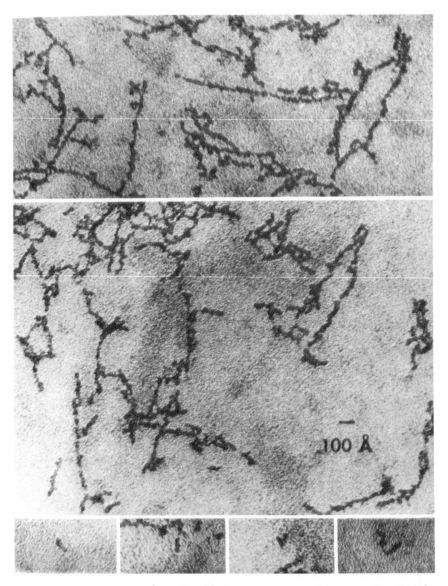

Fig. 12.18. Positive staining with 1% (w/v) aqueous uranyl acetate of a tRNA^{Ser} fraction which behaves as dimers on Sephadex. Note the presence of long threadlike structures showing some degree of branching. In some places short rods ~100 Å in length are observed. A few of these are shown in the lower part of the figure. [Taken from Ref. 35, with permission.]

TABLE 12.3

Unit Cell Dimensions for Crystalline tRNA

Species of tRNA[a]	Source	Unit cell dimensions[b]			γ (deg)	Suggested space group	No. of molecules in unit cell	References[c]
		a(Å)	b(Å)	c(Å)				
F-Met[d]	E. coli	53.2	43.2	118	90	$P222_1$	4	1
F-Met	E. coli	170	170	234	120	$P6_2$ or $P6_2 22$	~33	2
F-Met	E. coli	63.2	106.9	109.3	90	$C222$	8	3
F-Met	E. coli	55	25	120	90	$P222$	2	4
F-Met	Yeast	115	115	137	120	$P6_2 22$	12	5
Leu	E. coli	46	46	137	90	$P4_1$	4	5
Phe	E. coli	124	124	100	120	$P6_2 22$	24	6
Phe	Yeast	60.5	85	234	90	$C222_1$	32	7
Mixed	Yeast	53.0	44.7	120.3	90	$P222_1$	4	8

[a] Different crystalline forms have been obtained for the same tRNA, depending on the crystallization conditions.

[b] The unit cells listed here are either orthorhombic or hexagonal. Thus $\alpha = \beta = 90°$ in each case. A unit cell with rhombohedral space group $R32$ and dimension $a = b = c = 125$ Å, $\alpha = \beta = \gamma = 60.8°$ has been reported for E. coli tRNA^{Phe} (Ref. 8 below).

[c] Key to references:

1. Text Ref. 37.
2. S.-H. Kim and A. Rich, *Science* **162**, 1381 (1968).
3. S.-H. Kim, F. A. Schofield, and A. Rich, *Cold Spring Harbor Symp. Quant. Biol.* **34**, (1969).
4. Text Ref. 38.
5. J. D. Young, R. M. Bock, S. Mishimura, H. Ishikura, Y. Yamada, U. L. Raj Bhandary, M. Labanauskas, and P. G. Connors, *Science* **166**, 1527 (1969).
6. A. Hampel, M. Labanauskas, P. G. Connors, L. Kirregaard, U. L. Raj. Bhandary, U. L. Raj. Bhandary, P. Sigler, and R. M. Bock, *Science* **162**, 1384 (1968).
7. F. Cramer, F. von der Haar, K. C. Holmes, W. Saenger, E. Schlimme, G. E. Schultz, *J. Mol. Biol.* **51**, 523 (1970).
8. Text Ref. 36.

[d] Formyl methionine.

mixture of seven yeast tRNA's. These tRNA's have different chain lengths and sequences but can be accommodated in the same unit cell, which is also very similar to one of the unit cells obtained for purified *E. coli* tRNA[F-Met] (37).

One of the smallest unit cells is that obtained by Johnson *et al.* for *E. coli* tRNA[F-Met] (38), which is orthorhombic with dimensions $a = 120\,\text{Å}$, $b = 55\,\text{Å}$, and $c = 25\,\text{Å}$. The *P*222 space group requires that there be two molecules per unit cell. The *c* dimension of 25 Å corresponds approximately to the thickness of an RNA double helix. Thus the most probable shape for the molecule is that of a bulging rod, with two such molecules arranged in the unit cell as shown in Fig. 12.19 (38). The unit cell length of 120 Å corresponds

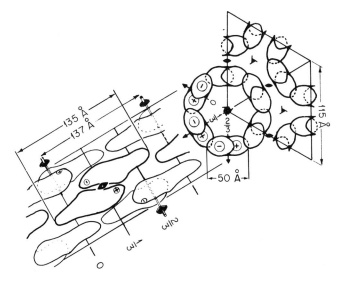

Fig. 12.19. A crystal-packing scheme postulated for the hexagonal tRNA[F-Met] crystals. The basic repeating unit is a dimer composed of two 90 Å long molecules. [Taken from Ref. 38, with permission of MacMillan and Co. Ltd.]

to the length of the dimer and matches the electron microscope dimensions discussed above. The other unit cells listed in Table 12.3 all contain more than two molecules, but their dimensions suggest similar rodlike structures in different three-dimensional arrays.

A rodlike conformation can be formed by arranging the arms of the cloverleaf in a variety of ways. Chemical evidence indicates that the *anticodon* and *dihydrouracil* loops are exposed, but the *TΨC* loop is in an inaccessible region (39). From the biochemical point of view, the anticodon must be

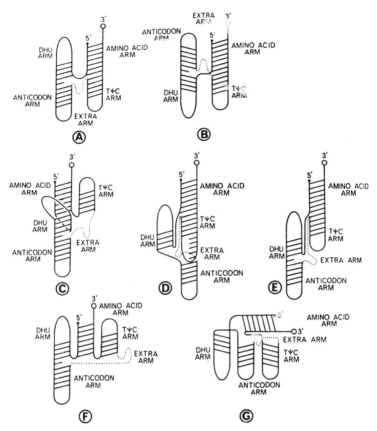

Fig. 12.20. Schematic of various proposed models of tRNA. The helical regions are shown as straight lines joined by the tilted base pairs, the variable region is dotted, and the amino acid is represented by a circle. [Taken from Ref. 40, with permission of MacMillan and Co., Ltd.]

exposed so that it can interact with the mRNA codon. It seems probable that the anticodon triplets in all tRNA's have the same conformation, which is likely to be a helical stack to facilitate base pairing with a similar arrangement of the codon section of mRNA. Seven of the structures that have been considered are shown schematically in Fig. 12.20 (40). The long axis of the molecule is vertical, and the arms of the cloverleaf are then wrapped about this axis. Model A appears to be one of the best with regard to the observed data and is shown in more detail in Fig. 12.21 (40). It seems likely that more detailed proposals for the conformation of tRNA will be available in the near future from the research groups working on the x-ray data.

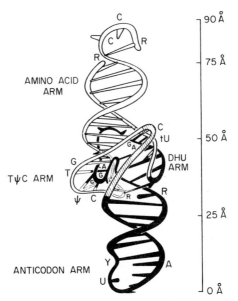

Fig. 12.21. A diagram of proposed tertiary structure for tRNA based on the sequence of tRNA$^{F\text{-Met}}$ from *E. coli*. [Taken from Ref. 40, with permission of MacMillan and Co. Ltd.]

Protein Synthesis and the Genetic Code

In general we have not discussed biological function in this book, except when we have been interested in, for example, proteins or polysaccharides as *structural* materials. In most cases we have avoided discussion of the *biochemical* roles of individual macromolecules. However, we feel that it is appropriate to include here a brief discussion of nucleic acids as the storage place for genetic information and their role in protein synthesis, which is the major motivation behind the structural studies of native and synthetic polynucleotides.

The information necessary for all functions of the cell is carried by a single strand of DNA in the cell nucleus, in the form of a specific sequence of nucleotides. The length of DNA containing the information necessary for the synthesis of a single protein is known as a *cistron*. The first step in protein synthesis is the transcription of the nucleotide sequence in the DNA to a molecule of messenger RNA (mRNA), which may contain one or more cistrons. This process probably takes place in a manner similar to that for DNA replication. The mRNA molecules migrate to the cytoplasm, where they form complexes with the ribosomes. The ribosomes themselves are complex particles made up of ribosomal RNA (rRNA) and protein, which serve as the site of protein synthesis.

Synthesis of a particular protein molecule requires translation of the message contained in the sequence of bases along the mRNA chain into a sequence of amino acids. The actual information is carried as a *triplet* code, with successive groups of three nucleotides corresponding to each amino acid residue. The translation takes place in two steps known as *trans I* and *trans II*. In the trans I process, the amino acids are linked by means of enzymatic reactions to their respective specific tRNA molecules. The code on the mRNA is then translated by matching the successive codon triplets with the tRNA anticodons—the trans II process. Two tRNA molecules interact with the ribosome and mRNA at any one time. Figure 12.22 shows this interaction schematically and represents any general stage during the synthesis other than initiation or termination. The growing protein chain is here attached to the first tRNA molecule. The second unit is an aminoacyl-tRNA, tRNA^{Ala} in this example, which has been recognized by linking the anticodon to one of the codons for alanine on the mRNA. When the binding to the mRNA and the ribosome have been effected, the peptidyl chain on the first tRNA is transferred to the amino acid residue of the second. The first tRNA is then released and the new peptidyl-tRNA, still attached to the alanine codon on the mRNA, moves to the first site on the ribosome. Now a third tRNA can occupy the aminoacyl-tRNA site, and when this is achieved by a tRNA matching anticodon to codon (in Fig. 12.22 this is a tRNA^{ser}) the process is repeated. Thus the peptide chain is synthesized by a series of steps of this type.

The sequence of amino acids in a protein is determined by the nucleotide sequence in the mRNA. Specifically, a triplet of nucleotides corresponds directly to a particular amino acid. For the four bases A, G, C, and U, there are $4^3 = 64$ possible triplet combinations. After exhaustive biochemical research, these triplets have all been assigned to their respective amino acids and are listed in Table 12.4. Reviews giving details of this research are listed in Ref. 41.

Confirming evidence for the code is that a large number of sequence variations and mutations reported for individual proteins can be predicted by postulating a change of a single nucleotide, e.g., a simple replacement or a frame shift in the mRNA. Nucleotide sequences have been determined for sections of some viral RNA's, which can be "read" directly with the aid of the code and correlated with known protein sequences.

Sixty-one of the 64 possible triplets code for individual amino acids. The remaining three, UAA, UAG, and UGA, are termination signals. When one of these signals is read, the protein chain is set free from the tRNA in site 1 on the ribosome, rather than joined to another amino acid. The signals for the start of synthesis are not completely understood. It is known that the AUG triplet for methionine and the GUG triplet for valine initiate synthesis,

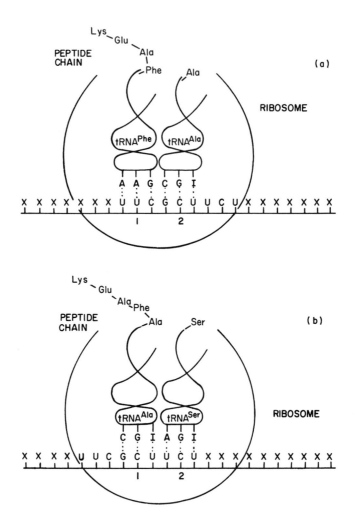

Fig. 11.22. Schematic diagram of protein synthesis in progress. (a) Site 1 on the ribosome contains a tRNA^{Phe}, with the peptidyl chain attached to the phenylalanine residue. Site 2 contains tRNA^{Ala}. These two t-RNA's are linked to the mRNA by base pairing of the anticodon to the codon. Note that the codon and anticodon sequences are antiparallel, as are the chains in DNA and RNA double helices. The complete peptidyl chain including the phenylalanine residue is linked to the alanine on tRNA^{Ala}, and the tRNA^{Phe} leaves site 1. (b) The peptidyl-tRNA^{Ala} moves to site 1 and the cycle is repeated. The next codon along the mRNA is for serine, and site 2 becomes occupied by a tRNA^{Ser}.

TABLE 12.4

The Genetic Code

Triplet	Amino acid	Triplet	Amino acid	Triplet	Amino acid	Triplet	Amino acid
UUU	Phe	UCU	Ser	UAU	Tyr	UGU	Cys
UUC	Phe	UCC	Ser	UAC	Tyr	UGC	Cys
UUA	Leu	UCA	Ser	UAA	Terminate	UGA	Terminate
UUG	Leu	UCG	Ser	UAG	Terminate	UGG	Try
CUU	Leu	CCU	Pro	CAU	His	CGU	Arg
CUC	Leu	CCC	Pro	CAC	His	CGC	Arg
CUA	Leu	CCA	Pro	CAA	Gln	CGA	Arg
CUG	Leu	CCG	Pro	CAG	Gln	CGG	Arg
AUU	Gln	ACU	Thr	AAU	Asn	AGU	Ser
AUC	Gln	ACC	Thr	AAC	Asn	AGC	Ser
AUA	Gln	ACA	Thr	AAA	Lys	AGA	Arg
AUG	Met	ACG	Thr	AAG	Lys	AGG	Arg
GUU	Val	GCU	Ala	GAU	Asp	GGU	Gly
GUC	Val	GCC	Ala	GAC	Asp	GGC	Gly
GUA	Val	GCA	Ala	GAA	Glu	GGA	Gly
GUG	Val	GCG	Ala	GAG	Glu	GGG	Gly

at least in bacteria. The mRNA is believed to be moved through the synthesis sites on the ribosome until an initiation triplet is reached, whereupon the corresponding tRNA is bound to the ribosome, and synthesis begins. However, the triplet may be preceded by a more complex sequence of bases which is part of the initiation signal. The amino acid that starts the chain, i.e., methionine or valine above, is probably removed some time later, as no individual amino acid is systematically in the first position in protein sequences.

Only methionine has a single code triplet; the remaining triplets can be divided into groups of two, three, or four, coding for the same amino acid. In these groups the first and second nucleotides are identical, and the only variation is in the third position. Crick proposed a *wobble* hypothesis to account for this degeneracy, whereby normal base pairs are formed successively at positions 1 and 2, whereupon some sort of hydrogen bonding, not necessarily Watson and Crick base pairing, would be formed at position 3. The anticodons from known sequences of tRNA are listed in Table 12.5. Nevertheless, the linkage between tRNA and mRNA plus the ribosome must be more complicated than simply base pairing, otherwise more mistakes in protein sequence would occur. More of the molecule is probably involved, and this may be another reason for the presence of so many minor bases.

TABLE 12.5

Anticodons in Known tRNA Sequences

tRNA	Source	Anticodon	tRNA	Source	Anticodon
Ala	Yeast	IGC	Phe	Wheat germ	GAA
Ser	Yeast	IGA	Ser	Rat liver	IGA
Val	Yeast	IAC	F-Met	*E. coli*	CAU
Tyr	Yeast	GΨA	Phe	*E. coli*	GAA
Phe	Yeast	GAA	Tyr(Su$_{III}^+$)	*E. coli*	CUA
Val	*T. utitis*	IAC			
Ile	*T. utitis*	IAU			

REFERENCES

1. N. Davidson, "The Biochemistry of the Nucleic Acids," 6th ed., Methuen, London, 1969.
2. E. Chargaff and J. N. Davidson, eds., "The Nucleic Acids," Vol. I–III, Academic Press, New York, 1950, 1955.
3. E. Chargaff, *Experientia* **6**, 201 (1950); see also E. Chargaff, *in* Ref. 2, Vol. I, p. 307, 1950.
4. J. D. Watson and F. H. C. Crick, *Nature* **171**, 737 (1953); **171**, 967 (1953).
5. W. T. Astbury and F. O. Bell, *Nature* **141**, 747 (1938); *Cold Spring Harbor Symp. Quant. Biol.* **6**, 109 (1938).
6. L. Pauling and R. B. Corey, *Proc. Nat. Acad. Sci. U.S.* **37**, 235 (1951).
7. W. Cochran, F. H. C. Crick, and V. Vand, *Acta Crystallogr.* **5**, 581 (1952).
8. L. Pauling and R. B. Corey, *Proc. Nat. Acad. Sci. U.S.* **39**, 84 (1953).
9. R. Langridge, H. R. Wilson, C. W. Hooper, M. H. F. Wilkins, and L. D. Hamilton, *J. Mol. Biol.*, **2**, 19 (1960).
10. D. A. Marvin, M. Spencer, M. H. F. Wilkins, and L. D. Hamilton, *J. Mol. Biol.* **3**. 547 (1961).
11. W. Fuller, M. H. F. Wilkins, H. R. Wilson, and L. D. Hamilton, *J. Mol. Biol.* **12**, 601 (1965).
12. J. Cairns, *J. Mol. Biol.* **6**, 208 (1963); A. K. Kleinschmidt, A. Burton and R. L. Sinsheimer, *Science* **142**, 961 (1963).
13. D. Voet and A. Rich, *Progr. Nucl. Acid Res. Mol. Biol.* **10**, 183 (1970).
14. K. C. Holmes and D. M. Blow, "The Use of X-Ray Diffraction in the Study of Proteins and Nucleic Acid Structures." Wiley (Interscience), New York, 1965; reprinted from *Methods Biochem. Anal.* **13**, 113 (1965).
15. W. Fuller, F. Hutchinson, M. Spencer and M. H. F. Wilkins, *J. Mol. Biol.* **27**, 507 (1967); S. Arnott, M. H. F. Wilkins, W. Fuller, and R. Langridge, *J. Mol. Biol.* **27**, 525 (1967); **27**, 535 (1967); S. Arnott, M. H. F. Wilkins, W. Fuller, J. H. Venable, and R. Langridge, *J. Mol. Biol.* **27**, 549 (1967).
16. S. Arnott, W. Fuller, A. Hodgson, and I. Prutton, *Nature* **220**, 561 (1968).
17. R. Langridge and A. Rich, *Nature* **198**, 725 (1963).
18. A. Rich, D. R. Davies, F. H. C. Crick, J. D. Watson, *J. Mol. Biol.* **3**, 71 (1961).
19. A. Rich, *Biochim. Biophys. Acta* **29**, 502 (1958).

20. R. F. Steiner and R. L. Beers, "Polynucleotides," p. 203, Amer. Elsevier, New York, 1961.
21. P. Doty, *Harvey Lect.* **55**, 103 (1961).
22. R. F. Steiner and R. L. Beers, *Biochim. Biophys. Acta* **29**, 189 (1958).
23. J. Fresco, *Trans. N.Y. Acad. Sci.* **21**, 653 (1959).
24. S. Timasheff, R. Brown, J. Cotter, and M. Davies, *Biochim. Biophys. Acta* **27**, 662 (1958).
25. G. Felsenfeld and A. Rich, *Biochim. Biophys. Acta* **26**, 425 (1957).
26. R. W. Holley, J. Apgar, G. A. Everett, J. T. Madison, M. Marquisee, S. H. Merill, J. R. Penswick, and A. Zamir, *Science* **147**, 1462 (1965); R. W. Holley, *Sci. Amer.* **214**, No. 2, 30 (1966).
27. H. G. Zachau, D. Dulling, and H. Feldman, *Hoppe-Seyler's Z. Physiol. Chem.* **347**, 212 (1966); H. G. Zachau, D. Dulling, H. Feldman, F. Melchers, and W. Karare, *Cold Spring Harbor Symp. Quant. Biol.* **31**, 417 (1966).
28. T. H. Jukes and L. Gatlin, *Progr. Nucl. Acid Res. Mol. Biol.* **17**, 303 (1971).
29. F. Cramer, *Progr. Nucl. Acid. Res. Mol. Biol.* **17**, 391 (1971).
30. S. Arnott, *Progr. Biophys. Mol. Biol.* **22**, 179 (1971).
31. D. D. Henley, T. Lindahl, and J. R. Fresco, *Proc. Nat. Acad. Sci. U.S.* **55**, 191 (1966).
32. C. R. Cantor, S. R. Jaskunas, and I. Tinoco, *J. Mol. Biol.* **20**, 39 (1966).
33. S. Arnott, F. Hutchinson, M. Spencer, M. H. F. Wilkins, W. Fuller, and R. Langridge, *Nature* **211**, 227 (1966); S. Arnott, M. H. F. Wilkins, W. Fuller, and R. Langridge, *J. Mol. Biol.* **27**, 535 (1967).
34. I. Pilz, O. Kratky, F. Cramer, F. von der Haar, and E. Schlimme, *Eur. J. Biochem.* **15**, 401 (1970); P. G. Connors, M. Labanauskas, and W. Beeman, *Science* **166**, 1528 (1969).
35. L. O. Froholm and B. R. Olsen, *FEBS Lett.* **3**, 182 (1969).
36. R. D. Blake, J. R. Fresco, and R. Langridge, *Nature* **225**, 32 (1970).
37. B. F. C. Clark, B. P. Doctor, K. C. Holmes, A. Klug, K. A. Marcker, S. J. Morris, and H. H. Paradies, *Nature* **219**, 1222 (1968).
38. C. D. Johnson, K. Adolph, J. J. Rosa, M. D. Hall, and P. Sigler, *Nature* **226**, 1246 (1970).
39. See Ref. 29 and references cited therein.
40. M. Levitt, *Nature* **224**, 759 (1969).
41. For further details on the genetic code see T. H. Jukes and L. Gatlin, Ref. 28; C. R. Woese, *Progr. Nucl. Acid Res. Mol. Biol.* **7**, 107 (1969); F. H. C. Crick, *Sci. Amer.* **215**, No. 4, 55 (1966).

13

GLOBULAR PROTEINS

X-ray Diffraction

Structure determination for macromolecules took on a new dimension following the publication of the structures of myoglobin by Kendrew and co-workers (1) in 1958 and hemoglobin by Perutz and co-workers (2) in 1961. These structures were determined by single-crystal x-ray methods, and progress since then allows us to consider the molecular conformation for some 20 globular proteins. These so-called globular proteins include the groups of molecules known as enzymes and hormones, and structures have been determined for a molecular weight up to about 70,000. The method of structure determination (see Chapter 3) requires crystallization of several isomorphous heavy-atom derivatives of the protein, usually from salt or buffer solutions. The x-ray intensities are obtained for each heavy-atom derivative and the phase angles determined by comparison of the various sets of data. A Fourier synthesis leads to an electron density contour map of the protein structure, and thus the conformation of the molecule is determined directly. Generally, the amino acid sequence is known prior to the structure determination and can be related directly to the conformation. In addition, structure determination has provided insight into the nature and mechanism of the action of certain enzymes, since cocrystallization of the enzymes with specific small molecules has shown up their interaction with the protein molecule, as well as any conformational changes that occur.

548

One limitation to the structure determination is that no x-ray data are available below a d-spacing of 2.0–1.4 Å. Thus, in general, individual atoms and chemical bonds cannot be resolved and, at best, only chemical groups can be identified. Accurate atomic coordinates cannot be determined but only postulated from known rules of stereochemistry on the basis of the shape of the chain. Consequently, a knowledge of the amino acid sequence is a valuable aid to structure determination. The cutoff in the observed d-spacings corresponds approximately to the limit of resolution for the structure. Thus, data cut off at $d = 2.0$ Å would lead to what is known as a structure at 2.0 Å resolution. In the early stages of structure determination only part of the data may be used, and a preliminary structure may be published at say 5.0 Å or 4.0 Å resolution. At 2.8 Å resolution the main chain of carboxypeptidase cannot be followed with certainty without knowledge of sequence, although with proteins for which homologous structures (i.e., having similar sequences and conformations) have previously been determined, the chain can be followed at 3.5 Å or 4.0 Å resolution, particularly if a significant fraction of the molecule is in the α-helical conformation, as is the case for myoglobin.

Under functional conditions a protein adopts an equilibrium conformation, and it is believed that this conformation is largely retained when it is crystallized. At first it was thought that the interchain forces within the crystal might force it into a different conformation. However, diffusion of substrates into the crystals has shown that enzymes retain their activity, albeit somewhat reduced, in the crystalline state. Furthermore, similar molecules such as different ribonucleases are found to have very similar conformations, even though they are crystallized under different conditions; crystals of ribonuclease S are obtained from salt solutions, while those of ribonuclease A are obtained from an alcoholic medium. Similarly, egg white lysozyme is found to have the same conformation in its different crystalline forms. The crystals of globular proteins contain a high proportion of solvent, often more than 50%, and are perhaps more like concentrated solutions than the more usual type of organic crystal. The protein molecule is surrounded by solvent molecules, and it is unlikely that these solvent forces are sufficiently different to produce a significant change from the solution conformation.

Some of the results of structure determination for globular proteins will be summarized in this chapter. Readers requiring further details are referred to the recent reviews by Davies (3), North and Phillips (4), Dickerson and Geis (5) and Blow and Steitz (6).

The first globular proteins whose structures were determined were myoglobin and hemoglobin. These molecules have more than 70% of the protein chains in the α-helical conformation and can be thought of as folded α helices. Subsequent work on a growing number of globular proteins has shown, however, that the globin structures are not typical and that the α-helical conformation is relatively rare. Lysozyme, for example, has only three short sections of distorted α helix involving less than 25 % of the molecule, and in chymotrypsin less than 5 % of the chain has the α form. However, the distribution of the amino acid side chains in myoglobin, where nonpolar side chains are usually on the inside of the folded molecule while the polar groups are usually exposed to the solvent, appears to be the general rule for globular protein structures. Charged side chains occur in the inner regions of the molecule only when they appear to be required for a particular purpose, i.e., when they are involved as functional groups at the active site. In addition, almost all the hydrogen bond donors are found in a convenient orientation to an acceptor or a bound water molecule.

The active sites for some enzymes have been located by cocrystallization with suitable substrates. In some cases, the polypeptide chain is arranged so that there is a deep cleft within the molecule, which contains the active site. Clefts of this type are present in lysozyme and ribonuclease. Otherwise, the active site may be no more than a shallow depression on the surface of an almost spherical molecule as, for example, in carboxypeptidase A.

Prediction of Chain Conformation for Globular Proteins

Efforts to predict the conformation of polypeptide chains is the subject of much of this book. It is not yet possible to predict the conformation of any but the simplest sequential polypeptide, much less that of a globular protein, simply from knowledge of the amino acid sequence. However, significant progress is being made toward this end, and various hypotheses have been formulated and checked against the conformations determined by x-ray analysis. It is hoped that, as more structures are determined for different globular proteins, and these data are taken together with those from further work on synthetic sequential polypeptides, these theories will be developed further, leading eventually to accurate prediction of the secondary and tertiary structures for globular proteins.

On the basis of conformational analysis of the type described in Chapter 2, Kotelchuck and Scheraga (7,8) have divided the 20 common peptide units into three groups: α-helix formers (*h*), α-helix breakers (*c*), and α-helix indifferent. The *h* units are Ala, Val, Leu, Ile, Met, Trp, Gln, Glu, Phe, Cys, and Arg; the *c* units are Ser, Thr, Asn, Asp, Tyr, Lys, and His. The peptide unit Pro is designated *h*; however, since proline cannot form an intramolecular hydrogen bond, it can be part of an α helix only if it is at the N-terminal end. Thus the peptide unit before Pro in the amino acid sequence is designated *c*. Glycine is considered helix indifferent and can be ignored, except where it is followed by proline, in which case it is designated *c*. A sequence of four *h* units is considered necessary to initiate formation (note that four peptides are necessary before an intramolecular hydrogen bond is formed), and the helix continues until the sequence contains two successive *c* units.

These theories were applied (8) to the peptide sequences of myoglobin, lysozyme, tosyl-α-chymotrypsin, and ribonuclease A, and the results were compared with the known conformations, as shown in Table 13.1 (7). The

TABLE 13.1

Comparison of the Observed and Predicted α-Helical Regions in Four Proteins, Using the Criteria of Kotelchuck and Scheraga[a]

Protein	Observed	Predicted
Myoglobin	3–18, 20–35, 36–42, 51–57, 58–77, 86–95, 100–118, 124–149	4–33, 5–61, 71–76, 83–91, 104–115, 127–131, 134–139
Lysozyme	5–15, 25–36, 88–99, 109–114, 120–124	2–17, 28–35, 107–115, 120–129
Tosyl-α-chymotrypsin	238–245	28–34, 65–73, 154–158, 179–188, 209–215, 237–245
Ribonuclease A	5–12, 28–35, 51–58	8–13, 54–60, 106–114

[a] Taken from Ref. 7, with permission.

structures of each of these proteins will be discussed below, and these predictions will then be considered in more detail. This hypothesis allowed 78% of the peptide units in these structures to be correctly assigned to their observed conformations. For the individual proteins, the percentages were 67, 78, 84, and 79 for myoglobin, lysozyme, chymotrypsin, and ribonuclease, respectively. Most of the observed helix regions are predicted by the model. However, the predicted helices are in many cases longer than those in the observed structures and some "random" regions are assigned as helical.

In a subsequent paper by Scheraga's group, Lewis *et al.* (9) assigned "helix probabilities" to the various amino acids based on a Zimm–Bragg formulation for the one-dimensional Ising model applied to denatured proteins. The amino acid residues are divided into three groups: Leu, Ile, Phe, Glu, Trp, Tyr, and Met are considered to be "helix formers"; Gly, Pro, Ser, and Asp are "helix breakers"; and Ala, Thr, Val, Gln, Asp, Cys, His, Lys, and Arg are "helix indifferent." Statistical weights were assigned to the amino acids, according to which group they were in, and "helix probability" profiles were calculated for 11 known globular protein structures. The probability profile for hen egg lysozyme is shown in Fig. 13.1 (9). The predicted helical regions for the 11 proteins are listed in Table 13.2 (8). Overall, some 68% of

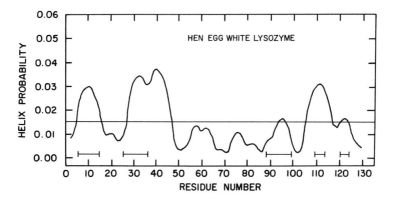

Fig. 13.1. Helix probability profile for hen egg white lysozyme, as calculated by Lewis *et al.* (9). The horizontal bars (⊢——⊣) denote those regions of the protein found to be in the right-handed α-helix conformation by x-ray methods. Predicted helical regions are those with probabilities above the horizontal line and are listed in Table 13.2. [Reproduced with permission of the National Academy of Sciences.]

TABLE 13.2

Comparison of Experimental and Predicted Helical Regions[a]

| | | | Helical regions | |
Protein	Number of residues	Residues correct (%)	Experimental	Predicted
Papain	211	68	26–41	27–53
			50–56	
			60–78	71–100
			116–126	104, 105, 112–113, 120–123, 125–126 131–144, 159–162, 171–172, 186–188
Hen egg white lysozyme	129	80	5–15	5–15
			23–36	27–46
			88–99	93–97
			109–114	106–116
			120–124	121–123
Horse oxyhemoglobin	141	56	3–18	7–14
			20–35	24–43
			36–42	
			52–71	55–58, 60–61
			80–89	85–113
			94–112	
			118–138	

(Continued)

TABLE 13.2 *(continued)*

Protein	Number of residues	Residues correct (%)	Helical regions	
			Experimental	Predicted
Subtilisin BPN	275	73	5–10	11–13
			14–20	25–30
			64–73	66–77
			103–117	89–96, 104–124
			132–145	134–144, 147–153
			223–238	227–257
			242–252	
			269–275	265–264
Chymotrypsin				
Chain A	13	50	None	5–10
Chain B	131	66	None	32–43
				64–95
Chain C	97	75		17–29
			90–97	79–96
Sperm whale myoglobin	153	52	3–18	8–18
			20–35	
			36–42	
			51–57	40–57
			58–77	67–72, 74–79
			86–95	91–87
			100–118	94–114
			124–149	131–146
Bovine ribonuclease A	124	64	5–12	4–11
			28–35	27–31, 33–36
			51–58	43–48, 55–67
				69–72
				94–111
Pig insulin				
Chain A	21	80	12–19	5–7
Chain B	30	70	9–18	11–25
Carboxypeptidase A	103	79	14–29	4–29
			72–88	71–88
			94–103	
Staphylococcal nuclease	149	80	54–70	61–78
			99–108	99–139
			121–134	
Horse ferricytochrome	104	47	None	8–20
				53–73
				80–100

[a] Data taken from Ref. 8 and references therein. [Reproduced with permission of the National Academy of Sciences.]

the residues were correctly assigned to their observed conformation. The comparable figure for these 11 proteins using the first method was 71 %.

Myoglobin

The first globular protein structure to be determined was that of sperm whale myoglobin by Kendrew *et al.* in 1958 (1). At this point the structure had been determined at 6.0 Å resolution and was followed in 1960 by a map to 2.0 Å resolution (10). Most of this molecule was shown to be in the α-helical conformation, and thus it was possible to follow the main chain almost completely at 6.0 Å resolution. The protein is roughly spheroidal and consists mainly of lengths of α helix folded together. A model of the myoglobin molecule based on the 6.0 Å map is shown in Fig. 13.2A (1). Myoglobins from

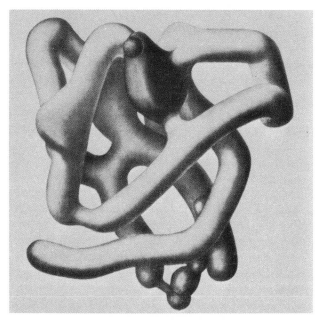

Fig. 13.2A. Structure of the myoglobin molecule proposed by Kendrew *et al.* (1). The α-helical sections of the chain appear sausagelike at this resolution; the disclike shape at the left center is the heme group. [Taken from M. F. Perutz, *Sci. 'Amer.* Nov. (1964), p. 64 with permission of W. H. Freeman Co.]

other sources have been found to have a similar shape. The 2.0 Å map showed that the α helices have right-hand sense and make up approximately 70 % of the molecule. The path of the main chain of the molecule is shown in Fig. 13.2B. There are eight sections of α helix, one of which has a distinct bend

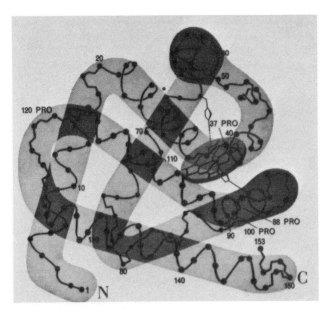

Fig. 13.2B. Path of the mainchain of sperm whale myoglobin. Every tenth amino acid subunit is marked, as are the proline subunits. [Taken from M. F. Perutz, *Sci. Amer.*, Nov. 1964, p. 64, with permission of W. H. Freeman Co.]

of ~7° in the middle of its axis. The remaining 30% of the molecule that is nonhelical does not develop any regular conformation. The molecule is very compact and, with a few exceptions, the polar residues are on the surface, while the interior consists of nonpolar residues. Most of the interactions are of the van der Waals type; there are a few saltlike interactions, and the majority of the groups that can form hydrogen bonds are in a position to bond either to other residues or to water molecules.

Myoglobin functions as an oxygen-storing protein in the blood. This is achieved by the formation of a complex between oxygen and the so-called "heme" group of the protein. The heme group is a planar ferrous organic base with the structure shown in Fig. 13.2C, and it fits into a pocket formed by the folded chain, as can be seen in Figs. 13.2A and B.

The amino acid sequence for sperm whale myoglobin (11) is listed in Table 13.3. The residues are numbered from 1 to 156 in order to facilitate comparison with the predictions in Tables 13.1 and 13.2 and, by convention, the eight helical sequences are labeled A to H, with the residues in each section numbered separately. The sections of chain between the sections of α helix are similarly labelled AB, BC, CD, etc. The infrequent occurrence of pairs of helix-breaking residues is apparent on inspection of the sequence. It

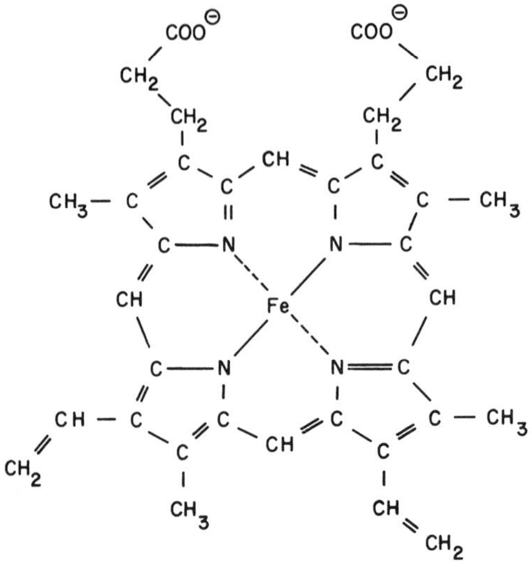

Fig. 13.2C. Chemical structure of the planar heme group.

is interesting that helices A and B are linked only by an alanine residue, which is normally an α-forming residue. Study of the primary sequences alone would have suggested continuing the helix at this point. It seems possible that the chain folds at this position for thermodynamic reasons, e.g., optimization of interfacial energetics, which also have to be considered in theoretical predictions. The final turns of helices A, C, E, and G are distorted to the tighter 3_{10} helix, while the last turn in both the F and H helices is more like the π-helix conformation.

Hemoglobin

The hemoglobins are oxygen-carrying blood proteins very similar to the myoglobins. Each hemoglobin unit consists of four separate protein chains: two α and two β chains (α and β here are merely labels and should not be confused with the α-helix and β-sheet conformations). Each component possesses a heme group (as in Fig. 13.2C) which complexes with oxygen. The amino acid sequences for the α and β chains of horse hemoglobin (11) are shown in Table 13.3, where the similarity to the myoglobin sequence can be seen.

The structure of horse oxyhemoglobin was determined to 5.5 Å resolution by Perutz *et al.* (2). The four units were found to be arranged tetrahedrally, as shown in Fig. 13.3, where each of the chains has a conformation similar to

TABLE 13.3

Amino Acid Sequences of Sperm Whale Myoglobin, and the α and β Chains of Horse Hemoglobin.

	NA1	2	3	A1	2	3	4
	1		2	3	4	5	6
Myoglobin	val	—	leu	ser	glu	gly	glu
Hemoglobin α	val	—	leu	ser	ala	ala	asp
β	val	gln	leu	ser	gly	glu	glu

	5	6	7	8	9	10	11
	7	8	9	10	11	12	13
Myoglobin	trp	gln	leu	val	leu	his	val
Hemoglobin α	lys	thr	asn	val	lys	ala	ala
β	lys	ala	ala	val	leu	ala	leu

	12	13	14	15	A16	AB1	B1
	14	15	16	17	18	19	20
Myoglobin	trp	ala	lys	val	glu	ala	asp
Hemoglobin α	trp	ser	lys	val	gly	gly	his
β	trp	asp	lys	val	asn	—	—

	2	3	4	5	6	7	8
	21	22	23	24	25	26	27
Myoglobin	val	ala	gly	his	gly	gln	asp
Hemoglobin α	ala	gly	glu	tyr	gly	ala	glu
β	glu	glu	glu	val	gly	gly	glu

	9	10	11	12	13	14	15
	28	29	30	31	32	33	34
Myoglobin	ilu	leu	ilu	arg	leu	phe	lys
Hemoglobin α	ala	leu	glu	arg	met	phe	leu
β	ala	leu	gly	arg	leu	leu	val

	16	C1	2	3	4	5	6
	35	36	37	38	39	40	41
Myoglobin	ser	his	pro	glu	thr	leu	glu
Hemoglobin α	gly	phe	pro	thr	thr	lys	thr
β	val	tyr	pro	trp	thr	gln	arg

	7	CD1	2	3	4	5	6
	42	43	44	45	46	47	48
Myoglobin	lys	phe	asp	arg	phe	lys	his
Hemoglobin α	tyr	phe	pro	his	phe	—	asp
β	phe	phe	asp	ser	phe	gly	asp

(continued)

TABLE 13.3 *(continued)*

	7	8	D1	2	3	4	5
	49	50	51	52	53	54	55
Myoglobin	leu	lys	thr	glu	ala	glu	met
Hemoglobin α	leu	ser	his	—	—	—	—
β	leu	ser	gly	pro	asp	ala	val

	6	7	E1	2	3	4	5
	56	57	58	59	60	61	62
Myoglobin	lys	ala	ser	glu	asp	leu	lys
Hemoglobin α	—	gly	ser	ala	gln	val	lys
β	met	gly	asn	pro	lys	val	lys

	6	7	8	9	10	11	12
	63	64	65	66	67	68	69
Myoglobin	lys	his	gly	val	thr	ala	leu
Hemoglobin α	ala	his	gly	lys	lys	val	ala
β	ala	his	gly	lys	lys	val	leu

	13	14	E15	16	17	18	19
	70	71	72	73	74	75	76
Myoglobin	thr	ala	leu	gly	ala	ilu	leu
Hemoglobin α	asp	gly	leu	thr	leu	ala	val
β	his	ser	phe	gly	glu	gly	val

	20	EF1	2	3	4	5	6
	77	78	79	80	81	82	83
Myoglobin	lys	lys	lys	gly	his	his	glu
Hemoglobin α	gly	his	leu	asp	asp	leu	pro
β	his	his	leu	asp	asn	leu	lys

	7	8	F1	2	3	4	5
	84	85	86	87	88	89	90
Myoglobin	ala	glu	leu	lys	pro	leu	ala
Hemoglobin α	gly	ala	leu	ser	asp	leu	ser
β	gly	thr	phe	ala	ala	leu	ser

	6	7	8	9	FG1	2	3
	91	92	93	94	95	96	97
Myoglobin	gln	ser	his	ala	thr	lys	his
Hemoglobin α	asn	leu	his	ala	his	lys	leu
β	glu	leu	his	cys	asp	lys	leu

	4	5	G1	2	3	4	5
	98	99	100	101	102	103	104
Myoglobin	lys	ilu	pro	ilu	lys	tyr	leu
Hemoglobin α	arg	val	asp	pro	val	asn	phe
β	his	val	asp	pro	glu	asn	phe

TABLE 13.3 *(continued)*

	6	7	8	G9	10	11	12
Myoglobin	105	106	107	108	109	110	111
	glu	phe	ilu	ser	glu	ala	ilu
Hemoglobin α	lys	leu	leu	ser	his	cys	leu
β	arg	leu	leu	gly	asn	val	leu

	13	14	15	16	17	18	19
Myoglobin	112	113	114	115	116	117	118
	ilu	his	val	leu	his	ser	arg
Hemoglobin α	leu	ser	thr	leu	ala	val	his
β	ala	leu	val	val	ala	arg	his

	GH1	2	3	4	5	6	H1
Myoglobin	119	120	121	122	123	124	125
	his	pro	gly	asn	phe	gly	ala
Hemoglobin α	leu	pro	asn	asp	phe	thr	pro
β	phe	gly	lys	asp	phe	thr	pro

	2	H3	4	5	6	7	8
Myoglobin	126	127	128	129	130	131	132
	asp	ala	gln	gly	ala	met	asn
Hemoglobin α	ala	val	his	ala	ser	leu	asp
β	glu	leu	gln	ala	ser	tyr	gln

	9	10	11	12	13	14	15
Myoglobin	133	134	135	136	137	138	139
	lys	ala	leu	glu	leu	phe	arg
Hemoglobin α	lys	phe	leu	ser	ser	val	ser
β	lys	val	val	ala	gly	val	ala

	16	17	18	19	20	H21	22
Myoglobin	140	141	142	143	144	145	146
	lys	asp	ilu	ala	ala	lys	tyr
Hemoglobin α	thr	val	leu	thr	ser	lys	tyr
β	asn	ala	leu	ala	his	lys	tyr

	23	24	HC1	2	3	4	5
Myoglobin	147	148	149	150	151	152	153
	lys	glu	leu	gly	tyr	gln	gly
Hemoglobin α	arg						
β	his						

Fig. 13.3. Hemoglobin molecule, as deduced from x-ray diffraction studies, is shown from above (top) and side (bottom). The drawings follow the representation scheme used in three-dimensional models built by Perutz and his co-workers. The irregular blocks represent electron density patterns at various levels in the hemoglobin molecule. The molecule is built up from four subunits: two identical α chains (light blocks) and two identical β chains (dark blocks). The letter N in the top view identifies the amino ends of the two α chains; the letter C identifies the carboxyl ends. Each chain enfolds a heme group (disc). [Taken from M. F. Perutz, *Sci. Amer.*, Nov. 1964, p. 64, with permission of W. H. Freeman Co.]

that of myoglobin. The small differences that were detected between the conformations of the α and β chains and myoglobin could be related to differences in the amino acid sequence. Notably, five residues in helix D of the β chain and myoglobin are absent from the sequence of the α chain. In addition, the carboxyl tail with nonhelical conformation is shorter in both α and β chains than in myoglobin.

An analysis to 2.8 Å resolution has now been achieved for hemoglobin (12). Perutz's first models, however, were based on the chain model for myoglobin, with the four chains arranged as indicated by the 5.5 Å map for hemoglobin. These models allowed conclusions to be drawn on the nature of the quaternary structure. The tetrahedral structure is maintained by salt bridges and hydrogen bonding between the pairs of α and β chains, i.e., α–α and β–β interactions. These types of bonding are also involved in the α–β interactions where, in addition, there is extensive nonpolar hydrophobic bonding. This is compatible with the observation that hemoglobin can be separated into symmetrical pairs of chains in acid or alkaline solutions, or at high concentration.

Perutz *et al.* (13) have surveyed a number of sequences of different globin molecules (including the myoglobins and the components of the hemoglobins) and have looked for relationships between the sequence and conformation. In spite of the very similar chain conformations, of the approximately 150 amino acid residues, only nine are invariant over the eighteen sequences surveyed. Of these, four are involved in linkage with the heme group and probably do not have conformationally directing effects. Of the remainder, four are located in the fold regions and probably are involved in determining the conformation; one is at a site of contact between two sections of the chain. However, these effects should be no more than minor contributions to the direction of the overall conformation. The ninth invariant residue appears to have no important function and could be a coincidence.

The proline residues always occur in the fold regions or near the beginning of the one of the helical sections. However, except for the proline at position C-2, these residues do not occur in invariant positions and cannot be the reason for the folds. Note also that there are no disulfide bridges in globin molecules which might direct the conformation.

These authors (13) also examined the sequences for regularity in the polar groups. Of the 43 ionizable groups on the surface of sperm whale myoglobin, only three are consistently basic, two are consistently acidic, and one is always acidic or basic. These effects seem scarcely more important than those mentioned previously. Similarly, there seems to be no appreciable constancy in the arrangement of nonpolar residues on the surface. However of the 33 internal sites, all but three are invariably nonpolar residues, and this seems to be the important factor controlling the conformation. This characteristic arrangement of the nonpolar residues is followed in all the other structures so far determined for globular proteins.

Most of the α-helical sections lie on the surface of the myoglobin molecule. Thus, a regular succession of polar followed by nonpolar residues can be detected in the amino acid sequence of the helical region, such that the polar side groups are on the outside and the nonpolar groups tend to be on the inside of the molecule. This sequence is broken up at the fold points of the chain. Figure 13.4 shows the sequence of polar and nonpolar residues in a globin chain. It is suggested that the presence of similar sequences in other proteins would indicate the presence of α helices on the surface of the molecule.

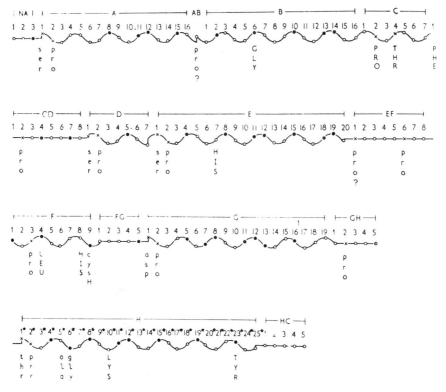

Fig. 13.4. Secondary structure of globin chains. The α-helical segments are represented as sine waves, and the nonhelical segments as straight lines. Black circles mark the sites where only nonpolar residues occur; crosses indicate where prolines or combinations of prolines with serine, threonine, aspartic acid, or asparagine occur. All other sites are represented by open circles. Residues in capital letters are invariant; those in small letters mark other residues of special interest, such as the reactive cysteine of the β chain, which must be on the surface. Note that this diagram shows the total incidence of proline in all species analyzed so far and that no single species contains proline at all the points indicated here. The sine waves are drawn so that the top of each wave points to the inside of the globin chain. [Taken from Ref. 13, with permission of Academic Press.]

Crystallographic methods have been used to investigate interactions between myoglobin and various small molecules. Once the phases are known for myoglobin they can be used in a difference synthesis to determine the positions of small molecules which form isomorphous complexes. These methods have been used to study deoxymyoglobin and various complexes, including azide ions, xenon, etc. (see, for example, Ref. 14).

Lysozyme

The first enzyme whose three-dimensional structure was determined was hen egg white lysozyme. Lysozyme functions as a catalyst in the cleavage of certain polysaccharide chains in the cell walls of bacteria. Crystallographic studies of lysozyme complexed with small saccharides have given a picture of the probable mechanism of cleavage.

Three groups of reseachers (15–17) have worked on lysozyme, and a clear picture of the chain conformation to 2.0 Å resolution has emerged (15,18). The path of the main polypeptide chain in lysozyme is shown in Fig. 13.5A (18). The electron density map is considered to be the clearest obtained at this resolution for any protein. The amino acid sequence was known almost completely prior to this work, and the remaining uncertainties were resolved from the map. The most difficult side chains to recognize were on the surface of the folded molecule, and it is possible that some of these side chains are free to adopt more than one conformation.

The distribution of side chains within the folded molecule of lysozyme follows the same pattern as in myoglobin and hemoglobin in that, with only one or two exceptions, the hydrophobic side chains are buried within the molecule (see Fig. 13.5A). The molecule is approximately ellipsoidal and is divided into two sections by a deep cleft in one side. One of these sections is rather sheetlike, with the thickness of only one chain; it consists mainly of hydrophilic residues which are thus in contact with the solvent molecules either on the outside of the molecule or in the cleft. The hydrophobic residues are found mainly in the other, larger section, where they form the central core. The conformation is stabilized by the presence of four disulfide bridges between the cystine residues.

In lysozyme, only 25% of the chain is in the α-helical conformation, which is considerably less than the figure of 70% for myoglobin. However, in the smaller of the two sections separated by the cleft, the chain has an extended conformation and, at one point, two antiparallel sections of the chain are hydrogen bonded together in an approximate antiparallel pleated-β sheet conformation. In addition, one short section adopts an approximate polyglycine II conformation. Nevertheless, these sections of regular confor-

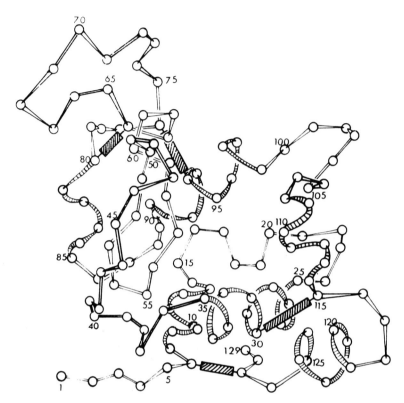

Fig. 13.5A. Schematic drawing of the main chain conformation of lysozyme. (By W. L. Bragg; taken from Ref. 18, with permission.]

mation are in the minority, and most of the molecule has an irregular " random " structure. The plot of the (ϕ,ψ) dihedral angles for lysozyme has been discussed in Chapter 2. Most of the (ϕ,ψ) combinations lie in the α, β, or polyglycine II regions; however, there is a regular progression along the chain of approximately constant (ϕ,ψ) angles only in the regions specified above.

The amino acid sequence for chick lysozyme (19) is given in Table 13.4, in which the sequences with approximately regular conformation are also shown separately. Helix A (residues 5–15) is a reasonably good Pauling and Corey α helix. Helix B (residues 24–34), however, is more distorted, with the N—H bonds pointing between the C=O bonds on the third and fourth residues along the chain. A similar distortion has been seen in the highest-resolution maps for myoglobin (20). This distortion amounts to a tightening of the B helix, which has a conformation intermediate between the α and 3_{10} helices. Helix C is also distorted from the ideal α conformation, but not

TABLE 13.4

Amino Acid Sequence of Chicken Lysozyme

1	2	3	4	5	6	7	8	9	10	11	12	13	14	15	16	17	18	19	20
Lys	Val	Phe	Gly	Arg	Cys	Glu	Leu	Ala	Ala	Ala	Met	Lys	Arg	His	Gly	Leu	Asp	Asn	Tyr

21	22	23	24	25	26	27	28	29	30	31	32	33	34	35	36	37	38	39	40
Arg	Gly	Tyr	Ser	Leu	Gly	Asn	Trp	Val	Cys	Ala	Ala	Lys	Phe	Glu	Ser	Asn	Phe	Asn	Thr

41	42	43	44	45	46	47	48	49	50	51	52	53	54	55	56	57	58	59	60
Gln	Ala	Thr	Asn	Arg	Asn	Thr	Asp	Gly	Ser	Thr	Asp	Tyr	Gly	Ile	Leu	Glu	Ile	Asn	Ser

61	62	63	64	65	66	67	68	69	70	71	72	73	74	75	76	77	78	79	80
Arg	Trp	Trp	Cys	Asn	Asp	Gly	Arg	Thr	Pro	Gly	Ser	Arg	Asn	Leu	Cys	Asn	Ile	Pro	Cys

81	82	83	84	85	86	87	88	89	90	91	92	93	94	95	96	97	98	99	100
Ser	Ala	Leu	Leu	Ser	Ser	Asp	The	Thr	Ala	Ser	Val	Asn	Asp	Ala	Lys	Lys	Ile	Val	Ser

101	102	103	104	105	106	107	108	109	110	111	112	113	114	115	116	117	118	119	120
Asp	Gly	Asp	Gly	Met	Asn	Ala	Trp	Val	Ala	Trp	Arg	Asn	Arg	Cys	Lys	Gly	Thr	Asp	Val

121	122	123	124	125	126	127	128	129
Gln	Ala	Trp	Ile	Arg	Gly	Cys	Arg	Leu

α-Helical regions (distorted)

	5	6	7	8	9	10	11	12	13	14	15
A 5–15	Arg	Cys	Gly	Leu	Ala	Ala	Ala	Met	Lys	Arg	His

```
                    127
                    Cys
```

	24	25	26	27	28	29	30	31	32	33	34
B 24–34	Ser	Leu	Gly	Asn	Trp	Val	Cys	Ala	Ala	Lys	Phe

```
                         115
                         Cys
```

	88	89	90	91	92	93	94	95	96
C 88–96	Ile	Thr	Ala	Ser	Val	Asn	Cys	Ala	Lys

```
                         76
                         Cys
```

so much as helix B. The presence of this distortion indicates that the formation of straight hydrogen bonds is not a necessary factor for the formation of the α-helical conformation. All the helical regions consist mainly of hydrophobic residues with hydrophilic units at the ends of the sequences. All three contain a cystine residue that is involved in a disulfide linkage.

The β-sheet sequence (residues 41–55) is shown in Fig. 13.5B and is made up almost entirely of hydrophilic residues. This section is part of the smaller of the two parts of the lysozyme molecule separated by the cleft and is in contact with water molecules on both sides of the sheet. Almost all the residues in this sequence are classified as "non-α-helix formers." The entire sequence for chick lyzozyme contains 12 sections where two or more such residues come together (ignoring the glycine residues), which matches the high proportion of nonhelical conformation in this molecule.

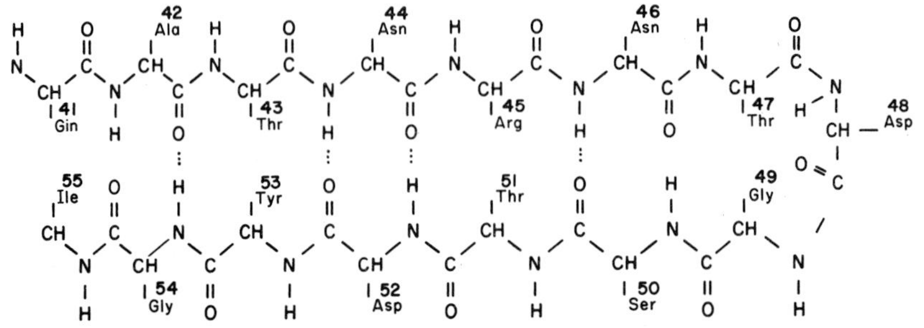

Fig. 13.5B. β-Sheet section of the lysozyme molecule showing the interchain hydrogen bonding. The actual structure is twisted and more distorted than shown here.

The nature of the active site of an enzyme can be characterized by determination of the structure of a suitable enzyme complex, which can also show up any conformational changes that occur during the interaction. Lysozyme is able to break the glycosidic linkages in the polysaccharide chitin (poly-N-acetyl-β-D-glucosamine) and in the alternating copolymer of N-acetyl-β-D-glucosamine and N-acetyl-β-D-muramic acid. However, monomers, dimers, and trimers involving N-acetylglusosamine and N-acetylmuramic acid act mainly as inhibitors of lysozyme. Blake *et al.* (18) determined the structure of lysozyme complexed with chitotriose, the trimer of β-(1,4)-linked N-acetyl-D-glucosamine. The trimer was found to lie in the cleft of the lysozyme molecule. This complex is stable, while the enzyme is known to attack higher oligomers, so the active site was thought to be away from the binding site for the trimer. When the trimer (residues A, B, and C) is converted to a chitin chain on a model, three more residues (D, E, and F) can be added to the reducing end of the triose before the chain passes out of the cleft. The geometry of the cleft requires the polysaccharide chain to have an approximate two-fold axis, as has been observed for chitin (see Chapter 11). In general, the contact distances between the enzyme and the polyhexose chain are good and include the possibility of hydrogen bonds. However, bad contacts occur between the enzyme and the CH_2OH side chain of residue D. It is believed that these bad contacts can be relieved if the ring of residue D is distorted into a "half-chair" conformation, which pushes the CH_2OH of the sugar residue into an axial position. Cleavage is thought to occur between D and E, while the chain is in the cleft. It is probable, therefore, that the distorted conformation required for residue D is no coincidence. Two charged residues, Glu 35 and Asp 52, are very close to this residue. It is suggested that the Glu 35 residue donates its proton to the glycosidic oxygen, and the negative Asp 52 residue promotes formation of a carbonium ion at C-1 of residue D, as the glycosidic bond is broken.

Ribonuclease

Ribonuclease is a cellular enzyme which cuts polyribonucleotide chains. It consists of a single chain of 124 amino acids (molecular weight 13,700), with four disulfide bridges, and it has a structure somewhat similar to that of lysozyme. The molecule is roughly ellipsoidal and is divided into two parts by a deep cleft. The cleft contains the active site, at which the RNA chain is broken by cleavage of the P—O-5′ bond. The amino acid sequence of bovine ribonuclease (21) is shown in Table 13.5.

TABLE 13.5

Amino Acid Sequence of Bovine Ribonuclease

```
  1    2    3    4    5    6    7    8    9   10   11   12   13   14   15   16   17   18   19   20
Lys-Glu -Thr-Ala -Ala -Ala -Lys-Phe-Glu- Arg-Glu -His -Met-Asp-Ser -Ser -Thr-Ser -Ala -Ala

 21   22   23   24   25   26   27   28   29   30   31   32   33   34   35   36   37   38   39   40
Ser -Ser -Ser -Asn-Tyr-Cys-Asn-Gln -Met-Met-Lys-Ser -Arg-Asn-Leu-Thr-Lys-Asp-Arg-Cys

 41   42   43   44   45   46   47   48   49   50   51   52   53   54   55   56   57   58   59   60
Lys-Pro-Val -Asn-Thr-Phe-Val -His -Glu -Ser -Leu-Ala -Asp-Val -Gln -Ala -Val -Cys-Ser -Gln

 61   62   63   64   65   66   67   68   69   70   71   72   73   74   75   76   77   78   79   80
Lys-Asn-Val -Ala -Cys-Lys-Asn-Gly -Gln -Thr-Asn-Cys-Tyr-Gln -Ser -Tyr-Ser -Thr-Met-Ser

 81   82   83   84   85   86   87   88   89   90   91   92   93   94   95   96   97   98   99  100
Ile - Thr-Asp-Cys-Arg-Glu -Thr-Gly -Ser -Ser -Lys-Tyr-Pro-Asn-Cys-Ala -Tyr-Lys-Thr-Thr

101  102  103  104  105  106  107  108  109  110  111  112  113  114  115  116  117  118  119  120
Gln -Ala -Asn-Lys-His -Ile - Ile -Val-Ala -Cys-Glu- Gly -Asn -Pro-Tyr-Val -Pro-Val -His -Phe

121  122  123  124
Asp-Ala -Ser -Val
```

α Helix

```
     2    3    4    5    6    7    8    9   10   11   12
A  Glu -Thr-Ala -Ala -Ala -Lys-Phe-Glu -Arg-Gln -His

    26   27   28   29   30   31   32   33
B  Cys-Asn-Gln -Met-Met-Lys-Ser -Arg
   |
   84
   Cys

    50   51   52   53   54   55   56   57   58
C  Ser -Leu-Ala -Asp-Val -Gln -Ala -Val -Cys
                                            |
                                           65
                                           Cys
```

β Sheet

```
                   42   43   44   45   46   47   48   49
                   Pro-Val -Asn-Thr-Phe-Val -His -Glu

 92   91   90   89   88   87   86   85   84   83   82   81   80   79   78   77   76   75   74   73   72   71
Tyr-Lys-Ser -Ser -Gly -Thr-Glu -Arg-Cys-Asp-Thr- Ile -Ser -Met-Thr-Ser -Tyr-Ser -Gln -Tyr-Cys-Asn

                   94   95   96   97   98   99  100  101  102  103  104  105  106  107  108  109  110
                   Asn-Cys-Ala -Tyr-Lys-Thr-Thr-Gln -Ala -Asn-Lys-His -Ile - Ile -Val -Ala -Cys
```

Crystallographic studies have determined the structures to 2.0 Å resolution
for both native ribonuclease, known as ribonuclease A (22), and the form
known as ribonuclease S (23), in which the bond between residues 20 and 21
has been cleaved enzymatically. In their crystalline structure, both the A and
S forms adopt essentially the same conformation. The most detailed structural
work so far is for ribonuclease S, and a skeletal model of this form is shown
in Fig. 13.6 (23). The short section of the chain (residues 1–20) is known as

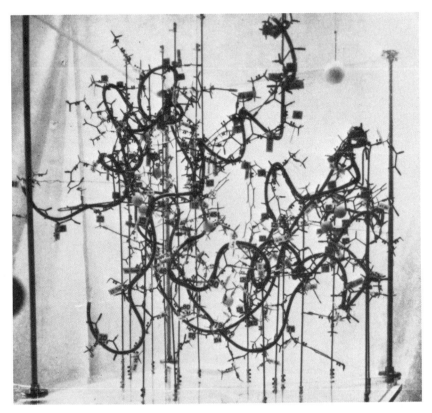

Fig. 13.6. Photograph of skeletal model of ribonuclease S. The break between the
S peptide (residues 1–20) and S protein (residues 21–124) is at the lower left-hand corner of
the photograph. The small balls represent sulfur atoms (of Cys and Met residues). [Taken
from Ref. 23, with permission of the American Society of Biological Chemists, Inc.]

the S peptide and remains part of the molecule even though there is no covalent linkage. The remainder of the molecule (residues 21–124) is known as
the S protein. Residues 20 and 21 are on the outside of the molecule and can
be seen in Fig. 13.6 to be separated by approximately 12 Å. In ribonuclease A,

these residues are linked by a covalent bond, and the conformation is un-changed except that neighboring parts of the chain are now drawn together.

Three short lengths of α helix are observed comprising 15% of the mole-cule, and their sequences are given in Table 13.5. These sequences are such that the hydrophobic side chains all come to one side of their respective helices and so can be packed on the inside of the molecule, as is the case with the helices of myoglobin, etc. Ribonuclease contains rather more extended β conformation (38%) than is found in lysozyme. Two long strands of anti-parallel β chains comprising residues 71–92 and 94–110 form a folded struc-ture, with residues 77 and 104 at the center of the fold (see Fig. 13.6). This structure runs the length of the molecule, on both sides of the cleft. A further β sequence (residues 42–49) is positioned along side residues 80–86. The amino acid sequences of the β sections are shown in Table 13.5. These se-quences contain a high proportion of the side chains that can form hydrogen bonds: Ser, Thr, Asn, and Gln. Much of the β sequence containing these hydrophilic residues forms the surface of the molecule, i.e., is in contact with the solvent.

Chymotrypsin

Enzymes that attack polypeptide chains are normally synthesized as in-active precursors, which are subsequently converted into the active form immediately prior to use. Three such enzymes will be considered here: α-chymotrypsin, papain, and carboxypeptidase A. α-Chymotrypsin is a di-gestive enzyme in the pancreas; it is formed from the precursor chymotryp-sinogen A, which consists of a single chain of 245 amino acid residues. Activation is achieved by enzymic removal of two dipeptides (14–15 and 147–148), leaving three sections (A, B, and C) of polypeptide chain linked by five disulfide bridges.

The structure of α-chymotrypsin has been determined at high resolution by Matthews *et al.* (24), and the path of the main chains for this molecule is shown in Fig. 13.7 (25). Two short sections of α helix are observed, both in the C chain; their amino acid sequences (26) are given in Table 13.6. The second sequence (234–245) contains a high proportion of α formers. How-ever, this is not true for the first section (164–170), and clearly more than the primary sequence is involved in determining the conformation at this point. This section is made up almost entirely of polar residues, and the helix occupies a very exposed position on the outside of the molecule. The rest of the molecule is made up of extended chains that are folded back on them-selves, so that, frequently, adjacent sections are antiparallel and separated by 4.5–5.0 Å for stretches of several residues, i.e., probably have the β con-formation. As usual, the central regions consist of nonpolar side chains,

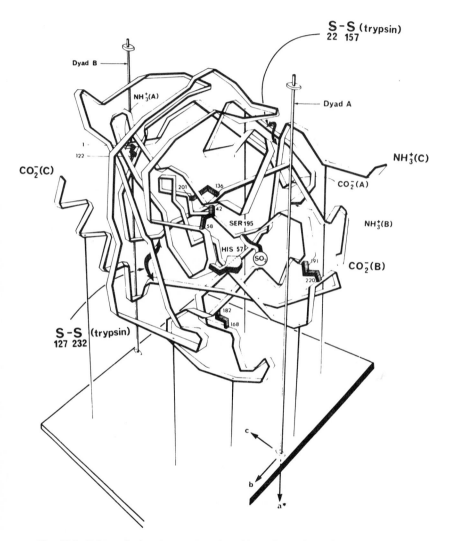

Fig. 13.7. Schematic drawing representing the conformation of the polypeptide chain in α-chymotrypsin. Disulfide bridges and the side chains of His 57 and the sulfonylated Ser 195 are shown. The positions of residues that form two additional disulfide bridges in trypsin are indicated. [Taken from Ref. 25, with permission of MacMillan and Co., Ltd.]

TABLE 13.6

α-Helical Sequences in Bovine α-Chymotrypsin

	164	165	166	167	168	169	170
A	Ser	Asn	Thr	Asn	Cys	Lys	Lys

	182
	Cys

	234	235	236	237	238	239	240	241	242	243	244	245
B	Leu	Val	Asn	Trp	Val	Gln	Gln	Thr	Leu	Ala	Ala	Asn

although some nonpolar groups are exposed on the surface. The molecule is roughly spherical, but with no obvious cleft for the active site, which is believed to be in the region of Ser 195.

Separate studies have revealed similar conformations for the precursor chymotrypsinogen A and also π-, δ- and γ-chymotrypsin (24, 27).

Papain

Papain, an enzyme isolated from the fruit papaya, consists of a single chain of 212 amino acid residues, with a molecular weight of 23,000. The structure as determined by Drenth *et al.* (28) shows a roughly ellipsoidal molecule, with a deep cleft containing the active site, as in lysozyme and ribonuclease. Four short sequences of α helix and a single section with the β conformation have been detected, and their amino acid sequences are listed in Table 13.7. The α sequences (A, B, C, and D) are made up mainly of

TABLE 13.7

Amino Acid Sequences of the α-Helical and β Sections of Papain

α Helix

	26	27	28	29	30	31	32	33	34	35	36	37	38	39	40	41
A	Trp	Ala	Phe	Ser	Ala	Val	Val	Thr	Ile	Glu	Gly	Ile	Ile	Lys	Ile	Arg

	50	51	52	53	54	55	56
B	Glu	Gln	Glu	Leu	Leu	Asp	Cys

	69	70	71	72	73	74	75	76	77	78
C	Trp	Ser	Ala	Leu	Gln	Leu	Val	Ala	Gln	Tyr

	116	117	118	119	120	121	122	123	124	125	126
D	Tyr	Asn	Glu	Gly	Ala	Leu	Leu	Tyr	Ser	Ile	Ala

β Chain

163	164	165	166	167	168	169	170	171	172
Ala	Val	Gly	Tyr	Asn	Pro	Gly	Tyr	Ile	Leu

α-forming residues, and there is an alternation of polar and nonpolar groups so that they can be concentrated on opposite sides of the helices. The β sequence contains mainly nonpolar residues and also a proline, which prevents an α conformation. The amino acid sequence for only certain sections was known before the x-ray analysis. The x-ray work determined the order of these sections and fitted in the remaining details.

Carboxypeptidase A

Carboxypeptidase A is another pancreatic enzyme, which digests polypeptide chains from the carboxyl end, as the name suggests. It consists of a single chain of 307 residues, with a molecular weight of 34,600. The structure has been determined to 2.0 Å resolution by Reeke *et al.* (29). The amino acid sequence was not known prior to this work and was determined from the x-ray analysis. The molecule is roughly ellipsoidal, with a small depression on the surface which contains the active site, including the zinc atom that is necessary for activity. Approximately 30% of the molecule is found to be helical, and the amino acid sequences (30) of the eight helical sections (A–H) are listed in Table 13.8. The distorted α-helical conformation seen in myo-

TABLE 13.8

α-*Helix Sequences in Bovine Carboxypeptidase A*

β Chain

    163 164 165 166 167 168 169 170 171 172
    Ala-Val -Gly -Tyr-Asn-Pro-Gly -Tyr - Ile - Leu

    14 15 16 17 18 19 20 21 22 23 24 25 26 27 28 29
A Thr-Leu-Asp-Glu - Ile - Tyr-Asp-Phe-Met-Asp-Leu-Leu-Val -Ala -Gln -His

    72 73 74 75 76 77 78 79 80 81 82 83 84 85 86 87 88
B Glu -Trp - Ile -Thr-Gln -Ala -Thr-Gly -Val -Trp-Phe-Ala -Lys-Lys-Phe-Thr-Glu

    94 95 96 97 98 99 100 101 102 103
C Pro-Ser -Phe-Thr-Ala - Ile - Leu-Asp-Ser -Met

    115 116 117 118 119 120 121 122
D Gly -Phe-Ala -Phe-Thr-His -Ser -Glu '

    174
E Val -Glu -Val -Lys-Ser - Ile - Val -Asp -Phe-Val -Lys

    215 216 217 218 219 220 221 222 223 224 225 226 227 228 229 230 231 232 233
F Asp-Lys-Thr-Glu -Leu-Asn-Gln -Val -Ala -Lys-Ser -Ala -Val -Ala -Ala -Leu-Lys-Ser -Leu

    254 255 256 257 258 259 260 261 262
G Ser - Ile - Asp-Trp-Ser -Tyr-Asn-Gln -Gly

    288
H Pro-Thr-Ala -Gln -Glu -Thr-Trp-Leu-Gly -Val -Leu-Thr - Ile -Met-Glu

globin and lysozyme, which requires bifurcated hydrogen bonding, is seen in some of the helical regions. These α helices are on the outside of the molecule and thus contain a high proportion of polar residues.

The most interesting feature of this molecule is that eight sections of β chain are arranged together to form a β sheet. The amino acid sequences and the interchain hydrogen bonds that are formed are shown in Fig. 13.8,

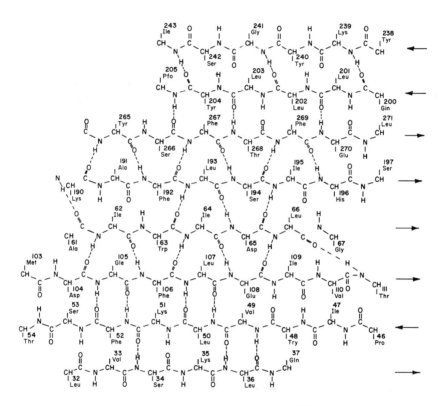

Fig. 13.8. Arrangement of the eight extended chains which form the β-sheet section on one side of the pocket in the carboxy peptidase a molecule, showing the interchain hydrogen bonding. The sheet is twisted, with the top chain rotated by 120° with respect to the bottom chain about a perpendicular axis in the plane of the paper, and is otherwise more distorted than shown here. [Diagram based on the structure due to Reeke *et al.*, (29).]

although the actual structure is more distorted than this diagram suggests. The top chain of the sheet is rotated approximately 120° with respect to the bottom chain about a perpendicular axis in the plane of the paper, the result being a twisted β sheet. The directions of successive chains are such that there are four pairs of parallel and three pairs of antiparallel chains. This is the

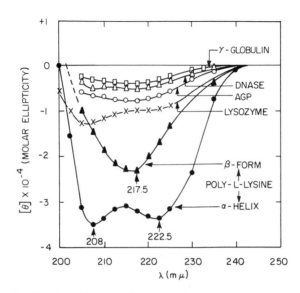

Fig. 13.9A. Circular dichroism of poly-L-lysine and several proteins. Poly-L-lysine (α helix at pH 11 and β-form at pH 11 after heating at 50°C for 10 min): conc., 1.24×10^{-3} M; path length, 2 mm; solvent, water. Lysozyme in 0.1 M potassium dihydrogen phosphate at pH 4.5: conc., 2.7×10^{-3} M, path length, 1 mm. α_1-Acid glycoprotein (AGP) in water: conc., 2.174×10^{-3} M, path length, 2 mm. DNase in water: conc., 4.5×10^{-3} M; path length, 1 mm. γ-Globulin in 0.01 M phosphate buffer plus 0.1 M sodium chloride: conc., 2.23×10^{-3} M, path length, 2 mm. The signal-to-noise ratio for poly-L-lysine was about 8:1 and that for the proteins was about 4:1; the possible error in the value of $[\theta]$ in the latter may be as large as $\pm 20\%$. An average residue weight of 115 was used in deriving the molar concentrations of proteins. [Taken from Ref. 31, with permission of the National Academy of Sciences.]

above effects do not contribute significantly in these cases. On the other hand, the ORD data (32) for ferricytochrome c show poor agreement with the x-ray results. The ORD curve shows the typical α-helix bands, and the calculated α content is 27%, but little or no α helix is observed in the x-ray structure at 4 Å resolution. The far-ultraviolet CD spectrum of carbonic anhydrase C is essentially featureless, and the ORD spectrum indicates little if any α helix (33). The equivalent spectra of the acid-denatured protein indicate some 20% α helix. However, the structure determined by x-ray crystallography has approximately 33% α helix in the natural form (34). It has been suggested that the contributions from aromatic residues may be overshadowing the spectra producing these apparent inconsistencies. Such side-chain effects have been observed, for example, with poly-α-tyrosine (see Chapter 6).

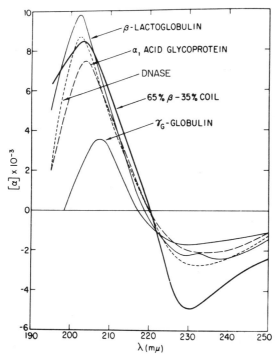

Fig. 13.9B. The ORD of several proteins in the far-ultraviolet region. γ-Globulin: conc., $1.9 \times 10^{-3}\,M$; path length, 1 mm. α_1-Acid glycoprotein: conc., $4.35 \times 10^{-4}\,M$; path length, 2 mm. DNase: conc., $4.37 \times 10^{-3}\,M$; path length, 0.13 mm. Solvents are same as in Fig. 13.9A. [Taken from Ref. 31, with permission of the National Academy of Sciences.]

The CD bands due to certain side chains have been identified for certain globular proteins and used to detect small conformational changes on binding of inhibitors. An example is the study of the conformational changes of lysozyme on binding of *N*-acetyl-D-glucosamine (35). Reagents that modify tyrosine and tryptophan residues produce a change in the CD spectrum, as does the binding of *N*-acetylglucosamine, which suggests that the aromatic residues are involved in binding of the sugar.

The heme proteins have received considerable attention by workers investigating the possibility of conformational changes between oxygenated or deoxygenated forms, as well as the changes involved when the heme is removed altogether (36). Differences among the spectra have been detected, which are difficult to interpret but are probably indicative of conformational changes. Deoxygenated hemoglobin has a trough at 233 mμ, which is about 5% deeper than that for the oxygenated form, However, interpretation of these results in terms of 5% more α helix in the deoxygenated form does not appear to be justified by x-ray work and is probably due to side-chain effects.

Infrared Spectroscopy

Assignment of conformation to particular proteins has often been made from the positions and complexity of the bands due to the vibration of the amide groups (see Chapter 5). In particular, the random and α-helical structures produce an amide I vibration at ~ 1650 cm^{-1}, while for antiparallel β chains, this vibration is at ~ 1630 cm^{-1} with an additional weak band at about 1690 cm^{-1}. Admittedly these assignments are based on observations for long chains of perfectly regular conformations, unlike the distorted α and β sequences in globular proteins. However, the bands for β sheets, for example, are given by the *Crysopa* egg stalk cross-β structure, where the fold period is of the order of the lengths of β chains in globular proteins, and use of the IR technique to investigate conformation in the latter is not unreasonable.

Most solution infrared work for globular proteins has been done in D_2O solution to avoid the band due to water (H_2O) in the amide I region. Myoglobin gives a sharp band at 1650 cm^{-1}, which is to be expected for a structure that is 75% α helix and 25% random coil. On the other hand, β-lactoglobulin shows a strong band at 1632 cm^{-1} with a shoulder at 1685 cm^{-1} (37), which would indicate the presence of a considerable percentage of β structure. This fits in with the similar interpretation of the ORD/CD spectra described above. Many proteins, however, do not give such clear-cut results. The value of the IR technique, however, is often found when it is used in conjunction with other techniques, such as ORD and CD, where a combination of results, indicating the *probable* conformation, begins to be convincing.

REFERENCES

1. J. C. Kendrew, G. Bodo, N. M. Dintzis, R. G. Parrish, H. W. Wyckoff, and D. C. Phillips, *Nature* **181**, 662 (1958).
2. M. F. Perutz, M. G. Rossmann, A. F. Cullis, H. Muirhead, G. Will, and A. C. T. North, *Nature* **185**, 416 (1960); A. F. Cullis, H. Muirhead, M. F. Perutz, M. G. Rossmann, and A. C. T. North, *Proc. Roy. Soc. London* **A262**, 15 (1961); **A265**, 161 (1962).
3. D. R. Davies, *Annu. Rev. Biochem.* **36**, 321 (1967).
4. A. C. T. North and D. C. Phillips, *Progr. Biophys. Mol. Biol.* **19**, 1 (1969).
5. R. E. Dickerson and I. Geis, "The Structure and Action of Proteins." Harper, New York, 1969.
6. D. M. Blow and T. A. Steitz, *Annu. Rev. Biochem.* **39**, 63 (1970).
7. D. Kotelchuck and H. A. Scheraga, *Proc. Nat. Acad. Sci. U.S.* **61**, 1163 (1968).
8. D. Kotelchuck and H. A. Scheraga, *Proc. Nat. Acad. Sci. U.S.* **62**, 14 (1969).
9. P. N. Lewis, N. Go, M. Go, D. Kotelchuck, and H. A. Scheraga, *Proc. Nat. Acad. Sci. U.S.* **65**, 810 (1970).

10. J. C. Kendrew, R. E. Dickerson, B E. Strandberg, R. G. Hart, D. R. Davies, D. C. Phillips, and V. C. Shore, *Nature* **185**, 422 (1960); J. C. Kendrew, H. C. Watson, B. E. Strandberg, R. E. Dickerson, D. C. Phillips, and V. C. Shore, *Nature* **190**, 666 (1961).

11. M. O. Dayhoff and R. V. Eck, "Atlas of Protein Sequence and Structure, 1967–68." National Biomedical Research Foundation, Silver Spring, Maryland, 1968.

12. M. F. Perutz, H. Muirhead, J. M. Cox, L. C. G. Goaman, F. S. Mathews, E. L. McGandy, and L. E. Webb, *Nature* **219**, 29 (1968); M. F. Perutz, H. Muirhead, J. M. Cox, and L. C. G. Goaman, *Nature* **219**, 131 (1968)

13. M. F. Perutz, J. C. Kendrew, and H. C. Watson, *J. Mol. Biol.* **13**, 669 (1965).

14. L. Stryer, J. C. Kendrew, and H. C. Watson, *J. Mol. Biol.* **8**, 96 (1964).

15. C. C. F. Blake, R. H. Fenn, A. C. T. North, D. C. Phillips, and R. J. Poljak, *Nature* **196**, 1173 (1962).

16. R. H. Stanford, R. E. Marsh, and R. B. Corey, *Nature* **196**, 1176 (1962).

17. R. E. Dickerson, J. M. Reddy, M. Pinkerton, and L. K. Steinrany, *Nature* **196**, 1178 (1962).

18. C. C. F. Blake, G. A. Mair, A. C. T. North, D. C. Phillips, and V. R. Sharma, *Proc. Roy. Soc. London* **B167**, 365 (1967); C. C. F. Blake, L. N. Johnson, G. A. Mair, A. C. T. North, D. C. Phillips, and V. R. Sharma, *Proc. Roy. Soc. London* **B167**, 378 (1967).

19. R. E. Canfield and A. K. Liu, *J. Biol. Chem.* **240**, 1997 (1965).

20. H. C. Watson, cited in Ref. 4.

21. D. G. Smyth, W. H. Stein, and S. Moore, *J. Biol. Chem.* **238**, 227 (1963).

22. G. Kartha, J. Bello, and D. Harker, *Nature* **213**, 862 (1967).

23. H. W. Wyckoff, K. D. Hardman, N. M. Allewell, T. Inarami, D. Tsernoglou, L. N. Johnson, and F. M. Richards, *J. Biol. Chem.* **242**, 3749 (1967); **242**, 3987 (1967).

24. B. W. Matthews, P. B. Sigler, R. Henderson, and D. M. Blow, *Nature* **214**, 652 (1967).

25. P. B. Sigler, D. M. Blow, B. W. Matthews, and R. Henderson, *J. Mol. Biol.* **35**, 143 (1967).

26. B. S. Hartley, *Nature* **201**, 1284 (1964).

27. J. Kraut, H. T. Wright, M. Kellerman, and S. T. Freer, *Proc. Nat. Acad. Sci. U.S.* **57**, 304 (1967).

28. J. Drenth, J. N. Jansonius, and B. G. Wolthers, *J. Mol. Biol.* **24**, 449 (1967); J. Drenth, J. N. Jansonius, R. Koekoek, H. M. Swen, and B. G. Wolthers, *Nature* **218**, 929 (1968).

29. G. N. Reeke, J. A. Hartsuck, M. L. Ludwig, F. A. Quicho, T. A. Steitz, and W. N. Lipscomb, *Proc. Nat. Acad. Sci. U.S.* **58**, 2220 (1967).

30. R. A. Bradshaw, L. H. Ericsson, K. A. Walsh, and H. Neurath, *Proc. Nat. Acad. Sci. U.S.* **63**, 1389 (1969).

31. P. K. Sarkar and P. Doty, *Proc. Nat. Acad. Sci. U.S.* **55**, 981 (1966).

32. D. W. Urry and P. Doty, *J. Amer. Chem. Soc.* **87**, 2754 (1965).

33. S. Beychok, J. McC. Armstrong, C. Lindblow, and J. T. Edsall, *J. Biol. Chem.* **241**, 5150 (1966); S. Beychok, *Annu. Rev. Biochem.* **37**, 437 (1967).

34. K. Fridburg, K. K. Kannan, A. Liljas, J. Lundin, B. Strandberg, R. Strandberg, B. Tilander, and G. Wiren, *J. Mol. Biol.* **25**, 505 (1967).

35. A. N. Glazer and N. S. Simmons, *J. Amer. Chem. Soc.* **88**, 2335 (1966).

36. R. F. Frankel and P. Doty, cited by S. Beychok, Ref. 33.

37. S. N. Timasheff and J. J. Susi, *J. Biol. Chem.* **241**, 2491 (1966).

AUTHOR INDEX

Numbers in parentheses are reference numbers and indicate that an author's work is referred to although his name is not cited in the text. Numbers in italics show the page on which the complete reference is listed.

A

Abe, A. R., *38*

Abu Shumays, A., *256*

Ackermann, Th., 338(67, 68), 339(67, 68, 70), 340(67, 68, 70), *342*

Adolph, K., 539(38), 540(38), *547*

Alexander, L. E., 84(1), 87, 94(1), *125*

Allewell, N. M., 568(23), *579*

Anderegg, J. W., 315, *341*

Anderson, J. M., 153(30), 156(30), 157(30), 158(33), *167*, 393(26), 394(26), 395(34), 399(45), *405, 406*, 414(7), *462*

Anderson, N. S., 69(50), *78*, 226(87), *228*, 290(58), *295*, 507(69), *513*

Anderson, S. O., 440(35), 443(35), *463*

Andries, J. C., 147(24, 25), 148(24), 149(25), 150(25), 151(28), 153, 154, 155(29), 156(30), 157(30), 160(24), 161(24), *166, 167, 380*, 395(34), 396(36), 397(43),

399(46), *405, 406*

Aoyagi, S., 275(42), *294*

Apgar, J., 535(26), *547*

Applequist, J., 245(18), *294*, 338(62), *342*

Archibald, W. J., 327, *342*

Armstrong, J. McC., 576(33), *579*

Arnon, R., 392(22), *405*

Arnott, S., 111(16), 113, 114, *125*, 523(15), 525(16), 526(16), 533(16), 535, 536(33), 537(30), *546, 547*

Asadourian, A., *380*

Astbury, W. T., *119*, 147, *166*, 517, *546*

Atkins, E. D. T., 118, 122(26), *125*, 147(23), *166*, 224, 225(84), *228*, 412(4), 413(4, *462*, 469, 492, 494, 505(64, 65), *511, 512, 513*

Auer, H. E., *380, 391*

Avigad, G., 289(54), *295*

Aylward, N. N., 194(43), 198(43), 200(47), 201(47), *227, 228*

SUBJECT INDEX

A

Actin, 454
  F-Actin, 454
  G-Actin, 454
Adenine, 15
  Raman spectrum, 195
Adenosine,
  Raman spectrum, 195
Agar, 507
L-Alanyl-L-proline, conformational
    analysis, 56
Alginic Acids, structure of, 505
Amide Vibrations, 172–175
  Fermi resonance, 172
  fibrous proteins, 190
Amino Acids, 1
Amorphous Polypeptides, 159
Amylopectin, 499
  infrared spectroscopy of, 224
Amylose, 499
  A-amylose, 499
  B-amylose, 68, 499
    structure of, 503
    x-ray diffraction pattern of, 504
  C-amylose, 499

KBr-amylose, 69
  conformational analysis, 69
V-amylose,
  complexes of, 500
  conformational analysis, 67
  infrared spectroscopy of, 224
  single crystals of, 502
  structure of, 501
  unit cell parameters of, 501
  x-ray diffraction pattern of, 501
  7_1 helix, 69
Amylose Triacetate, 503
  conformational analysis, 69
  x-ray diffraction of, 109
Anticodon, 537, 540–541, 542, 543

B

Base Stacking, 237
Bases (in nucleotides), planar structure, 69
Bessel functions, in x-ray diffraction
    theory, 102
β sheet conformation for polypeptides
  ORD and CD spectra, 575
  in solution, 383–384
  definition, 29

Molecular Biology

An International Series of Monographs and Textbooks

Editors

BERNARD HORECKER

Department of Molecular Biology
Albert Einstein College of Medicine
Yeshiva University
Bronx, New York

NATHAN O. KAPLAN

Department of Chemistry
University of California
At San Diego
La Jolla, California

JULIUS MARMUR

Department of Biochemistry
Albert Einstein College of Medicine
Yeshiva University
Bronx, New York

HAROLD A. SCHERAGA

Department of Chemistry
Cornell University
Ithaca, New York

HAROLD A. SCHERAGA. Protein Structure. 1961

STUART A. RICE AND MITSURU NAGASAWA. Polyelectrolyte Solutions: A Theoretical Introduction, *with a contribution by Herbert Morawetz.* 1961

SIDNEY UDENFRIEND. Fluorescence Assay in Biology and Medicine. Volume I–1962. Volume II–1969

J. HERBERT TAYLOR (Editor). Molecular Genetics. Part I–1963. Part II–1967

ARTHUR VEIS. The Macromolecular Chemistry of Gelatin. 1964

M. JOLY. A Physico-chemical Approach to the Denaturation of Proteins. 1965

SYDNEY J. LEACH (Editor). Physical Principles and Techniques of Protein Chemistry. Part A–1969. Part B–1970. Part C–in preparation

KENDRIC C. SMITH AND PHILIP C. HANAWALT. Molecular Photobiology: Inactivation and Recovery. 1969

RONALD BENTLEY. Molecular Asymmetry in Biology. Volume I–1969. Volume II–1970

JACINTO STEINHARDT AND JACQUELINE A. REYNOLDS. Multiple Equilibria in Protein. 1969

DOUGLAS POLAND AND HAROLD A. SCHERAGA. Theory of Helix-Coil Transitions in Biopolymers. 1970

JOHN R. CANN. Interacting Macromolecules: The Theory and Practice of Their Electrophoresis, Ultracentrifugation, and Chromatography. 1970

WALTER W. WAINIO. The Mammalian Mitochondrial Respiratory Chain. 1970

LAWRENCE I. ROTHFIELD (Editor). Structure and Function of Biological Membranes. 1971

ALAN G. WALTON AND JOHN BLACKWELL. Biopolymers. 1973

In preparation

A. J. HOPFINGER. Conformational Properties of Macromolecules

R. D. B. FRASER AND T. P. MACRAE. Conformation in Fibrous Proteins

WALTER LOVENBERG (Editor). Iron-Sulfur Proteins

OSAMU HAYAISHI (Editor). Molecular Mechanisms of Oxygen Activation